W9-BCW-379

The Ever-Changing Sky provides a comprehensive and uniquely non-mathematical guide to spherical astronomy. In a clear and lucid text, the reader is guided through terrestrial and celestial coordinate systems, time measurement, and celestial navigation, to the prediction of the rising and setting of the stars, Sun, and Moon. The book focuses on the geometrical aspects of the night sky *without* the use of complex trigonometry (which is saved to a handy appendix).

The book progresses to a general study of the Earth and sky, including the stars and constellations (with useful star maps provided), the motions and appearance of the Moon, tides and eclipses, the orbits of the planets, and the smaller bodies of the Solar System (asteroids, meteors, meteorites, and comets). Finally, there is a brief overview of atmospheric phenomena (including rainbows and halos).

This text will be invaluable to students taking courses in naked-eye astronomy, amateur and professional astronomers, as well as more general readers wanting to know how the night sky changes.

The Ever-Changing Sky

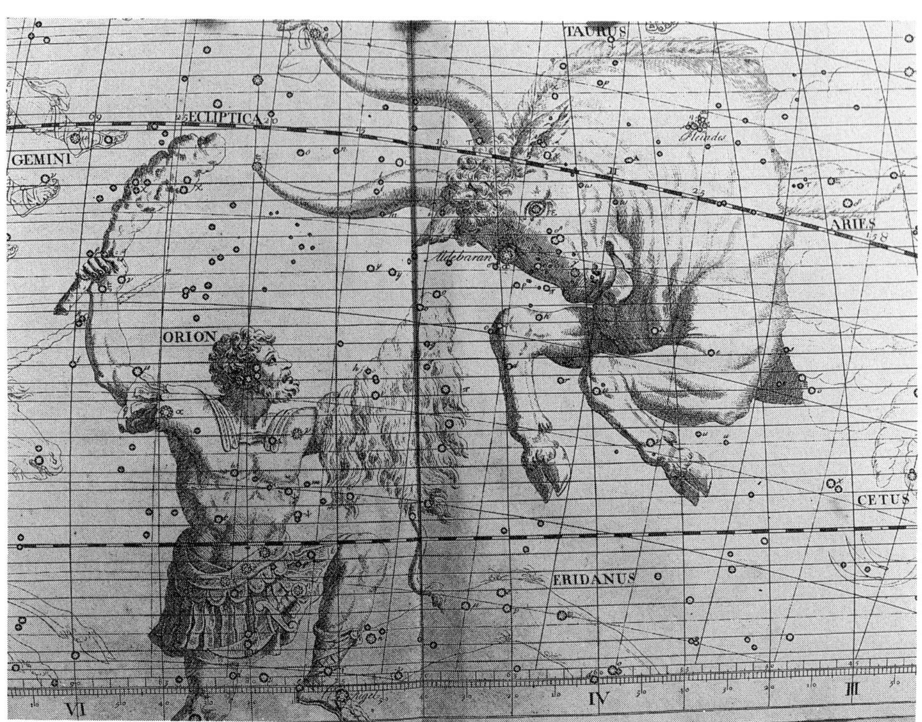

Our changing perceptions of the sky are epitomized by John Flamsteed's *Atlas Coelestis* of 1729, which shows ancient mythologies (the great hunter Orion in combat with the zodiac's Taurus) superimposed on a scientifically precise star atlas.

The Ever-Changing Sky

A Guide to the Celestial Sphere

JAMES B. KALER
University of Illinois

Published by the Press Syndicate of the University of Cambridge
The Pitt Building, Trumpington Street, Cambridge CB2 1RP
40 West 20th Street, New York, NY 10011-4211, USA
10 Stamford Road, Oakleigh Melbourne 3166, Australia

First published 1996

Printed in Great Britain by Biddles Ltd., Guildford and King's Lynn

A catalogue record for this book is available from the British Library

Library of Congress cataloguing in publication data

Kaler, James B.
The ever-changing sky : a guide to the celestial sphere/James B. Kaler.
p. cm.
Includes bibliographical references.
ISBN 0 521 38053 7
1. Spherical astronomy. 2. Celestial sphere. I. Title.
QB145.K33 1966
522′.7–dc20 95-8507CIP

ISBN 0 521 38053 7 hardback

WV

To my teachers. Thank you Gertrude Rosenberg, Benjamin Boss, Hazel Losh, Dean McLaughlin, Hermann Zanstra, Albrecht Unsöld, and Lawrence Aller.

Contents

Tables

Preface

As the subject of astrophysics has grown over the years, it has gradually replaced much of the traditional material once commonly found in elementary textbooks. As a result, an entire body of knowledge revolving primarily around terrestrial and planetary motions and aspects of the celestial sphere is becoming less accessible. It is not that these subjects are unimportant but that there is simply less and less available text space.

The problem was first made clear to me when I was asked to give a seminar to a graduate anthropology class that was exploring subjects in South American ethno-astronomy. My task was to explain some basic astronomical principles – the rising and setting of the Sun and stars for example – needed to help the students make sense of Incan building alignments. This discussion, plus a series of related experiences (I was even helping a doctoral candidate in agronomy who was doing a project on ancient agriculture and needed information on planting times as indicated by the rising of star groups), demonstrated that there was little available in the way of qualitative, relatively non-mathematical reading material on the subject: I was recommending out-of-print, and often out-of-date, 1930s and 1940s textbooks.

At that point I decided to fill the gap by writing what was initially intended to be a cross-disciplinary guide to spherical astronomy for anthropologists, archaeologists, classicists, etc. However once this venture was underway I could begin to see a more general purpose: a complete non-mathematical treatment of all aspects of the sky, useful not only to those in other professions who need to know these elements of astronomy, but to astronomers as well, whether student, professional, or amateur.

The chief theme of the book is the exploration of the things that move, or seem to move, across the heavens, with the emphasis on what can be seen with the naked eye. I introduce a variety of phenomena, and then explain them in a generally non-mathematical way. Mathematics – its usefulness and grandeur – is indeed constantly a presence, but it is largely silent: there are few equations, but much geometry, couched in non-technical terms and, where necessary, explained from the beginning. The reader should finish the book with a firm conceptual foundation of the science of spherical astronomy and celestial motions, and would be well equipped to move on to more technical treatments.

The text begins (*Chapter 1*) with an introduction to the sky, where I define the basic architecture of the celestial sphere, including the fundamental concepts of circles, angles, and arcs. In *Chapter 2* the reader sees how the stars appear to move when the Earth is set turning and the observer sent traveling around the globe. Here we explore the rising and setting of stars, and how the observer's position affects what he or she sees.

With these basics in place, *Chapter 3* shows what happens in the sky as the Earth moves through space in its orbit about the Sun, causing that body to appear to travel through the stars. The chapter defines the ecliptic (the solar path), explains the seasons, establishes why the length of the day and the positions of sunrise and sunset change with the time of year and the observer's location, and demonstrates what happens when the observer travels to the arctic regions of the globe and to the tropics. More technical matters are then developed in an examination of celestial coordinate systems, which was begun in Chapter 2. These geometrical concepts provide a foundation for subsequent discussions.

In *Chapter 4* the reader is placed under the nighttime sky and begins the study of the subject that in the popular mind defines astronomy, the stars and constellations. The text examines the history and significance of constellation lore, the Milky Way, and various non-stellar objects. Extensive tables of the stars and constellations are given in conjunction with basic but fairly detailed star maps that are presented in *Appendix 2*. *Chapter 5* adds further to the complexity of the explanations of celestial motions that were developed in the first three chapters with a detailed account of precession and its attendant movements, and uses the material of Chapter 4 to make the slow motion seem more real.

With the basics of celestial motions in place, *Chapter 6* develops the subject of time measurement, including the practicalities of keeping precise time on a world-wide basis. This material leads naturally into *Chapter 7*, in which we see how the times and positions of sunrise and sunset can be found, and into *Chapter 8*, which examines the science of star positions and celestial navigation.

In the next four chapters we turn specifically to a study of the bodies of the Solar System. *Chapter 9* takes up a detailed examination of the motions and phenomena of the Moon, beginning with the most fundamental concepts of the lunar phases, and successively adding complexities that lead to the development and definition of the elements of orbital mechanics and the niceties of gravitational perturbations. Here we also examine the special problems of moonrise and moonset. *Chapter 10* continues the discussion to include tides, eclipses, and calendars.

In *Chapter 11*, this subject is extended to the planets, to examine their observed characteristics and apparent motions. The material introduced in Chapter 9 is expanded to encompass the history and development of the laws of planetary motion and the underlying concept of gravity, that elegant construction that explains how the planetary system works. A brief telescopic view of the planets is included and the chapter closes with a critical look at astrology, which can then be intelligently juxtaposed against the science that both parallels and opposes it. *Chapter 12* then turns to the other bodies of the Sun's family, the comets, meteors, and meteorites, phenomena to which even a casual observer can easily relate.

Finally, *Chapter 13* moves the reader back toward the Earth to examine atmospheric phenomena that are played out against the backdrop of the starry or sunlit sky. The chapter begins with elementary optical principles and uses the workings of telescopes as practical examples. We then explore the extraordinary richness of atmospheric optics: rainbows, halos, sundogs, sunsets, the green flash, aurorae, the twinkling of stars, and a variety of other topics.

The book terminates with three appendices. The first, initially referred to in Chapter 3, presents various graphs and tables that involve the times and positions of the rising and setting of the Sun and Moon. The second relates to Chapter 4, and consists of a set of star maps that are sufficiently detailed to illustrate the positions of all the constellations and a few non-stellar objects. *Appendix 3* develops the basic rules of spherical trigonometry, allowing the student of this subject to make a variety of astronomical calculations and providing a transition to more mathematically oriented texts. The ultimate goals throughout the book are to interest the reader in the myriad aspects of the sky and to provide clear, useful explanations of them.

I am indebted to R. Tom Zuidema for starting me on this project and for his interest and enthusiasm. Deep thanks also go to David Bright and Ray White, who read, commented on, and improved the manuscript, to Dennis D. McCarthy, who instructed me on various matters of time and the Earth's rotation, and to Lauren Kaler Brewer, who checked the clarity of the mathematics in Appendix 3. I also thank Simon Mitton for his patience while the manuscript was being completed, Bob MacFarlane for his excellent draftsmanship, Jack Gladin for his fine photographic work, and Maxine for continuous encouragement over the years of preparation.

Chapter 1

The Earth and the celestial sphere

Step outside on a clear dark night and be welcomed by a lavish celestial display. Quite possibly the Moon will be visible, and it will surely be the first thing you notice. If it is not too bright, you will see a profusion of stars arranged in distinctive patterns or *constellations*. At first the number of stars looks countless; some few are quite bright, but the dimmer ones are far more numerous. If you had the patience, though, and if the Moon were not present, you would find you could count about 2000. There might also be one or more very bright bodies that you would recognize as planets. Finally, if the night were very dark, you might see a wide luminous encircling band, the Milky Way, the combined light of millions of faint stars.

If you watch for perhaps 15 minutes, you will see that the sky is not stationary. Stars set below the western horizon and new ones take their places in the east. Over the course of the night, the heavens change considerably in appearance. Then the eastern horizon begins to glow and the stars start to fade away. As sunrise approaches, they disappear altogether and the sky becomes dominated by the westward-moving Sun. You can now see that the whole celestial vault – stars, Moon and Sun – appears to turn about points that are fixed in the north and south.

If you were to watch the sky for many days, you would see more subtle effects and slower changes. The most obvious would be that the Moon appears to move to the east against the stellar background. On successive nights it rises and sets steadily later, changing its shape as it goes. You would also note that the stars are never found in the same place at any given time from night to night. Each evening they are displaced a little to the west, setting progressively earlier until they are lost in twilight. Simultaneously, new ones arrive in the east, until after several months of watching, the entire aspect of the sky has changed. Observations made over several days or weeks also show that the planets are discriminated from stars not by their brightness but by their seemingly erratic movements, shifting east then west but on the whole drifting eastward against the starry display, very roughly following the path of the Moon.

In the daytime the position of the Sun also changes. Your first viewing might be on a warm summer day somewhere in northern climes. The Sun passes nearly overhead at noon and sets far to the northwest, providing long daylight hours. But as you watch, it appears to slip south day by day. Sunrise and sunset move to the southeast and southwest, and the days become colder and shorter until you might think that the Sun will disappear forever from view. But to your relief, the southerly movement eventually stops and the Sun starts northward again, the whole performance to be repeated year after year.

If you could now watch not just for days or even a few years but instead for centuries, you could detect subtler movements in the heavens. The whole sky would appear to be shifting slowly in an independent manner as stars creep past the fixed points of daily rotation. Stars and constellations never before seen would appear above the southern horizon while some others would disappear altogether from view. The familiar constellations of each season would slowly change. The sight of Leo in the evening announces Spring for northerners. Many thousands of years from now it will proclaim Autumn. We cannot see these changes in a mere lifetime, but they are obvious over many generations, and we have our written records, which can give us humans the effect of having lived for the required millennia.

On a still longer time scale of thousands of centuries even the constellations would start to dissolve, the result of the motions of the stars themselves and our changing perspective, as the Sun, with its family of planets, pursues a relentless path through space. Thousands of millennia hence, the sky will still be filled with twinkling stars, but almost none you see today will be visible. We live in a world of constant and complex motion, a truly dynamic universe. That is the point and message of the tale that follows.

In the succeeding sections and chapters we will examine the appearance and motions of the sky in detail, approaching the subject from two perspectives. First, the sky will be described as it appears to us as observers on the Earth. As we watch, it does not seem that our planet is moving at all, but that it is the *sky* that is in motion. Second, the reasons behind the actions will be explained. That is, we will deal with a given problem not just from an observer's standpoint, but from a physical one as well. In this way the reader will be able to make the generalizations necessary for the understanding of a given phenomenon as seen from any point in time or any place on Earth.

1.1 Angles on a plane

To understand the mechanism of the heavens, we must deal with spheres. Indeed, the term *spherical astronomy* is commonly used to describe much of the subject matter to be covered. The Earth is a very good approximation to a sphere, and the sky can be thought of as an exact sphere. We must be able to define *position* on a sphere so that we can locate London with respect to Buenos Aires, the constellation Orion (the famed celestial hunter: see Chapter 4) relative to the Big Dipper (or the Plough, as it is known in Britain), and Orion with regard to the terrestrial observer.

To measure position, we use the *angle*, which is gauged at the intersections of the line pairs in Figure 1.1. We can measure the angle either interior to the two lines (a, b, c), or exterior to them (a', b', c'). The sum ($a + a'$ etc.) is always 360°, the maximum

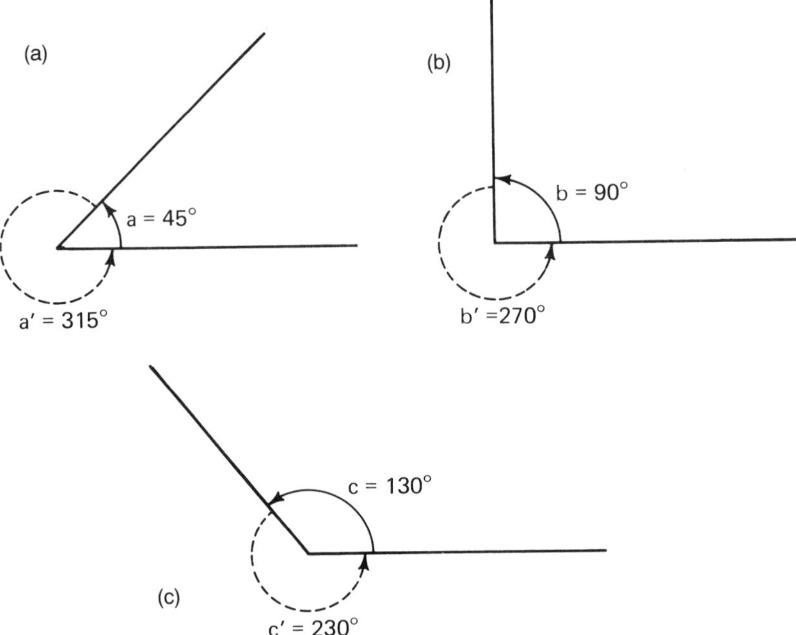

1.1. Various angles, measured at the intersections of two lines. The solid and dashed arrows respectively denote interior and exterior angles, which must sum to 360°. Angles *a* and *c* are respectively acute and obtuse. The 90° angle (*b*) formed by the perpendicular lines is a right angle.

angle that can be measured through or around a full circle. Why 360°? Whence comes this odd number? Why not divide the circle by 100 or 1000 units (as is actually done in some instances)? The reason immediately illustrates the ancient and abiding influence of the sky on daily life. There are 365 days to the year, and 360 is the closest number that is easily divisible into parts. With this definition, the Earth then orbits simply and steadily through an angle centered on the Sun of about 1° every day. Or, more significantly, the Sun, reflecting the Earth's motion, *appears* to move about 1° per day against the backdrop of the constellations.

Certain angles, especially those that quarter the circle, are of special significance. In particular, the 90° angle, formed by two perpendicular lines (Figure 1.1b), is called a *right angle*. Its exterior complement is obviously 270°. Angles less than 90° (Figure 1.1a) are called *acute*, those greater (Figure 1.1c) *obtuse*. The 180° angle is sometimes used synonymously for "opposite." Others – 45°, 60°, 30° – do not carry specific names but are commonly used in setting examples and are important in astronomy's parallel in the pseudoscientific world, astrology.

Angles are also measured around the circumference of a circle. Figure 1.2 shows two lines intersecting to form an angle called *a*. Draw a circle centered on the intersection

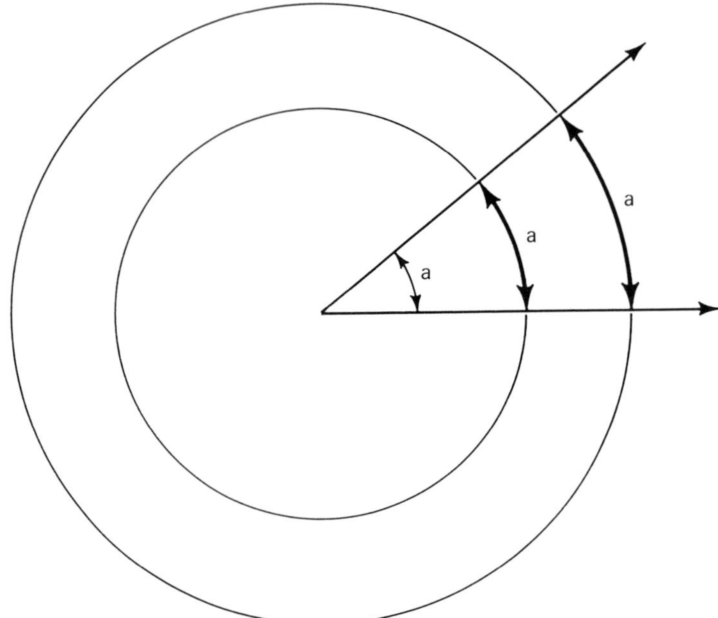

1.2. Arc measurement on a circle. The angle *a* is 40° and the sections of the circles between the lines have arcs of 40° as well. The length of an arc in degrees does not depend upon the physical radius of the circle.

and extend the two lines outward until they intersect it. The lines that form angle *a* define a part of the circle's circumference called the *arc* of angle *a*. We can then measure that section in *degrees of arc*, there being 360° of arc on the entire circle. We say that the arc *a subtends* (is opposite to) the angle *a*, or vice versa. Arc measurement does not depend on the physical dimension, or radius, of the circle, only on the fraction of its circumference defined by the lines. Two circles of different radius are drawn in Figure 1.2 and it is obvious that angle *a* subtends arcs *a* in both cases.

We can now define positions along the circle not by inches or centimeters, which would be dependent on the radius, but through angular measurement. To do so we must define an *origin* or starting point. The origin may be picked with some basis in mind, or it may be selected arbitrarily. Point *P* on the circle (Figure 1.3) can be located precisely and uniquely by specifying its arc around the circle from the origin and the direction of measurement.

The human eye can easily notice an angle of 1°, and specialized instruments can do much better. The degree is subdivided into 60 divisions called *minutes* (written 60′) and the minute into 60 *seconds* (60″). Do not confuse these terms with the more common time units of minutes and seconds. One word is used to express two concepts and the meaning must be taken from context. We might specify, for example, that in

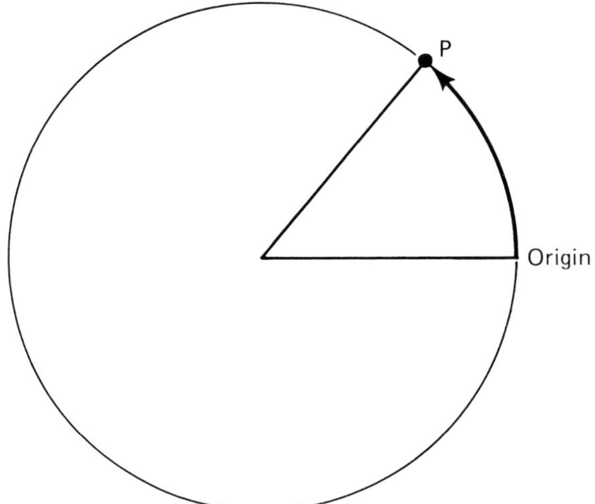

1.3. Position by angle or arc on a circle. P is about 50° from the origin, measured counterclockwise.

Figure 1.3 point P is 50° 16′ 32″.2 from the origin, as measured counterclockwise, or that it is 360° – 50° 16′ 32″.2 from the origin as measured clockwise.

In the subtraction one must always subtract seconds from seconds, minutes from minutes, and degrees from degrees. The larger unit must first be converted into a form to make the subtraction possible. Since 1° equals 60′, 360° can be written as 359° 60′, and similarly this number can be expressed as 359° 59′ 60″. Thus

$$\begin{array}{r} 359^\circ\ 59'\ 60''.0 \\ \underline{-50^\circ\ 16'\ 32''.2} \\ 309^\circ\ 43'\ 27''.8, \end{array}$$

the remainder being the obtuse angle in Figure 1.3.

1.2 Angles on a sphere

These concepts can be transferred, with some modification, from the plane to the sphere. The sphere must be drawn on the flat page: with practice, the perspective will become familiar. Examine the sphere drawn in Figure 1.4. The small dot (labeled C) represents its center, and A and B are two points arbitrarily located on its surface with the singular qualification that they are not on a line with the center. Next draw a circle on the sphere around the center through the points. The circle defines a plane that is the same as those in Figures 1.2 and 1.3. The only difference is that in Figure 1.4 we are looking at it not from overhead but at an angle, so that the circle appears distorted. You can get the same effect in either of the two previous figures if you place your eye

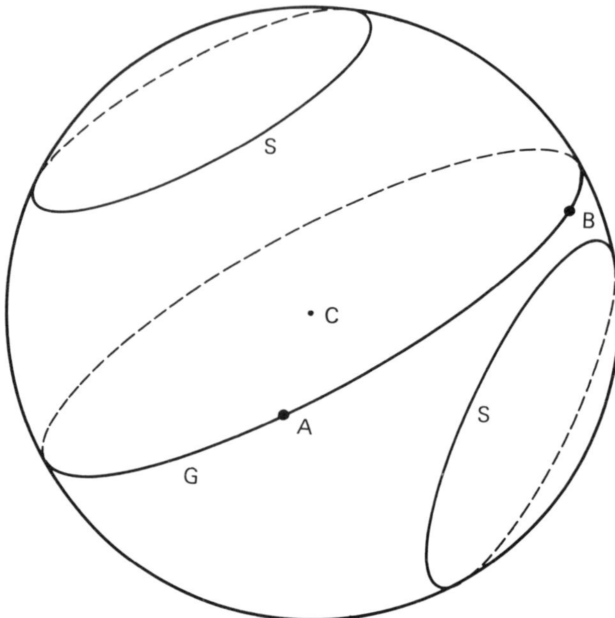

1.4. Great and small circles on a sphere. *C* is the center of the sphere. A great circle (G) is defined as a circle through arbitrary points *A* and *B* that also has *C* as its center. The small circles (S) are not centered on *C*.

near the edge of the page. The sphere is a solid object that may or may not be transparent. The half of the circle through *A* and *B* that passes around the back is shown as a dashed line that may in practice be faint or actually invisible. The result is a three-dimensional visualization of a sphere.

It is obvious that we can draw one *and only one* circle on the sphere through *A* and *B* that has as its center the center of the sphere. If you have any doubts about this restriction try it on a real globe, one of the Earth, or perhaps a basketball. This kind of circle is special. It is called a *great circle* and is marked G in Figure 1.4. By definition a great circle is a circle on a sphere whose center (*C*) is coincident with the center of the sphere. A great circle is a circle of largest radius that can be drawn on a sphere. It is a diametric slice that always divides the sphere into two equal hemispheres. The shortest distance on the surface of the sphere between any two points *A* and *B* is along the great circle connecting them.

Any two points on a sphere, as long as they are not opposite each other, define a unique great circle. There is an infinite number of point pairs, and consequently an infinite number of great circles. Any circle that is not a great circle (one whose center is not coincident with that of the sphere) is called a *small circle*; two of these are labeled S in Figure 1.4.

Measurement of angle on the sphere is the same as it is for the circle. The sphere with the great circle through *A* and *B* reappears in Figure 1.5 with radii

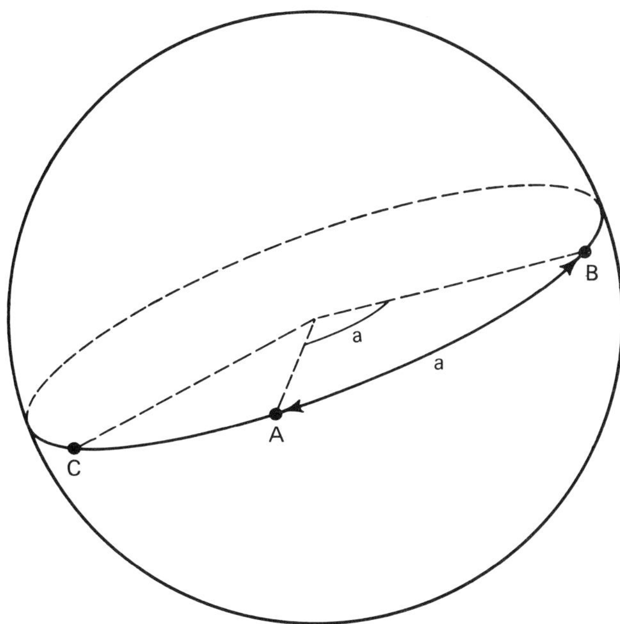

1.5. Angular measurement on a sphere. The angle *a* at the center, defined by the radii to *A* and *B*, is the same as the arc between *A* and *B* measured along the great circle. The great circle is identical to the circle in Figure 1.3 except that it is viewed at an angle. *AB* is about 90°, *AC* about 60°.

extended from the center to each of the points. The plane defined by the circle is identical to that in Figure 1.3 except that we see it from a tilted perspective. The radii define angle *a* at the center, which is numerically equivalent to the arc along the great circle from *A* to *B*.

We might then express the distance between two cities on the Earth in degrees along the great circle that connects them. Try it on a globe. Select London, England, and Montreal, Quebec. Fix a piece of string tightly end-to-end around the globe to construct the great circle through them. Estimate (or measure) the fraction of the string that lies between the cities, multiply by 360° (the full circle) and you find an arc of about 45°.

A globe is rarely readily at hand and it would be cumbersome to use one constantly as you read. In this book we must consequently rely on eye estimates of arc and angle from the perspective representations. In Figure 1.5 the distance between *A* and *B* is about one-half of the solid-line semicircle on the sphere's front half. The angular length of the semicircle is 180°, so the arc between *A* and *B* (and the angle *a*) must be about 90°; that between *A* and *C* is roughly 60°. This kind of estimation is not meant to be precise but only approximate for purposes of learning.

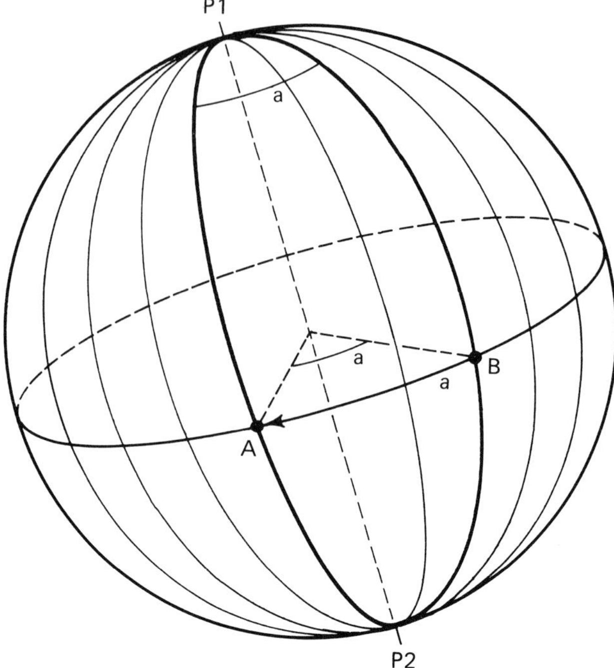

1.6. The poles and secondaries of a great circle. The poles (named *P*1 and *P*2) of the great circle defined by *A* and *B* are the points of intersection between the sphere and the line through its center perpendicular to the circle's plane. Several secondary great semicircles are drawn through the poles perpendicular to the primary great circle defined by *A* and *B*. The specific secondaries through *A* and *B* meet at the poles with the polar angle *a*, which is numerically equal to the angle *a* at the center and the arc *a* along the primary.

1.3 Poles and secondaries

The practice sphere in Figure 1.5 is repeated in Figure 1.6 with point *B* somewhat shifted. Run a line through the sphere's center perpendicular to the circle's plane. It will exit the sphere at two opposite points that are called the *poles* of the great circle, each of which is in every direction 90° of arc from its parent great circle, that is, the poles are centered on the hemispherical surface created by the great circle. We distinguish between the poles by assigning names, here just *P*1 and *P*2. The most familiar example of this concept is the equator of the Earth. It is clearly a great circle as is evident from a globe. The poles of the equator are simply the north and south poles of the Earth, which define our planet's rotation axis. Every great circle has two opposite poles associated with it, and we will look at many additional examples in the pages to come.

Note that Figure 1.6 is not a true perspective construction. The great circle is seen tilted, yet the poles are drawn *on* the circle that represents the sphere, and are not tilted forward and backward as they would be in real perspective. Such a stylized, schematic rendering of a sphere is traditional in classical astronomy, the perspective being distorted for ease in drawing and learning.

Pick pole *P*1 and point *A* in Figure 1.6. A great circle passed through these two points must be perpendicular to the *primary great circle* (that through *A* and *B*) and must pass through the opposite pole as well. This second circle is called, appropriately, a *secondary great circle*. Every point on the primary defines another secondary, and consequently there must be a whole family – an infinite number – of them. Several are lightly drawn in Figure 1.6, with those through points *A* and *B* singled out as special. Only the front halves are indicated to avoid confusion. Every great circle on the sphere must have its own poles and family of secondaries.

We now have a third way of defining the angle between *A* and *B*. The two secondaries through these points converge at either pole, and make an angle (the *polar angle*) that is also equal to *a*, the angle at the center, and the arc *a* between *A* and *B*.

1.4 Coordinates

We may now use these great and small circles, and the angles that have been defined, for the measurement of place or position on a sphere. On the sphere in Figure 1.7 we draw a great circle that we will designate the *fundamental great circle* (or FGC). This circle may have some physical basis or it may be chosen arbitrarily. The poles are again defined by the perpendicular to the circle's plane. Since we are considering only a generic sphere, they are distinguished by calling them *P* and *Q*; north and south or plus and minus would be equally valid. As before, a set of secondaries must run from pole to pole perpendicular to the FGC.

We are now prepared to construct a *coordinate system* that will generate a pair of numbers and will define any point on the sphere. As an example, we wish to define the coordinates of a point *S* in Figure 1.7. We first establish a *zero point*, or origin, on the FGC from which to measure, as required in Figure 1.3. Again the point may have some significance, or it may be arbitrary. The secondary great circle (or SGC) through the zero point is now a special case and is called the *prime secondary*. From here on, we deal with secondaries only as semicircles. That is, only the semicircle through the origin is called the prime secondary; the part of the circle opposite the origin is not included. We also draw another secondary (semi)circle through point *S*, which intersects the FGC at point *R*. We have now defined two arcs suitable for angular measurement. One is on the FGC between zero and *R*, and is analogous to the arc between *A* and *B* in Figures 1.5 and 1.6. We can obviously go between zero and *R* in two directions, to the right or left of the zero point, so we must specify which direction we want. The other arc is between the FGC at *R* and point *S* along the SGC. Here we have to specify on which side of the FGC we find *S* – toward pole *P* or pole *Q*. We refer to the two arcs as *x* and *y* in Figure 1.7. Every point *S* on the sphere now has a unique set of *x* and *y*, specified according to direction. In the diagram, *S* has coordinates $x = 80°$ (measured clockwise from zero, looking down from pole *P*), and $y = 45°$, measured toward *P*. Note again that direction *must* be specified.

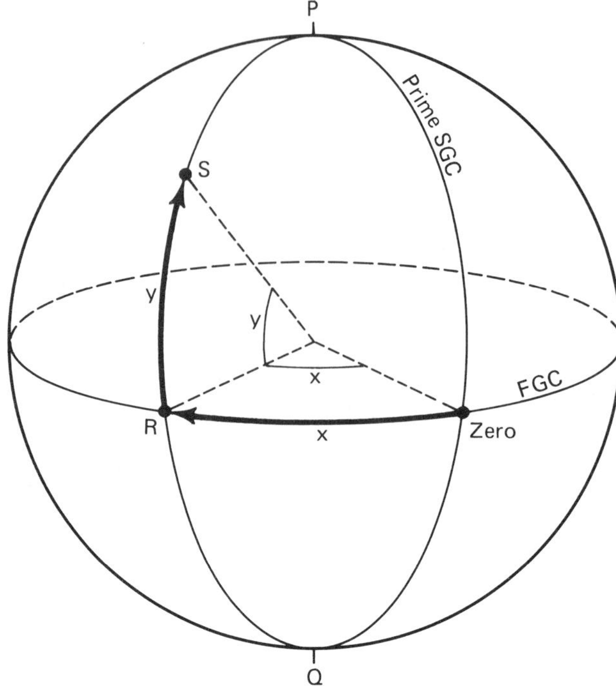

1.7. A general system of spherical coordinates. The FGC is the fundamental great circle, which then defines poles *P* and *Q*. The zero point is arbitrarily chosen. Secondary great circles run from the poles through zero and point *S*. The one through zero is the prime secondary. Arc x is measured along the FGC between zero and point *R*, the intersection of the secondary through *S* and the FGC. Arc y is measured along the secondary between *R* and *S*. Any point on the sphere can now be located by the unique pair of numbers x and y plus their directions of measurement.

All points on the FGC have coordinate $y = 0°$. The poles *P* and *Q* do not have a coordinate x, since all the secondaries converge there. That is, x does not exist at the poles. For the poles, specifying that $y = 90°$ (in the appropriate direction) is sufficient.

1.5 The Earth

The most important example of a coordinate system is the one used to specify position on Earth. The Earth is not a perfect sphere but an *oblate spheroid* with a diameter through the equator (12 756 km) that is 41 kilometers greater than that through the poles. For our purposes here, however, we will assume that the Earth is indeed perfectly spherical, the deviation from the sphere to be addressed in Section 5.2.

How do we indeed know that the Earth is a sphere? At first glance it looks to be flat, and for centuries was considered so. The most obvious evidence is that if we look at an object that is sufficiently distant – the classic example is a ship out at sea – we only

(a)

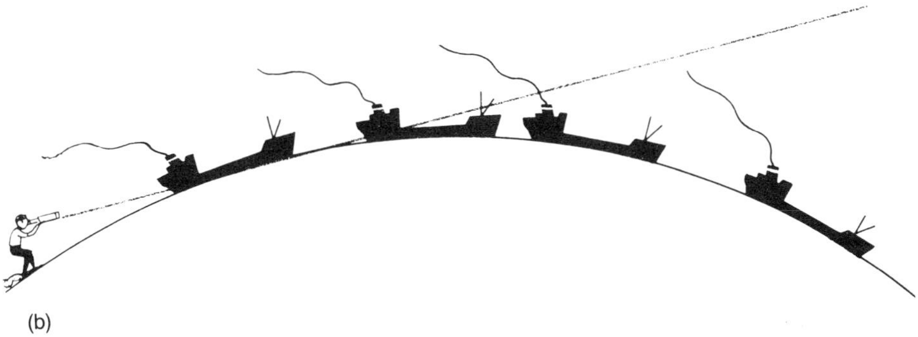

(b)

1.8. The curvature of the Earth. Only the upper parts of a distant ship are visible (a), the lower hull rendered invisible by the curvature of the Earth (b). ((a) Author's photograph; (b) from *Principles of Astronomy: A Short Version*, S. P. Wyatt and J. B. Kaler, Allyn and Bacon, Boston, 1981.)

see its upper parts (Figure 1.8). When it sails far enough away, it disappears altogether. But that observation only shows Earth to be curved, not that it is round and closed back in on itself. The best of the ancient proofs (known to Aristotle) is that the Earth's shadow – cast on the Moon in a lunar eclipse (see Chapter 10) – is always circular, and the only figure whose cross-section is always a circle is a sphere. Now, of course, we can circumnavigate the globe and even photograph it from space. There is no doubt about its figure. (Remarkably, however, "flat Earth" societies still exist.)

The Earth is symbolized in Figure 1.9. As for any sphere, we could choose any great circle we wished as the FGC for a coordinate system. The one that runs between Mt. Everest and the Cape of Good Hope might be good since it would be defined by visible

Table 1.1 *The Greek alphabet*

Letter	Upper-case	Lower-case	Letter	Upper-case	Lower-case
alpha	Α	α	nu	Ν	ν
beta	Β	β	xi	Ξ	ξ
gamma	Γ	γ	omicron	Ο	ο
delta	Δ	δ	pi	Π	π
epsilon	Ε	ε	rho	Ρ	ρ
zeta	Ζ	ζ	sigma	Σ	σ
eta	Η	η	tau	Τ	τ
theta	Θ	θ	upsilon	Υ	υ
iota	Ι	ι	phi	Φ	φ
kappa	Κ	κ	chi	Χ	χ
lambda	Λ	λ	psi	Ψ	ψ
mu	Μ	μ	omega	Ω	ω

features. It would not be very convenient as a basis for actual measurement, however. It is far better to choose a circle that has some physical significance with reference to the Earth as a whole, namely one that is based on the rotation of the planet. The Earth spins once a day on an axis that passes through its center and exits the sphere at the *poles of rotation*. The poles are the only points on the terrestrial surface that remain stationary. Midway between the poles is a great circle around the Earth, the *equator*. The equator is the circle that is farthest away from the rotational axis.

Since the equator is defined by the Earth's spin, we make it the FGC. The poles of rotation then become the poles of the coordinate system as well. We call one *north* (N) and the other *south* (S) to distinguish one from the other. The north pole is defined such that when one looks down on it from overhead the Earth appears to turn in the counterclockwise direction (and, obviously, in the opposite direction if the south pole is so examined).

The zero point on the Earth's equator is selected arbitrarily by choosing a prime SGC that everyone agrees upon. The terrestrial semicircular secondaries are called *meridians*. The *prime meridian* is chosen to pass through Greenwich, near London, England, and is also called the *Greenwich meridian*. The intersection of this prime SGC and the equator is the origin, or zero, which lies in the Atlantic Ocean off the coast of Africa.

Terrestrial coordinates are called *longitude* and *latitude* and are analogous to x and y in Figure 1.7. The lower-case Greek letter "lambda" (λ) is generally used to denote longitude; latitude will be referred to by the Greek letter "phi" (ϕ). (The Greek alphabet, which will frequently be encountered, is given in Table 1.1.) Longitude is an arc measured along the equator in either the clockwise or counterclockwise directions, not exceeding 180°. Counterclockwise, in the direction of the Earth's rotation, defines *east*, and clockwise, *west*. East and west are always directed perpendicular to the secondaries.

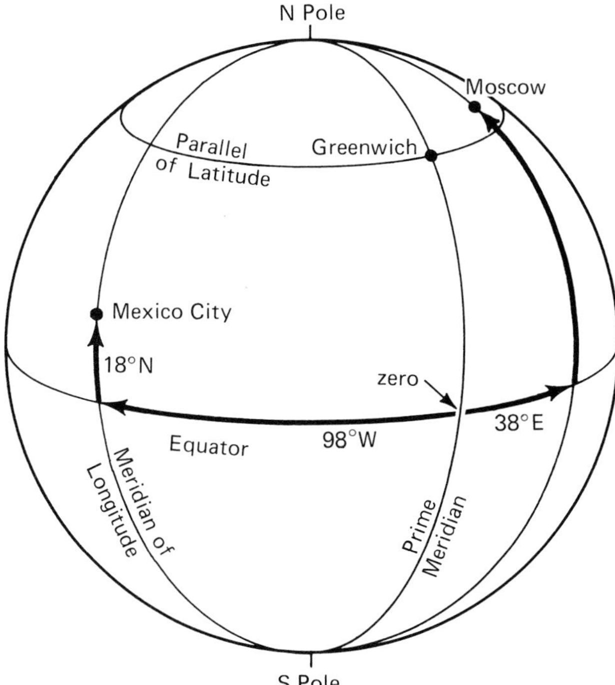

1.9. Latitude and longitude. The north and south poles of the coordinate system are defined by the rotation of the Earth. The equator then becomes the FGC. The prime secondary, or prime meridian, passes through Greenwich, and its intersection with the equator defines "zero." Longitude is measured along the equator east and west of the Greenwich meridian to the secondaries (meridians) through the cities; latitude is measured north and south of the equator along a meridian. Greenwich itself has a latitude of 51° N. Its parallel of latitude is a small circle parallel to the equator.

Latitude is the arc measured perpendicular to the equator along any SGC, or *north* and *south* toward the north or south rotation poles.

A pair of examples illustrate the concept. In Figure 1.9, select two locations, Mexico City and Moscow. First draw a secondary through each point. Mexico City is closer to the prime meridian if measured west (W). Its longitude, measured along the equator (to the intersection of the secondary), is 98° west of Greenwich. Moscow is closer if measured east (E). Its longitude is 38° E. The latitudes, measured along the appropriate secondaries, are 18° and 57° north (N) of the equator respectively. Points south of the equator are designated S. For example, Santiago de Chile has a latitude of 33° S. Occasionally the symbols "+" or "–" are substituted for N and S (e.g., +38° instead of 38° N). All points that lie at the same latitude form a small circle parallel to the equator. These small circles are called *parallels of latitude* as opposed to the *meridians of longitude*. East and west are always directed along parallels, north and south along meridians.

Every point on the Earth's surface is now uniquely defined by a set of numbers giving latitude and longitude. The prime meridian has a longitude of 0° and the equator a latitude of 0°. Note again that longitude is always measured along the zeroth parallel (the equator). The poles are again defined by only 90° N or 90° S. Longitude does not exist at the poles. Since there are no parallels or meridians at the poles (the meridians all converge together to a point), neither can there be any east or west. All directions at the north pole are south, and all at the south pole are north.

The circumference of the Earth is about 40 000 kilometers (25 000 miles), which corresponds to a distance of about 112 kilometers (70 miles) per degree of latitude, or per degree of longitude measured along the equator. The length of one minute of arc along the equator, 1853 meters = 6080 feet, is defined as one *nautical mile*. Since the meridians become closer north and south of the equator, a degree of longitude corresponds to smaller and smaller distances as the latitude is increased or decreased from zero.

What is so special about Greenwich, England that the whole system of longitude should be based upon it? Why not Rome or Berlin? In fact there is nothing physically distinctive about the location. It is easy to find old maps in which longitude is expressed according to the meridian of Paris for example. (Use caution when using old maps, and pay attention to the legends and captions.) England long stood as the world's great sea power, however, and the English explored and conquered extensively. Consequently the meridian of London superseded all others. It is now adopted by common convention among all nations.

The determination of longitude is an astronomical art. It is, as we will see, measured from the positions of the stars in the sky. The prime meridian is based on an astronomical telescope located within the Old Royal Greenwich Observatory (Figure 1.10). A section of the prime meridian is marked there by a brass strip set in concrete: it is real, not just some mathematical abstraction.

1.6 The sky, or celestial sphere

We now move from the Earth to the sky and to begin to examine the circles and points that can be drawn upon it. It is necessary to look at the sky from the point of view of the observer, that is, to see how the sky *appears*, not necessarily the way it actually *is*. If you stand in a field on a clear summer day you will see the sky as a hemispherical bowl over your head. It is an illusion. The blue sky is not solid but is caused by sunlight scattering in air. Everything in space is so far away that all sense of distance is lost, and the stars, planets, Moon, and Sun appear to be tacked upon this illusory bowl. The sphere of sky actually extends all the way around: you see only the upper half because the Earth gets in the way. This apparent surface is termed the *celestial sphere*. It is a fiction, but a necessary one. It is not a vague concept, however: just look outside to see it.

Start with the Earth in space (Figure 1.11a) in the same orientation as Figure 1.9. The observer is located at an arbitrary point, in the example at a latitude of 45° N. The longitude is, for now, irrelevant. The person appears tipped on his or her side;

1.10. The Old Royal Greenwich Observatory. The prime meridian runs through the telescope in the building to the lower right and is visible as a brass strip at lower left. The eastern hemisphere of the world, longitude east, is below it, the western is above. The two hemispheres meet at 180° longitude on the opposite side of the globe. (National Maritime Museum, London.)

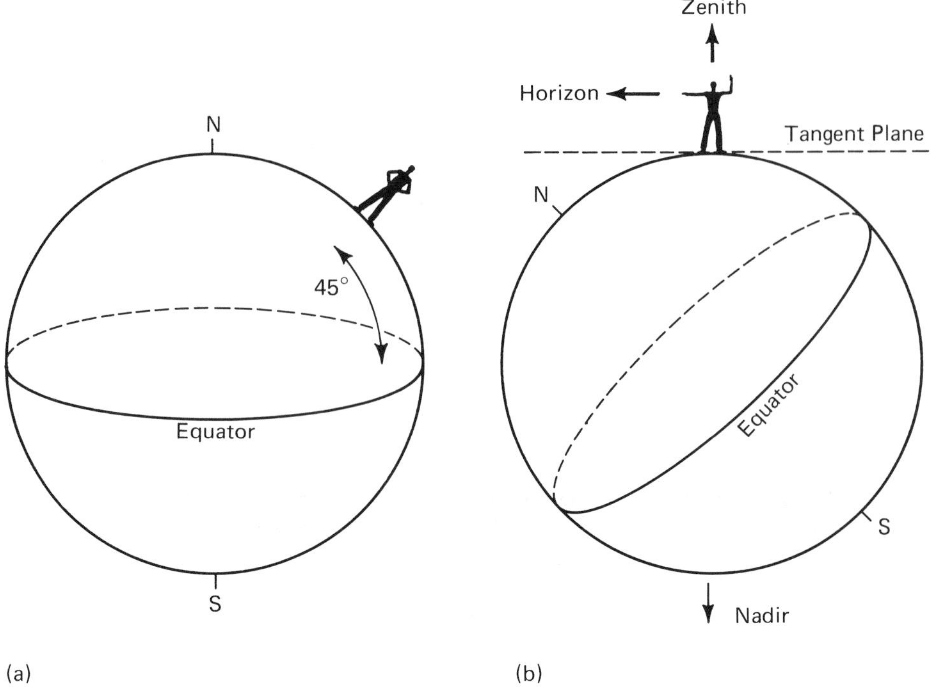

1.11. The observer in space. (a) The Earth is shown with an observer at 45° N latitude. (b) The Earth has been rotated 45° counterclockwise, rendering the observer apparently upright. The person points both to the zenith and to the horizon. The nadir is in the direction opposite the zenith.

however that is not the way it would look to the individual. As far as we are concerned, if we are at 45° or any other latitude, we seem to be standing on top of the globe. We know, of course, that there is no up or down in space, but because of our planet's gravity we always appear to ourselves to be vertical, standing straight up and down. And in Figure 1.11b that is exactly the way in which the Earth is drawn. If the person in Figure 1.11a is 45° north of the equator, and the equator is 90° from the pole, he or she must then be 90° – 45° = 45° south of the pole (i.e. halfway from the equator to the pole). Figure 1.11b is then just Figure 1.11a rotated 45° counterclockwise to bring the individual upright. We see now that the apparent orientation of the Earth's equator, axis, and poles in space *with respect to the observer* depends upon latitude.

The sky, or celestial sphere, is not yet drawn around the Earth in Figure 1.11b. Before that is done it would be helpful to define a pair of concepts. First the person has one arm pointing straight up, directly away from the center of the planet, toward a point in the sky called the *zenith*. It is the point on the celestial sphere directly overhead, and is actually and properly defined by the direction opposite the pull of the Earth's gravity, which is toward the center. The zenith is a personal property: everyone

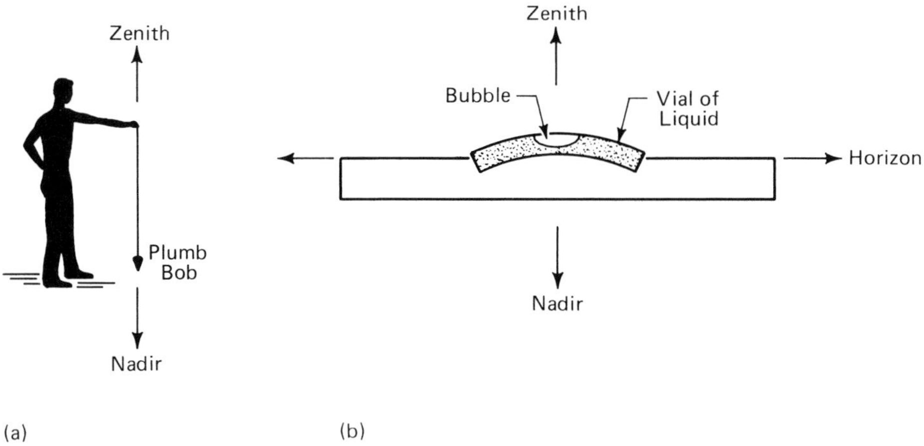

1.12. Plumb line and level. (a) The person holds a plumb line weighted with a plumb bob. The line hangs toward the center of the Earth and establishes the direction to the zenith and nadir. (b) The level holds a curved tube of liquid that is set with precision into a long bar. The bubble rises to the top of the curve. When the bar is rotated so as to place the bubble between two markers, it points exactly to the horizon and is exactly perpendicular to the plumb line.

has his or her own. Since the globe is continuously curved, any other individual is to some degree around the curvature of the Earth from you, and must have a different zenith. It matters not if your friend is standing nearby on a flat floor: the individual will have a slightly different direction of gravity from you and must have a different zenith. The difference may at first seem too trivial to be concerned with, but in fact differences in latitude and longitude, and in the position of the zenith relative to the stars in the sky, can be noticed over only a meter or so.

The opposite point, 180° away, lies directly beneath one's feet, and is called the *nadir.* Since it and the zenith are defined by gravity, a weight on a string (a carpenter's plumb bob, from the Latin "plumbum" for the heavy metal lead, Figure 1.12a) will point right at it. A *plumb line* generated by the bob stretches between zenith and nadir. The nadir is a point in the sky, not one on the Earth. Since the Earth is in the way you can never see your nadir. But you can be comforted by knowing that your zenith is ever in view.

Then note that the person in Figure 1.11b is also pointing off to your left in the direction perpendicular to the plumb line toward the *horizon.* Figure 1.11b is severely distorted, as the individual stands there about 2000 kilometers tall. But to draw to proportion would preclude seeing the curvature of the Earth. Use your imagination to shrink the person to size. Then you see the observer to be pointing along the dashed line, which represents the plane tangent to the Earth at his or her feet ("tangent" meaning touching the Earth perpendicular to the plumb line). The person cannot see any part of the sky below the horizon as the Earth is in the way.

Step outside and you can see the horizon where the Earth appears to meet the blue heavens. The visible horizon may not be a very accurate rendition of the true one (the *astronomical horizon*) defined by the perpendicular to the plumb line, but it will give you the idea. The best and closest fit between the visible and ideal horizons will be at sea, where the ocean makes a good tangent plane. There you will view almost exactly half the sky. On land you can easily locate the direction to the horizon with a simple carpenter's level. A bubble trapped in a curved vial of liquid will float to the top (Figure 1.12b), showing you where to sight toward the true horizon. Note from these simple tools that this direction must be exactly perpendicular to the string of the plumb bob. Buildings are actually constructed from these two venerable devices along astronomical sightlines. Just as your zenith and nadir are your personal property, so is your horizon. Since your neighbor is just around the curvature of the Earth from you, his or her horizon will be different also.

With this background, we can draw the celestial sphere around the Earth (Figure 1.13), enclosing Figure 1.11b. The various things on the sphere, stars etc., are so far away that the sphere's radius is very large compared to the Earth's, and is effectively infinite. That is, our planet is tiny, a mere mote, compared with the sky. However, the Earth and the celestial sphere cannot be drawn together in proportion on a page. In Figure 1.13 the size of the Earth is greatly exaggerated to enable the relations between it and the celestial sphere to be seen. You will have to use your imagination again and shrink the Earth to the actual point it is to avoid distortions. Another complication is that we are in reality enclosed by this apparent sphere, but it obviously cannot be drawn that way. In reading about the celestial sphere we are in the position of standing outside the sphere looking in, and we have to imagine ourselves inside, looking out. With some practice, one becomes accustomed to this viewpoint rather readily.

Finally, we can place the zenith, the nadir, and the horizon onto the celestial sphere. The zenith and the nadir are directly above and below the observer, and the horizon is a great circle around the observer, the direction to which is everywhere perpendicular to the line to the zenith. The horizon cuts the celestial sphere into visible and invisible hemispheres (or into its "sky" and "Earth" portions) whose surfaces are centered on the zenith and nadir. Since your zenith and nadir are at right angles to the horizon, they must be its poles.

1.7 The celestial poles and equator

The horizon, zenith, and nadir are all related to the observer. We now define points and circles that are independent of the observer and that are defined by the Earth alone. In Figure 1.14, extend the Earth's *axis* (the line that passes through the north and south poles, labeled NP and SP) into space in both directions. The line must intersect the celestial sphere at two points, analogous to those on the Earth, called the *north celestial pole* (NCP) and the *south celestial pole* (SCP). The NCP is directly above the Earth's north pole, and the SCP directly above the south pole. If an observer were standing directly on the north pole of the globe, the north celestial pole would be

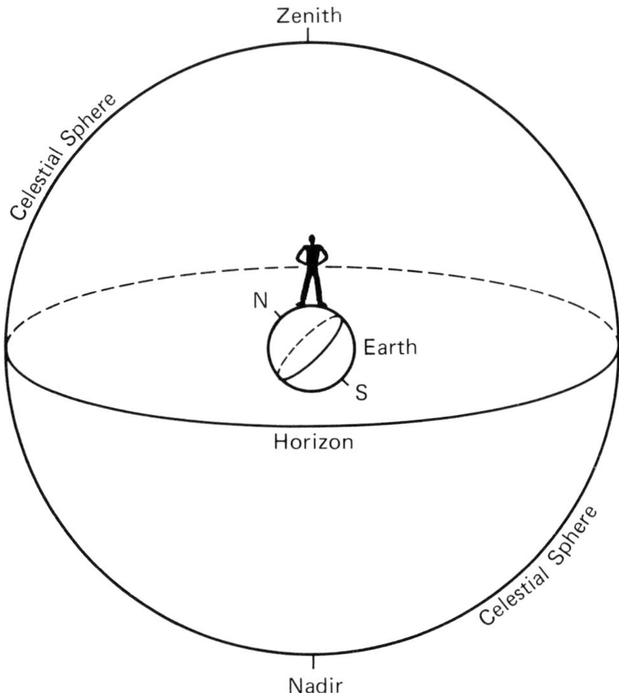

1.13. The celestial sphere. The sphere of the sky is drawn around the Earth with a center common to both. The observer appears to be on "top" of the Earth as in Figure 1.11b. You are on the outside looking in, but must imagine yourself to be on the inside looking out. The sky is infinitely large compared to the Earth. The size of the Earth is greatly exaggerated, and you must imagine it to be very small. The great circle of the horizon is the intersection between the celestial sphere and the plane tangent to the Earth at the observer's feet. The zenith is a point on the sphere directly over the observer's head; the nadir is opposite.

directly overhead, at the zenith. The celestial poles are to the sky what the terrestrial poles are to the Earth. Just as the planet turns around its poles, the sky appears to rotate oppositely about its celestial counterparts. We will look at the subject of motions in detail in the next chapter.

Note that although the Earth's north pole is not visible to the observer in Figure 1.14, the NCP *is* visible, since it is in the sky above the horizon. The SCP is *not* visible since it is below the horizon. As the SCP and NCP are opposite one another, only one can be seen by any observer: a person in the northern hemisphere will see the NCP only, and vice versa.

The NCP and the zenith are two points on a sphere, and as such define a great circle, in fact the circle drawn on the page to represent the sphere itself. This great circle is called the *celestial meridian*. Since the nadir and the SCP are respectively opposite the zenith and NCP, they must fall upon it as well; the NCP and zenith or

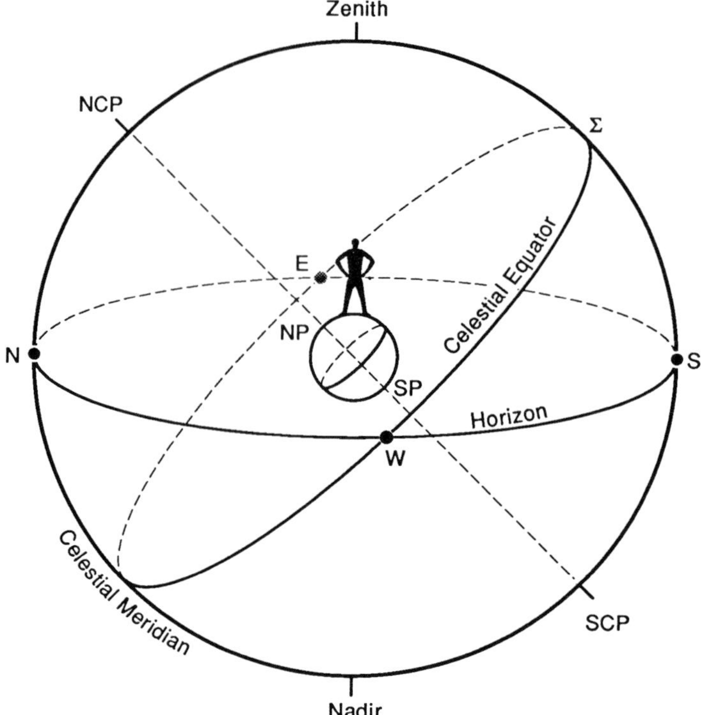

1.14. The celestial poles and compass points. NP and SP mark the Earth's north and south rotation poles. The extended rotation axis intersects the celestial sphere at the north and south celestial poles, NCP and SCP. The meridian is the great circle drawn through the zenith, nadir, NCP, and SCP, and intersects the horizon at the north and south points (N and S) respectively. The celestial equator is formed by the intersection of the celestial sphere with the extended plane of the Earth's equator. It intersects the horizon at east (E), in the back of the sphere, and at west (W), in the front, and intersects the meridian (above the horizon) at the equator point (Σ).

the SCP and nadir actually define the "meridian" (as it is loosely called) just as well. Do not confuse it with terrestrial meridians used to determine longitude. The same basic idea is present, but for any individual there is only *one* celestial meridian.

One more circle belongs on the celestial sphere, the counterpart to the Earth's equator, the *celestial equator*. To define it extend the Earth's equatorial plane in Figure 1.14 outward in all directions until it intersects the celestial sphere, which it must do in a great circle centered on the Earth. The celestial equator must at all points be 90° from both the north and south celestial poles, that is, it divides the sky into equal northern and southern hemispheres with the NCP and SCP as its poles. Just as the

NCP and SCP lie above the Earth's poles, the celestial equator lies above its terrestrial version.

1.8 North, south, east, and west

There are now three great circles on the celestial sphere, the horizon, the celestial meridian, and the celestial equator. Each must intersect the other at two opposite points. Look first at the intersections of the celestial meridian and the horizon in Figure 1.14. The one nearest the NCP is called *north* (N) and that nearest the SCP, south (S). "North" is actually found on the horizon by locating the NCP and dropping a plumb line from it to the horizon. From Figure 1.14 we see that if we face the north celestial pole and walk toward it, we must also be walking toward the globe's north rotation pole, and consequently along a terrestrial meridian, or due north.

Because the poles of the equator and horizon are both on the meridian, the intersections between the equator and the horizon must each be halfway between north and south, at the *east* (E) and *west* (W) points, 90° from north and south; the intersections of the equator and horizon in fact *define* east and west. If we face north, east must be along a parallel of latitude exactly to our right and west to our left; the east and west points of the horizon are the poles of the meridian. These four points, labeled in Figure 1.14, are on the horizon and the celestial sphere. Exact direction on Earth can be found only by astronomical means.

Finally, the equator and meridian must cross once above the horizon and once below. The point of intersection above the horizon is called the *equator point* (or *sigma point*), Σ. In this diagram, viewed from the northern hemisphere, it is seen to the south. That will not always be true and will depend on the location of the observer (Chapter 2). Alas, the other point, the one below the horizon, carries no name: out of sight, out of mind. Note the beautiful symmetry. There are three principal great circles, and the poles of each one lie on one of the others. Moreover, each divides the others exactly in half, so that half the meridian and equator are above the horizon and half below.

The reader should try to locate all these points and circles in the actual sky. The zenith and nadir are easy. As a preview of the last section of this chapter (stated for the moment without proof), note that the angular elevation of the celestial pole above the horizon in degrees equals your latitude. If you live in the north, just look up your latitude on a map, face north, then look upward that many degrees to find the NCP; if you reside in the southern hemisphere look to the south instead to find the SCP. Once you have the pole, connect a circle with it and the zenith to trace out the meridian. Next, point one arm at the pole, and move the other in a circle always at right angles to it. You are then drawing the celestial equator in the sky.

How in practice are the celestial poles located to find north and south? Just as the terrestrial poles are the only stationary points on Earth, the celestial poles are the only fixed positions in the sky. Looking ahead to the next chapter we see that the Earth's spin causes the stars to appear to move across the heavens, and the Sun to rise and set. We merely find the spot in the sky that does not move. How this point is located exactly will be saved for later, but it is easy to walk outside and approximate it by just looking for a few hours. Residents of the northern hemisphere are lucky in that the NCP is closely marked by a star – the legendary *north star* or *Polaris* – only half a degree removed from it. If you know your constellations (it is at the end of the handle of the Little Dipper: see Chapter 4) face Polaris and walk north: we thereby encounter the first rule of celestial navigation. Those living in the south have no such handy guide, but may still estimate the position of the SCP from its known location among the stars.

Oddly, although these directions are called "compass points," a magnetic compass does not locate them very well. The needle of a compass points to locations in Canada and Antarctica respectively called *magnetic north and south* that are fairly close to, but not *at*, the rotation poles. A correction must always be made to a direction found magnetically. This subject will be expanded in Chapter 13.

The horizon and the cardinal compass points (N, S, E, W) are subdivided to indicate finer directions. The position exactly between north and east, for example, is called *northeast* (NE); that between north and northeast is *north by northeast* (or north-northeast, NNE); the direction between northeast and east is east by northeast (ENE), and so on. These compass points are often beautifully and artistically laid out on a mariner's compass, shown in Figure 1.15. The terms are also used in a vague general sense. That is, "east" does not necessarily mean *exactly* east, but can also indicate the general direction somewhere to the right of north, or between north and south, as opposed to the general "west." The meaning will usually be clear from the context.

1.9 Horizon coordinates: altitude and azimuth

With these definitions in place it is now possible to establish the first of what will eventually be several sets of coordinates in the sky. The simplest of them uses the only visible great circle, the horizon, as the fundamental, or FGC. From Section 1.6 the zenith and the nadir are the poles of the horizon, and consequently of the *horizon coordinate system* as well, which will serve to locate any point in the sky.

Figure 1.16 has a star named *S* in the southwestern sky, on the front side of the celestial sphere. Follow the rules of Section 1.5 for the establishment of coordinates. A secondary circle is first passed from the zenith through the star, perpendicular to the horizon, to the nadir. The secondaries in the horizon system are called *vertical circles* (sometimes *azimuth circles*). We now need only choose a zero point on the horizon and the system is complete. The zero point used in astronomy is north (N). The coordinate measured on the horizon is called *azimuth*, *A*. (The word comes from the Arabic *al-sumut*, meaning "the way." There are a great many Arabic terms in astronomy stemming from a vast heritage from those lands.) By common convention, azimuth is always measured clockwise as seen looking down from the zenith, that is, from north through east, south, and west. Since east is 90° around the horizon from north, its azimuth is clearly 90°; that of south and west are respectively 180° and 270°. The vertical circle through star *S* crosses the horizon roughly in the southwest, so its azimuth must be

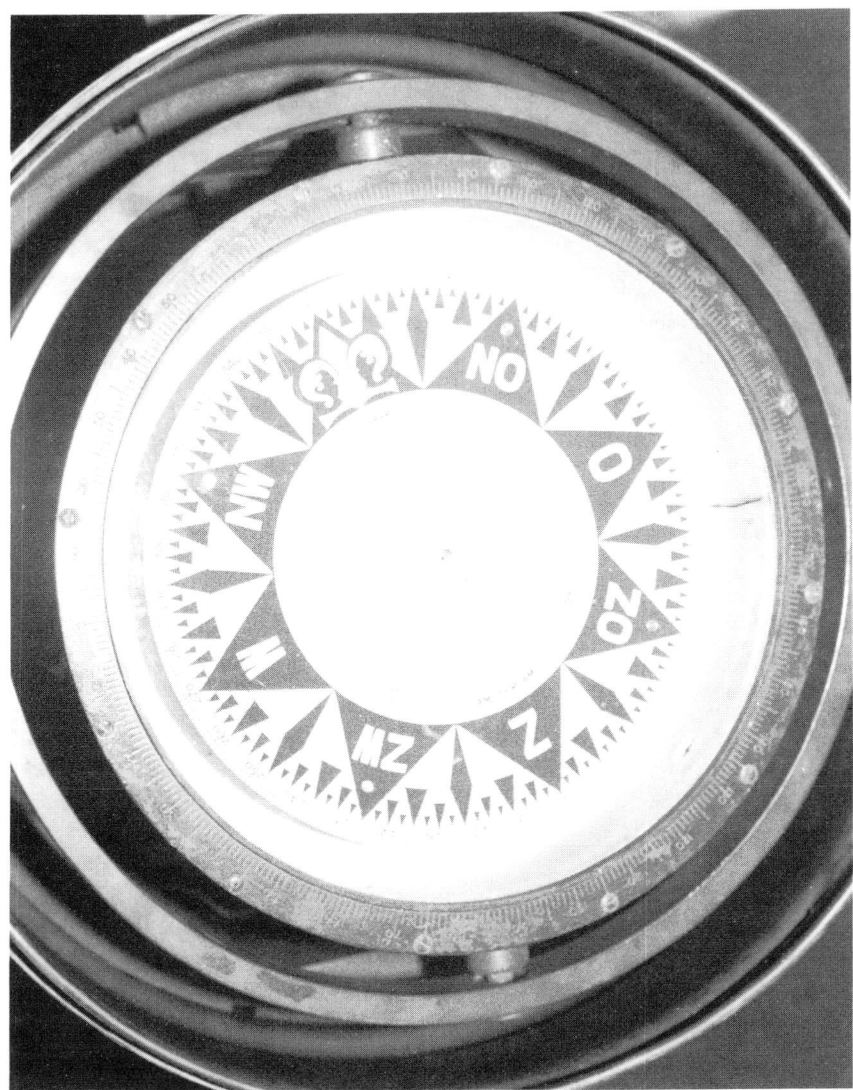

1.15. A mariner's compass. This is the plate from a compass that rode at sea aboard a Dutch ship and provided directions for the helmsman. "O" in Dutch stands for east (*oost*), and "Z" for south (*zuid*). North, south, east, and west are subdivided into finer divisions, southeast, east-southeast, etc. (Author's photograph, courtesy of John Schmitt of Ned Kelly's Steakhouse, Urbana, Illinois, USA.)

somewhere between 180° and 270°. From Figure 1.16 (allowing for perspective), we might estimate that $A = 240°$.

The angle measured along the secondary perpendicular to the FGC is called *altitude*, h. (Do not confuse it with physical altitude in meters: the term as used here refers to

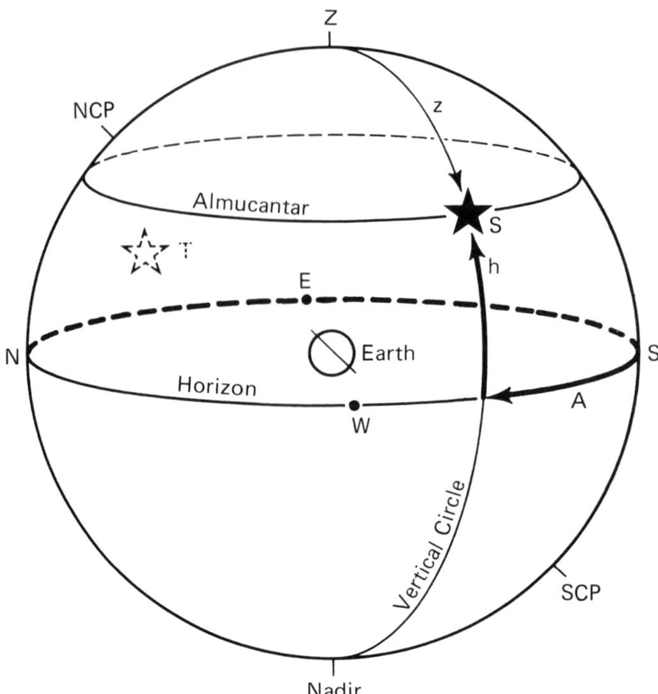

1.16. Horizon coordinates. The horizon is the FGC. Secondary circles called vertical circles run from the zenith (Z) to the nadir. Azimuth (A) is measured from north through east on the horizon to the foot of the vertical circle through the star; altitude (h) is measured along a vertical circle up or down from the equator. The almucantar is a small circle of constant altitude. The zenith distance, z, is the complement of h: $z = 90° - h$.

angle; h is sometimes also called the *elevation*.) It is always measured from the horizon, positive toward the zenith, negative toward the nadir (the "plus" is ordinarily dropped). Any star with a negative altitude must be below the horizon and cannot ordinarily be seen. The horizon and points N, E, S, W have altitudes of zero degrees. That of the zenith is 90°, and of the nadir −90°. The altitude of star S in Figure 1.16 is about 40°. Every point on the celestial sphere can be identified by specifying its azimuth and altitude, A and h. As a further example, star T on the back side of the sphere has coordinates $A = 30°$, $h = 20°$.

The *complement* of altitude is quite often used instead of altitude itself. The complement of an angle is 90° minus that angle. Since the altitude of the zenith is 90°, the angular distance of the star from the zenith must be 90° minus altitude. The angle between the star and the zenith is called the *zenith distance*, z, or $z = 90° - h$ (and conversely, $h = 90° - z$). The position of a star in the sky is often specified by A and z instead of by A and h.

Finally, look here at some special points and circles. Just like longitude at the poles, azimuth disappears at the zenith and nadir: we need only specify an altitude of plus or minus 90°. The celestial meridian, since it passes through the zenith and nadir, is obviously a vertical circle, the northern half with azimuth 0°, the southern with 180°. The vertical circle that runs from east to west (not indicated in the figure: draw it in yourself) is called the *prime vertical*. (Consistency is not always a strong point in astronomy: the "prime" meridian on Earth denotes a longitude of 0°, whereas the "prime" vertical designates azimuths of 90° *and* 270° depending upon whether you look east or west.) Parallels of altitude, small circles of constant altitude (or zenith distance) parallel to the horizon, are called *almucantars*, from the Arabic *al-muqantarat* meaning "the arched bridges." The prime vertical and almucantars have specific uses in positional measurement and navigation.

1.10 The orientation of the celestial sphere

Now that altitude has been properly introduced we can establish formally how the celestial sphere is oriented for the fixed observer. It is obvious from Figures 1.10 through 1.16 that if the observer stands somewhere in the northern hemisphere, the north celestial pole will be visible above the northern horizon. But where to find it? And more to the point, when found, what will it tell us about our relationship to the Earth? Examine Figure 1.17. It is a stripped-down version of Figure 1.14 in which there is no attempt to show three-dimensional structure. The figure is simply a slice of the sky and the Earth, shown as two concentric circles, in the plane of the celestial meridian. The Earth's equator is now represented by a line through the center, the equatorial plane that cuts our planet into northern and southern hemispheres; its outward extension strikes the meridian at the equator point, Σ. The horizon, from which altitudes are measured, is represented by a horizontal line through the middle of the Earth, since the Earth should actually be a mere point at the center.

The observer still stands atop the circle of the Earth at a northern latitude ϕ, which means (Section 1.5) that the arc measured from the person to the equator is $\phi°$ long. That is, if we could view the observer and the equator from the Earth's center the angle between them would be $\phi°$. By extension, the line from the center of the globe to the zenith (Z) intersects the line to the equator point also by $\phi°$. So far, this part of the figure looks just like Figure 1.2, where it is shown that the length of an arc defined by two lines radiating from the center of a circle is independent of the size of the circle. It is then clear that the zenith distance, z, of the equator point must always be identical to the observer's latitude, that is, z of sigma equals ϕ, or in mathematical shorthand,

$$\boxed{z\,(\Sigma) = \phi} \quad .$$

The altitude (90° – z) of the equator point must therefore be 90° – ϕ, or

$$\boxed{h\,(\Sigma) = 90° - \phi} \quad .$$

The quantity (90° – ϕ) is a special, occasionally used number called the *co-latitude*, which is the observer's angular distance in degrees from the terrestrial pole rather than from the equator, and indeed is referred to as the *polar distance* (remember, degrees,

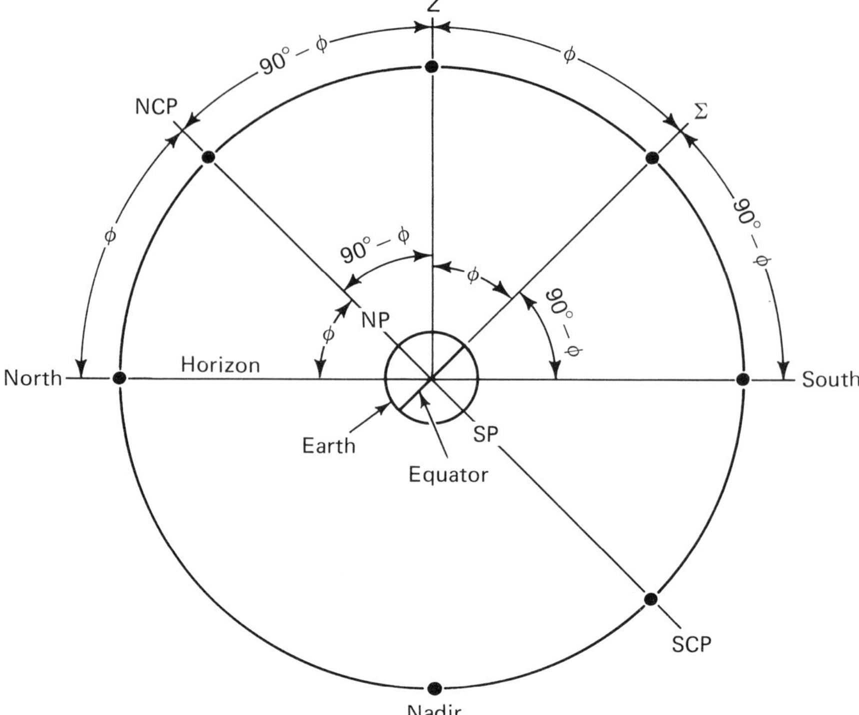

1.17. The altitude of the pole. The two concentric circles represent cross-sections of the celestial sphere and Earth sliced along the celestial meridian. NP and SP are the Earth's north and south poles. The Earth should be mentally shrunk to a point at the center. The lines from the center of the Earth to the zenith (Z) and to the equator point (Σ) intersect at an angle ϕ, the latitude, and the lines from the zenith and the NCP intersect at ($90° - \phi$), the co-latitude. Since the zenith is 90° from the horizon, the NCP must be $90° - (90° - \phi)$ from the horizon. Thus ϕ, simultaneously the altitude of the pole and the zenith distance of the equator point, equals the latitude of the observer.

not kilometers). No matter where you are in the northern hemisphere you can find the highest arch of the celestial equator, the equator point, by looking up by a number of degrees equal to your co-latitude. (From symmetry, you would look north from a southern hemisphere location; we will travel south shortly.)

From the same argument used above, we see from Figure 1.17 that the zenith distance of the NCP is equal to the co-latitude, or

$$\boxed{z\,(\mathrm{NCP}) = 90° - \phi} \quad .$$

That is, the arc between the NCP and Σ is 90°, hence the arc NCP to the zenith must be the complement of the arc ZΣ, which is $90° - \phi$. Since the altitude of the zenith is 90°, the arc between the north point of the horizon and the NCP must be 90° minus the co-latitude, or 90° – (90° – ϕ), and we simply recover ϕ. That is, *the altitude of the north celestial pole equals the latitude of the observer*,

$$\boxed{h(\mathrm{NCP}) = \phi} \quad .$$

Once you know this rule, the sky is oriented for you, and you can find the stars. Conversely, if you know your stars, *you know your latitude*, one-half of your position on Earth. You are now becoming a celestial navigator.

Return to Figure 1.16. The latitude of the observer was set in Section 1.7 at 45°, halfway between the equator and the north pole. Therefore the altitude of the NCP is 45°, or halfway up the sky. If you are in Havana, Cuba, latitude 23° N, you look up in the north by 23°, and from Stockholm, ϕ = 59°, up by *that* amount. The rule works in the southern hemisphere as well. If you are in Santiago, Chile, 33° S, or –33°, h(NCP) = –33°, which means that the NCP is 33° below the northern horizon, and consequently that the SCP (180° removed from the NCP) must be 33° above the southern horizon: symmetry in action.

You now have a way of beginning to find constellations if you are not already familiar with them. Just look up your latitude in an atlas. If you live in the north, sight up from the northern horizon by that many degrees, and you will see a modestly bright star, Polaris. Because of its fame, it is commonly assumed it must be brilliant. It is not, but it is easily seen even from town. There will be no ambiguity, since there are no comparable stars near it. The star forms the end of the handle of the so-called Little Dipper, or the end of the tail of Ursa Minor, the small bear. The Dipper is made of some faint stars, but from the dark countryside on a moonless night, you can pick out its outline, which is shown on the north polar star map in Appendix 2. Your outstretched hand would about cover it.

This approach does not work well in the southern hemisphere as the SCP lies atop no bright marker. But if you look around a circle 30° away from the pole you can at least locate the bright *Southern Cross*, known also as Crux. As another example that can be used almost anywhere, the belt of great Orion, the hunter – three bright stars in a row, only 3° from end to end – lies very nearly along the celestial equator. During his season of visibility, northern winter or southern summer, just locate the celestial equator (east to west, 90° from the pole, $z(\Sigma) = \phi$) and you will see him easily. Once you find one figure, you can use it to locate others. We will look at the subject of stars and constellations in Chapter 4, but we still need passing familiarity with at least a few to get on with the story of their motions in the sky, which we will address in the next chapter.

Chapter 2

The moving Earth and the traveling observer

In the last chapter, the sky was examined as it would appear in a snapshot, as it looks from one time and one place. This chapter delves into what happens when time is allowed to proceed and the Earth rotates, and when the observer is free to wander over its surface.

2.1 Declination and hour angle

In the celestial sphere in Figure 2.1 we, the observers, still stand at a latitude of 45° N. Remember that we appear to be on top of the Earth and that we must imagine ourselves to be in the inside looking out. The Earth, however, has now been shrunk to a dot at the center.

Remember also that the sky does not contain just circles and points that are artifices of the mind but also contains real observable objects. In the daytime the Sun and sometimes the Moon are to be found on the sphere, and at night it is covered with thousands of stars. We must create order out of this seeming chaos. A start was made with the horizon system of coordinates, wherein any star can be located by altitude and azimuth. But this system, as neat as it appears, is not really analogous to the terrestrial coordinates of latitude and longitude. Since the celestial poles and equator are formed by the outward extensions of their terrestrial counterparts, it would be more useful, in relating Earth to sky, to have a system based directly upon them: a "latitude and longitude" of the celestial sphere. Furthermore, as is evident to anyone who watches the sky, the stars and their companions move, rising above the horizon and setting below it. They do not stay fixed, and consequently their azimuths and altitudes change continuously, and in a complex way.

As a result, astronomers long ago established an *equatorial system* of celestial coordinates. There are actually two of them, one fixed to the observer, one to the stars. We will explore the first one here, and defer the other to Chapter 3 after we have had a chance to look at how the Sun and stars actually appear to move. The basic concept is outlined in Figure 2.1, and is essentially the same as the terrestrial system. The celestial

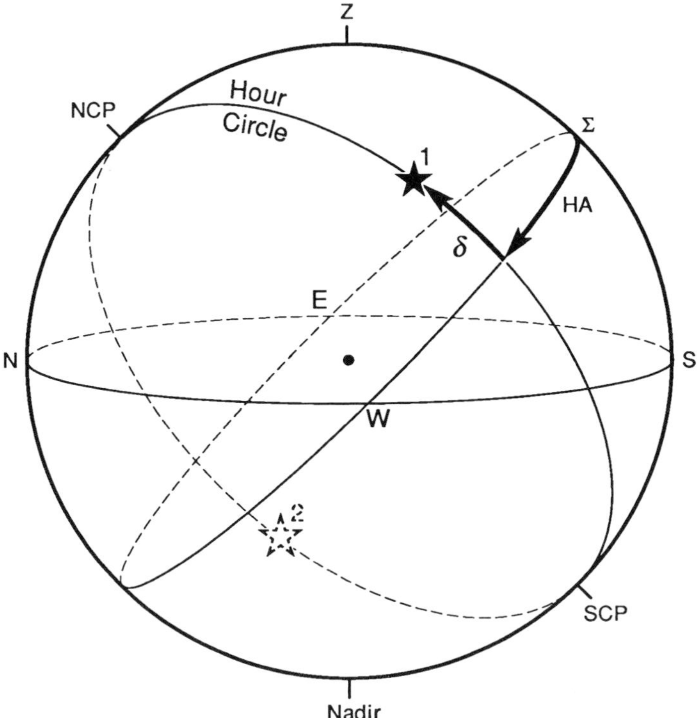

2.1. Hour angle and declination. The celestial sphere is still set for 45° N latitude. It is no longer necessary to label the celestial equator and horizon. Hour circles are passed from pole to pole through the stars. Hour angle (*HA*) is measured from the equator point westward along the celestial equator to the intersection of the hour circle. Declination (δ) is measured north and south of the equator along the hour angle to the star. Star 1 has *HA* = 45°, δ = +20°; star 2 has *HA* = 225°, δ = −20°.

equator (which no longer needs explicit labeling) is the FGC and the poles are the NCP and SCP. Note star 1 in the figure and draw the secondary great (semi)circle from the NCP through the star to the SCP. This secondary is called an *hour circle* and is analogous to a meridian of longitude. To complete the system we must choose a zero point on the equator. There is more than one way of defining its location, giving rise to the two kinds of equatorial coordinates. For this particular system the equator point (Σ) is used. We now have two arcs we can measure: one along the equator from Σ to the intersection of the hour circle and the equator, called the *hour angle* (*HA*), and the other along the hour circle north or south from the equator, called the declination (δ). Hour angles (unless otherwise designated) are measured to the west from Σ. Declinations, like latitudes, must be designated north (N, +) or south (S, −). In the figure, the star has an hour angle of roughly 45° and a declination of about +20°.

The coordinates of the equator point are obviously *HA* = 0°, δ = 0°. The east and west points of the horizon have declinations of 0° and hour angles of 90° and 270°

respectively. The NCP and SCP are at declinations of plus and minus 90°, and just as longitude disappears at the poles, so does hour angle. As another stellar example, estimate the coordinates of star 2, which is on the sphere's back (or eastern) side. The hour circle is the dashed line. Now we must go from Σ through the west point, past the meridian at the bottom, and back up. The hour angle is about 225°. We could also express it as 360° − 225° = 135° *east* of the meridian. But in that case 135° E (or −135°) must be specified; the default is always *west*. This star is south of the celestial equator (closer to the SCP than the NCP) and the declination is about 20° S. Note that star 2 would be invisible to the observer, since it is below the horizon.

2.2 The rotating Earth: daily paths

The Earth is not a stationary body. It has a variety of motions, its *rotation* the most obvious. Our planet spins on the axis through its center once a day (one rotation being the definition of a day). Our speed around the axis depends on latitude. If you are at one of the poles, you would simply turn on your body's center line like a spinning skater or dancer. But at the equator you would be 6500 kilometers from the axis and would travel in a circle 40 000 kilometers in circumference in 24 hours, at an extraordinary speed of 1700 kilometers per hour (nearly half a kilometer per second). Even at 45° latitude the speed is still 1200 kilometers per hour. Yet we feel nothing. The air turns with us so that no such fierce wind blows and the motion is so uniform that we do not perceive it.

We are not aware of the Earth's rotation except that it is reflected in the apparent motion of the sky. We witness a simple example of relative movement. To the driver of a northbound car, the landscape appears to slip south. If you turn clockwise on one foot in a circle, the room appears to spin counterclockwise. The Earth turns counterclockwise, west to east, around the north and south poles. This motion makes the sky – including the Sun, Moon, planets, and stars – *appear* to rotate in the opposite direction, or clockwise around the NCP and the SCP, as viewed from above the NCP.

Now look again at stars 1 and 2 redrawn in Figure 2.2. As the Earth rotates east, the stars will appear to move toward the west. Since the sky rotates about the celestial poles, the NCP and SCP must stay in their fixed positions, that is, the altitude of the pole must always equal the observer's latitude regardless of where the Earth is in its rotation (which will later be equated with the time of day). The celestial equator is always 90° from the NCP and SCP, so that it and the equator point must appear to be stationary too. Since the angles of the stars from the equator are specified and constant, they appear to trace out paths in the sky parallel to the equator. These apparent tracks are called the *daily* or *diurnal paths* (*diurnal* is to daily as *annual* is to yearly). Unless a star is exactly on the celestial equator, its daily path will be a small circle, which is followed every day.

Daily paths can easily be traced. We have all watched the Sun move across the sky as it tracks a path from east to west parallel to the celestial equator. (Do *not* attempt actually to look at the Sun since it is very dangerous; still it is easy to know where it is without staring at it.) The Moon serves better. Watch it during the night or day

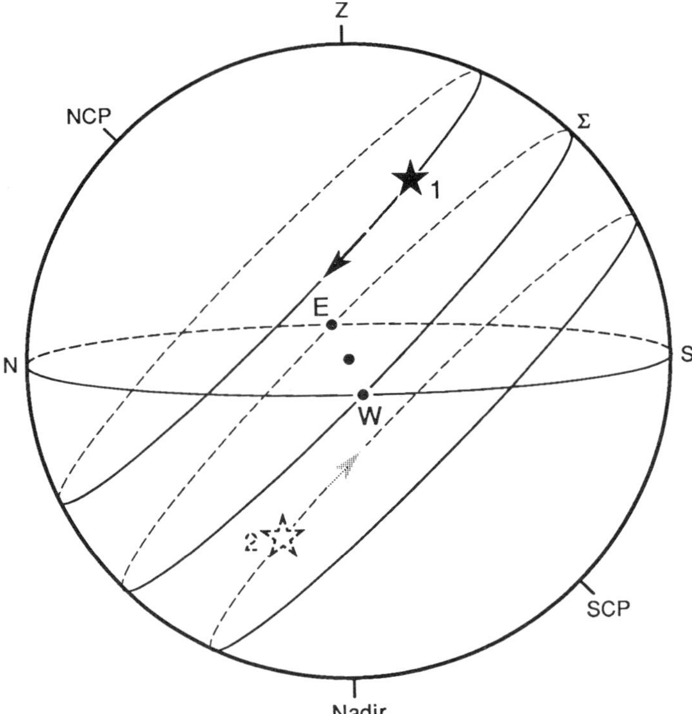

2.2. Daily paths of stars. The daily paths are small circles parallel to the celestial equator. Only if a star is exactly on the equator will its path be a great circle, that of the equator itself. Since declination is fixed, the stars like 1 and 2 in the drawing must maintain the same angular separations from the poles and the equator.

when it is above the horizon. The daily motion is relatively slow, however. It is better seen if a camera is opened to the sky for a long interval. The stars then simply take their own pictures, and their motions cause them to leave streaks across the film, vividly revealing their daily paths (Figure 2.3).

As stars 1 and 2 move along their great circles they must cross the horizon twice a day, when they *rise* (in the east) and *set* (in the west), where the altitude is 0°. Note that if a star is close enough to the pole it will not cross the horizon and will be perpetually visible or invisible depending upon which pole it is near. We will address this subject in detail in Section 2.5. A star must also cross or *transit* the meridian twice a day, once at 0° hour angle, and again at 180°. When it is at 0°, it must also be at maximum altitude, and the crossing is called *upper culmination* (or *upper transit*). *Lower culmination* (*lower transit*) occurs at 180°, below the pole, and at minimum altitude. The upper culmination of star *1* in Figure 2.2 takes place at h = 68°; lower culmination at −24°.

The effect of the star's apparent motion on its coordinates is obvious from Figure 2.2. Look at star 1. It is in the western hemisphere of the sky, and its altitude will

2.3. Real daily paths. Stars of Ursa Major rising over Kitt Peak trace out sections of their daily paths in a time exposure, showing the effect of the rotating Earth. The large dome houses the 4-meter telescope and the smaller building just to the right contains a 2.3-meter instrument belonging to the University of Arizona. (National Optical Astronomy Observatories photograph.)

decrease and its azimuth will increase. Star 2's altitude and azimuth are both at the moment increasing. The changes are complex because the daily paths are tilted relative to the horizon. The equatorial system of declination and hour angle simplifies the picture, however, as declination is constant. And since the daily paths are perpendicular to the meridian and parallel to the celestial equator, the stars move in the same direction in which hour angle is measured, and the change is smooth and uniform as the Earth rotates. In the next chapter we will build on this scheme to construct a system in which the coordinates do not (so far as we are concerned at the moment) change at all.

2.3 Units of time

This book has a great deal to do with the matter of *time*: what it means, how it is measured, and how it is used. The subject will arise in one form or another, culminating in Chapter 6. The time of day is simply a measure of where the Earth stands in its daily rotation, taking one day of 24 hours to make a full turn. In principle we pick a celestial body and look to see where it is in the sky as the Earth turns. It will take one day to appear to move all the way around. There are various definitions of the day and

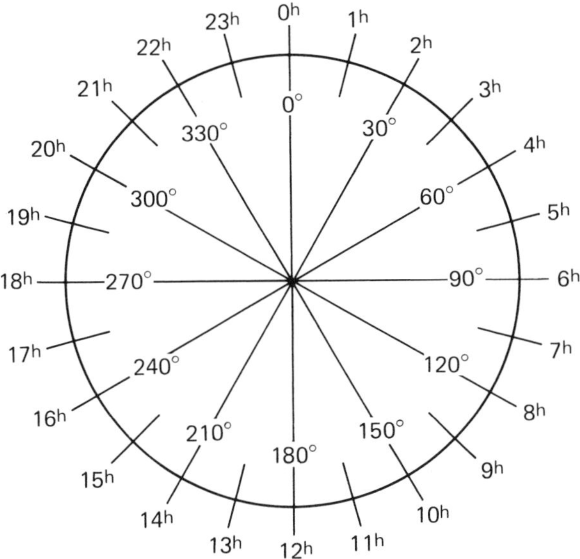

2.4. The circle divided into hours and degrees. Time units can also be used to measure angle, at 15° per hour. A right angle is both 90° and 6 hours.

of kinds of time, some having to do with the stars and some with the Sun; these will remain undifferentiated for the moment. It is the purpose of a *clock* to keep track of the rotational position of the Earth or of the apparent location of the Sun and stars in the sky. The clock is a *model* or representation of the sky and as such the hands move to the right, or *clockwise*, in the same direction as the Sun as seen from the northern hemisphere (where our clocks were invented). By definition then, the Earth turns *counterclockwise*, opposite the apparent movement of the heavens.

We will nearly always deal here with a 24-hour clock: 12-hour clocks, in which the hour hand revolves twice a day, are banned from the discussion until later, when they will appear only briefly. Think of a clock dial divided into 24 equal units (Figure 2.4), a circle divided into 24 parts. There is no reason why we cannot use these units instead of degrees for angular measurement. We simply say that the circle has 24 "hours" in it rather than 360°. Note that these hours involve *angular* measurement, in which the units have the same names as *time* measurement with which they will later be mated. There is a subtle but real difference between the two concepts.

Since 24 hours (of angle!) equates with 360°, one hour must be equivalent to $^{360}/_{24}$ or 15°. An hour of time is divided into 60 minutes, and so is an hour of angle; one minute is similarly divided into 60 seconds. It was noted before, however, that one degree is split into 60 minutes of arc, and one minute into 60 seconds of arc. We have one word used for two different units. Although it is arcs and angles that are being measured by these time units, we still call them minutes and seconds of *time* in order to differentiate them from the subdivisions of a degree, which are called minutes and seconds of *arc*. Minutes of *time*, abbreviated min or m, are not the same as minutes of

Table 2.1. *Conversion between time and arc units*

$1 \text{ hour} = 15° = 60^m$
$1° = 4^m = 60'$
$1^m = 15' = 60^s$
$1' = 4^s = 60''$
$1^s = 15''$
$1'' = {}^1/_{15}{}^s = 0^s.0667$

arc, abbreviated with a single prime (′), and seconds of time (sec or s) are not the same as seconds of arc (double prime, ″). The two concepts are well separated by their different abbreviations.

The relations between the two sets of units are easy to establish. Fifteen degrees equals 1 hour, which is 60 minutes of time. Consequently, one degree is $^{60}/_{15}$ or 4 minutes of time. But one degree is also 60 minutes of arc. So *one* minute of time is $^{60}/_4$ or 15 minutes of arc. The same reasoning can be applied to seconds, resulting in the summary of Table 2.1. Why is such apparent confusion necessary? Because the relations tell us how far in angle or arc a celestial object moves in a specific amount of time as the Earth rotates. Whether the modern scientist likes it or not, he or she is confronted with ancient units that must be reconciled. The rationalists of the French revolution tried decimalizing arc and time units: it did not work; no one would accept them.

Now return to Figure 2.2 and the two stars. Star 1 has an hour angle of 45°/15 = 3 hours. The second star's hour angle is 225°/15 = 15^h, (or 24 – 15 = 9^h E). As we watch the sky turn, it moves through 15° or 1 hour of angle in 1 hour of time, or 1° in 4^m, and so forth according to Table 2.1. Watch for yourself. The angular diameter of the full moon is ½°, so it will appear to move through its own angular width in two minutes. Line it up with a post, and time it. It is clear that the turn of the sky can be measured with a clock. Consequently, it is simpler to use time units in the measurement of hour angle. Say it is 5 o'clock (rather 05 hours time) in Figure 2.2. Then at 06 hours, star 1's hour angle will be 4^h, and star 2 will be at 16^h (8^h E). At 10 o'clock, 4 hours later still, the hour angles will have increased to 8^h and 20^h respectively. Note that eventually (at 14 hours time) the hour angle of star 2 reaches 24 hours. It then crosses the meridian at upper culmination, and we start again at zero: 24 hours is equivalent to 0 hours.

It is obvious now why "hour circles" and "hour angles" bear the names they do. Hour angles increase uniformly with time and will be used (later) to *measure* time. Measurement of hour angle in degrees is never wrong, it is just not always convenient. Declination, however, is *always* measured in degrees.

2.4 The rising and setting of stars

In Section 2.2 we briefly examined the passage of stars on their daily paths across the meridian and the horizon. This section now deals specifically with how these stars rise

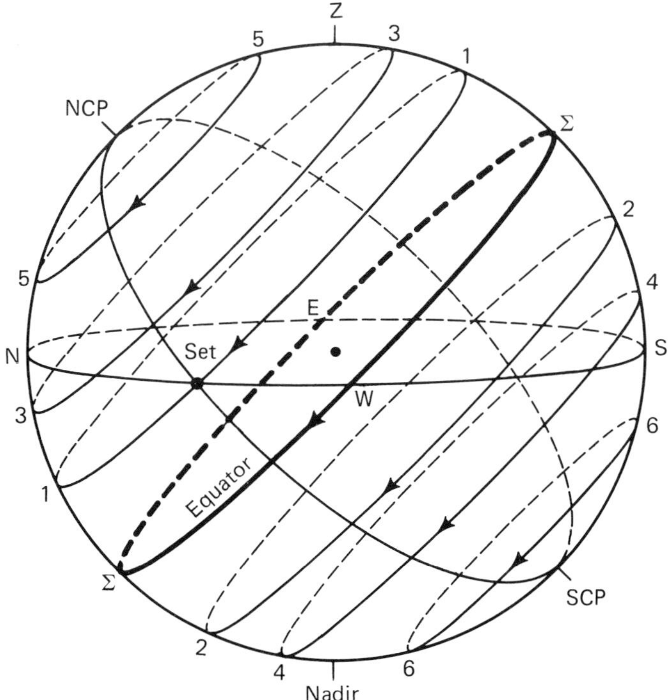

2.5. The rising and setting of stars. The heavy circle is the celestial equator. The daily paths of stars 1 and 2 are the same as before. The northern one rises and sets (crosses the horizon) in the northeast and northwest respectively, the southern one in the southeast and southwest. The farther north or south the star (stars 3 and 4) the closer to the north and south points are the risings and settings. If the star is closer to the pole than the pole is to the horizon (δ greater than $90° - \phi$), it (star 5) cannot set; if δ is less than $-(90° - \phi)$ (star 6) it cannot rise. A star on the equator is up for 12 hours and down for 12 hours. The more northerly the star, the longer it is above the horizon and vice versa. The hour circles of the setting point of star 1 and of the rising point of star 2 are drawn and give hour angles of about 7.5 hours west and 4.5 hours east respectively. The northern star is then up for 15 hours, the southern for only 9.

and set. We will look at the subject here only from our now standard point of view in the northern hemisphere, at 45° N, and take up the matter of what happens when we travel in Sections 2.8, 2.10, and 2.11.

Figure 2.5 shows the daily paths of stars 1 and 2 of Figures 2.1 and 2.2 and adds four more at different declinations. First, examine again the daily paths of stars 1 and 2. Neither rises and sets exactly east or west. Star 1 descends to the horizon in the northwest and climbs above it in the northeast. Only if a star is exactly on the equator will it rise and set due east and west. The rising and setting points of a star depend upon its declination. Star 2 has a southerly declination, and as it moves on its daily path, it rises in the southeast and sets in the southwest. The farther north the star (increasing positive declination) the farther toward the north point it rises and sets.

Star 3 has a declination of +35° and sets not far from the north point. Alternatively, the greater the southerly declination of a star the farther toward the *south* point it rises and sets: star 4, at $\delta = -35°$, does not rise very far above the horizon.

If the star is far enough north of the equator, like star 5 in Figure 2.5, at $\delta = +60°$, it can pass beneath the NCP and go through lower culmination without setting at all. Stars of this sort are always present above the horizon at a given latitude (except at the equator: see Section 2.10) and are called *circumpolar*. Since the altitude of the NCP is equal to the observer's latitude (Section 1.11), any star that is numerically closer to the pole than the latitude must be circumpolar. A star that is $\phi°$ (the latitude) from the pole will exactly graze the horizon at lower culmination. Since the pole is 90° from the equator, this star's declination is $90° - \phi$, another way of saying that the declination of the north point of the horizon in Figure 2.5 is $90° - \phi$. The criterion for a celestial body to be circumpolar is that δ must be greater than the co-latitude $(90° - \phi)$. The example used in the figures so far has placed the observer at $\phi = 45°$. Therefore any object with $\delta >$ (greater than) 45° will never set: the observer could see it any time of the day or night. (The stars are always there in the daytime. We do not see them because of the brightness of the sky, and we will pretend that they can be seen in spite of the solar glare; the brighter ones are actually visible with telescopes.) Familiar to these northern observers are such configurations as the Big and Little Dippers and the constellation of Cassiopeia, which are always in view. Figure 2.6 shows another time exposure photograph, this one centered on the north celestial pole. The circumpolar stars simply spin around it, over and under, never setting below the horizon.

The SCP is below the observer's horizon by an arc that is also equal to the latitude. Consequently, any star that is closer to the SCP than the latitude will never *rise* at a given northern location. A star that just barely makes it to the edge of the southern horizon at upper culmination in Figure 2.5 must have a declination of $90° - \phi$ south (i.e. the south point's declination is $-(90° - \phi)$. Consequently, any star whose declination is below (more negative than) $(90° - \phi)$, like star 6 in Figure 2.5 at $\delta = -60°$, will never appear to the observer. At our selected value of $\phi = 45°$ N that includes all stars south of $\delta = -45°$ and such famous sights as the Southern Cross and the nearest star, Alpha Centauri.

The northern parallel of declination that defines the circumpolar stars is infrequently called *the circle of perpetual apparition* (even astronomers will usually look puzzled upon hearing the term); the other that defines the invisible stars is in parallel named *the circle of perpetual occultation* (from *occult* for hidden). "Circumpolar" is commonly reserved for the perpetually visible zone of the sky; the other hidden region has no similar name (like the antipode of the equator point). The sizes of these zones obviously change with latitude as the poles become higher or lower in the sky. We will look at how, and in particular what happens when we reverse hemispheres (i.e. go to Australia), in Section 2.11.

Not only do the rising and setting points change with declination, but so do the times that stars spend above and below the horizon. First, it is clear from Figure 2.5 and previous discussions (Section 1.8) that half the celestial equator is above the horizon and half below. Thus, a star on the equator will not only rise exactly in the east and

2.6. Daily paths of circumpolar stars. The position of no movement in this time exposure of the northern sky marks the north celestial pole, about which all the other stars spin. The bright star near the center is Polaris, whose circle around the pole is nearly a degree across. (Lick Observatory.)

set in the west, but will be above and below the horizon for equal intervals, namely 12 hours at a time. (For now we neglect the distinction between the day as defined by the stars and that defined by the Sun, which we take up in the next chapter). Then, look at one of the daily paths in Figure 2.5, again that of star 1. The star moves uniformly along its apparent track as the Earth spins. No matter what the declination, it takes the star a full rotation of the Earth, or one day, for it to return to its starting point. The time interval between rising and setting is therefore the fraction of the daily path above the horizon times 24 hours. In other words, that interval is the eastern hour angle of the rising point plus the western hour angle of the setting point, since hour angles are directly related to clock times. However, the meridian divides the sky into equal hemispheres, so the rising and setting portions of a daily path are exactly symmetrical about the pole, and consequently the period of visibility of a star is just double the hour angle of its setting point. The hour circle of this point is drawn in Figure 2.5; the hour angle is estimated to be about 7.5 hours. The star must therefore be above the horizon for 15 hours and invisible for 24 – 15 or 9 hours.

For the observer in Figure 2.5 (who, remember, is in the northern hemisphere), the greater the northerly declination of a star the larger will be the hour angle of its setting point and the greater will be the duration of its visibility. Finally, when δ is greater than $90° - \phi$, the duration is infinite, i.e. the star is always above the horizon. As we look to southerly stars, the duration of visibility becomes less and less until eventually the star is never seen to rise. The hour circle of the rising point of Star 2 is also drawn in Figure 2.5. The rising hour angle is about $4^h.5$ east, which means that the star is up for only 9 hours.

2.5 Horizon–equatorial relations: the astronomical triangle

Here we examine the relations between the horizon and equatorial coordinate systems. Look first at the general case. Figure 2.7 shows a northern star set in the western hemisphere. Both secondaries (vertical and hour circles) are drawn through the star. Sections of three great circles connect the star, the zenith, and the pole in a triangle called *the astronomical triangle*. It is an example of a *spherical triangle* in which all the sides are arcs of great circles. Unlike plane triangles, each side (as well as each angle) is measured in degrees. The arc from the zenith to the star is the zenith distance (z) of the star (Section 1.9). That from zenith to pole is 90° minus the latitude, or $z(\text{NCP}) = 90° - \phi$ (Section 1.10). Since the NCP is 90° from the equator, the arc between the pole and the star, the star's polar distance, is $90° - \delta$.

The vertical circle and the meridian come together at the zenith in the *zenith angle*, called Z (the same symbol, but italicized, that is used for the point marking the zenith itself; do not get it confused with the zenith *distance*, $z = 90° - h$). From Section 1.3 and Figure 1.6, the zenith angle must be numerically the same as the arc from the north point to the intersection of the vertical circle and the horizon and is obviously directly related to azimuth. If the star is in the eastern hemisphere, Z is identical to azimuth A; if in the west, Z is $360° - A$. The intersection of the hour circle and meridian form an angle t that is, similarly, the hour angle. If the star is in the eastern

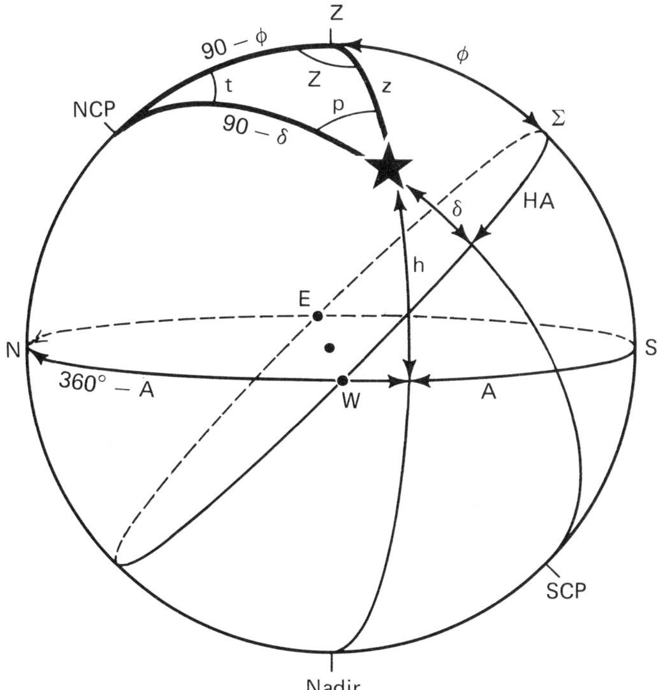

2.7. The astronomical triangle. The star is in the west. The arcs between the zenith and star (on the vertical circle), pole and star (on the hour circle), and zenith and pole (on the celestial meridian) form a spherical triangle with sides z = (90° − h), (90° − δ), and (90° − ϕ) respectively. The zenith angle Z is either azimuth (A) or (360° − A) (as it is in this drawing); t is the hour angle, HA, or if the star is in the eastern hemisphere, it is HA (east); p is the parallactic angle. If the star is fixed by A and h (255° and 45° respectively) then t and δ can be estimated from the drawing to be $2^h.5$ and +25° respectively: see the text. Trigonometric relations connect the sides and angles allowing for coordinate transformations: see Appendix 3.

hemisphere, t is hour angle *east*, 24^h − HA, where the HA is measured *west*. Finally, the vertical and hour circles make the *parallactic angle*, p, which is only infrequently used.

Once we have a triangle, we have a means of coordinate conversion through the laws of trigonometry. They will not be stated here in this chapter, which is meant to be non-mathematical. There are equations that connect the three sides and any angle, or that proportionately relate angles and their opposite sides. For example if we know Z, h, and ϕ, we can calculate δ and t, and knowing t, δ, and ϕ, we can recover Z and h. Thus, knowing the position of a star in one set of coordinates and the latitude allows the calculation of the other set. The equations themselves and their use are discussed in detail in Appendix 3. Alternatively, if we know Z (or A), h, and δ, we can find ϕ. So the lesson on navigation is extended a bit further. Eventually, we will see how the positions of the stars in the sky can exactly determine terrestrial locations.

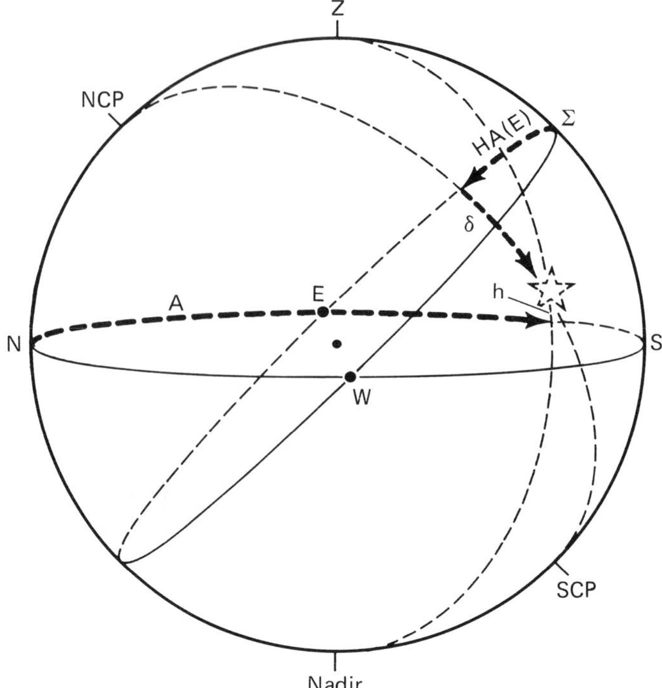

2.8. A coordinate conversion. The star is placed by its hour angle of $21^h.5 = 2^h.5$ east and its declination of 30° S. The vertical circle is drawn through the star establishing the astronomical triangle, from which we estimate that A is about 150° and h about 10°, not far from the exact solutions given in the text.

Even if we do not employ explicit equations here, we can still estimate coordinate conversions from drawings of the celestial sphere. Try a pair of examples. First, use Figure 2.7. Adopt a star with $A = 255°$, $h = 45°$. The star is located by drawing the vertical circle at the proper azimuth. Then by placing the hour circle through the star, we might estimate that $\delta = +25°$, $t = HA = 2^h.5$. The exact solution found by solving the equations for the astronomical triangle is $\delta = +21°45'$, $t = 47°15'$ = (from Table 1.1) $HA = 3^h\ 09^m$, not far off the eye estimate.

The procedure is reversed in Figure 2.8, where we adopt a star with equatorial coordinates $HA = 21^h.5$, $\delta = -30°$. Now the hour circle is drawn first. At $21^h.5$ it is on the eastern (back) side of the sphere, and HA (east) is 2.5 hours. The star is then along it, 30° south of the equator. Estimating from the figure, $A = 150°$ and $h = 10°$. The exact solution is found to be $A = 147°52'$, $h = 7°36'$.

The estimates from the figures are clearly not exact. In the first example, the hour angle is off by over half an hour, and δ by nearly 5°. The values estimated will depend on the way in which the viewer perceives curvature and perspective on the sphere. Nevertheless, the results are close enough for instruction, and with care, even for

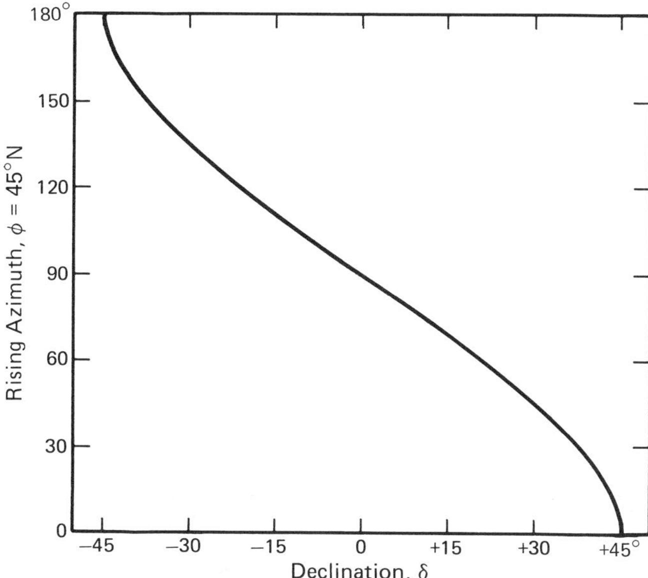

2.9. The rising azimuths of stars at 45° N latitude. The graph shows that at $\delta = 0°$ the star rises exactly east ($A = 90°$). As δ increases, A_r decreases until at $\delta = 45°$ the azimuth becomes 0° and beyond that declination there is no solution because the star does not set. Below 0°, A_r increases and at −45° becomes 180°: below that declination the star does not rise. The setting azimuth, A_s, is just $360° - A_r$.

practical use. Note also that the real answers (from the equations in Appendix 3) can be expressed with precision.

Next let us determine the azimuths and hour angles at which the stars rise and set. First, try it yourself. Draw in the daily paths of the two stars in Figures 2.7 and 2.8, or trace them with your finger. The astronomical triangle now goes from the zenith to the pole to the setting point, where $h = 0°$. In Figure 2.7 *A*(set) and *HA*(set) are estimated to be about 310° and 8^h respectively; the trigonometrically correct answers are 302° and $7^h\ 34^m$. The star is therefore aloft for $15^h\ 08^m$. In Figure 2.8 *A*(set) and *HA*(set) are estimated to be 230° and 3.5 hours, compared to the exact solutions of 229° and $04^h\ 07^m$. The differences again are caused by perspective effects.

Figure 2.9 shows a graph of how the rising azimuth, A_r, changes with a star's declination for $\phi = 45°$ N, calculated for the exact astronomical horizon, $h = 0°$. To use it, pick a declination on the bottom line (horizontal axis). Then read upward to the curve and look to the left to find A_r. If $\delta = +40°$, $A_r = 25°$; for $\delta = +10°$, $A_r = 76°$; at $\delta = 0°$, $A_r = 90°$ or the star rises at the celestial equator. If δ is over 45° N there is no solution and the star is circumpolar; and if less than −45°, it does not rise, so again A_r does not exist. Generalized curves for every 10° of declination (and for special solar and lunar cases) are given in Appendix Figure A1.1. Note again that rising and setting are exactly symmetrical (Section 2.4). The azimuth differences between the rising and setting

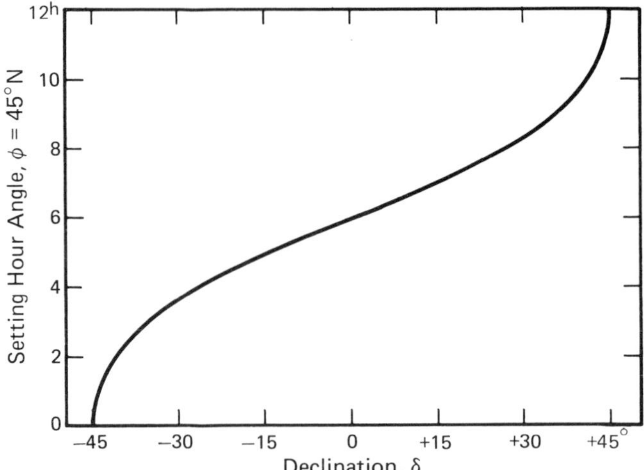

2.10. The hour angle of the setting point at 45° N latitude. At $\delta = 0°$, the hour angle is 6 hours. The duration above the horizon is double the setting hour angle, and is thus 12 hours. Above 0° declination the hour angle increases to 12^h at 45°: above that the star is up for 24 hours. Below 0° declination the hour angle decreases to 0^h at -45°, and below that the star does not rise.

points and the north point of the horizon are the same, or if A_r = azimuth of rising and A_s = azimuth of setting, $A_r = 360° - A_s = Z_s$. As a corollary, the difference between A_r and the east point is the same as that between A_s and the west point.

Figure 2.10 plots the hour angle of the setting point (HA_s) against the declination, again for 45° north latitude. The rising hour angle is just HA_s measured to the east, or $24^h - HA_s$. The curve goes to an hour angle of 12^h at $\delta = 45°$ N, meaning that the star grazes the horizon. Above that there is no solution: the curve shoots straight up, the duration above the horizon is infinite, and the star is circumpolar. At $\delta = -45°$, $HA = 0^h$, and the star culminates the meridian at the southern horizon; below this declination the star never rises. Full solutions for HA_s (and by symmetry for HA_r) for every 10° of declination (and again for special solar and lunar cases) are plotted against latitude in Appendix Figure A1.2. Take careful note of the caption. Locate the two stars of Figures 2.7 and 2.8 on the graph in Figure A1.2 by their declinations, and the curves will lead you to the values for the setting azimuths and hour angles given above (relative, of course, to a flat or astronomical horizon).

2.6 Stars on the meridian

If stars are on the meridian, we can examine exact relationships between latitude, altitude (or zenith distance), and declination. Under these circumstances, the hour circle falls directly on the meridian, and $HA = t$ must be zero. The astronomical triangle becomes a straight line, or "degenerate." Note again that a star must cross or transit the meridian twice a day, once at upper culmination at maximum altitude ("above the

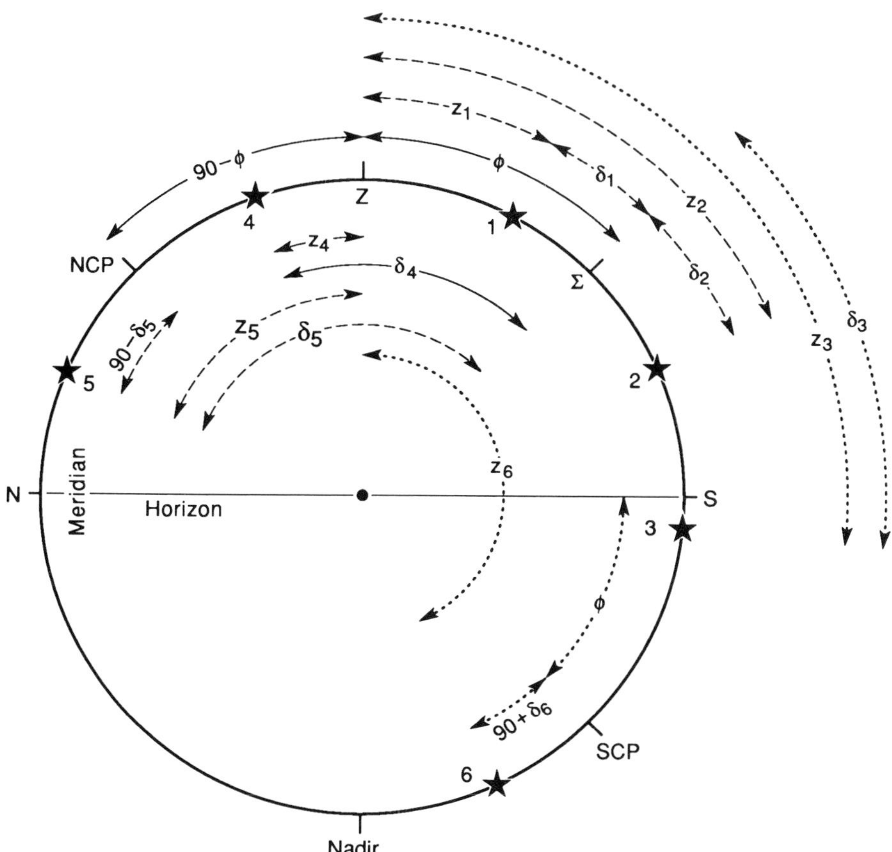

2.11. Relations between latitude, declination, and zenith distance on the meridian, or at culmination. The diagram is simplified into two dimensions, and only the celestial meridian is shown. The arcs for stars 1 and 2 are represented by dashed lines outside the sphere; those for star 3 by dotted lines outside; those for stars 4, 5, and 6 by solid, dashed, and dotted lines inside the sphere respectively. The relations are different for the different arcs Z–SCP, Z–NCP, NCP–nadir, nadir–SCP.

pole") and again at minimum altitude ("below the pole"). Both transits will be visible if the star is circumpolar, and neither will be seen if it is too close to the pole that is below the horizon.

The relations are sketched in Figure 2.11, and depend on where the star stands relative to the zenith and pole. First, look at star 1, which is transiting south of the zenith. The zenith distance of the equator point is the latitude, $z(\Sigma) = \phi$ (Section 2.1). But that arc is equal to the sum of the star's zenith distance and declination. Therefore, for stars south of the zenith at upper transit, observed from the northern hemisphere,

$$\boxed{\phi = \delta + z} \quad .$$

The altitude, h, is $90° - z$. The relationship also works for stars (star 2) south of the equator, for which δ is negative. Here is a more general way of obtaining latitude. We can easily determine the altitude (and consequently the zenith distance) of a star with precision. Just project lines of sight toward the horizon beneath the star and to the culminating star itself and measure the angle between them. There are a variety of devices with which to make such a measurement; see Chapter 8. Declinations of stars are available in many different catalogues; look up that of your star, find $z = 90° - h$, and just add the two together. If $z = 0$, then we see that the declination of the zenith is simply the latitude, the corollary to the latitude also being the zenith distance of the equator point (Σ). To summarize,

$$\boxed{\phi = h(\mathrm{NCP}) = z(\Sigma) = \delta(\mathrm{Z})} \quad .$$

Alternatively, we can employ the relation to derive stellar declinations. If the latitude is known, measure h or z at culmination and find $\delta = \phi - z$. As examples, fix ϕ again at 45°. If star 1 transits at $z = 27°$, as in the drawing, $\delta = 45° - 27° = 18°$. If we take a southern object, star 2, for which $z = 65°$ ($h = 25°$), $\delta = 45° - 65° = -20°$, or 20° S. To complete the picture, $z = \phi - \delta$. If δ is more negative than $90° - \phi$, z will be greater than 90°, and the star cannot rise above the horizon: that is, if δ is −50° (star 3), $z = 45° - (-50°) = 95°$, $h = -5°$, and the star culminates 5° below the southern horizon. We then recover our earlier rule for perpetually occulted stars (see Section 2.4).

For stars at upper transit that pass *north* of the zenith (we are still in the northern hemisphere), the relation is reversed. For star 4 in Figure 2.11 we see that $90° - \phi$, the zenith distance of the NCP, equals the star's zenith distance plus its polar distance, $90° - \delta$. Consequently, $90° - \phi = z + 90° - \delta$, or

$$\boxed{\phi = \delta - z} \quad .$$

Put in some numbers for that star: if $z = 19°$, $\delta = 45° + 19° = +64°$.

Lower transit is a bit more complicated. Look at star 5 below the NCP. Zenith-to-pole is $90° - \phi$, and pole-to-star is $90° - \delta$, which when summed equals z, or $90° - \delta + 90° - \phi = 180° - \delta - \phi = z$, or

$$\boxed{\phi = 180° - \delta - z} \quad .$$

This relation serves for stars transiting under the pole but north of the nadir. We can rewrite it as $z = 180° - \delta - \phi$. For z to be less than 90°, or visible at lower transit (or circumpolar), δ must be equal to $90° - \phi$ or larger, and we recover the criterion for circumpolarity. The only section now left is that between the nadir and the south celestial pole, which is of no particular interest. But for the sake of completeness, start at the zenith and work southward to star 6: zenith-to-horizon is 90°, horizon-to-SCP is ϕ, SCP-to-star is $90° + \delta$ (plus because δ is negative). Consequently, $z = 90° + \phi + 90° + \delta$, or

$$\boxed{\phi = z - \delta - 180°} \quad .$$

This relation always produces zenith distances greater than $90° + \phi$, or altitudes less than $-\phi$. All four relations are summarized in Table 2.2. The two upper-transit relations

Table 2.2. *Relations between ϕ, z, and δ for an observer in the northern hemisphere*[a]

	Section of meridian	Relation
Upper transit	Zenith–SCP	$\phi = \delta + z$; $z = \phi - \delta$
	Zenith–NCP	$\phi = \delta - z$; $z = \delta - \phi$
Lower transit	NCP–Nadir	$\phi = 180° - \delta - z$; $z = 180° - \delta - \phi$
	SCP–Nadir	$\phi = z - \delta - 180°$; $z = 180° + \delta + \phi$

[a]Reverse the signs for an observer in the southern hemisphere.

give the same answer at the zenith, i.e. if $\delta = \phi$, $z = 0°$; similarly at the nadir both lower relations yield $z = 180°$ for $\delta = -\phi$.

2.7 Maximum and minimum azimuths

As a conclusion to the discussion of the astronomical triangle and to our view of the sky as fixed observers, examine the changes that take place in azimuth as a star rises, transits, and sets. Any star that crosses the meridian at the zenith must have a declination equal to latitude (Table 2.2), and if it crosses at the nadir ($z = 180°$), $\delta = -\phi$. Between $\delta = +\phi$ and $-\phi$ (going through zero), the star's azimuth will pass through the entire circle, through 360°, in its daily path, from 0° at lower transit, to 180° at upper, and then to higher values to 360° (exemplified by stars 1 and 2 in Figure 2.5).

If a star transits north of the zenith, however, it cannot reach an azimuth of 180°. Look at star 5 in Figure 2.5. Its azimuth is 0° at lower transit. Azimuth thereafter rises to a maximum and then decreases again to 0° (or 360°) at upper transit. Subsequently, it decreases from 360° to a minimum in the west, before increasing again to 360° = 0° at lower transit. Star 5 only reaches about 40° azimuth in the east (A_E), goes back to 0°, then plunges only to about 320° in the west (A_W: note the symmetry). The maximum A_E and minimum A_W depend on declination; Polaris, only ½° from the pole, will reach an azimuth of only ½° in the east, and drop to one of 359½° in the west. As δ approaches ϕ, A_E approaches 90°, A_W, 270°. A star with $\delta = \phi$ has no azimuth at the zenith, but just before transit, A must just approach 90°; just after transit it leaves the meridian at 270°, that is, the change is *discontinuous*. If $\delta <$ (less than) ϕ, the changes are smooth and A goes through all angles.

A star with $\delta < -\phi$ (star 6 in Figure 2.5) behaves oppositely. At upper transit, which will be below the horizon at $\phi = 45°$, $A = 180°$. As hour angle increases, A increases to a maximum, then decreases to 180° at lower transit. As it subsequently climbs toward the horizon, A will go through a minimum.

2.8 The traveling observer: to the north pole

In the previous discussion, the observer has been at a latitude $\phi = +45°$, but it is obvious that the various relations must hold if the observer steps north or south of that parallel. For example if you move to 40° N, the equator point rises by 5°, the NCP falls by 5°, you see fewer circumpolar stars, and you can watch a few drift above the southern horizon that you have never before seen. In this section we leave our 45° home and begin to travel both north and south and watch the sky take on some very different aspects.

The key to understanding the changing appearance of the sky is always to realize that if you move in one direction, the heavens will appear to move oppositely. For example, as the Earth carries you east, the sky appears to move to the west. You would see the same effect were the Earth not rotating and you flew east in a speeding airplane. If you flew west, counter to the Earth's spin, the angular rates of motion of the stars on their daily paths would slow down, and if we moved west relative to the ground below at exactly the Earth's rotational speed they would stand still, that is, we would be stationary relative to the sky. If we were to fly west faster than the Earth's rotation speed (quite possible for an orbiting satellite), the stars would reverse their apparent directions of motion, and move west to east, that is, they would rise in the west and set in the east. If we were to turn our airplane and travel north, the sky would begin to spin south, and vice versa.

This principle was used by the third century BC Alexandrian Greek astronomer Eratosthenes to determine the size of the Earth. He knew that the Sun shifted by 1/50th of a circle to the south upon going from Syene in southern Egypt to Alexandria. The Earth's circumference therefore had to be 50 times the distance between the two locations. From our best estimate of the length of the Greek distance unit, Eratosthenes was off by only a few percent.

Look next at Figure 2.12a, again at latitude 45° N, and picture yourself as the observer. Two daily paths are drawn on the sphere at declinations of 30° N and 20° S. As you begin to walk north, you move around the curvature of the Earth, and your zenith will appear to follow along, always staying over your head. Our planet, however, is large and the curvature gentle, so you would really not be aware of your true path. Furthermore gravity always directs you to the Earth's center, and no matter where you stand, you always appear to be on the globe's top. So instead of considering your own movement, think that you are stationary and that the Earth is moving beneath your feet in the opposite direction. The appearance is that you are on a treadmill, always walking, but with your feet causing the world and the celestial sphere to rotate to the south and to slip behind you. That is not the way it *is*, but the way it *looks*. The NCP is aligned with and defined by the Earth's rotational axis, so that as the north pole is approaching you, the NCP appears to rise higher in the sky, becoming closer to your zenith.

You started at 45° N latitude. By the time you reach 70° N – perhaps near Kaktovik, on the north Alaskan coast, or near Tromso, Norway – the sky looks as it does in Figure 2.12b, with the altitude of the celestial pole now 70°. The zone of circumpolar stars has enormously enlarged: all those north of 20° declination can be seen all the time. The celestial equator, always 90° from the pole, has dropped in the sky by the same amount, and the altitude of the equator point is only 20°. You have also lost a great many of the southern stars you saw before: only those with declinations greater than −20° (i.e. less negative than −20°) can make it above the south point.

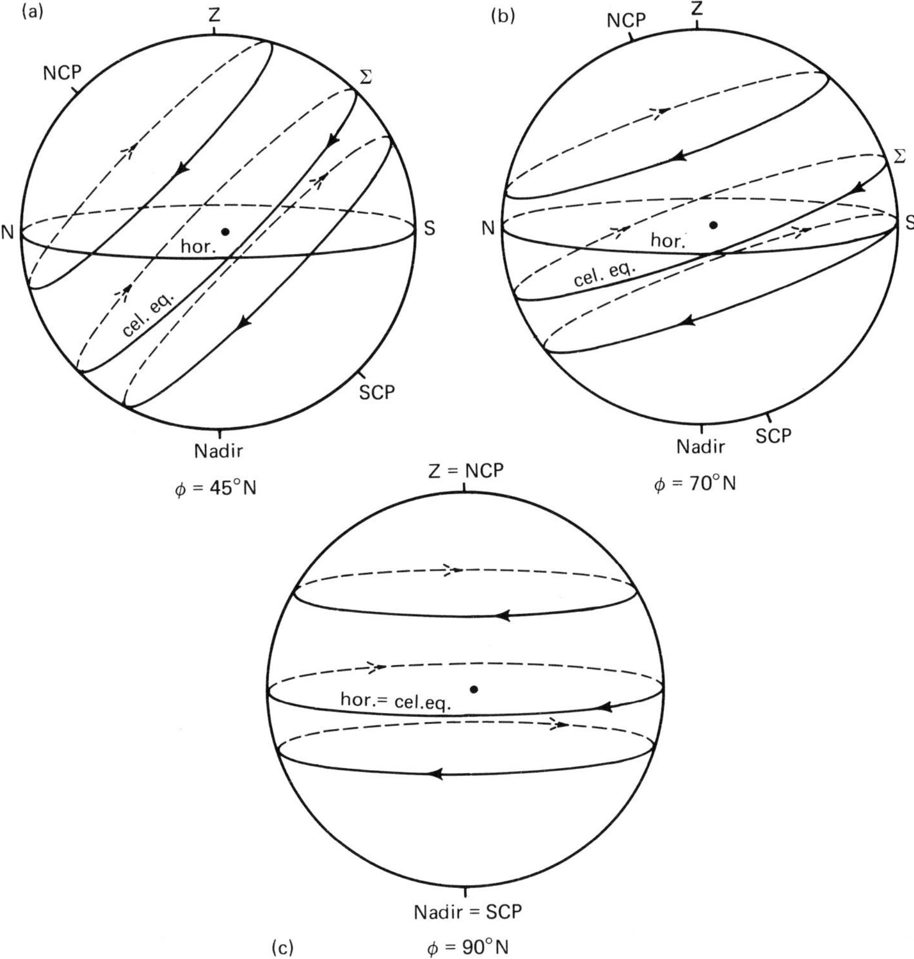

2.12. The sky from different northern latitudes. Each shows the daily path of a star at declinations +30° and −20° for: (a) ϕ = 45° N, which has been the standard reference latitude; (b) ϕ = 70° N; (c) ϕ = 90° N, the observer at the north pole. As the observer goes north, the daily paths flatten out and become parallel to the horizon at the pole, where the horizon and equator are equivalent. There the stars move to the right as the Earth spins to the left. The circumpolar zone increases as we proceed north. At ϕ = 70° N, the northern star is circumpolar and the southern one only grazes the southern horizon. At the pole, the entire northern hemisphere is circumpolar and none of the southern hemisphere can be seen.

Moreover, the celestial equator now makes a much flatter angle with the horizon. At 45° N the equator crossed the east and west points at a 45° tilt. This angle is always 90° − ϕ and now it has fallen to 20°. All the parallel daily paths have flattened as well, which produces a change in their rising and setting points. Those north of the equator move toward the north and those south of it move south, in both cases away from east and west. The length of stay of a star above (or below) the horizon changes accordingly. Only a star on the equator will always be up for 12 hours and down for 12 hours. Those in the north will stay aloft progressively longer as we move to the north pole, those to the south progressively shorter, a corollary to our seeing more and more area of circumpolarity.

Now pass this frigid latitude of 70° N and keep trekking north. The solid ice will allow you to stand exactly on the north rotation pole in the middle of the Arctic Ocean (it has been done – with great difficulty – many times). When you finally arrive at the north pole, $\phi = 90°$, the NCP must be at your zenith, and $h(\mathrm{NCP}) = 90°$. The Earth's axis passes directly upward through your body. If the NCP is in the zenith, the celestial equator must lie exactly on the horizon, as shown in Figure 2.12c. Longitude has disappeared, and with it, azimuth and the celestial meridian. The equator has no intersection point with the horizon, and east and west vanish. You can go no farther north: there is only south. The daily paths of stars, which are parallel to the equator, are now also parallel to the horizon. Stars simply spin around the sky at constant altitude, always toward the right, ever riding their almucantars. The entire sky is circumpolar! Nothing rises and nothing sets; everything you can see is always up. None of the southern stars are visible. Orion, who straddles the equator, shows only his upper body, his legs walking below the horizon: only his smaller dog, marked by the star Procyon to his right (your left) is seen; the larger dog, with brilliant Sirius, is always hidden. The north pole is an odd but fascinating place, and it will seem even more so when we place the Sun in the sky in the next chapter.

2.9 Proof that the Earth spins: the Foucault pendulum

Having arrived at the north pole, we have a grand opportunity to examine our world's rotation. Standing at home, viewing the sky, you really have no way of knowing what is actually moving. You see the heavens wheel around you, and perhaps, as was thought centuries ago, that is truly the case: maybe the stars *are* on a crystal sphere that spins about the stationary Earth. So far, you have taken the author's (or some teacher's) word that it is the Earth that turns. Now, you might say, prove it.

Start with a "thought experiment." At the Earth's north pole erect a large tripod and from its peak hang a heavy ball on a string, as in Figure 2.13. The string is attached to the tripod by a frictionless bearing so that motion of the weight cannot be transferred to the support, and vice versa. Now pull the ball back and line it up with a star traveling its daily path parallel to the horizon. Let it go and watch it swing at the star. You will see a remarkable sight: the plane of the pendulum's swing will continue to point at the star as the star travels to the right, that is, the direction of swing will slowly creep clockwise and go through a full rotation in one day. It appears as if there must be some mysterious force that causes the pendulum to align itself with the heavens; and you could likely convince someone that such was the case. But what actually is happening is that it is *not* the sky that moves, pulling the pendulum, but the Earth that moves in the opposite direction. The pendulum, free of its tripod on its frictionless bearing, is also free of the world. Once you set it swinging, it just keeps going in the same direc-

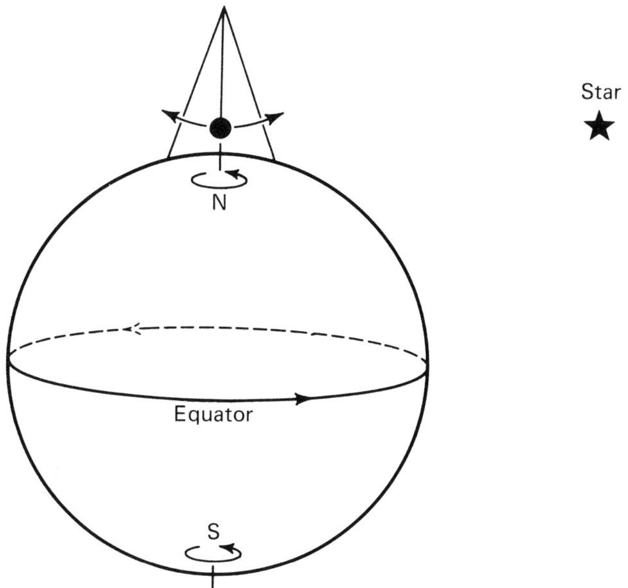

2.13. The Foucault pendulum at the north pole. The pendulum swings free of the Earth and its swing-plane always points toward the star. The Earth turns beneath it, so that the swing-plane appears to rotate once in 24 hours.

tion. There is no force that can move it, and the Earth, with you on it, turns below, making the swing-plane only *appear* to move.

The experiment is run on these pages at the north pole because it is conceptually easy there (obviously it works just as well at the south pole). At the equator, it will not work at all, since there is no actual rotation beneath the pendulum. At intermediate latitudes, where most of us live, the pendulum will not actually follow a star, since daily paths are not parallel to the horizon and the swing cannot be made perpendicular to them. Still our planet's spin produces some effect of rotation beneath the pendulum that can easily be calculated. The plane will still rotate, but with a period greater than 24 hours, the length dependent on latitude (24 hours divided by the trigonometric sine of the latitude). At 45° N (or 45° S), the period is about 35 hours, and at the equator it becomes infinite (i.e. zero rotation).

The experiment was actually first performed in the dome of the Pantheon in Paris in 1851 by Léon Foucault, whose name is forever connected with the device (Figure 2.14). It is not a simple thing to rig. Truly frictionless bearings are not possible, but with care in suspending a cord or wire, we can come close. However, the cable must be long and the weight very heavy to overcome residual frictional effects so as to keep the weight swinging for a sufficiently long time. Foucault used a heavy iron ball and a wire nearly 60 meters in length! On the underside of the ball was a pin that grazed a pile of sand at each swing. Slowly, it was seen to cut different paths, rotating always clockwise.

2.14. Léon Foucault's pendulum experiment. It consisted of a heavy ball swinging on a wire some 60 meters long. At the latitude of Paris, $\phi = 49°$ N, the pendulum's plane takes 32 hours to swing through 360°. (From an old French newspaper.)

Several museums have Foucault pendulums on permanent exhibit. Note the direction of its swing, go look at something else, then come back. It will have changed. The Earth rotates!

2.10 South to the equator

Now return briefly to our "home" at 45° N in Figure 2.12a and begin a new journey toward the south and the terrestrial equator. The zenith will now move *away* from the NCP and toward the equator or equator point. To you, it would appear that the NCP was sinking toward the northern horizon and that the equator and Σ were climbing higher. The size of the circumpolar zone shrinks, and new stars, never seen before, rise

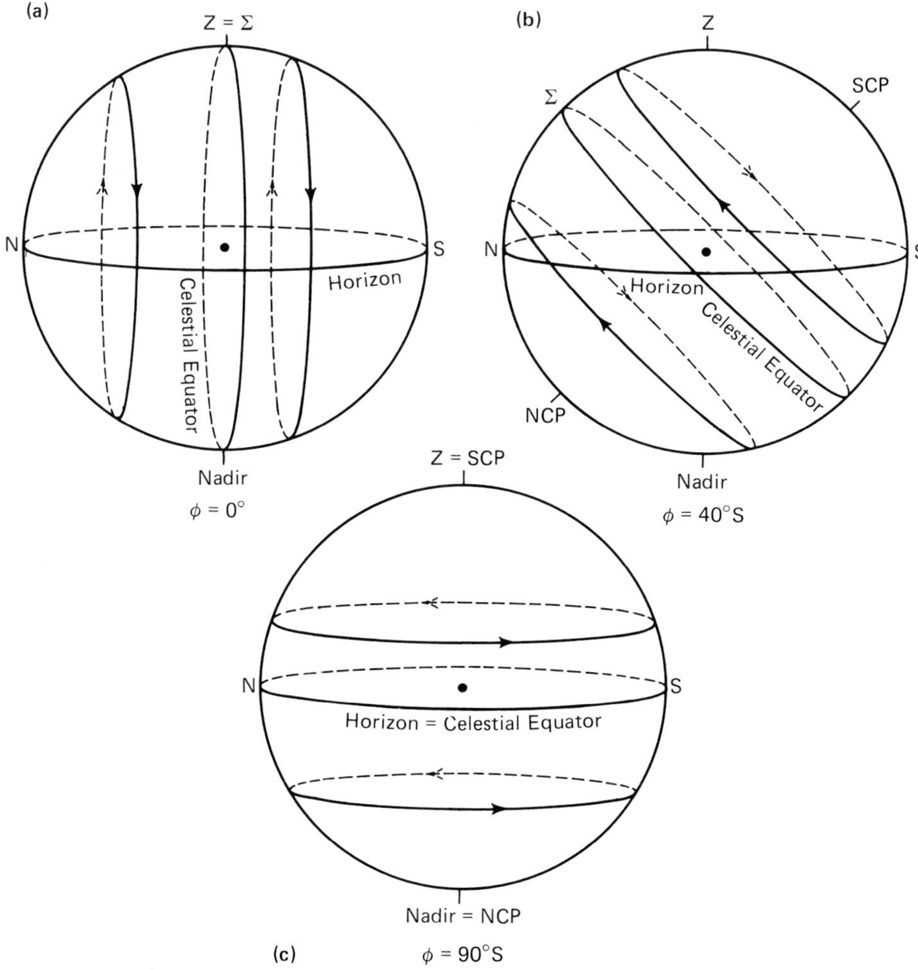

2.15. The sky from the equator and southern latitudes. Daily paths for stars at 30° N and 20° S declination are shown for several latitudes. (a) $\phi = 0°$ (at the equator). There, the poles lie on the horizon, and we can see all the stars over the course of a day. The daily paths are perpendicular to the horizon, and the stars are up for 12 hours and down for 12 hours. (b) $\phi = 40°$ S. The SCP is now up and the NCP below the northern horizon. (c) $\phi = 90°$ S, the south pole. The stars again move parallel to the horizon, but now to the left. Here, the entire southern hemisphere is circumpolar, and none of the northern can be seen. Compare these drawings with those in Figure 2.12.

in the southeast. The pitch angle of the equator increases with the decline of latitude, as do the angles of the daily paths. Now the rising and setting points of the stars move farther from the poles and closer to the east and west points.

Finally, we arrive at the Earth's equator, $\phi = 0°$, and the NCP must be on the northern horizon, at $h = 0°$ (Figure 2.15a). The celestial equator, always above its terrestrial analogue, runs from east to west through the zenith, perpendicular to the horizon. There are no longer any circumpolar stars. Since the SCP is opposite the NCP it lies exactly at the south point, so no star can stay hidden. We now observe from that remarkable place – perhaps Quito, Ecuador, or central Borneo – where we can see the entire sky. All the stars rise and set and all daily paths (stars at declinations 30° N and 20° S are again shown) are perpendicular to the horizon. Consequently, all stars are up for 12 hours and down for 12 hours. Even Polaris pops up and culminates the meridian at an altitude of ½°. The azimuths of the rising and setting points of stars are now also as close to the east and west points of the horizon as possible. In addition, no star can pass through 360° of azimuth. A star exactly at $\delta = 0°$ rises at azimuth 90° then passes through the zenith and discontinuously switches to 270° for the descent.

2.11 Into the southern hemisphere

We now cross the Earth's equator into the southern hemisphere, where the traveler may be amused by several reversals. Readers who already live there may feel a bit neglected that we have so far confined ourselves to northern climes. But we had to start somewhere, north or south, and since the author is a native northerner, the choice is not surprising. The starting point, though, is irrelevant. The principles are what count, that is, the comparisons that can be made between north and south, between Australia and England.

As we step across the equator, the SCP rises and the NCP is at last lost from our view (at half a degree south, Polaris is always invisible). Now we see the same phenomena witnessed earlier as we went north. In Figure 2.15b we arrive at latitude 40° S, perhaps Valdivia, Chile, or King Island in Bass Strait between Australia and Tasmania. The SCP is now 40° above the south point; the daily paths are tilted in the other direction, and the rising and setting points again move away from east and west. The rule about ϕ(observer) = h(NCP) is still true, because h(NCP) is now negative and so is the latitude.

If we face the NCP in the northern hemisphere, the stars move around it counterclockwise; when we face away from it, they move from the east, at left, toward the right. In the southern hemisphere, the apparent daily motions are reversed. If we face away from the SCP, east is on the right and stars (and the Sun) move to the left. If we face the SCP, the apparent rotation is clockwise instead of counterclockwise.

We continue on, cross the Antarctic Ocean, and fight our way to the Earth's south pole, an extraordinary, forbidding place at $\phi = -90°$. In Figure 2.15c we see the analogue to the sky at the north pole, now with the SCP at $h = 90°$, and the NCP at $h = -90°$. All of the southern hemisphere is circumpolar, and none of the northern can be seen. The stars all track parallel to the horizon, upon which lies the equator, but now they move to the left rather than to the right. Orion plods the horizon again, and all we see are his legs and feet. Since he was invented by northerners, he appears to be upside-down.

Now return to the rules in Section 2.6 and Table 2.2 that relate ϕ, z, and δ for stars on the meridian. Similar rules must hold in the southern hemisphere. Some alterations must be made, however. Since ϕ is now negative, we must change all the signs, plus for minus and vice versa. Alternatively, we could consider ϕ always as positive, and declination positive when it has the same sign as ϕ and negative when the signs differ (viz., the declination of a northern star viewed from the southern hemisphere would be considered negative).

2.12 A summary: rules of rising and setting

Start at the north pole, where all stars with northern declinations are circumpolar and all with southern declinations are invisible. As we proceed south, stars nearest the equator begin to rise and set. A star with a given northern declination, δ_N, first sets when we reach a latitude of $90° - \delta_N$. This rule is the analogue to the one that says a star must have a declination greater than $90° - \phi$ before it is circumpolar. That is, if $\delta = 90° - \phi$, then $\phi = 90° - \delta$. Similarly, a star with a given southern declination, δ_S, first becomes visible at latitude $90° + \delta_S$ (where δ_S is negative). The points A_s and A_r (azimuths of the setting and rising points) approach the east and west points and the duration above the horizon of the northern stars decreases, while that of the southern stars increases. The daily paths of the stars become more and more perpendicular to the horizon. When we reach the equator, A_s and A_r are as close to east and west as they can be. As we continue south, the southern stars begin to become circumpolar by exactly the same rules. A star with declination δ_S will become circumpolar when we reach latitude $-(90° + \delta_S)$. A_s and A_r approach N and S for stars with northern and southern declinations respectively. When A_s and A_r reach the north and south points the stars become invisible and circumpolar respectively. The duration above the horizon for northern stars continues to decrease while that of southern stars continues to increase. By the time we reach the south pole, all southern stars are circumpolar, and all northern stars are invisible.

The effects described above are presented graphically in Appendix Figures A1.1 and A1.2. Figure A1.1 shows the value of A_r as latitude changes for stars of different declinations. Pick a northern latitude, say 40° N. Then look upward along the line until you find the declination of a northern star you have chosen, perhaps one at 30°. Look to the left to find its rising azimuth, which we read to be 49°. The setting azimuth is then 360° minus that value, or 311°. If there is no solution, that is if the line along ϕ does not intersect a δ line, then the star is circumpolar (which we see for stars north of $\delta = 50°$ for $\phi = 40°$, as expected). If the star has a southern declination, use the right-hand scale. Then, if the latitude line does not intersect a declination curve, the star does not rise. Figure A1.1 will work whether ϕ is north or south. The only difference is that stars that would be circumpolar for an observer in the north are those that will not rise for one in the south, and vice versa. It is easy to interpolate lines for declination that fall between the 10° intervals. The special curves for the Sun and Moon will be discussed in later chapters.

Figure A1.2 shows the hour angle of the setting point (one-half the time above the horizon) again for stars of different declination for a continuous change of latitude. Use the left-hand axis if latitude and declination have the same sign (both plus or both minus), and use the right-hand one if the signs are opposite (e.g. northern latitude and southern declination). Stars for which there are no intersection points are circumpolar if ϕ and δ have the same sign, or are always invisible if the signs are opposite. As an example, again pick $\phi = 40°$ N and a star with $\delta = +30°$. It sets at an hour angle of 7^h 55^m, and its duration above the horizon must be 15^h 50^m. The $\phi = +40°$ line does not intersect the $\delta = +50°$ curve, and any star at that declination must be circumpolar. If $\delta = -30°$, use the right-hand axis, and find that $HA = 4^h$ 05^m, so it is up for 8^h 10^m. Note that the two hour angles are exactly symmetric about 6^h, the setting hour angle of the celestial equator.

We have now laid the foundation necessary to examine the motion of the Sun in the sky, to which we next turn our attention.

Chapter 3

The orbital motion of the Earth

Two great astronomical facts command everyone's attention. The first is the rising and setting of the Sun, the periodic change of day to night, which controls the flow of our lives. The other is the much slower but just as inevitable change of the seasons that are accompanied by the northerly and southerly motion of the Sun. The warm summer days begin to cool. As winter approaches, the air chills, and the noontime passage of the Sun becomes lower in the sky. It looks as if the Sun may disappear altogether, until – to the relief and celebration of all – the motion is reversed. The Sun thereupon climbs the sky again and the days start to warm, the cycle repeated throughout the aeons.

3.1 The sky at night

The clue to this once mysterious motion of the Sun can be seen by anyone in the seasonal passage of the stars at night. In early northern midwinter evenings, the constellation Orion pursues his hunt high in the sky near the celestial meridian. But by spring, he is seen only briefly in the west, not to appear again until the following autumn. If we look carefully each night at the same time, we see that the stars shift just a bit under one degree toward the west for every turn of the Earth.

It was apparent even 2500 years ago that the most rational explanation requires that the Earth move about the Sun. It is the combination of *revolution* (the proper term for orbital motion, the movement of one body about another) and *rotation* (the correct word for spin, the motion of a body about its axis) that makes the sky appear to move as it does. In the last two chapters the viewer was held fixed and the celestial sphere allowed to rotate. Now let us tie this fiction a bit better into the real world. Figure 3.1 shows the rotating Earth, centered as usual in an enormous surrounding celestial sphere. To the left of the Earth, on the apparent sphere, is the Sun, represented by its universal symbol, $\odot$. The Sun must illuminate one-half of the Earth at all times. Consequently, as the planet rotates, an observer will alternately pass through daylight and then night, each (on the average) about one-half day, or 12 hours, long.

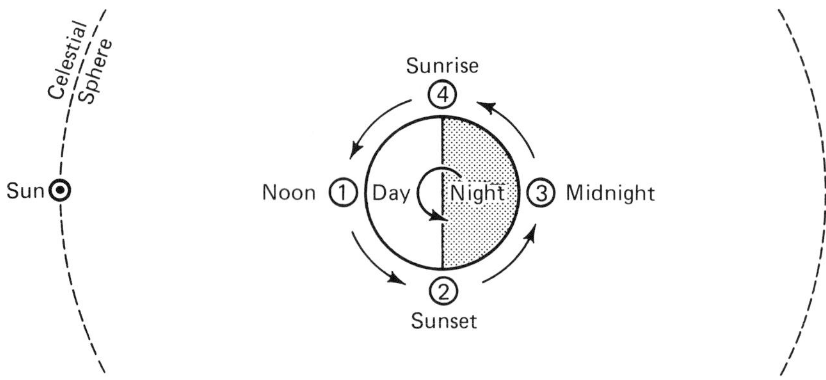

3.1. Day and night on Earth. At position 1 the observer sees the Sun overhead at the zenith. The Earth rotates counterclockwise carrying the person to position 2 where the Sun appears to set. At position 3, midnight, the Sun is at the observer's nadir, and a quarter turn later, at position 4, the viewer enters daylight again. As the Earth spins the Sun appears to move in the opposite direction.

Follow the observer during a rotation. At position 1 the Sun is opposite the center of the Earth, and it must be on the celestial meridian, that is, it is *noon*. Then, as the person is carried toward position 2, the Sun must appear to move to the west, and at 2 it will be on the western horizon and setting. The observer subsequently enters darkness, where the Sun cannot be seen, and a quarter-turn later it crosses the meridian again (this time below the horizon) at position 3, when it is *midnight*. Eventually the observer is carried to the "top" of the figure, at position 4, where the Sun appears to rise in the east, followed by a return in daylight to noon. To this observer, the Sun would appear to go about the Earth on a daily path. Although there are always stars in the sky, the viewer would actually see them only between positions 2 and 4, when the sky is dark. As a general rule, you see the stars in the sky that are more or less opposite the Sun.

Next, Figure 3.1 is transformed into the broadened view of Figure 3.2, which shows the Earth revolving about, or orbiting, the Sun on a nearly circular path. The radius of the orbit, called the *astronomical unit*, is 150 million kilometers, but that need not concern us yet since we are now only interested in the orbital geometry. Our planet orbits once a year, moving in the counterclockwise direction as viewed roughly from above the north pole. The revolution is in the same direction as the rotation. In Figure 3.2, the plane of the Earth's path is in the plane of the paper. The rotation axis does not actually project perpendicularly out of it, although it is drawn that way here and in Figure 3.1 for simplicity. The axial tilt will be examined in Section 3.3.

The infinite celestial sphere, which holds the stars and the constellation patterns they form, now surrounds *both* the Earth and Sun, enclosing the orbit. The directions to four constellations, Sagittarius, Pisces, Gemini, and Virgo, each about 90° from the other, are indicated in Figure 3.2. These are selections from a special set that is in the

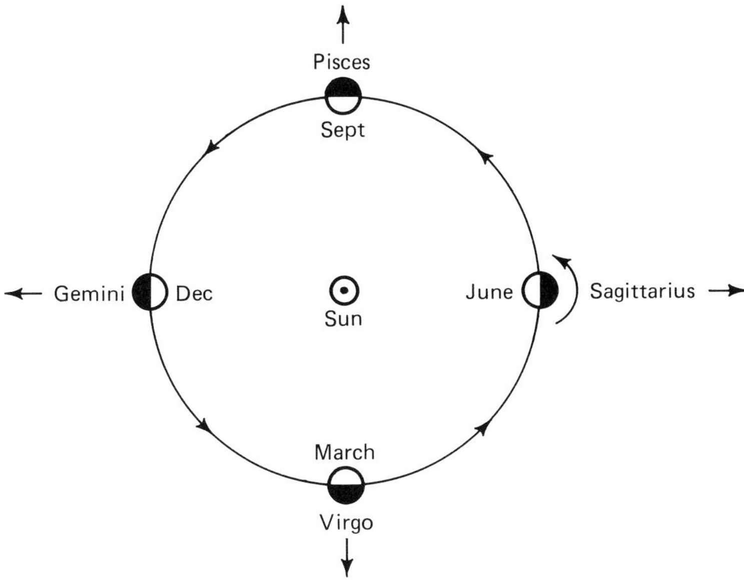

3.2. Change in the nighttime sky. As the Earth goes about the Sun the observer will see at night the part of the celestial sphere directed away from the Sun, illustrated by a principal constellation of the zodiac. The entire celestial sphere will be visible at night over the course of the year. The Sun will appear to move along the ecliptic through the constellations of the zodiac at a rate of about 1° per day, and if viewed at the same time at night, the stars will appear to shift west by 1° per day.

plane of the Earth's orbit. Since the part of the sky in view at night depends on the direction to the Sun, the constellations we see must change as the Earth moves in its orbital circle. In June, for example, we will find Sagittarius on the meridian at midnight. Three months and a quarter-orbit later we will see Pisces there at this time, and Sagittarius will be setting in the west. By December our nighttime view takes us to the other side of the Sun where at midnight we can admire Gemini, with Pisces setting and Virgo rising, and another three months later Virgo is similarly on display.

The change of the constellations at night does not occur in three month jumps, but smoothly. Since there are 365 days in a year, each night the Earth is about 1° farther along in its orbit, and consequently Sagittarius will appear to be 1° farther west toward its setting each successive midnight. During the year, the entire celestial sphere (except for those stars that do not rise at a specific latitude) will be visible. Keep in mind that the sky now reflects not one but *two* distinct motions. Every night the stars will rise, cross the heavens on their diurnal paths, then (except for the circumpolar ones) set. The orbital motion causes the stars to be displaced one degree westward in their daily paths *if we observe at the same time each night*. (For reference the full moon is ½° across and Orion's belt is 3° wide.) Night-to-night the change may be hard to see, but the shift amounts to 30°, or two hours of hour angle, in a 30-day month, noticeable to the most casual of observers.

3.2 The ecliptic path

The consequence of the Earth's motion is that the Sun must appear to move against the background of the stars, also in the counterclockwise direction, or to the east. If you walk around a table, the table changes its direction with respect to the other objects in the room: it appears to orbit around *you* in the same direction in which you revolve about *it*. When Sagittarius is on the meridian at midnight in Figure 3.2, the Sun must appear among the stars of Gemini, and when Pisces rides high in the midnight sky, the Sun hovers within Virgo. Just as the stars (and the Sun) have apparent daily paths in the sky caused by Earth's rotation, the Sun must *in addition* have an *annual* apparent path produced by revolution. The Sun is centered in the Earth's orbit. Therefore, if you could stand on it and watch our planet revolve, the Earth would trace out a great circle against the celestial sphere; from our point of view however, the Sun instead *appears* to move in a great circle about *us*.

The Earth's orbit is very stable and almost permanently fixed in space. Therefore the apparent path that the eastward-moving Sun traces out through the background stars is fixed as well. This great circle path is central to astronomy, and is called the *ecliptic*, so named because that is where eclipses of the Sun and Moon (Chapter 10) take place. From the Sun, the Earth would also follow the ecliptic, the two bodies always 180° apart. Note again both motions of our planet. The Sun must always appear to track its daily path, rising in the east, setting in the west. The easterly motion caused by terrestrial revolution is slow, only 1° per day.

The ecliptic passes through a set of 12 ancient classical constellations collectively called the *zodiac*, of which 4 are mentioned in Figure 3.2. The complete collection is listed in Table 4.1. These have always been special to humanity because they hold the life-giving Sun, and as we will see, present the backdrop for the Moon and planets as well. The figures form the foundation of the pseudoscience of *astrology*, the practitioners of which attempt to tell the fortunes of people and even of nations. We will look at this subject in more depth in Chapter 11. Astronomically, and of much greater significance, the zodiac marks for us the paths of the bodies of our Solar System.

Like any great circle, the ecliptic plane has two poles, the points where the perpendicular to the ecliptic pierces the celestial sphere. The one nearest the NCP is called the *north ecliptic pole* (NEP) and the other the *south ecliptic pole* (SEP). Figure 3.2 is viewed from the north ecliptic pole, which renders the orbit of the Earth a circle. Since the ecliptic is fixed in space, so are the ecliptic poles: the northern one is in the constellation Draco (a celestial dragon), the southern in Dorado, a modern figure that represents a peacock.

3.3 The tilt of the ecliptic

If the Earth's rotational axis were perpendicular to the plane of its orbit, the NCP would be coincident with the NEP; that is, the terrestrial north pole would be centered in the drawings of the Earth in Figure 3.2. The Sun's ecliptic path would then lie on the celestial equator. We would live in a never-changing environment. The Sun would

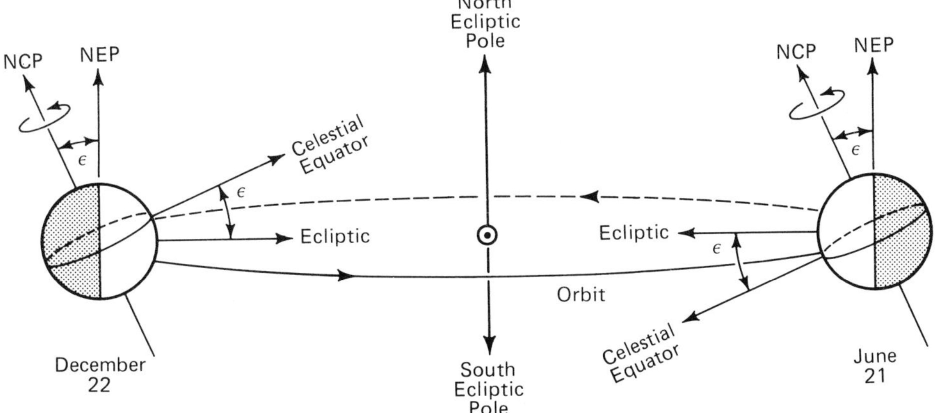

3.3. The tilt of the Earth's axis. The orbital axis of the Earth, the perpendicular to the orbital plane, points to the north and south ecliptic poles. The Earth's rotational axis is tilted by an angle, ϵ, of 23½° relative to the orbital axis. This angle is also that between the equatorial and ecliptic planes. On the left the Sun shines overhead in the southern hemisphere. Six months later it is overhead in the north. In between it would be overhead at the equator.

always rise and set directly east and west, and we would always have 12 hours of daylight, 12 hours of night. But by chance of nature, our axis is *not* perpendicular to the ecliptic plane, and thus arises an interesting complication that gives a great deal of variety to our lives.

The Earth's axis is tilted with respect to the orbital perpendicular by 23½° (more precisely, 23°26′), which displaces the celestial poles by 23½° from their companion ecliptic poles. There is no specific reason that we know of for the tilt; it happened in the violent, early days of the Solar System. All planets have such tilts that are peculiar to them, and most, like that of the Earth, are under 30°. This angle of tilt is called the *obliquity of the ecliptic* and is represented by the Greek letter epsilon, ϵ. It causes the plane of the ecliptic to be tilted through the same angle of 23½° with respect to the plane of the celestial equator. Therefore, instead of remaining *on* the celestial equator, the Sun moves alternately 23½° north and then 23½° south of it.

The orbital configuration is shown in Figure 3.3 where we look at the Earth's path almost edge-on. In the left-hand drawing of the Earth, the axis is tipped away from NEP in the plane of the page, and as a consequence, the Sun shines overhead in the southern hemisphere. The angles between the lines to the NCP and equator and to the NEP and ecliptic are (by definition) both 90°. So if you tilt the NCP 23½° away from the NEP and the Sun, you must tilt the celestial equator 23½° northward of the ecliptic, which puts the Sun at its limit 23½° *south* of the ecliptic, or at a declination of −23½°. As the Earth orbits the Sun, its axis maintains a constant orientation in space, and in the drawing it will remain tilted to the left. (There is a very slow motion, called *precession*, that does change this orientation, which we will take up in Chapter 5, but for the present discussion we can ignore it.) Six months later, in the right-hand

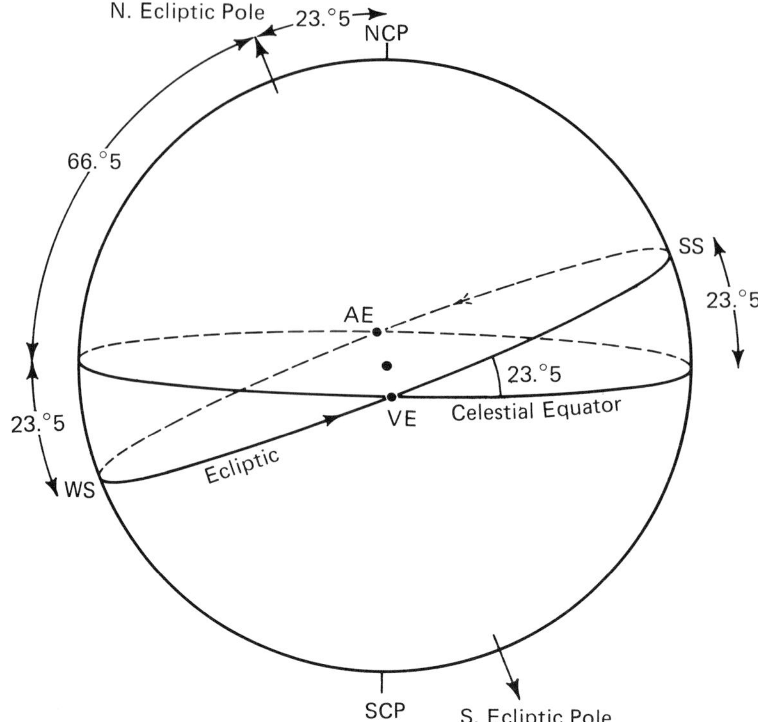

3.4. The celestial equator and the ecliptic. The Sun's path, the ecliptic, crosses the celestial equator at two points at an angle of 23½°. AE, VE, SS, and WS are respectively the autumnal and vernal equinoxes and the summer and winter solstices. The Sun is at the vernal equinox on March 21, and in succession at the SS, AE, and WS on June 21, September 23, and December 22.

drawing of the Earth, the Sun is similarly seen overhead in the north at an extreme declination of 23½° N. As the Earth moves from left to right, the Sun must appear to move smoothly from south to north, and then back again. At the two positions in between, the axis, though still tipped to the left, must be perpendicular to the line to the Sun, which would then be directly *on* the celestial equator, at $\delta = 0°$. That is, for the Sun to go from 23½° south declination to 23½° north, it must pass through 0°, and at that moment, the Earth's axis must be perpendicular to the solar sightline.

Figure 3.4 next shows the configuration of the ecliptic and the equator on the celestial sphere. Imagine yourself again at the center of this sphere, at the central dot that represents the Earth, looking out. The horizon has been removed to concentrate attention on the relationship between the ecliptic and the equator; there is no reference to the observer's latitude. The ecliptic represents the Earth's orbital plane and the Sun moves counterclockwise around it once a year. The axis of the Earth's rotation, defined by the NCP on the celestial sphere, is tilted by 23½° from the NEP, which marks the orbital perpendicular; the ecliptic and equator are thus also tilted relative to each other, and their circles intersect at an angle of 23½°. Once a year the Sun reaches a declination of $23^1/_2°$ N, and once a year $23^1/_2°$ S. The NCP is $23^1/_2°$ from the NEP, and therefore the latter has a declination of $90° - 23^1/_2° = 66^1/_2°$ N. The SEP's declination is $66^1/_2°$ S.

3.4 The ecliptic and the seasons

As the Sun moves on the ecliptic it must cross the celestial equator twice per year, once when it is moving north and again when it is traveling south. Since the equator of the sky lies everywhere above that of the Earth, at these two times the Sun must be seen in the zenith at the terrestrial equator sometime during the day. Because the equator cuts the horizon at east and west everywhere on Earth, the Sun must also rise and set exactly at those positions, and be above and below the horizon for 12 hours at a time for all of us. The intersections of the ecliptic and celestial equator are therefore appropriately called the *equinoxes*. They are points in the sky set among the stars, and like the stars rise and set once per day. The point at which the Sun moves into the northern celestial hemisphere is called the *vernal equinox*, and marks the position of the Sun on the first day of northern hemisphere spring (*ver* in Latin.) The Sun is there on March 20 or 21. The other point is called the *autumnal equinox*, and marks the Sun's position on September 23, the usual first day of northern hemisphere autumn (the 22nd is also possible). After the Sun passes the vernal equinox, it continues to move to the north until about June 21, when it lies at a maximum angle from the equator. This point, marking the first day of northern hemisphere summer, is called the *summer solstice*. Analogously, on December 22 (or possibly the 21st), the first day of northern hemisphere winter, and three months past autumnal equinox passage, the Sun stands $23^1/_2°$ south of the equator at the *winter solstice*. The solstices are also fixed among the stars, and like them have their own daily paths around the observer. The Sun appears to move north or south relatively rapidly when it is near the equinoxes. But as it goes past the solstices, where the ecliptic becomes momentarily parallel to the equator, the northerly or southerly motion becomes fairly insensible; hence the name "solstice": *sol*, Latin for "Sun," and *sistere*, meaning "to stand still."

Spring, as a season, refers to the time when the Sun enters and then moves deeper into the celestial analogue of the sky to the hemisphere in which you live. It is terminated by the start of summer, when the Sun reverses its direction and heads toward the equator. Autumn is the inverse of spring, that is, the Sun enters and moves deeper into the *opposite* hemisphere. It is then halted by winter, when the Sun begins its trek back towards your side of the equator. Spring and autumn are reversed in the two terrestrial hemispheres: as spring begins in the north, autumn commences in the south. Summer and winter are reversed as well. Nevertheless, the astronomical use of the term "vernal equinox" refers to the solar passage across the equator to the north even to a southerner whose autumn is just beginning. The names of these four principal points will remain the same no matter what the observer's hemisphere to avoid confusion.

Because these four positions are fixed in the sky they must always appear in the same direction relative to the background stars. The vernal equinox lies among the stars of the constellation Pisces, the pair of fishes; the summer solstice, autumnal equinox, and winter solstice belong respectively to Gemini, Virgo, and Sagittarius, the four figures noted in Figure 3.2. Always we must strive to relate the abstract points and circles that we cannot see to the concrete: real stars, real places in the sky. All four are positioned in the heavens on the star maps in Appendix 2.

The cultural significance of these four points is deep indeed: witness the placement of our oldest holidays. We find Christmas (which replaced a much older celebratory holiday) around the winter solstice, Easter and Passover near the vernal equinox, and Rosh Hashanah and Yom Kippur close to the autumnal equinox. Even the summer solstice – Midsummer Night – is celebrated by Shakespeare. In between the four major dates lie the *cross-quarter days*: *Candlemas*, or in the United States *Ground-Hog Day*, on February 2; *Halloween* (All Saints' Day Eve) on October 31; *May Eve*, the night before May Day, May 1 (*Beltane*, an old Scots–Irish feast of fertility); and the more obscure *Lammas Day* (celebrating the first ripe grain in what is now Great Britain) on August 1. We cannot escape our astronomical roots.

3.5 The ecliptic and the horizon

The introduction of the horizon complicates the picture. The problem is that the sky rotates daily around the celestial poles parallel to the equator, whereas the ecliptic is tilted. Therefore, the ecliptic is continually changing its relation to the horizon and equator. At all times an observer will see exactly half of the ecliptic, but the half seen depends on the time of day and year. Sometimes just the southern half can be seen, at other times only the northern, and most commonly, parts of each.

Start first with the simple case of Figure 3.5, which illustrates the ecliptic and equator from the viewpoint of a real observer who has a real horizon and a specific position on the Earth at the usual 45° N latitude. The drawing is identical to Figure 1.14 except that the ecliptic and its four reference points have been included, and it is done for the specific time of day when we see the autumnal equinox setting in the west, and the vernal equinox, 180° away, rising in the east. The vernal equinox is now indicated by the symbol that represents the constellation Aries, or stylized set of ram's horns, ♈. (The vernal equinox was once found in Aries, as explained in Chapter 5.) Remember that the equinoxes are defined by the equator, and must rise and set exactly in the east and west. In Figure 3.5, the two solstices, each 90° from the equinoxes, must be found on the meridian. If we ignore daily motion, the Sun must creep to the east, or counterclockwise as viewed from the NEP, at a rate of 1° per day. From the autumnal equinox it moves a quarter-turn to the winter solstice, which is south of the equator. Therefore, the observer sees the winter solstice at upper culmination at a declination of 23½° south. The summer solstice, opposite at 23½° N declination, is therefore at lower culmination. The person watching the sky at this instant sees only the southern half of the ecliptic. From the previous section, the constellation Sagittarius is on the upper meridian, Pisces is rising, and Virgo setting. Because of the symmetry of the configuration, with the equinoxes on the horizon, the ecliptic poles have to be on the meridian, here with the NEP at upper transit, 23½° from the NCP, and the SEP at lower transit.

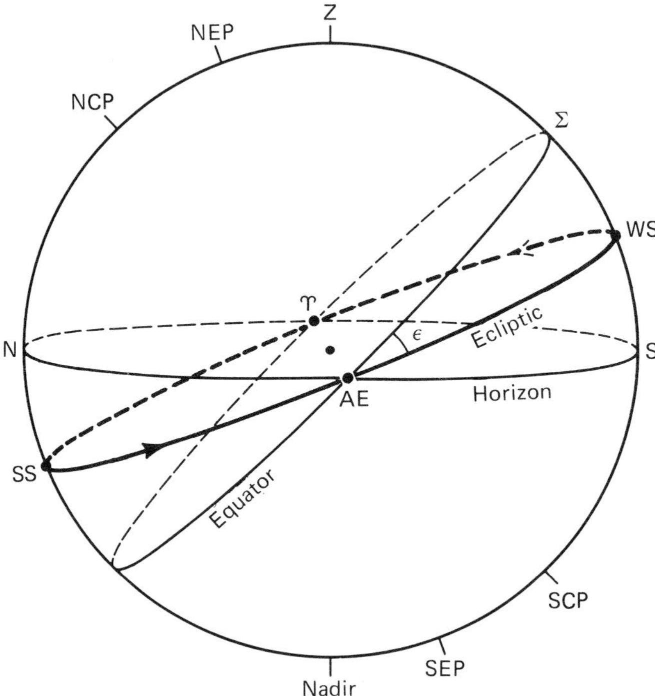

3.5. The orientation of the ecliptic. The celestial sphere shows how the ecliptic is oriented for an observer at 45° N latitude at the moment when the vernal equinox is rising and the autumnal equinox is setting. The vernal equinox is now represented by the symbol ♈. The winter and summer solstices then must be culminating. The observer now sees only the southern half of the ecliptic. The arrows show the annual direction of solar motion.

Finally, we could place the Sun on the diagram (it is not drawn in). The time of year, which is given by the Sun's location relative to the four ecliptic points, has not been specified in the drawing. If the date is December 22, the Sun is at the winter solstice, and in Figure 3.5 it is also on the meridian; the Sun is at its highest point in the sky for that day, and it must be noon. If it is June 21, the Sun is at the summer solstice, and it must be midnight, if March 21 it is sunrise, and so on. The date, the time of year, is tied to the position of the Sun on the ecliptic relative to the four points, specifically to the vernal equinox. Our standard western calendar (described in Section 6.7) is set up so that on the average the Sun passes it on March 20 or 21 (depending on closeness to a leap year). That way the seasons and the calendar stay synchronized.

3.6 The changing ecliptic

The orientation of the circles in Figure 3.5 exists for but a moment. The equinoxes are not always rising and setting and the solstices are not in perpetual transit. These points are fixed among the stars and must move about the sky with them. A few minutes after the time of Figure 3.5, the autumnal equinox will be below the horizon, the vernal equinox above it, and we will begin to see some of the northern section of the ecliptic.

The orientation of the ecliptic relative to the horizon thus depends upon the rotational position of the Earth relative to the stars, or as we will see, on both the time of year and the time of day. Mentally rotate Figure 3.5. Every point on it must move along a daily path about the celestial poles. The key to understanding the apparent motion of the ecliptic, and that of the Sun in the sky, is in remembering that the solstices and equinoxes (and every point on the ecliptic as well as the ecliptic poles) are fixed among the stars and have their own daily paths around the sky parallel to the celestial equator, rising and setting (unless circumpolar) once every day.

Figure 3.6 shows this rotation. Think of the four drawings as a crude motion picture with a frame taken every time the Earth turns through 90° (or 6 hours). Figure 3.6a reproduces Figure 3.5, and Figures 3.6b, c, and d show what happens at 6, 12, and 18 hours past the instant marked by Figure 3.6a. The daily paths of the equinoxes always trace out the equator, staying on opposite sides of the sky. Unless they are exactly rising and setting an observer can see only one at a time. The paths of the winter and summer solstices are small circles respectively 23½° south and north of the celestial equator. They are also always opposite each other and are perpetually 90° (6 hours) in hour angle from the equinoxes. In Figure 3.6b, the equinoxes have moved through one-quarter of the circle and are now on the celestial meridian; the winter solstice has set, and the summer solstice has risen and is seen in the east. Note that since the winter solstice is south of the equator, at an hour angle of 6 hours, it is below the horizon. And because the summer solstice is in the north, at 18 hours, it will be *above* the horizon. To draw in the ecliptic we simply connect all four dots with a great circle. With the vernal equinox at culmination, the observer sees the ecliptic tilted from northeast to southwest, angling across the sky; somewhat over half the northern part and a little less than half the southern can be seen.

Next, in Figure 3.6c, the sphere has rotated through one-half a turn from Figure 3.6a. The equinoxes have reversed their locations, and now it is the vernal that is setting and the autumnal rising. The solstices, too, have flipped their positions, but remember again that since the solstices are not on the equator, their daily paths are *small* circles; each solstice remains in its respective hemisphere and does not cross the celestial equator. The summer solstice culminates the meridian high in the sky in Figure 3.6c, but the winter one transits low, below the equator, in Figure 3.6a. As a result, the ecliptic has reversed its orientation relative to the equator. The observer now sees only the northern half, which climbs the sky *above* the equator rather than below it.

Finally, in Figure 3.6d, the sphere has turned three-quarters of the way from 3.6a, and the equinoxes are once again on the meridian. Figure 3.6d is also 12 hours in time past Figure 3.6b, and the two are reversals of one another. The summer solstice is in the front and the winter one in the back. The hour angle of the summer point is now 6^h, and it is above the horizon because it is in the north, along with the observer; the winter point, at 18^h hour angle, has not yet risen. The ecliptic now cuts southeast to

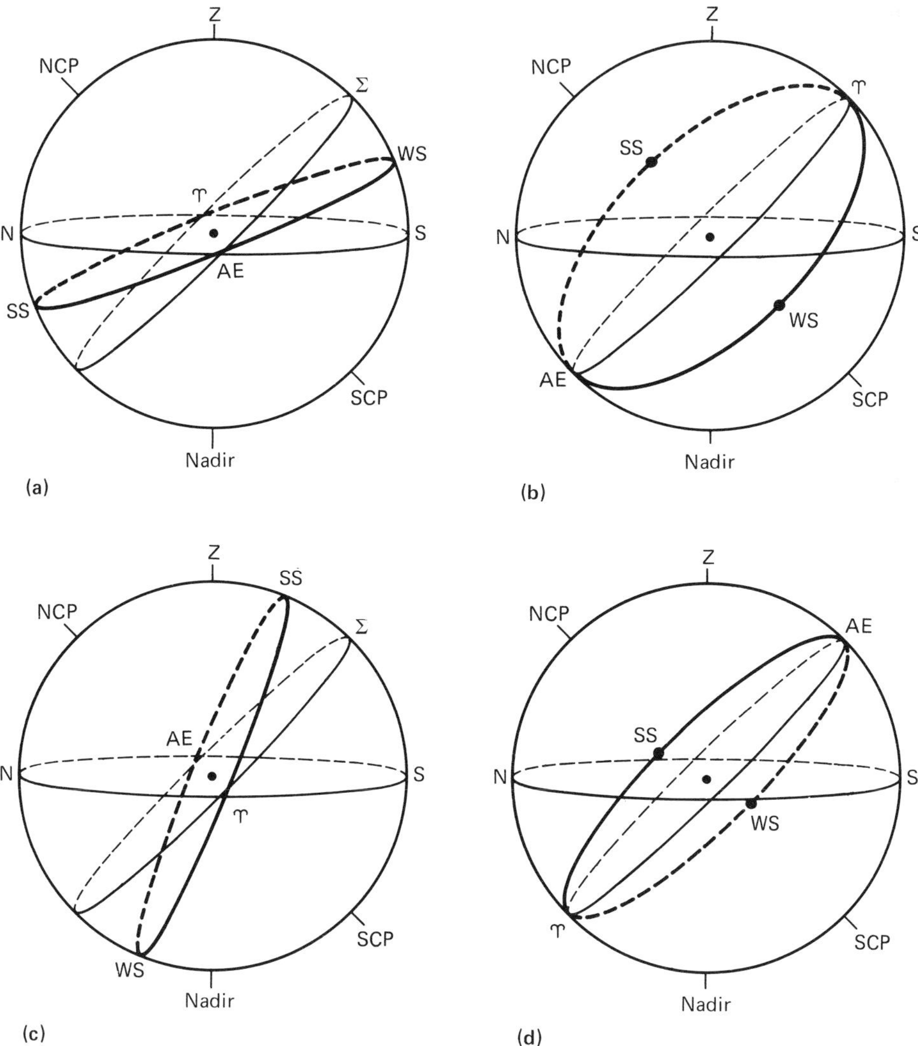

3.6. The daily motion of the ecliptic. The celestial equator is drawn with a lighter line, the ecliptic with a heavier one. Each part, drawn for a latitude of 45° N, is six hours later than the preceding one ((d) precedes (a)). (a) reproduces Figure 3.5 and shows the vernal equinox rising. The equinoxes and solstices all have their own daily paths around the celestial sphere. As the sky rotates we see different parts of the ecliptic come into view, so that it keeps changing its orientation relative to the horizon.

northwest, and again we see a bit over half the northern ecliptic and somewhat less than half the southern.

The view from the north is clearly parochial. In Figure 3.7 the rotation of the sphere and the orientation of the ecliptic are seen from a latitude of 45° S. Each pair (Figures 3.6a and 3.7a etc.) are set for the same moment and are viewed from the same longitude,

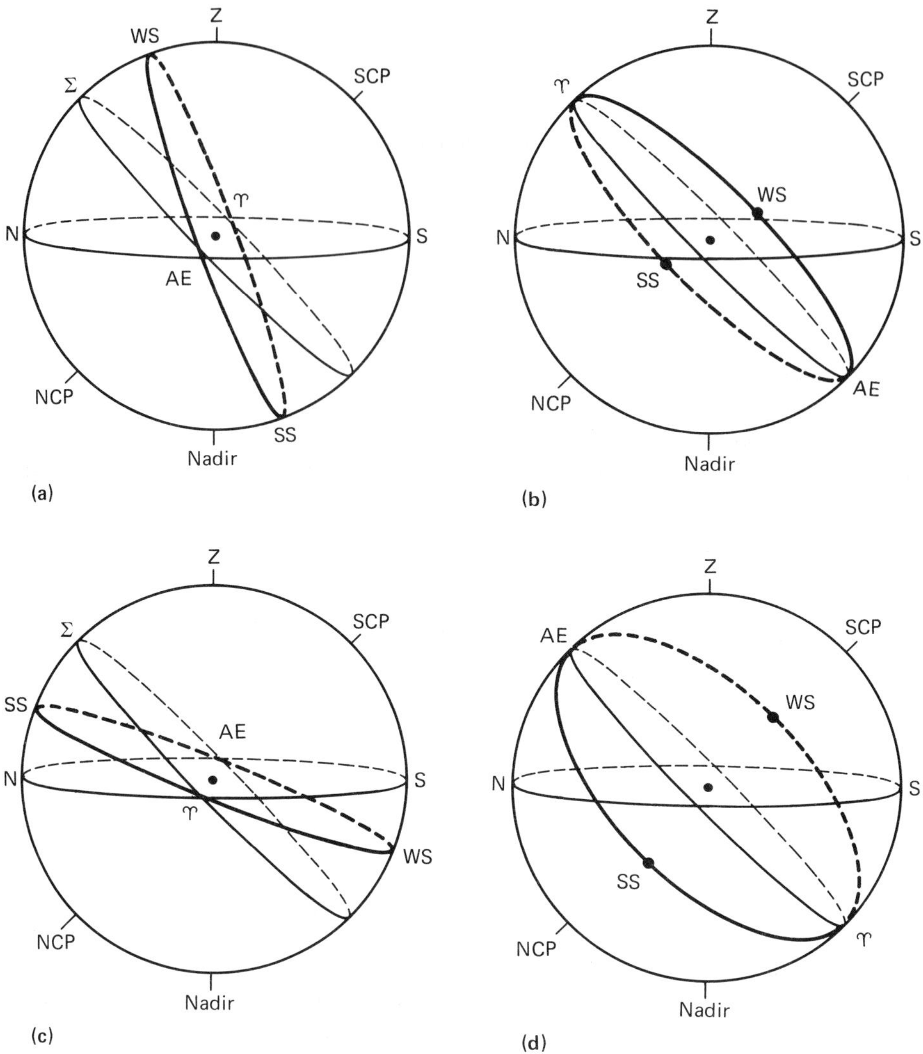

3.7. The daily motion of the ecliptic from the southern hemisphere. The drawings are set for the same longitude as those in Figure 3.6, but for a latitude of 45° S.

but at latitudes that are 90° apart. Consequently, the apparent orientations are reversed: if the ecliptic is high in the sky from the northern hemisphere, it will be low from the southern.

In both cases, north and south latitude, try to envision the positions that occur between the drawings. Between Figures 3.6a and 3.6b the winter solstice must set. Since the summer solstice is diametrically opposed to the winter, it must be rising. *Now* the observer sees exactly half the northern ecliptic and half the southern, but the vernal equinox has not yet culminated. In Figure 3.7, the winter solstice will set between Figures 3.6b and 3.6c, *after* the vernal equinox has made its lower transit.

With some study, the changing picture presented by the ecliptic should become apparent from these drawings. It helps to have a three-dimensional celestial globe at hand. Set the axis of the sphere to any desired latitude (the altitude of the pole above the horizon) and spin it to see how the ecliptic changes its orientation. If you do not have one, take a ball and mark out a great circle (as the equator) and its poles. Draw another great circle inclined to the first at roughly 23°. Tilt the axis relative to a desk top, which can serve as your horizon, and spin the ball around its poles. Compare the view with the figures.

If you know any of the constellations of the zodiac, go outside and trace the ecliptic with your finger. See how it changes as the sphere rotates about you. Even if you do not know them, the Moon will roughly trace out the ecliptic path over the course of the month; sometimes the Moon and a couple of bright planets will even explicitly mark it like a dotted line in the sky. Watch how the ecliptic tilts relative to the equator at different times of the night or year.

The ecliptic is actually rather simple to draw on diagrams of the kind used here. Only one of the four major points need be located. From west to east (counterclockwise) we encounter the vernal equinox on the equator, the summer solstice 90° around the equator at 23½° N declination, the autumnal equinox at 180° around and back on the equator, and the winter solstice at 270° and 23½° S declination. Locate the points and just connect them, as is done in all the drawings.

3.7 Solar time

The next step is to place the Sun on the ecliptic to show how its orientation relates to the date and the time of day. The date, as described in Section 3.4, is closely tied to the location of the Sun on the ecliptic relative to the equinoxes and the solstices. On March 21 the Sun is at the vernal equinox; on the 22nd it is 1° east, and after about 3 months, by June 21, it has gone 90°, and so on.

The time of day is of equal significance since it is related to where the Earth stands in its daily rotation and to the rising, setting, and culmination of the equinoxes and solstices. Almost everyone has some awareness that the time is related to the position of the Sun. It is somewhere near 6 hours (6 AM) when the Sun rises, noon when it is highest, and 18 hours (6 PM) or so when it sets. Fewer know that clock time is in fact strictly locked into the solar position, more specifically to the Sun's hour angle. *Solar time* is defined very simply as the hour angle of the Sun plus 12 hours. The extra 12 hours is added so that the new day (0 hours) starts at midnight instead of noon. (The 12 hours were added in 1925. Until then, the astronomical – not the civil – day did start at noon.)

This kind of simple solar time is that read by a *sundial* (Figure 3.8). In its simplest form, a stick (called a *gnomon*) pointed at the visible celestial pole casts a shadow onto a horizontal or curved plate that is graduated in hour angle. If you have ever used one,

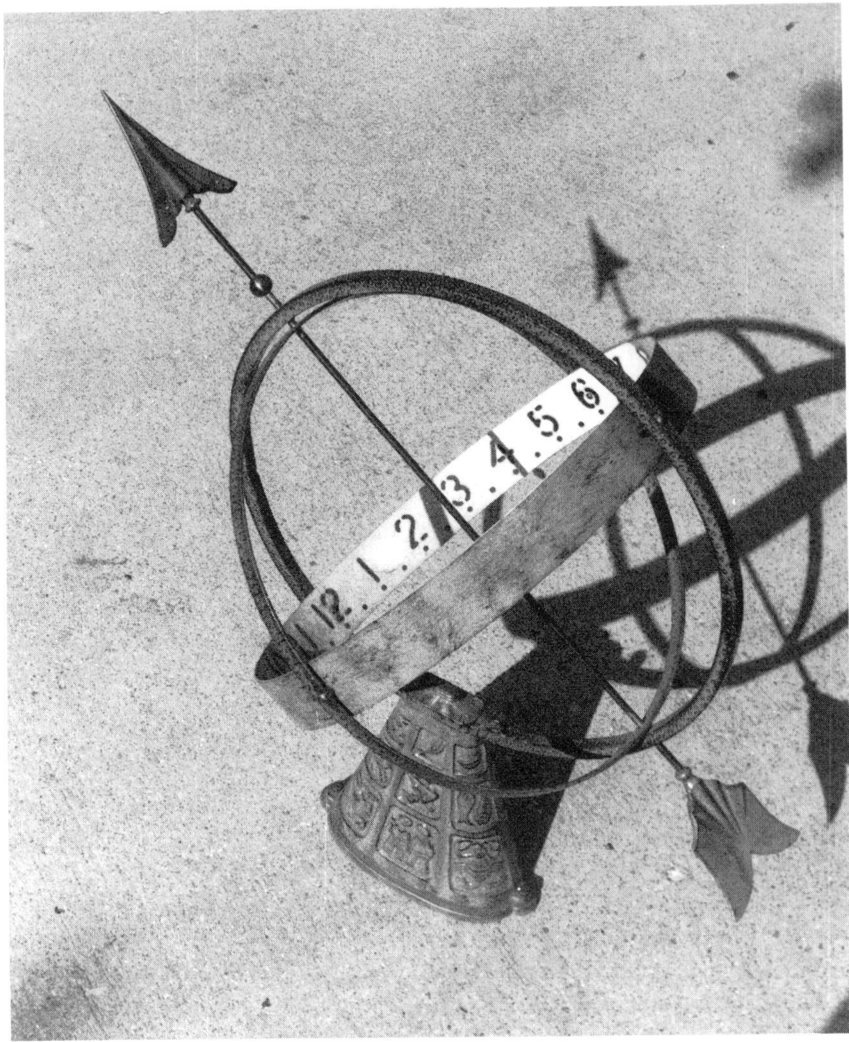

3.8. A sundial. The gnomon, in this case an arrow, points at the celestial pole and casts a shadow on a graduated dial from which local apparent solar time can be read directly. Here the local apparent solar time is a few minutes past 4 PM or 16 hours. (Author's photograph.)

though, you probably noticed that the time it told differed, perhaps considerably, from that read on your watch. Time measurement is in fact quite complex because of variations and irregularities in the Earth's motions that have not yet been discussed. There are, for example, two kinds of "Suns" that are used, the real or *apparent* Sun and an average or *mean sun* that smooths out the apparent Sun's somewhat variable motion. Moreover, since motion on the Earth's surface is reversely reflected into motion in the sky, as you move east the Sun appears to move west (and vice versa), which changes

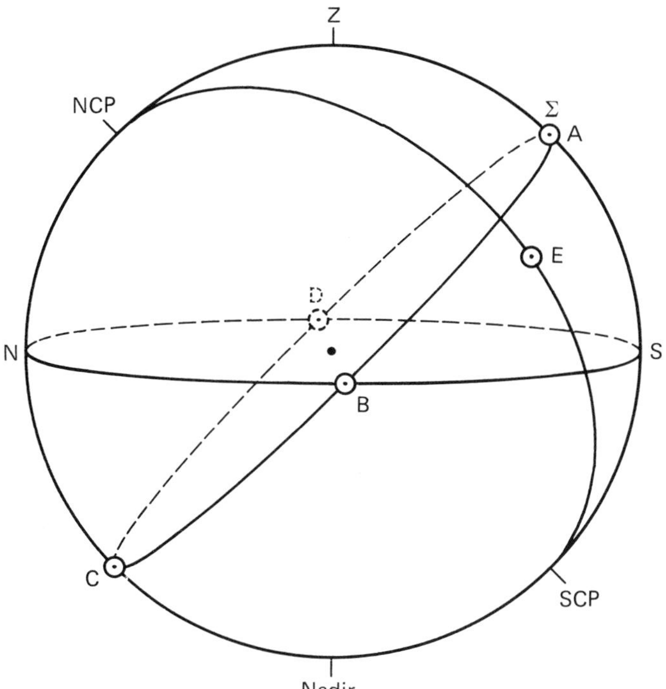

3.9. Solar time. The equatorial Sun is shown at four hour angles *A* through *D*, 0^h, 6^h, 12^h, and 18^h respectively. The solar times for each are 12^h, 18^h, $24^h = 0^h$, and 6^h. At position *E* the Sun is placed south of the equator during mid-afternoon and the time is about 14^h.

the solar hour angle. The time therefore depends on your longitude. These distinctions will be taken up in Chapter 6. Here we will define only a generalized solar time based on the apparent Sun, more properly called *local apparent solar time*, since it depends (as we will see) on longitude.

A variety of examples is presented in Figure 3.9, which shows the sky again at latitude 45° N and the Sun on the celestial equator. First, when the Sun is on the meridian, position *A*, and the hour angle is 0 hours, the time is $0^h + 12^h$ or 12 hours. This view accords well with experience. Twelve hours is called noon, and the Sun is at upper culmination, at maximum altitude. If the Sun is setting exactly at the west point, position *B* (which, since the Sun is on the equator, would occur only on March 21 or September 23), the hour angle is 6^h and the time is $6^h + 12^h = 18$ hours. Normally, a 24 hour clock is used here. But a 12-hour system is also allowed wherein the Sun is referred to the upper meridian. Thus 18^h is also 6 PM, where "PM" means *post* (Latin for "after") *meridiem* (midday, from which comes our word "meridian"). Similarly, "AM" means *ante meridiem*, or "before midday."

At *C* the Sun is at lower transit and the hour angle must be 12 hours. The time, $12^h + 12^h = 24$ hours, is identical to 0 hours, that is, the end of one day is also the beginning of the next. This time is popularly known as midnight. Neither noon nor midnight can have AM or PM attached to them. There is no 12 AM or 12 PM, since the terms have no meaning *on* the meridian: the proper term is 12 noon or 12 midnight. A 24-hour clock eliminates the confusion, as 12 midnight becomes "zero hours." After another quarter turn the Sun arrives at the east point (*D*) and rises. The hour angle is now 18 hours, and the time is $18^h + 12^h = 30$ hours. However, clocks only go to 24 hours and then start cycling from zero. So if the time exceeds 24 hours, a full day can be subtracted, and hence the time of sunrise is $30^h - 24^h = 6$ hours. As a fifth example, the Sun is placed south of the equator and somewhere in the western celestial hemisphere at position *E* (the date must now be between September 23 and March 21). What time is it? First draw in the hour circle. The hour angle is about 2^h, so the time is 14 hours (or 2 PM).

It is well to reiterate that a clock is in fact a model of the Earth's rotation, or rather of the placement of the Sun in the sky. If it is nighttime or cloudy, we cannot use a sundial to measure the time of day. We then represent the Sun by an accurate mechanical device whose hands move in the same direction, one that we can bring indoors to read the solar hour angle. No machine is perfect, nor is the Earth's rotation. Clocks – even the most modern – must be continually calibrated by direct observation of the sky, returning us to our ancient astronomical roots.

3.8 The position of the Sun

The next step is to combine the above concepts and place the Sun on the celestial sphere to relate time and date simultaneously to the orientation of the ecliptic. In Figure 3.10a the sky is again rotated so that the vernal equinox is rising, one possible configuration out of many. We may locate the Sun anywhere on the ecliptic we like: the date is then given by where it is relative to the equinoxes or solstices, as described briefly in Section 3.7. If we put it on the winter solstice it is December 22, and so forth; if we place it just midway between the autumnal equinox (which it passes on September 23) and the winter solstice (where it will arrive on December 22), it must be around November 6. To a reasonable approximation, all you need do is count the number of degrees counterclockwise from the vernal equinox to the Sun, call that the number of days, and (with the aid of a calendar or Table A1.1, which gives day numbers) add it to March 20 or 21. Of course since there are 365 days in a year, you will be off by 5 days if your date is, say, March 19. For better accuracy, count degrees forward and backward between the vernal and autumnal equinoxes. Better yet, just count degrees forward and backward up to 45° between the Sun and any of the four principal ecliptic points. For example, if the Sun is 20° in front of the winter solstice, that is, it has 20° to go to get there, the date must be about December 2.

Once the Sun is positioned, the time of day can quickly be told from its hour angle as per the last section. Figure 3.10a is set for December 22 at noon (12 hours), March 21 at 6 hours, June 21 at 0 hours, and September 23 at 18 hours. The configuration of the ecliptic in Figure 3.10a will actually occur once a day as the Earth spins (the equinoxes must rise and set daily), but at a different time for every day of the year. There are 361 possibilities other than those given above. For example, if it is December

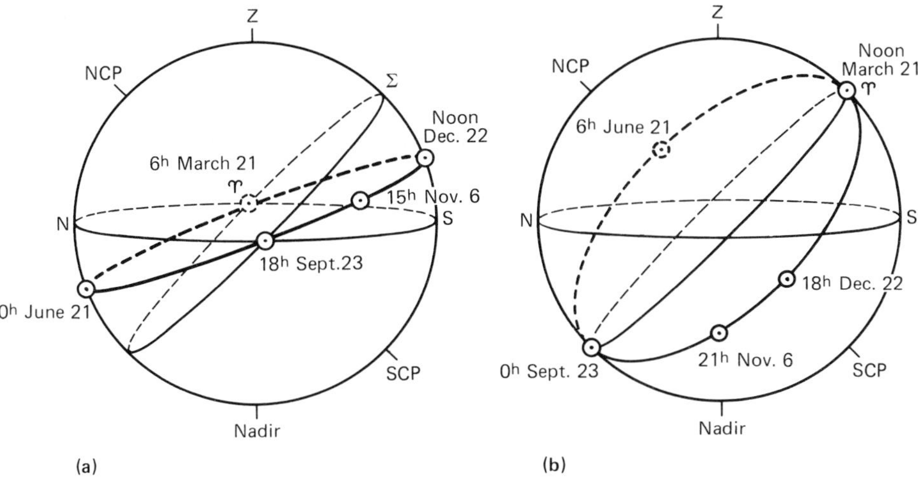

3.10. Position of the Sun at different times of the day and year. The drawings are the same as for Figures 3.6a and 3.6b except that the Sun is placed on them at different positions so as to define the day and date. (a) shows the sky for March 21 at 6^h solar time, June 21 at 0^h, December 22 at 12^h or noon, and September 23 at 18^h (or at interpolated points between, e.g. November 6 at 15^h or March 22 at 5^h 56^m, the last not shown). (b) is progressively 6 hours later than (a) for the same date or 3 months later for the same time.

23, the Sun would be 1° east of the meridian and the time would be 11^h 56^m. Or if it is November 6 in Figure 3.10a, the Sun's hour angle is about 3 hours and the time about 15 hours.

Now look at the next drawing in Figure 3.10b. If the Sun is at the vernal equinox, the date is March 21, and the time must be 12 noon. This configuration is three months later than that in Figure 3.10a. The Sun positioned in succession at the other three ecliptic points yields June 21 at 6 hours (note that though the Sun is above the horizon the hour circle through it cuts the east point, rendering the hour angle 18 hours), September 23 at 0 hours (midnight) and December 22 at 18 hours. Comparing the two diagrams we see that they are six hours apart for the same *date* or they are three months apart for the same *time*. We have returned now to the description that opened this chapter. The comparison between Figures 3.10a and b shows the same phenomenon as Figure 3.2: over a period of three months (a bit over 90 days) the celestial sphere will slip a quarter turn, 90°, to the west, if viewed at the same time every day. Return to Figure 3.6 and mentally place the Sun in Figures 3.6c and 3.6d, and note the same differential. Figure 3.6c is 12 hours or, for the same time, 6 months from Figure 3.6a; Figure 3.6d is 18 hours or 9 months from 3.6a.

3.9 The daily path of the Sun

Like the equator, the daily paths of the stars observed from a given latitude are fixed in the sky. Citizens of New York City see the brilliant northern star Vega trace out

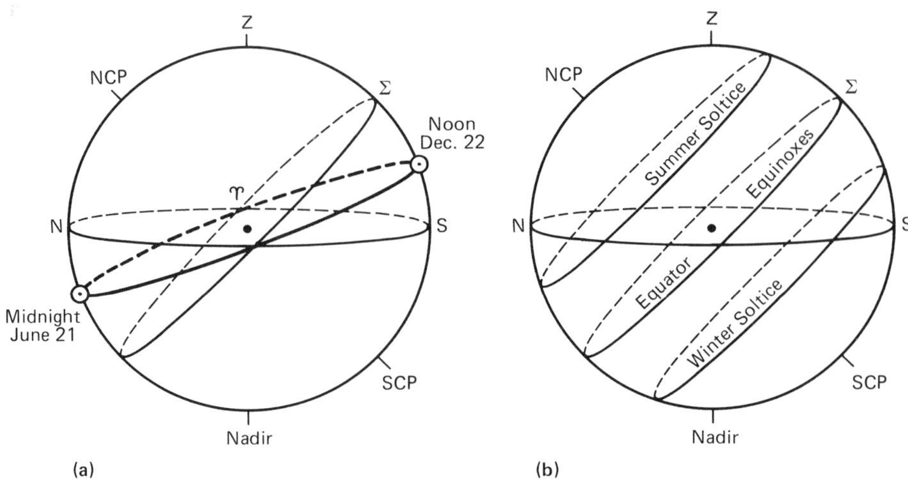

3.11. The daily path of the Sun at the solstices and equinoxes. (a) shows the ecliptic at ϕ = 45° N with the vernal equinox rising. When the sphere is rotated (b), the solstices trace out daily paths according to their declinations, just as they would were they stars. The daily path of the equinoxes is the equator.

precisely the same circuit day after day, year after year, perpetually culminating 2° south of the zenith (the declination of Vega is 38¾° N, the latitude of New York is 40½° N). But since the Sun rides the ecliptic, and consequently must pass from north to south and back again, its daily path is a little different each day. Look at Figure 3.11a and place the Sun at the winter solstice: it is now (as we saw in the last section) noon on December 22. As the Earth turns and the sky rotates above the observer to the west, the Sun will trace out a daily path 23½° south of the equator, as in Figure 3.11b. Refer to Section 2.4: the Sun will set in the southwest, and rise in the southeast. We also saw there that the duration of visibility for a star in the hemisphere opposite to that of the observer is always less than 12 hours. The Sun, as seen from Figure 3.11b, is below the horizon for a greater time than it is above it, and it will transit the meridian low in the sky. Winter sunlight will be at a minimum, but astronomers will be delighted with the long hours of darkness.

By June 21, the configuration is reversed. If we place the Sun at the summer solstice in Figure 3.11a it is midnight, 0 hours solar time. The daily path now lies 23½° *north* of the equator, and from Figure 3.11b the Sun will rise in the northeast and set in the northwest. It will cross the meridian high in the sky, and will be above the horizon for a longer duration than it will be below it. Daylight hours are long and leisurely, but astronomers must rush to complete their nightly tasks. If in Figure 3.11a we place the Sun at the vernal equinox it is 6 hours solar time, and if we place it at the autumnal equinox it is 18 hours. In either case, the Sun rises exactly east, sets exactly west, and rides the celestial equator during the day, that is, the equator *is* the daily path. Since the horizon exactly bisects the celestial equator, days and nights are of equal duration.

The Sun takes six months to go from declination 23½° N to 23½° S, and six months to return. Figure 3.11b shows only the extremes, but obviously, the Sun will at some time of the year be found in between the limits and the equator. It is an easy matter for the astronomers to calculate the Sun's declination for any date: tables of declination are readily available in a variety of publications (see Appendix Figure A1.3). For example on April 18 and August 24 δ(Sun) will be +11°, and the daily path will lie in between that for the summer solstice and the equator.

In the above descriptions, terrestrial rotation has been decoupled from orbital revolution. That is, the Sun is assumed to have a certain declination on any one day, which locates a daily path parallel to the equator. The next day it is shifted a bit north or south. In fact, however, the motions are simultaneous, and the Sun's true daily path is never really parallel to the celestial equator as is that of a star. Since the Sun is always moving eastward along the inclined ecliptic, the actual apparent daily solar path is a slow helix. As the Sun appears to circuit about the Earth once a day, it is always moving either north or south, completing the full cycle only after the passage of a year.

3.10 The origin of the seasons

It is commonly believed that the different seasons are caused by the changing distance of the Earth from the Sun. Even casual observation refutes this idea. If such were the case, the entire Earth chill at one time and the seasons would not alternate between north and south. Moreover, if we moved far enough away from our heating source to produce a winter, the Sun would look smaller in the sky. However, the solar angular diameter stays nearly the same year-round. (But note: do *not* attempt to verify this by looking casually at the Sun. It is *only* safe to do so with specialized filters or when the Sun can be comfortably viewed on the horizon through thick haze.)

The sole cause of the change of the seasons is the obliquity of the ecliptic. In mid-latitude winters, the Earth receives the Sun's heat only for an abbreviated part of the day. More importantly, the Sun is never far above the horizon. The line to the Sun makes a small angle with the ground; sunlight must spread itself out over a larger surface area than it does in summer, and each square meter of the Earth receives less solar radiation than it would if the Sun were overhead (Figure 3.12). If the obliquity, ϵ, were 0°, then with the Sun always on the equator there could be no seasons at all. If ϵ were larger than it is now, the seasons would be more severe.

Because it takes time for the Earth to absorb and release heat and because of the slow change of solar declination near the solstice, the highest temperatures of summer lag behind the highest solar altitude (which occurs at the solstice) by a month or more. Conversely, the coldest period of winter will occur a month to six weeks after lowest solar culmination (the winter solstice passage in the northern hemisphere or the astronomical summer solstice in the southern hemisphere).

Near the Earth's equator the four-season phenomenon disappears. Citizens of northern South America and central Africa have the celestial equator nearly over their heads, and the Sun goes back and forth from somewhat north of the zenith to somewhat south. The variation in solar heating is much less, so temperature variations are minimized as well.

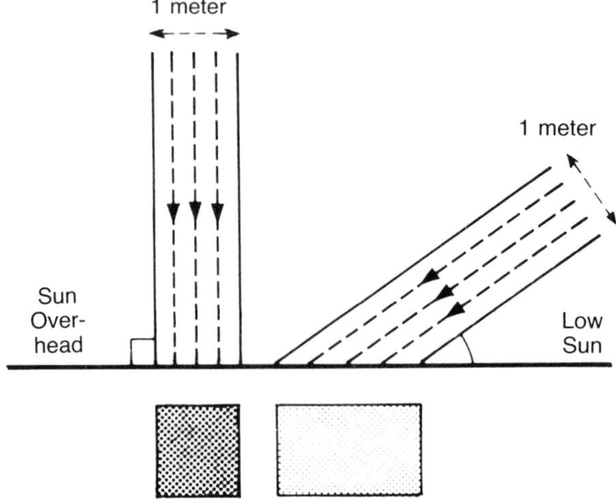

3.12. The origin of the seasons. When the Sun is overhead (or nearly so) in summer, the solar energy reaching a given area in one minute is a maximum. When the Sun is low in the sky, in winter, this same amount of energy is spread out over a large area, resulting in a lower heating rate. (From *Principles of Astronomy: A Short Version*, S. P. Wyatt and J. B. Kaler, Allyn and Bacon, Boston, 1981.)

In fact, the Earth does change its distance from the Sun. Our planet's orbit is slightly elliptical, the distance varying by a bit under 2% over the year. The change has little sensible effect on the seasons, however, since the resulting 3.5% variation in solar heating is quite lost in the effects produced by the obliquity. The Earth is actually closest to the Sun (149 million kilometers) about January 2, in the dead of northern hemisphere winter, which shows what a small effect the orbital eccentricity has. The Earth is most distant from the Sun (152 million kilometers) about July 4. All things being equal, southern hemisphere winters should be more severe than northern, but even that small effect is swamped by climate differences caused by the different distribution of land masses and oceans between the two hemispheres.

3.11 Ideal sunrise and sunset

In Section 2.4, we examined how the rising and setting points of a star and its daily visibility all depend on its declination. The farther north a star, the closer to the north point it will set, or the greater the azimuth of its setting point (and the smaller the azimuth of its rising point), and (in the northern hemisphere) the longer it will stay above the horizon. Now we apply the same logic to the Sun. It must behave in the same way, clearly evident from the paths in Figure 3.11b. If in our usual mid-northern example ($\phi = 45°$ N) we watch the Sun over the course of the year, we see that on the first day of winter, December 22, it will set toward the southwest. As the year progresses, the point of sunset moves toward the west point of the horizon, crossing it on

March 21, and continues to the northwest to a maximum azimuth on June 21, after which it swings slowly southward again.

These rising and setting points must also change in response to changes in the observer's latitude, as described in Sections 2.8 and 2.10 through 2.12 and seen in the succession of drawings in Figures 2.12 and 2.15. Just consider the Sun as the point in question rather than a star. Adopt a northern summer day as an example. The farther north the observer travels, the closer sunrise will be to the north point of the horizon. At a high enough latitude, the Sun can be circumpolar (see Section 3.12). At the Earth's equator, the Sun's daily path is perpendicular to the horizon. Rising and setting azimuths as well as the duration of daylight and night change in accordance with the summary of Section 2.12.

The angles of interest can be estimated from hand-drawn celestial spheres or calculated exactly from the astronomical triangle and a knowledge of spherical trigonometry (see Section 2.5 and Appendix 3). These angles include the azimuths of sunrise and sunset, the hour angles of these points (which define the time of sunrise and sunset and the duration of daylight), and the altitude of the Sun as it crosses the meridian. Indeed we can obviously calculate the azimuth and altitude of the Sun exactly at any time of the day and year from its declination and hour angle, and vice versa.

To quantify this exploration of the solar motion, we must be precise about the Sun's position relative to the equator over the course of the year, that is, we need to know how solar declination ($\delta_\odot$) changes with the date, or more precisely, how $\delta_\odot$ changes with the time elapsed since the exact moment the Sun crossed the equator at the vernal equinox. Given the numerical value of the obliquity of the ecliptic, we can always calculate the perpendicular arc from the Sun to the celestial equator for any position along the ecliptic. At the time of the solstices, the declination equals the obliquity, $23^1/_2°$ north or south. The average declination of the Sun for the date at Greenwich (the prime meridian) is shown in Figure A1.3. Because the length of the year is not exactly equal to an integral number of Earth days, spring does not begin at the same time each year, and $\delta_\odot$ for a given date will change by a small fraction of a degree from one year to the next. In addition, your date (in decimal units) may differ by as much as half a day from that at Greenwich.

From the solar declination in Figure A1.3 and the relations in Table 2.2, we can immediately find the altitude of the Sun at noon. The azimuth of sunrise, A_r (or sunset, $A_s = 360° - A_r$), can also be found for any latitude and any day of the year from Figure A1.1. Try a pair of examples. First, what is the azimuth of sunrise at 35° N latitude on May 21? From Figure A1.3, $\delta_\odot = +20°$. On Figure A1.1, find the latitude on the lower axis and read up to the 20° declination curve. Then read to the left to find an azimuth of 65°. The setting azimuth must be 360° – 65° or 295°. Next, what is A_r for November 12 at that latitude? On November 12, Figure A1.3 shows $\delta_\odot$ to be –17°.5. On Figure A1.1 we read to the right since declination is negative. We must also interpolate between the 10° and 20° declination curves, which we can do reasonably by eye estimate. The azimuth of sunrise is about (looking to the right) 110°; that of sunset would be 250°. The solstice curve in Figure A1.1 is given as a special case and shows the minimum and maximum rising azimuths for any latitude, that is, it will tell just how close to north and south sunrise (or sunset) can be.

We can similarly find the hour angle of sunset or sunrise for any latitude and any date from Figure A1.2, for which the declination limit of the solstice Sun is again singled out there as special. The hour angle gives the local apparent solar time of sunset (see Section 3.7) and twice the hour angle of sunset gives the duration of daylight. Consider the above two examples. On May 21 (at 35° N latitude) the hour angle of the setting Sun (read to the left on the figure) is 7^h. The apparent solar time of sunset is then 19^h (or 7 PM) and daylight lasts for 14 hours. On November 14, the hour angle (read to the left, since latitude and declination have different signs) is $5^h\ 15^m$, so the apparent solar time is $17^h\ 15^m$ (5:15 PM), and daylight endures for only $10^h\ 30^m$. Reading along the curve of the solstice Sun will give the apparent solar time of sunrise and sunset and the maximum and minimum daylight hours for any latitude.

Solar time as discussed so far – apparent solar time – is limited in its usefulness. We do not live by apparent solar time, but by standard time, which involves averaging the variability of apparent time and making corrections for longitude. In addition, the calculation of daylight duration and sunset and sunrise times and azimuths is complicated by the extended angular diameter of the Sun – it is not a point like a star – and by a phenomenon called *atmospheric refraction*, the bending and distortion of light in the Earth's atmosphere. The effects of these complications on the ideal world described in this chapter will be discussed in Chapter 7 after the subject of time has been thoroughly examined in Chapter 6.

3.12 The tropics, arctic, and antarctic

From Figures 1.17 or 2.11 and the relations in Table 2.2, the declination of a star passing through the observer's zenith is the same as the observer's latitude. Because the Sun is constrained to declinations between 23½° north and south, it can never be overhead north or south of latitude 23½°. At our usual latitude of 45° N, for example, the solar zenith distance on June 21 is 45° – 23½° = 21½°. At 23½° N latitude (actually, 23°26′, the precise obliquity of the ecliptic), however, the Sun passes overhead when it reaches the summer solstice (Figure 3.13a); at 23½° S, it is overhead on December 22 (Figure 3.13b). These two parallels of latitude define the *tropics* (from Latin for "turning points"). The northern limit is called the *tropic of Cancer* and the southern the *tropic of Capricorn*. These names reflect the time when astronomy and its terminology were first formulated in the modern sense (some two millennia ago) and the summer solstice was in the zodiacal constellation of Cancer and the winter solstice was in Capricornus. They are consistent with the astrological signs for the two solstice dates that can (sadly) be found in almost any newspaper. That the solstices are actually now located one constellation to the west, in Gemini and Sagittarius, is a consequence of the third motion of the Earth, precession, which will be examined separately in Chapter 5. These anachronisms are the analogues to the use of the symbol ♈ that denotes Aries and the vernal equinox.

The generic term "tropics" is defined as the zone or set of latitudes on Earth within

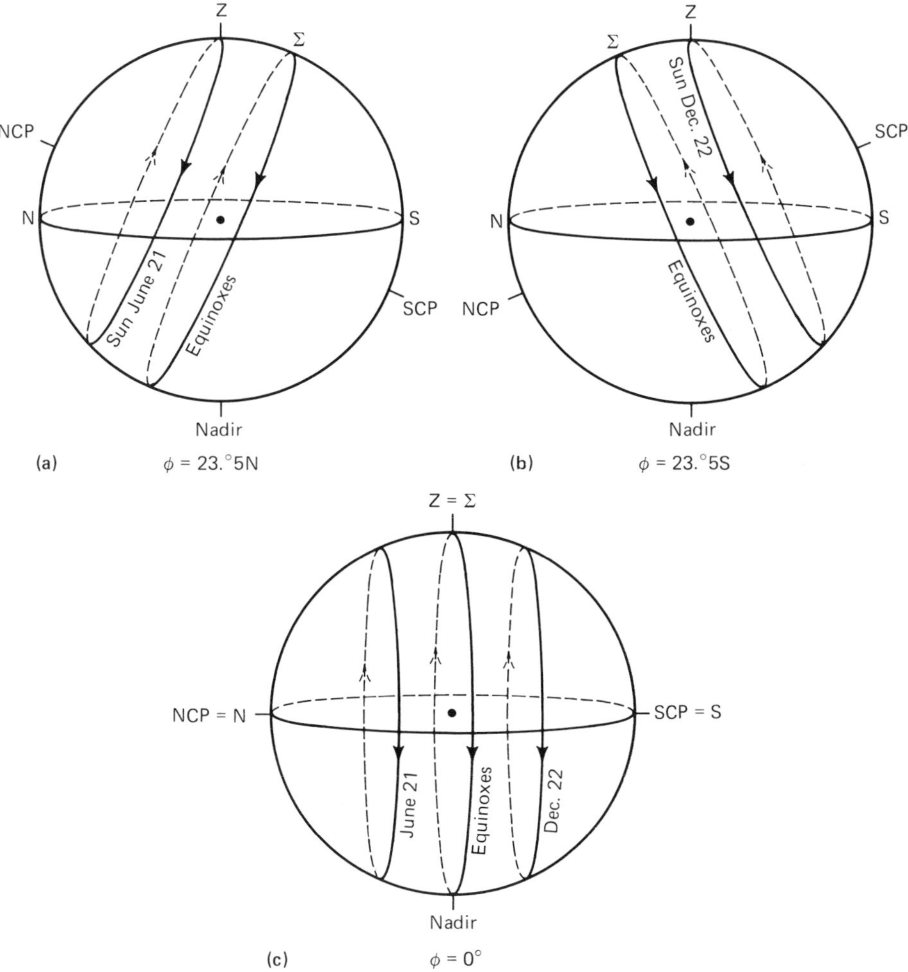

3.13. The Sun's daily path at the tropics. (a) and (b) show the Sun passing through the zenith: (a) the time of the summer solstice, June 21, at the tropic of Cancer, 23½° N latitude; (b) the time of the winter solstice, December 22 at the tropic of Capricorn, 23½° S. (c) This drawing, set at $\phi = 0°$, shows that the Sun must appear overhead twice during the year between $\phi = \pm\ 23½°$.

which the Sun will shine at the zenith some time during the year: the tropics must therefore be the hottest place on the planet. Except for the tropics of Cancer and Capricorn themselves, the Sun must pass overhead *twice* each year, once going north, and again going south (Figure 3.13c).

Next, return to the discussion of circumpolar stars in Section 2.8. In tropical and middle latitudes the Sun is seen always to rise and set. But if we get sufficiently close to one of the poles of the Earth, the altitude of the visible celestial pole can be greater than 90° minus the Sun's declination, and the Sun can become circumpolar (Figure

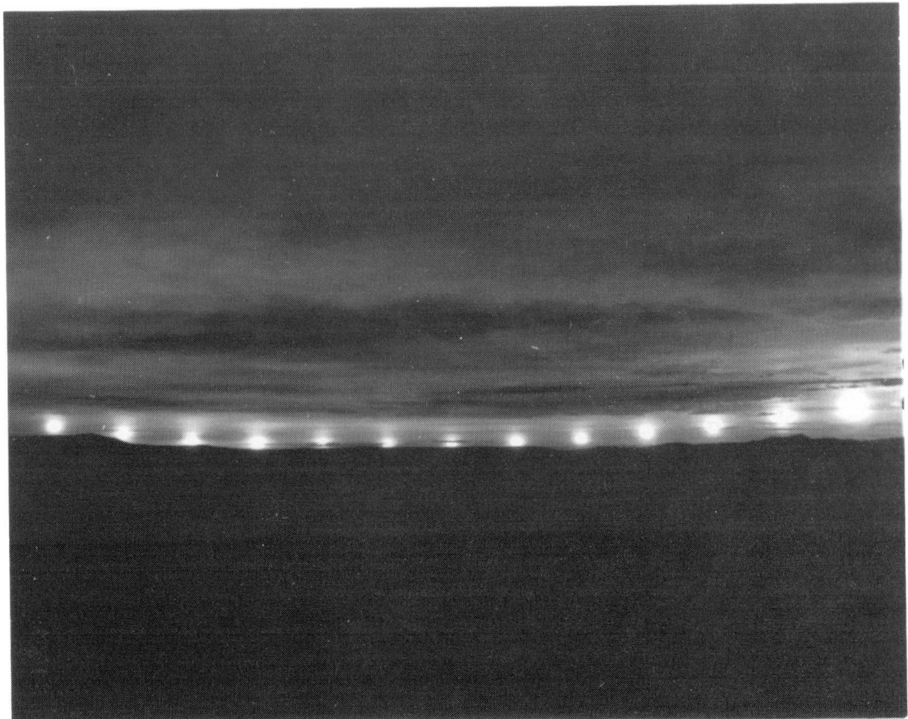

3.14. The midnight Sun. A multiple exposure made from northern Alaska shows the Sun passing lower culmination beneath the north celestial pole in June. (Mario Grassi.)

3.14). Alternatively, the Sun can be far enough below the observer's equator so that it could not rise on a given day. Since the maximum solar north declination is 23½° on June 21, the Sun must be circumpolar at a latitude of 90° − 23½° = 66½° N (the precise obliquity actually yields 66°34′ N) on that date, as seen in Figure 3.15a. This parallel is called the *arctic circle*. South of it (in the northern hemisphere) the Sun can never be circumpolar.

If we were to travel north of the arctic circle we would experience successively more circumpolar days, a time of continuous sunlight, centered on June 21. Figure 3.15b shows the sky at 80° N latitude. People of Spitsbergen, Severnaya Zemlya, and Ellesmere Island see the solstice Sun clear the northern horizon by 13½° at midnight. When the Sun is 80° from the NCP, or has a declination of +10°, it is still circumpolar. Figure A1.3 shows that the Sun reaches this declination on April 16, and descends below it on August 27, at this latitude yielding 133 days of continuous sunlight!

Finally, at the north pole, where the northern celestial hemisphere is always circumpolar (Figure 3.15c), the Sun is above the horizon from March 21 to September 23, for six full months of daylight. The Sun just keeps going around and around the sky always to the observer's right at very nearly constant altitude. The daily path is actually

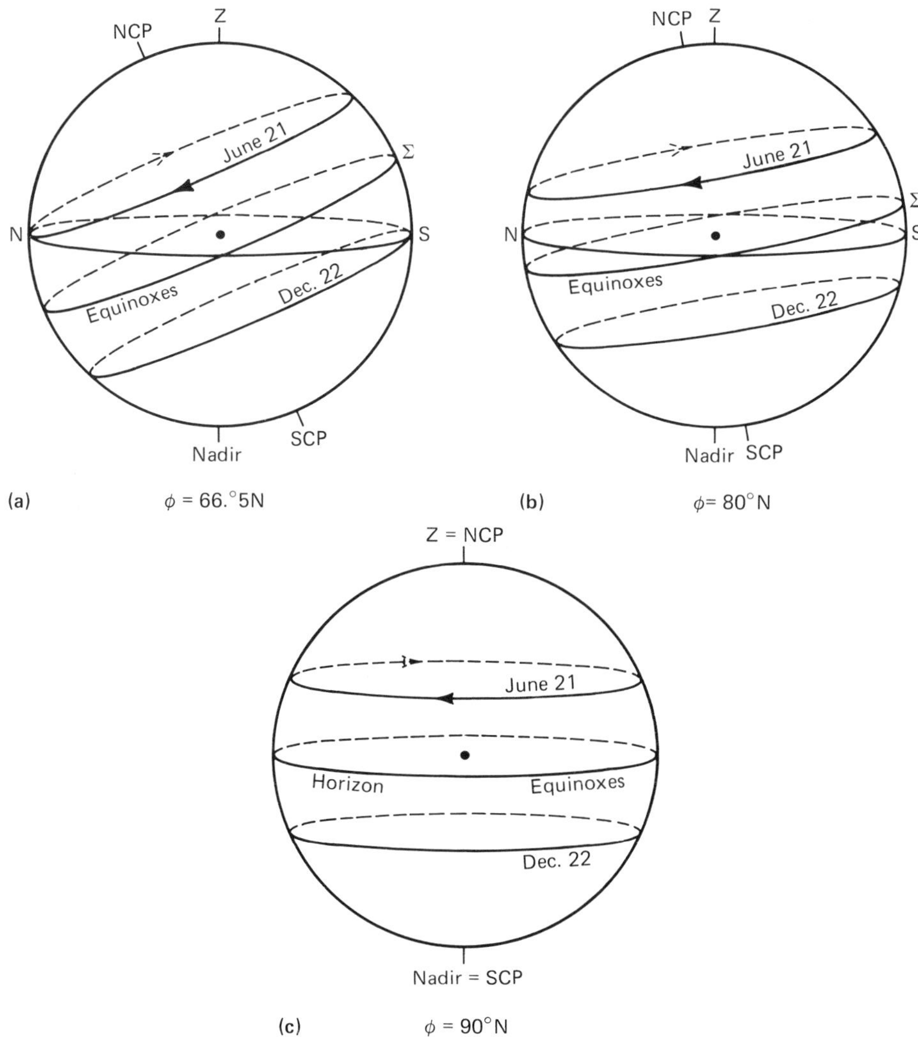

3.15. The Sun's daily path from the arctic circle to the pole. (a) shows the daily paths for the arctic circle, $\phi = 66.5°$ N, at the time of the Sun's solstice and equinox passages. On June 21 the Sun is barely circumpolar and on December 22 it just grazes the southern horizon but does not rise. In (b), 80° N latitude, the Sun is circumpolar when $\delta > 10°$ N, between April 16 and August 27. On June 21 it clears the northern horizon by 13½°. However it does not rise between October 20 and February 23. At the pole (c), 90° N, the Sun is circumpolar between March 21 and September 23 and does not rise between September 23 and March 21.

a slow helix, upward between March 21 and June 21, then downward to the autumnal equinox, when the Sun becomes extinguished below the horizon.

Analogously, at the arctic circle, the Sun will not rise at the time of the winter solstice, December 22, when it is 23½° south of the equator (again, Figure 3.15a). At 80° N latitude (Figure 3.15b) the Sun will not clear the horizon after it drops below 10° S declination, or (from Figure A1.3) between October 20 and February 23. Then at the north pole (Figure 3.15c), the Sun is below the horizon between September 23 and March 21, when it is in the southern celestial hemisphere.

Exactly the same phenomena can be seen in the southern hemisphere, but six months out of phase with the northern. The *antarctic circle* is at latitude 66½° south. There, the Sun is circumpolar on December 22 and does not rise at all on June 21. When the penguins south of the antarctic circle see the midnight Sun, it will be night to the reindeer north of the arctic.

The real world is again complicated by the solar angular diameter and refraction, which will change the visibility of the Sun in arctic regions. Moreover, real night does not set in as soon as the Sun disappears as there is always a period of twilight caused by sunlight scattered downward from the Earth's atmosphere. Consequently, the frigid polar astronomer would see the sky and the frozen ocean become light well before "spring" (such as it is in such an extreme environment) begins, and a glow would linger well after the start of "autumn." We will return to these matters in Chapter 7.

3.13 Solar and sidereal days

Up to this point, the concept of the day has been loosely defined. We know from childhood that it represents the interval between successive sunrises, or more accurately, between successive midnights or noons. What we are really defining is the rotation period of the Earth. To do so, we must refer it to some outside source that does not participate in the rotation. Our civil day is necessarily based upon the Sun, but that is not the only celestial source available: one can also use the stars. Because the Earth goes around the Sun, and the Sun is moving along the ecliptic relative to the stars, the use of the Sun or the stars as the outside reference must result in days of different lengths.

Hold a wristwatch in your hand, and slowly rotate on your axis. Use the watch to measure your period of rotation. You cannot do so unless you look for a reference point, say a tree. Whenever the tree sweeps past your central vision, note the time, and derive the interval. But what if your reference were moving? If you picked a dog that was walking past the tree, the reference would be in a different location every time it swept past your field of view, and your period relative to the animal would necessarily be different than it would be for the tree. Why, you ask, would anyone rational actually use a moving body as a reference? Because it is overwhelmingly the brightest thing in the sky and gives us life. The tree is a star and the dog is the Sun, which appears to move through the sky because *we* move about *it*.

Place the Sun and a star together on the celestial meridian on a certain day. The

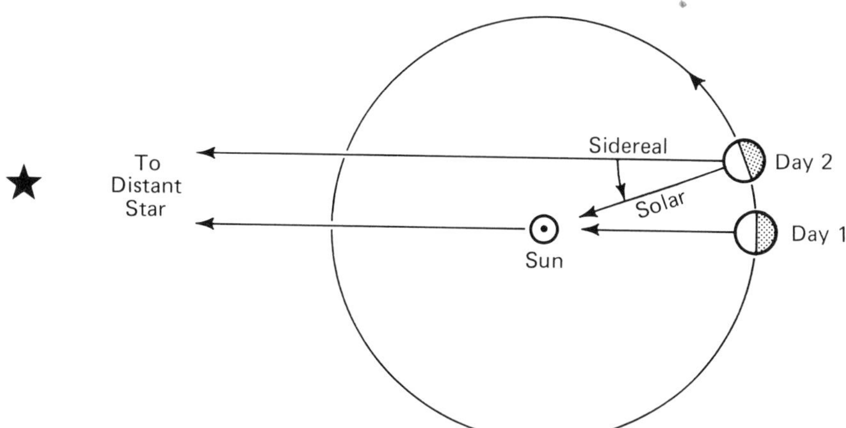

3.16. Sidereal and solar days. On day 1 the Sun is lined up with a star on the meridian. By day 2 the Earth has moved in its orbit (by a very exaggerated amount), and the star returns to the meridian in advance of the Sun. Since the Earth moves about 1° per day, the sidereal day is about 4 minutes shorter than the solar day.

alignment is shown from space in Figure 3.16 and from the perspective of a terrestrial observer on March 21 in Figure 3.17. As the day progresses, the Sun will move to the east and slightly north of the star as each follows its daily path toward the west. When the star returns to the meridian on the next day, the Sun will be almost 1° east of it and will require just under 4 more minutes (see Table 2.1) to complete *its* circuit back to the meridian. By then, the star will be nearly 1° to the west of the meridian, which is the same phenomenon we saw at night in Section 3.1. We now have the basis on which to define two different kinds of day. The *sidereal day* (from the Latin word *sidereus* for star) is the interval between successive culminations of the meridian by any star, and a *solar day* is the analogous interval for the Sun. The solar day is obviously the longer by nearly 4 minutes.

The exact relation between the two kinds of day is easy to compute. The length of the year, equinox to equinox, is actually 365.2422. . . solar days. During that interval, the Sun must pass through 360°, so that on the average (there is some daily variation due to orbital eccentricity: see Chapter 6) it moves ($^{360}/_{365.2422}$) = 0.985 6473 degrees per day. Consequently, by the time the Sun returns to the meridian the star will have an hour angle of 0°.985 6473, or since 1° = 4^m, of 3^m $56^s.56$. In the time it takes the Sun to move through 24 hours of hour angle, the star has moved through 24^h 03^m $56^s.56$ = $24^h.065$ 71, which is 1.002 738 sidereal days. That is, one solar day is 1.002 738 sidereal days long, or the duration of one sidereal day is $^1/_{1.002\,738}$ = 0.997 27 solar days. If 1.002 738 is multiplied by the number of solar days in the year, 365.2422, the result is 366.2422 sidereal days. There is exactly one more sidereal day in a year than there is solar.

The orbital motion of the Earth adds an additional sidereal day. Think of a table in the middle of a room as the Sun and the other fixtures in the room as stars. Walk

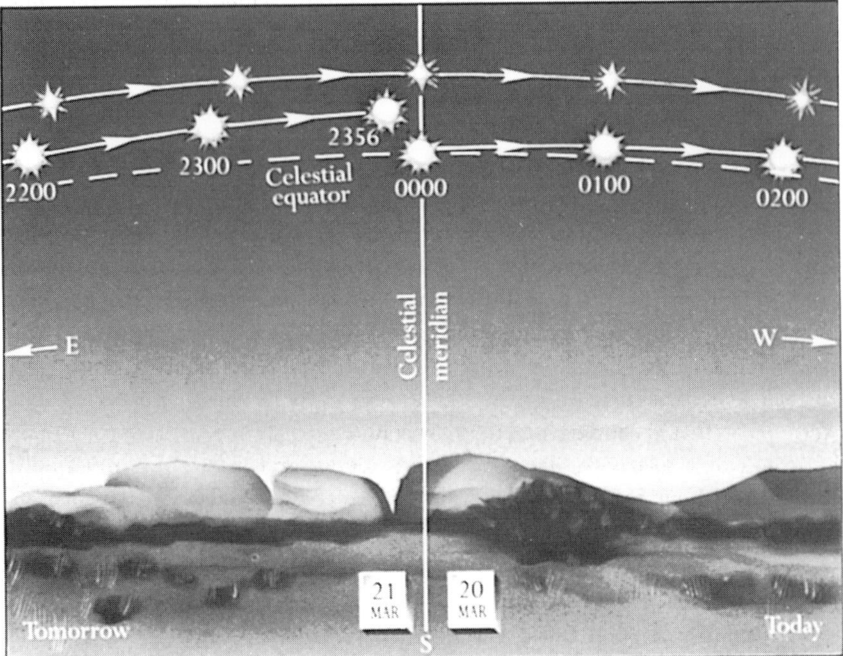

3.17. The observer's view of sidereal and solar days. The observer looks south and sees the Sun and a star lined up on the celestial meridian on March 20 at $00^h\ 00^m$ solar time. As the day progresses and the bodies move along their daily paths, the Sun moves to the east and slightly to the north of the star. By the time the star has returned to the meridian (an interval of a sidereal day), the Sun is nearly a degree to the east of it. It will take about another 4 minutes for the Sun to return to the meridian. The solar day is therefore about 4 minutes longer than the sidereal. (From *Stars*, J. B. Kaler, Scientific American Library, Freeman, NY, 1992.)

around the table, always facing it. Your "sun" will always be on your nose, or meridian, and will neither rise nor set. But each object in the room will rise over the horizon of your vision, cross your meridian, and then set once each orbital revolution. In every "year" you will have zero solar days, but one sidereal day. The act of orbital motion itself introduces the extra sidereal day, whether you are rotating with respect to the Sun or not. Any planet that rotates in the same direction in which it revolves will have one more sidereal day than solar. A planet that rotates opposite the direction of revolution (Venus, Uranus, and Pluto: see Chapter 11) will have one fewer.

How long is the sidereal day? In one *sidereal* day the Sun moves $^{360}/_{366.2422}$ degrees, or at an average rate of $0.982\,956° = 3^m.9318 = 3^m\ 55^s.91$ per sidereal day. After a full sidereal day the Sun will be shy of the meridian, or have an eastern hour angle, of $3^m\ 55^s.91$. Thus the length of each sidereal day is $24^h - 3^m\ 55^s.91$ or $23^h\ 56^m\ 04^s.09$. The sidereal day is actually the true rotation period of the Earth. It is what would be measured by an outside observer, a hypothetical Martian or a Venusian. For obvious reasons, we live our life by the solar day. But the sidereal day is the fundamental astronomical time unit.

3.14. Sidereal time

These two definitions of the day form the bases for two different kinds of time, *solar time* and *sidereal time*. Solar time was defined in Section 3.7 as the hour angle of the Sun plus 12^h. Sidereal time (properly *local sidereal time*, LST, since it too depends on longitude) is merely the *hour angle of the vernal equinox* (the 12 hours is *not* added). The vernal equinox is (for our purposes here) fixed to the stars, and therefore makes an excellent external reference for sidereal time. We traditionally divide the solar day into 24 units, each called an hour, as the basis of normal civil timekeeping. These units are technically called *solar* hours. We do the same to the sidereal day, and define a *sidereal hour* that is shorter than its solar counterpart. If a sidereal day is 0.997 27 times the length of a solar day, the ratio of the sidereal and solar hours must be the same. The sidereal hour is then $59^m\ 50^s.2$ long in solar time, and the sidereal minute has a length of 59.84 solar seconds.

Build two clocks. One makes a complete 24-hour circuit each solar day; the other must make its circuit in the shorter sidereal day. The sidereal clock must tick faster to compress its 24 sidereal hours into 23 hours 56 minutes 4.09 seconds of solar time. Since the average solar day is 1.002 738 sidereal days long, after a full solar day the sidereal clock must have gone through $24^h \times 1.002\ 738 = 24^h\ 3^m\ 56^s.56$. (If we start the Sun and the equinox together on the meridian in Figures 3.16 and 3.17, after a solar day, the hour angle of the equinox will be $3^m\ 56^s.56$.) Thus if the two clocks both read 12 hours on a certain day, by 12 hours on the solar clock the next day, the sidereal clock, moving the faster, will read $12^h\ 3^m\ 56^s.56$. It gains $3^m\ 56^s.56$ per day on the solar clock, and over the course of the year, it will gain $3^m\ 56^s.56 \times 365^d.2422 = 1440$ minutes, which is 24 hours, or one full day. The differential of $3^m\ 56^s.56$ is equal to $236^s.65$. Thus the sidereal clock gains $236^s.65/24 = 9^s.86$ per hour, or 1 second about every 6 minutes. It is fascinating to watch sidereal and solar clocks working side by side. You can literally see one pulling ahead of the other: you are watching the Earth revolve about the Sun.

Now look at the absolute relation between the two kinds of time. They must be the same when the Sun passes the autumnal equinox on September 23, as illustrated in Figure 3.18. If the vernal equinox is at upper culmination, the sidereal time is 0^h. The autumnal equinox and the Sun must be at lower culmination. The hour angle of the Sun is therefore 12^h, and the solar time is $12^h + 12^h = 24^h = 0^h$, the same as the sidereal. The sidereal clock will afterward move ahead of the solar clock by just under four minutes – $3^m\ 56^s.56$ – for each day past September 23. In a full year of 365.2422 solar days, the sidereal clock will have gained $365.2422 \times 3^m\ 56^s.56 = 24$ sidereal hours, or a full day, again illustrating that there are 366.2422 sidereal days in a year.

You can easily make a crude mental calculation of sidereal time. What, for example, is the sidereal time at noon on April 10? If the clocks read the same at the time of the

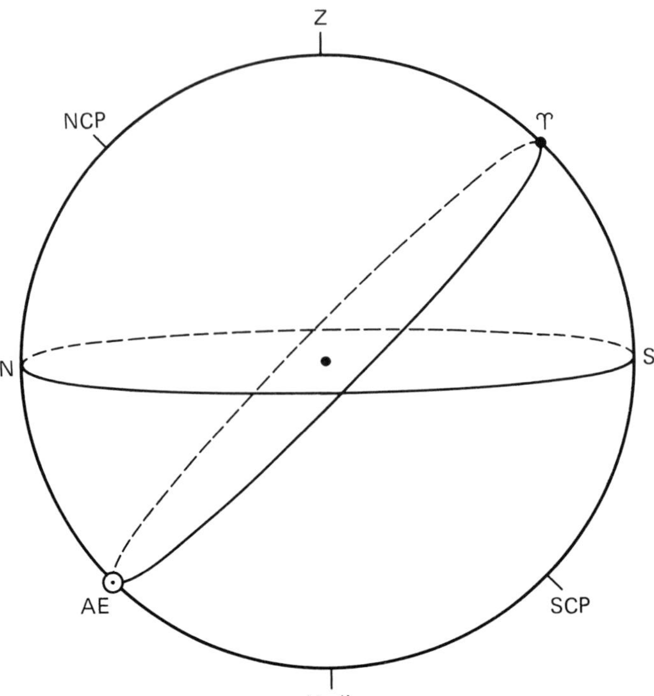

3.18. Zero hours sidereal and solar time. The hour angle of the vernal equinox is zero as is the sidereal time. The hour angle of the Sun is 12^h, so the solar time is also 0^h. Since the Sun is opposite the vernal equinox it must be at the autumnal equinox, and thus the date is September 23.

autumnal equinox, and the sidereal gains a full day over the solar in a year, they must be about 6 hours apart on the date of the winter solstice, December 22, 12 hours on March 21, and 18 hours on June 22. April 10 is 20 days past the date of the vernal equinox, so the sidereal clock must be 12 hours plus 20 days times 4 minutes or 12^h $80^m = 13^h\ 20^m$ ahead of solar. If the solar time is 12^h, the sidereal is $25^h\ 20^m = 1^h\ 20^m$. In this chapter, the concept of sidereal days and sidereal time is introduced in part to illustrate the relation between the Earth's rotation and its orbital revolution. Although sidereal time is not used directly in civil affairs, it is important in the establishment of modern timekeeping systems and it is vital to the astronomer. The purpose of a clock is to represent the heavens. An ordinary civil clock tells us where to find the Sun; a sidereal clock tells the astronomer where to find the stars by locating the vernal equinox.

3.15 Right ascension and declination

Section 2.1 introduced a system of equatorially based coordinates, declination and hour angle. Declination has been used extensively to describe the rising and setting of stars and the Sun, and hour angle has been used to define the duration of visibility of a star.

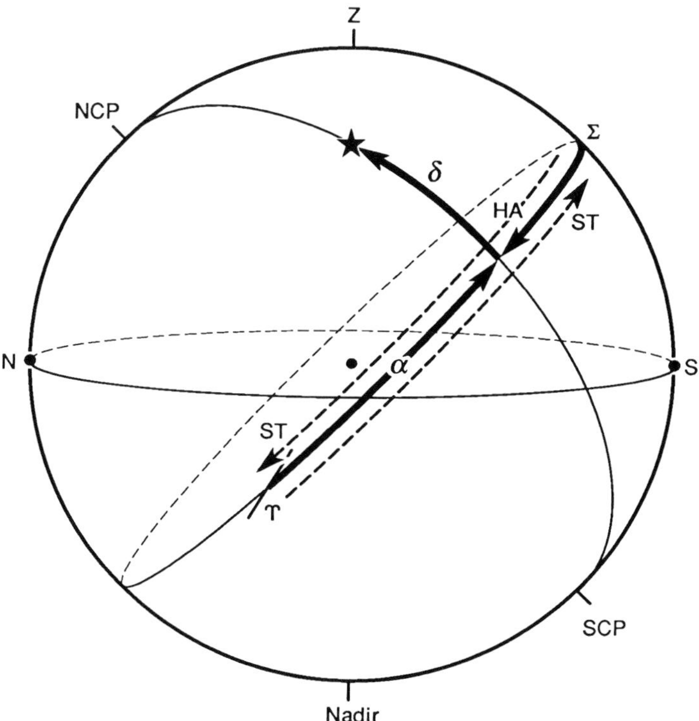

3.19. Right ascension and declination. Only a short segment of the ecliptic is shown at the vernal equinox (♈). Right ascension (α) is measured along the celestial equator eastward (counterclockwise) from the vernal equinox to the hour circle. Declination (δ) and hour angle (*HA*) have been previously defined. Hour angle increases uniformly with sidereal time, but right ascension is constant since the equinox (for the present purposes) is fixed among the stars. The sidereal time is the hour angle of the vernal equinox or the right ascension of the meridian, and is the sum of the star's hour angle and right ascension.

Declination stays constant, but the hour angle of a star steadily changes, increasing uniformly with time; as seen above, hour angle is in fact used as the basic measure of the passage of time. For many purposes, however, including time measurement, we must have a system in the sky similar to latitude and longitude, in which the coordinates of a star do not rapidly change. Hour angle is measured from the celestial meridian, which is fixed to the *observer*. We must have a system that is fixed to the *stars* and rotates with them.

The solution is to use the vernal equinox as a zero point. We define a coordinate called *right ascension* (α), which is measured *eastward* (opposite to hour angle), or counterclockwise, from the vernal equinox along the equator to the hour circle through any specific star (Figure 3.19). Since the origin of right ascension moves on its daily path with the same rate as the stars, right ascension remains constant as the day proceeds. Like hour angle, right ascension is counted in time units; degrees are not incorrect, only cumbersome.

The line that establishes the terrestrial zero point for longitude is called the prime meridian (see Section 1.6). The analogue in the sky's equatorial coordinates, the hour circle at zero hours right ascension, is the *equinoctial colure*. The difference in the two is that the prime meridian is established by a point at high latitude (Greenwich) that is used to locate zero on the equator. The equinoctial colure, which actually goes all the way around the sky through 12 hours, is established after the fact from the vernal equinox, which is already on the equator. A similarly named hour circle, the *solstitial colure*, runs through the solstices at 6^h and 12^h.

All three equatorial coordinates are illustrated in Figure 3.19. The star's hour angle is 3 hours (west of the meridian), its right ascension is 6 hours (measured oppositely), and its declination is +30°. The coordinates α and δ are the true celestial analogues of latitude and longitude. Since the equinox rotates with the stars, they do not change. With right ascension and declination, we can easily keep track of all the celestial bodies: the coordinates of some 15 million stars have been measured. We now also know where to find the Sun, Moon, or any of the planets, because astronomers can calculate its *ephemeris*, a table of how its α and δ change with time. For example, the right ascension of the Sun at the beginning of spring (by its very definition) is 0^h; the summer solstice is at $\alpha = 6^h$, the autumnal equinox 12^h, and the winter solstice 18^h.

Longitude is created by selecting an arbitrary point on the Earth's surface and passing a reference, or prime, meridian through it. Why not just select a bright reference star, and pass a "prime hour circle" through it to determine the zero point on the celestial equator? The problem is that the stars actually have slow, but very measurable, motions among one another, as they and the Sun orbit the center of our galaxy. The zero point would then move, and in an indefinable way; we would be measuring the positions of all stars against that of a moving reference. It would be as if Greenwich and England floated on the ocean on an unknown path. (They *do*, of course, because of continental drift, but that motion is sufficiently slow to be ignored.) The vernal equinox, however, is precisely defined by the planes of the Earth's rotation and revolution, and it gives us an excellent external reference standard. In fact, it too moves because of the phenomenon of precession, which we will examine in Chapter 5. But the rate of motion is known with great accuracy, allowing the exact position of the vernal equinox to be located.

3.16 Right ascension, hour angle, and sidereal time

The two equatorial systems – the moving and the fixed – are linked through the sidereal time. The sidereal time is defined as the hour angle of the vernal equinox (see Figure 3.19), that is the arc from Σ to ♈. It can also be defined as the right ascension of the meridian, since the angle measured from ♈ to Σ is the same as that measured from Σ to ♈. That is,

$$\boxed{\text{LST(local sidereal time)} = HA(♈) = \alpha(\Sigma)} \quad .$$

The arc between Σ and ♈ is broken into two parts by the hour circle through the star:

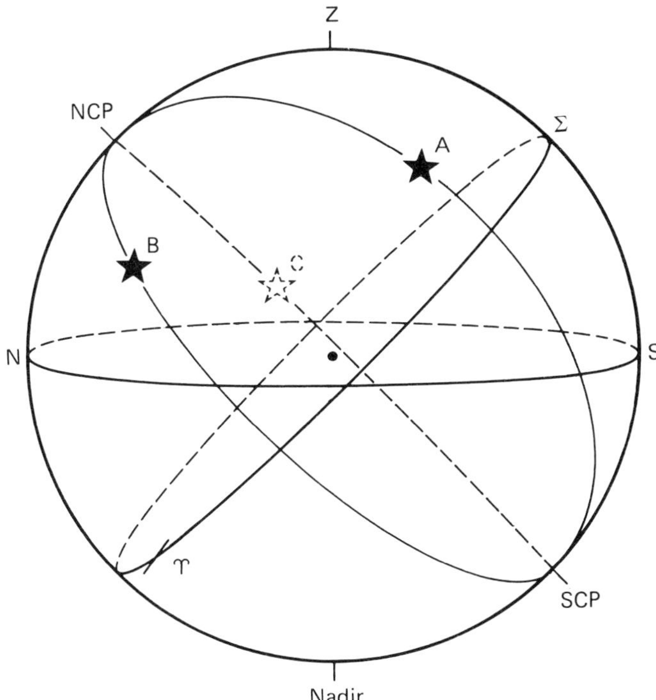

3.20. Sidereal time, hour angle, and right ascension. Stars *A*, *B*, and *C* respectively have hour angles of 2, 8, and 18 hours and right ascensions of 9, 3, and 17 hours. All three give a sidereal time (right ascension plus hour angle) of 11 hours. If the sidereal time is known, the hour angles can be computed.

the part from Σ to the hour circle is the star's hour angle, and that from ♈ to the hour circle is its right ascension. The whole being equal to the sum of its parts,

$$\boxed{\text{LST (Local Sidereal Time)} = HA + \alpha} \quad .$$

This equality is shown for three stars in Figure 3.20. The hour angles of stars *A*, *B*, and *C* are 2, 8, and 18 hours respectively, and their right ascensions are 9, 3, and 17 hours. For *A* and *B* the two sum to 11 hours, which is obviously the hour angle of ♈. For *C*, $HA + \alpha = 35^h$. Twenty-four hours can always be added or subtracted, and $35^h - 24^h = 11^h$, the same as for the other two.

All observatories are equipped with a sidereal clock. To locate any star, look up the α and δ in one of several readily available catalogues. Then, since LST = $HA + \alpha$,

$$\boxed{HA = \text{LST} - \alpha} \quad .$$

If the LST in Figure 3.20 is 11 hours, and we read the right ascensions of the three stars to be 9, 3, and 17 hours, the differences, or hour angles, are 2, 8, and −6 hours. The minus sign means that the star is 6^h east of the meridian, and $-6 + 24 = 18$ hours.

Use the constellation Orion as another example. Adopt the usual latitude of 45° N. Assume the solar time to be 15^h (3 PM) on March 21, so that the sidereal and solar times are 12^h apart. The Sun is then at the vernal equinox, which means that both must have hour angles of 3^h, which is also the local sidereal time. Orion's right ascension is roughly 5 hours. Thus $HA = 3 - 5 = -2$ hours. Orion, as indicated several times already, rides the celestial equator. You can then go outside, draw the equator with your finger, locate the equator point (its zenith distance equals your latitude), then just go 2^h east (22^h west), or one-third the way from Σ to the east point. There is Orion on that time and date. You cannot see it because the sky is bright or cloudy. But it is surely there.

The star Vega is used as yet another example. Its α and δ are $18^h\ 35^m$ and +39° respectively. If the sidereal time is 3^h, $HA = \mathrm{ST} - \alpha = 3^h - 18^h\ 35^m = 27^h - 18^h\ 35^m = 26^h\ 60^m - 18^h\ 35^m = 8^h\ 25^m$. So now proceed westward from Σ to roughly 8½ hours, which takes you below the horizon, draw in the hour circle, and go north by 39°. Vega is close to setting in the far northwest; without even looking you know where it is. With a sidereal clock, the sky is at your command.

3.17 Setting the astronomical telescope

To find a bright planet or a faint constellation in the sky with the unaided eye all you need do is to look up. Even binoculars or low-power telescopes can be directed by eye alone. However, how do you point a large instrument to locate faint stars below the level of human vision? Equatorial coordinates and the sidereal time provide the ability to roam across the sky and to find anything you want.

An astronomical telescope is designed to move on two axes so as to be able to point to any position in the sky. In the most common type of mounting, called an *equatorial*, one axis, the *polar axis*, points at the visible (north or south) celestial pole (Figure 3.21). The telescope can be moved in hour angle by rotating it around the polar axis, and it can be moved in declination by rotation about the *declination axis*, which is perpendicular to the polar axis. Older telescopes come equipped with *setting circles*, graduated rings affixed directly to the rotation axes that give hour angle and declination. Modern instruments have electronic devices that can sense the degree of axial rotation and can send the hour angle and declination to a computer for visual display.

If we wish to look at a star, we first look up the right ascension and declination, α and δ, in a star catalogue. The declination axis can immediately be set. We know the sidereal time from the observatory's sidereal clock, and need only subtract α from LST to find the hour angle, whereupon we can set the other axis. The star will then appear in the field of view of a small *finder telescope* affixed to the main instrument. We center the star in the finder and there it is in the field of view of the bigger telescope. A sidereal clock mounted on and mechanically linked to the polar axis then drives the telescope in hour angle to allow the instrument to track a star along its daily path. A computer-controlled telescope will have a built-in sidereal clock. The observer merely enters α and δ from a keyboard and the electronics do the rest, commanding both the axes to rotate by the correct amount to find the star. Such telescopes can be made so

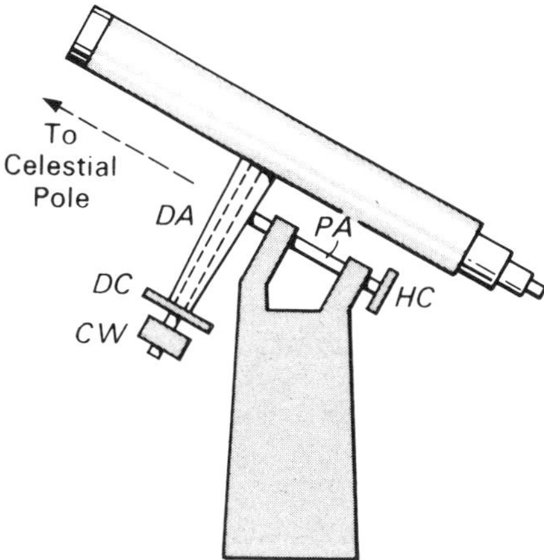

3.21. A telescope mounting. The polar axis is labeled PA; it carries the hour angle setting circle, HC, for setting the desired hour angle, and it also carries the declination axis, DA. The declination axis carries the telescope, a counterweight, CW, and the declination setting circle, DC, for setting the desired declination. From *Principles of Astronomy: A Short Version*, S. P. Wyatt and J. B. Kaler, 2nd ed., Allyn and Bacon, Boston 1981.)

precise that the object almost magically appears in the field of view, usually on a television monitor. The sidereal clock then controls the electronics and tracks the star.

With computer capability we need not even construct an instrument along the plane of the celestial equator. It is much cheaper to build a telescope with axes that move in altitude and azimuth (an *alt-azimuth mounting*). We feed the computer α and δ. It solves the astronomical triangle for h and A, then directs our view to the proper place. Daily motion – simple equatorially, but complex in altitude and azimuth – is done by continuously solving the astronomical triangle to increment h and A appropriately: impossible to do in "real time" by hand.

Other calculations and a variety of corrections are needed to set a telescope properly. These will be dealt with in Chapter 5. We will look at the optical design of telescopes in Chapter 13.

3.18. Celestial, or ecliptic, coordinates

It may seem as if astronomers glory in the construction of coordinate systems. Section 1.5 opened the subject with terrestrial latitude and longitude. Section 1.9 followed with horizon coordinates, altitude and azimuth. Then Section 2.1 introduced the fixed set of equatorial coordinates, hour angle and declination, and Section 3.15 added the moving equatorial set of right ascension and declination.

Here is yet another one in which the ecliptic, rather than the equator, is the fundamental great circle, and the NEP and SEP (Section 3.2) are the system's poles. The coordinates are called *celestial longitude* (λ) and *celestial latitude* (β), or *ecliptic longitude*

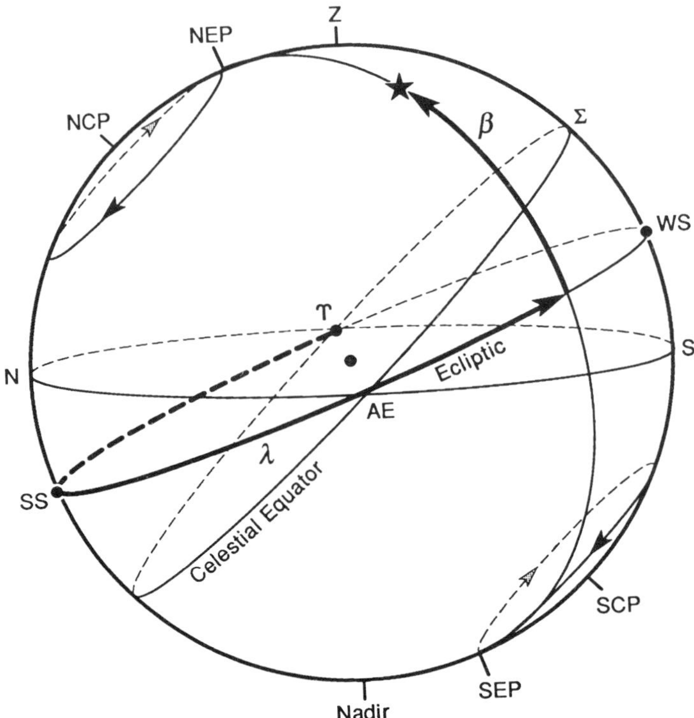

3.22. Celestial, or ecliptic, coordinates. The celestial and ecliptic poles are all marked. With the vernal equinox rising, the NEP and SEP are transiting the meridian. Celestial longitude (λ) and latitude (β) are based on the ecliptic, λ measured counterclockwise from ♈, β measured perpendicular from the ecliptic toward the north or south ecliptic poles. The λ and β for the star are 230° and +50° respectively. With the equinoxes at the horizon, the meridian forms the secondary of each pole in the other system. In equatorial coordinates, α(NEP) = 18^h, δ(NEP) = +66½°, α(SEP) = 6^h, δ(SEP) = −66½°. In ecliptic coordinates, λ(NCP) = 90°, β(NCP) = +66½°, λ(SCP) = 270°, β(SCP) = −66½°.

and latitude (Figure 3.22). The zero point is again chosen to be the vernal equinox. In the example, a secondary is passed from one ecliptic pole to the other through the star. Celestial longitude is always measured counterclockwise – to the east, just like right ascension – from ♈ to the intersection of the secondary and the ecliptic. It is always measured in degrees, not hours. Celestial latitude, as expected, is measured north (+) or south (−) of the ecliptic along the secondary. Note especially that celestial longitude and latitude are *not* the analogues to terrestrial longitude and latitude, in spite of the fact that the same words are used: the ecliptic is always tilted relative to the celestial and terrestrial equators.

In Figure 3.21, the star's celestial longitude is about 230°, and β is +50°. The equinoxes and solstices all have β = 0° since they are all on the ecliptic. The longitudes of the four principal points, vernal equinox, summer solstice, autumnal equinox, and

Table 3.1. *Summary of coordinate systems*

Name	Terrestrial	Horizon	Equatorial I	Equatorial II	Celestial (Ecliptic)	Galactic I (Old)	Galactic II (New)
FGC	terrestrial equator	horizon	celestial equator	celestial equator	ecliptic	galactic equator	galactic equator
SGC	meridians	vertical circles	hour circles	hour circles	. . .	. . .	. . .
Prime SGC	prime meridian	celestial meridian	celestial meridian	equinoctial colure[a]	. . .	. . .	. . .
Zero point	intersection of equator and prime meridian	north point of horizon	sigma point	vernal equinox	vernal equinox	intersection of galactic and celestial equators	center of galaxy
FGC coordinate	longitude, λ	azimuth, A (or Z)[b]	hour angle, HA	right ascension, α	celestial longitude, λ	galactic longitude, l^{I}	galactic longitude, l^{II}
Units	degrees/hours	degrees	hours/degrees	hours/degrees	degrees	degrees	degrees
Direction	east and west to 180°	east	west (clockwise), or east[c]	east (counter-clockwise)	east (counter-clockwise)	north	north
SGC coordinate[d]	latitude, ϕ	altitude, h	declination, δ	declination, δ	celestial latitude, β	galactic latitude, b^{I}	galactic latitude,b^{II}
Pole	north, south pole	zenith, nadir	north, south celestial poles	north, south celestial poles	north, south ecliptic poles	north, south galactic poles	north, south galactic poles

[a]the hour circle through the solstices is called the solistial colure

[b]zenith angle, Z, is measured in both directions from north to 180°

[c]hour angles measured east are negative

[d]small circles for which this coordinate is constant are called *parallels*; parallels of altitude are called *almucantars*.

winter solstice are just 0°, 90°, 180°, and 270° since they quarter the circle. The latitudes of the NEP and SEP are +90° and −90°, where as usual, longitude disappears.

Just as it is possible to go back and forth from horizon to equatorial coordinates, conversions can readily be made between equatorial and ecliptic: β and λ can be calculated precisely from α and δ, and vice versa, by means of trigonometric formulae (see Appendix A3.4). Estimates can also be made from drawings of celestial spheres once the ecliptic, equator, and both sets of poles are located, although the drawings can become rather complicated. In Figure 3.22, the ecliptic poles are placed at culmination for simplicity, so that the hour circle for the NEP is the upper meridian, and that for the SEP is the lower. Remember also that the ecliptic poles, being fixed among the stars, must have daily paths, as shown, around the celestial poles. We see that the right ascensions of the NEP and SEP are 18^h and 6^h respectively. The ecliptic poles must be 23½° away from the celestial poles (see Section 3.3) and their declinations are thus +66½° and −66½°. The celestial meridian in Figure 3.21 at the moment also serves as a secondary in the ecliptic system. By measuring along the ecliptic we see that λ(NCP) = 90° and λ(SCP) = 270°. By the same reasoning as above, the celestial latitudes of the NCP and SCP must also be +66½° and −66½°. As an example, the right ascension of the star in Figure 3.22 is about 16^h and the declination is near +30°.

Since the declination of the NEP is +66½°, it is circumpolar everywhere north of the tropic of Cancer and culminates the zenith at the arctic circle. Above the tropic of Cancer, the south ecliptic pole does not rise. The SEP becomes circumpolar south of the tropic of Capricorn, where the NEP is invisible, and culminates the zenith at the antarctic circle. Only between 23½° N and 23½° S latitude will an observer be able to see both ecliptic poles, and both will rise and set.

These systems is not merely academic exercises. We need coordinates that are appropriate to the problems at hand. When we are dealing with the position of the Sun in its relation to position on the Earth, we use α and δ. But while ecliptic coordinates are not now in common use, they do simplify some problems, i.e. those that pertain strictly to the Solar System. The right ascension and declination of the Sun change with time in a complex way, but by its very definition, the celestial latitude of the Sun is always 0°; we need consider longitude only, which increases simply and uniformly with time as the year progresses. The solar celestial longitude is 0° when the Sun crosses the vernal equinox. It then increases at an average rate of 0°.9856 per solar day, completing 360° in 365.2422 days. Since the planets move roughly parallel to the ecliptic the system is useful for plotting their motions as well.

3.19. A summary of coordinates

The last three chapters have introduced a total of five coordinate systems. One more with two subdivisions will be defined in Chapter 4. Table 3.1 presents a summary of all of them, including the names of the fundamental and secondary great circles (FGC and SGC), the names of the poles, the definitions of the zero points, how the various arcs and angles are measured, and other data.

Chapter 4

Stars and constellations

The Sun is part of our daily lives. We are so familiar with it and its travels across the sky that we usually do not even think of it as an astronomical body. The real fascination with astronomy comes with sunset, when the sky darkens enough to reveal the myriad other sights of the Universe. Only after we begin to study astronomy and become familiar with the Moon, stars, and planets does our curiosity return again to the Sun and its role as an astronomical entity rather than a life-sustaining one.

In our modern western civilization, few people are really aware of the nighttime sky. From the environs of a big city, only the brightest stars and the Moon and planets are readily seen. We spend most of our nighttime hours indoors, and have little time outside to watch the motions and changes that these bodies exhibit. It comes as a surprise to many that the Moon and stars rise and set just like the Sun.

Our technology has largely removed the aspects of the sky from our daily lives. But such was not always the case, and in many parts of the world is still not. People in a society without artificial lighting, those living an outdoor existence for example, will become intimately familiar with the stars. In older cultures, the celestial bodies represented – and for some still do represent – gods and heroes. Perhaps earthly happenings and the dispositions of the gods could be read through aspects of the skies. On a homelier level, the stars were used as a reminder of stories, epics, and creation myths, the sky in a sense serving as the written word. All groups, everywhere, name the stars.

Knowledge of the skies was, and still is, also of immense practical importance. Human activities are synchronized by the daily motion of the Sun across the celestial vault. A modern society uses a mechanical or electronic clock to keep time, but the clock still represents the Sun. The Moon and stars were vital in the development of calendars. The word for "month" in many languages is derived from the word for "moon." The very concept of 12 months in a year comes from the Moon's 12-plus orbits of the Earth in a year. We can keep track of where the Sun is among the stars – the time of year – by observing the positions of the stars after the Sun goes down. The stars can be used to mark our seasons: we might know that harvest should begin with our first glimpse of a certain one in the east before dawn.

We have long known that a mythology is not needed to explain the appearance of, and occurrences in, the sky. We have learned what the stars are and have good ideas of how they came to be. At the same time, we have removed ourselves from study of the skies in our daily lives. But the stars and the patterns they appear to make – the *constellations* – are still there for our appraisal, just as they were in ancient times. They are our most direct link with the ancient past, and tell us much about our history.

4.1. The constellations

Common opinion regards the astronomer as an individual who knows all the stars and their constellations by heart, who keeps track of them and their changing seasonal positions, and who watches over the movement of the Sun, Moon, and planets among them. On the contrary, modern astronomy focuses on the physical construction of stars, galaxies, and the Universe, and the majority of professional research astronomers would be hard pressed to identify any but the most prominent of constellations.

Nevertheless, the constellations are still significant in modern astronomy. Although their importance is not now what it was to the ancients, they still have a useful role. A look at the heavens on a clear, dark, moonless night at first presents a bewildering sight – some two thousand or more visible stars scattered largely at random across a seemingly enormous sky. With a second look, however, we begin to make patterns and figures out of the stars. With each view we see the same groupings, and so gain familiarity. These constellations serve to partition the sky into small, recognizable areas. They are a simple aid to communication. An astronomer can specify exact coordinates of a newly found bright comet; but if the discoverer says simply that the visitor is in Leo, we all know where to look.

An examination of the star names and constellations of the world's cultures would require a treatise of hundreds of pages. Here, we will take only a summary look at those of our western culture to provide some initial familiarity with the skies. The world's astronomers now recognize 88 "official" constellation figures. By an agreement reached at the 1922 convention of the International Astronomical Union, the sky is divided into 88 regions with well-defined rectangular boundaries. These zones largely take the names from, and conform to, the positions of the older myth-related constellation patterns. Every star in the sky lies within formal constellation boundaries delineated by right ascension and declination.

The 88 constellations are broadly divided into two groups: the *ancient* and the *modern*. There are 48 old figures whose origins are largely lost to antiquity. The modern constellations were formed from the seventeenth century onward to fill in the gaps, mostly in the southern hemisphere, left over from the ancient 48. Some of the early figures have been subdivided, and many other modern ones were invented that never achieved acceptance, or were used for a while and were subsequently dropped or subsumed by a neighboring stellar pattern.

4.2 The ancient 48

The oldest groups may have originated in their present form among the peoples of Mesopotamia perhaps 4000 years ago or even earlier. Some may have been infused from ancient Egypt. The old figures were accepted by the ancient Greeks, who added new ones and wove many detailed myths about them. Several are connected with numerous, often conflicting, stories. The reference to "ancient" really means before classical times. The Greeks of the first millennium BC and later accepted the constellation figures as we do, and like us speculated about their origins. The ancient groups were codified for us by Eudoxus of Cnidos about 375 BC, by the writer Aratos (about 270 BC) in his poem *On the Phaenomena* and by Hipparchus a century later, and finally were cast into their present form by Ptolemy (second century AD) in his *Syntaxis*. These figures were subsequently handed down to us through the Arabians, such as Ulugh Begh, who lived in the fifteenth century AD.

It is a disappointment to many that the constellations only rarely look like what they represent. How could they, since they are truly random scatterings of stars. They are meant to *represent*, not to *portray*. But with good humor and relaxed imagination, the viewer may begin to see the figures in the mind as they move gracefully across the sky.

The early storytellers did not feel compelled to fill the sky with constellation figures. A look at the heavens will show a number of outstanding patterns – the Big Dipper (in Britain the Plough, or Wagon, Figure 4.1), Orion (frontispiece and Figure 4.2), and others – with spaces between in which only a few faint stars are visible. The ancients used the prominent patterns and the bright stars to represent and honor their heroes and gods, and to tell their stories. The spaces between were called by the Greeks the *amorphotoi*, the unformed, or the scattered. Sometimes an unformed region "belonged" to a given constellation, most times not. It was believed that the gods themselves organized the stars and constellations to their own desires. They could elevate a great man to immortality in the heavens by making a constellation of him, like Hercules or Orion, or commemorate a brave woman, as they did for Ariadne, by placing her crown (Corona Borealis) in the sky. Presumably, the unformed regions were simply the spaces still to be used.

The accepted constellation figures are listed in Tables 4.1 (the ancient) and 4.2 (the modern). The tables give a variety of information: first, the Latin name, and the meaning or identification; next, the approximate right ascension (α) and declination (δ) of the center of the figure so that it can be quickly located on the maps in Appendix 2; then the Latin possessive of the constellation name, used for the naming of stars (see Section 4.7); the modern official abbreviation; and finally, references to notes at the end of the table.

For a variety of reasons a number of separate constellations may be related. By far the most prominent and important of these companies is the set of constellations through which the ecliptic passes, the zodiac (Table 4.1a), first introduced in Section 3.2. The traditional zodiac consists of 12 constellations, one for each month of the year. Their number really comes from the Moon, which annually orbits the Earth some 12½ times. Roughly speaking, the Sun is in a different zodiacal constellation for every passage of the lunar phases, that is, at every full (or new) Moon. With modern constellation boundaries, the Sun spends widely different times in each of these figures (Table 4.3), and although the ecliptic actually goes through the modern version of ancient Ophiuchus (Table 4.1e), it is never considered as part of the zodiac.

The zodiac traditionally begins with Aries, the ram (Figure 4.3), since that constel-

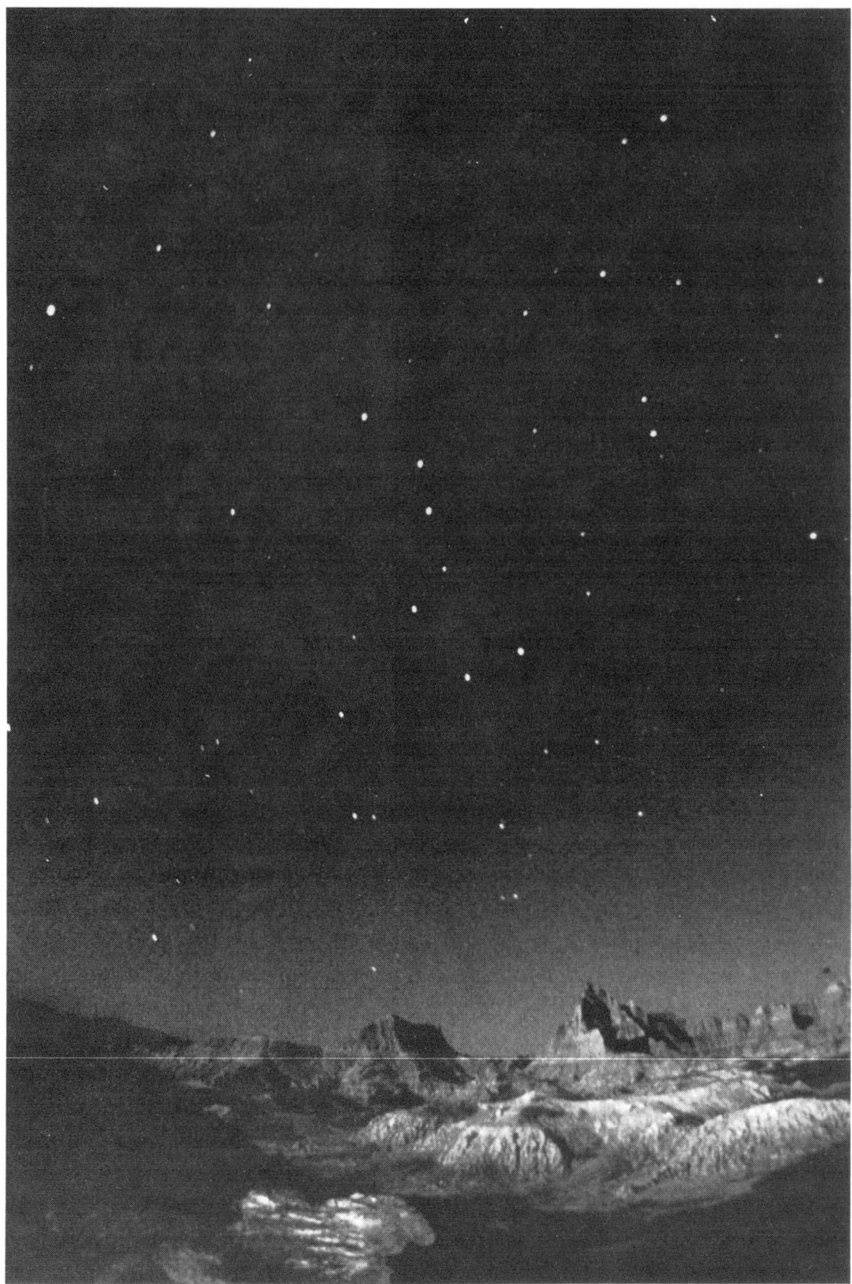

4.1. The northern sky. The Big Dipper or Plough, the hindquarters and tail of Ursa Major, rides over the Black Hills of South Dakota. The entire bear is visible below the Dipper's bowl. Follow the two front bowl stars, the Pointers – up and to the right toward Polaris, which is at the end of the Little Dipper's handle. The Little Dipper stretches up and to the left to the bright "Guardians of the Pole." Draco winds between the Dippers, and Canes Venatici is represented by a pair of stars below the curve of the Big Dipper's handle. Follow the curve of the handle to Arcturus and Boötes at the left. Denebola in Leo is not quite halfway up at the extreme left. Compare with the star maps in Appendix 2. (Rick Olson.)

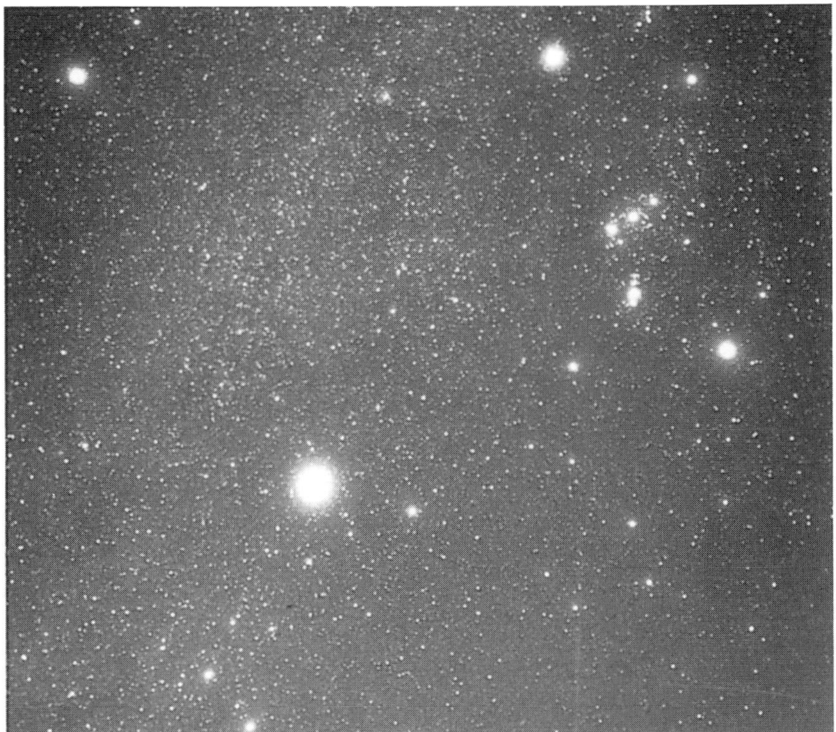

4.2. Orion, the hunter, and his retinue. The three-star belt (up and to the right) rides the celestial equator; Betelgeuse and Rigel, both first magnitude, are respectively at upper left and lower right in the pattern. Compare the photo with the frontispiece taken from the Flamsteed atlas. Down and to the left is Sirius and the whole figure of Canis Major, the large dog, and at upper left is Procyon in Canis Minor. Lepus, the hare, is directly below Orion. (Akira Fujii.)

lation contained the vernal equinox in classical times. Notice, moving down in Table 4.1a from Aries, that all the figures but one represent living creatures. The word *zodiac* comes from a Greek phrase that means "circle of animals"; it is formed from the same Greek root word that produced "zoo" and "zoology." The exception is Libra, the scales. However, even that at one time was part of the "double sign" of Scorpius, and represented the claws of the scorpion. (Note in Table 4.7 the names of the two prominent stars in Libra.) The distinct concept of the scales originated in recorded classical times.

The next six groups (Tables 4.1b through 4.1g) are arranged roughly according to renown and prominence. The three figures associated with the bears – Ursa Major (the larger bear), Ursa Minor (the smaller), and Boötes (the herdsman) – are listed in Table 4.1b; all appear in Figure 4.1. Ursa Minor (Figure 4.4), the bearer of the North Star, is not of truly ancient origin, but much like Libra was invented in classical Greek times. Table 4.1c presents Orion, the great hunter, and his companions, his two dogs Canis

Table 4.1. *The ancient constellations*

Name	Identification	α	δ	Possessive	Abbreviation	Notes
			a. The Zodiac			
Aries	ram	3	+20	Arietis	Ari	
Taurus	bull	5	+20	Tauri	Tau	
Gemini	twins	7	+20	Geminorum	Gem	
Cancer	crab	8.5	+15	Cancri	Cnc	
Leo	lion	11	+15	Leonis	Leo	
Virgo	virgin	13	0	Virginis	Vir	
Libra	scales	15	−15	Librae	Lib	
Scorpius	scorpion	17	−30	Scorpii	Sco	
Sagittarius	archer	19	−25	Sagittarii	Sgr	
Capricornus	water goat	21	−20	Capricorni	Cap	1
Aquarius	waterman	22	−10	Aquarii	Aqr	1
Pisces	fishes	1	+10	Piscium	Psc	1
			b. Ursa Major			
Ursa Major	large bear	11	+60	Ursae Majoris	UMa	
Ursa Minor	small bear	16	+80	Ursae Minoris	UMi	
Boötes	wagoner, bear driver	15	+30	Boötis	Boo	
			c. Orion			
Orion	hunter, giant	6	0	Orionis	Ori	
Canis Major	large dog	7	−20	Canis Majoris	CMa	
Canis Minor	small dog	8	+ 5	Canis Minoris	CMi	
Lepus	hare	6	−20	Leporis	Lep	
			d. Perseus			
Andromeda	chained woman	1	+40	Andromedae	And	
Cassiopeia	queen	1	+60	Cassiopeiae	Cas	
Cepheus	king	22	+65	Cephei	Cep	
Cetus	whale	2	−10	Ceti	Cet	
Pegasus	winged horse	23	+20	Pegasi	Peg	
Perseus	Perseus	3	+45	Persei	Per	
			e. Argo (ship), Jason			
Carina	keel	9	−60	Carinae	Car	2
Puppis	stern	8	−30	Puppis	Pup	2
Vela	sails	10	−45	Velorum	Vel	2
Hercules	hero; kneeler	17	+30	Herculis	Her	
Hydra	water serpent	12	−25	Hydrae	Hya	
Aries	ram-fleece	3	+20			3

Table 4.1. (*cont.*)

Name	Identification	α	δ	Possessive	Abbreviation	Notes
			f. Centaurus			
Centaurus	centaur	13	−45	Centauri	Cen	
Lupus	wolf	15	−45	Lupi	Lup	
Ara	altar	17	−55	Arae	Ara	
			g. Ophiuchus			
Ophiuchus	serpent bearer (Asclepius)	17	0	Ophiuchi	Oph	
Serpens	serpent	17	0	Serpentis	Ser	4
			h. Single constellations			
Aquila	eagle	20	+ 5	Aquilae	Aql	
Auriga	charioteer	6	+40	Aurigae	Aur	
Corona Australis	southern crown	19	−40	Coronae Australis	CrA	5
Corona Borealis	northern crown	16	+30	Coronae Borealis	CrB	6
Corvus	crow, raven	12	−20	Corvi	Crv	
Crater	cup	11	−15	Crateris	Crt	
Cygnus	swan	21	+40	Cygni	Cyg	
Delphinus	dolphin	21	+10	Delphini	Del	1
Draco	dragon	15	+60	Draconis	Dra	7
Equuleus	little horse	21	+10	Equulei	Equ	
Eridanus	river	4	−30	Eridani	Eri	
Lyra	lyre	19	+35	Lyrae	Lyr	
Piscis Austrinus	southern fish	22	−30	Piscis Austrini	PsA	1
Sagitta	arrow	20	+20	Sagittae	Sge	
Triangulum	triangle	2	+30	Trianguli	Tri	

Notes

1. Constellations of the wet quarter.
2. Carina, Puppis, and Vela are modern parts of the original ancient Argo.
3. See the zodiac.
4. Serpens is divided in two parts on either side of Ophiuchus (Serpens Caput, head, and Serpens Cauda, tail), but is still a single constellation.
5. Sometimes considered as Sagittarius's crown.
6. Ariadne's crown.
7. Contains the north ecliptic pole.

Table 4.2. *The modern constellations*

Name	Identification	α	δ	Possessive	Abbreviation	Inventor[a]	Notes
Antlia	air pump	10	−35	Antliae	Ant	Lacaille	
Apus	bird of paradise	16	−75	Apodis	Aps	Keyser/de Houtman	
Caelum	engraving tool	5	−40	Caeli	Cae	Lacaille	
Camelopardalis	giraffe	6	+70	Camelopardalis	Cam	Plancius	
Canes Venatici	hunting dogs	13	+40	Canum Venaticorum	CVn	Hevelius	
Chamaeleon	chameleon	10	−80	Chamaeleontis	Cha	Keyser/de Houtman	
Circinus	compasses	15	−65	Circini	Cir	Lacaille	
Columba	dove	6	−35	Columbae	Col	Plancius	
Coma Berenices	Berenice's hair	13	+20	Comae Berenices	Com	. . .	1
Crux	southern cross	12	−60	Crucis	Cru	Royer	2
Dorado	goldfish/ swordfish	6	−55	Doradus	Dor	Keyser/de Houtman	3
Fornax	furnace	3	−30	Fornacis	For	Lacaille	
Grus	crane (bird)	22	−45	Gruis	Gru	Keyser/de Houtman	
Horologium	clock	3	−55	Horologii	Hor	Lacaille	
Hydrus	water snake	2	−70	Hydri	Hyi	Keyser/de Houtman	
Indus	indian	22	−65	Indi	Ind	Keyser/de Houtman	
Lacerta	lizard	22	+45	Lacertae	Lac	Hevelius	
Leo Minor	smaller lion	10	+35	Leonis Minoris	LMi	Hevelius	
Lynx	lynx	8	+45	Lyncis	Lyn	Hevelius	
Mensa	table	6	−75	Mensae	Men	Lacaille	
Microscopium	microscope	21	−40	Microscopii	Mic	Lacaille	
Monoceros	unicorn	7	0	Monocerotis	Mon	Plancius	4
Musca (aust.)	(southern) fly	12	−70	Muscae	Mus	Lacaille	5
Norma	square	16	−50	Normae	Nor	Lacaille	
Octans	octant	–	−90	Octantis	Oct	Lacaille	
Pavo	peacock	20	−70	Pavonis	Pav	Keyser/de Houtman	
Phoenix	phoenix	1	−50	Phoenicis	Phe	Keyser/de Houtman	
Pictor	easel	6	−55	Pictoris	Pic	Lacaille	
Pyxis	compass	9	−30	Pyxidis	Pyx	Lacaille	6
Reticulum	net	4	−60	Reticuli	Ret	Lacaille	

Table 4.2. (*cont.*)

Name	Identification	α	δ	Possessive	Abbreviation	Inventor[a]	Notes
Sculptor	sculptor's workshop	1	−30	Sculptoris	Scl	Lacaille	
Scutum	shield	19	−10	Scuti	Sct	Hevelius	
Sextans	sextant	10	0	Sextantis	Sex	Hevelius	
Telescopium	telescope	19	−50	Telescopii	Tel	Lacaille	
Triangulum Australe	southern triangle	16	−65	Trianguli Australis	TrA	Keyser/de Houtman	
Tucana	toucan	0	−65	Tucanae	Tuc	Keyser/de Houtman	7
Volans	flying fish	8	−70	Volantis	Vol	Keyser/de Houtman	
Vulpecula	fox	20	+25	Vulpeculae	Vul	Hevelius	

[a]Source: *The Origin of Our Constellations*, Ian Ridpath, *Mercury*, November/December 1990.

Notes

1. Star cluster, many very old references to it, but not one of the ancient 48. Referred to as Ariadne's hair (see Corona Borealis) by Eratosthenes.
2. Originally a part of Centaurus.
3. Contains Large Magellanic Cloud and the south ecliptic pole.
4. Of older origin.
5. Previously a "bee" invented by Bayer.
6. Grouped with Argo.
7. Contains Small Magellanic Cloud.

Major and Canis Minor, and some of his prey, Lepus the hare. The whole group is pictured in Figure 4.2.

The six old constellations of the Perseus myth, perhaps the most famous told by the constellation figures, are in Table 4.1d. Cassiopeia, queen to King Cepheus, boasted that her daughter Andromeda was lovelier than the sea nymphs. As punishment, Neptune had Andromeda chained to a rock on the coast to be devoured by Cetus the sea monster or whale. Perseus (Figure 4.5), arriving on the flying Pegasus, used the Medusa's head to turn Cetus to stone, thereby rescuing the young lady. Another famous story, that of Jason, is represented by six constellations in Table 4.1e. Carina, Puppis, and Vela (Figure 4.6) are recent subdivisions of Jason's ship, Argo, all of which are counted as one in a listing of the ancient 48. Jason set off in his ship accompanied by various heroes, including Hercules, to find the golden fleece, represented by Aries, which does double-duty as a zodiacal figure.

The last two groups are less well known. Centaurus (Table 4.1f and Figure 4.7), which contains the closest star, Alpha Centauri, is the classic centaur, half-man half-horse. He is sometimes depicted as sacrificing Lupus the wolf upon Ara the altar.

Table 4.3. *Dates of passage of the Sun through the constellations of the zodiac*

Aries	April 19 – May 14
Taurus	May 15 – June 20
Gemini	June 21 – July 20
Cancer	July 21 – August 10
Leo	August 11 – September 16
Virgo	September 17 – October 30
Libra	November 1 – November 24
Scorpio[a]	November 25 – December 17
Sagittarius	December 18 – January 19
Capricornus	January 20 – February 16
Aquarius	February 17 – March 11
Pisces	March 12 – April 18

[a]Scorpio is combined here with the non-zodiacal constellation Ophiuchus, through which the Sun passes between November 30 and December 17.

Ophiuchus (Table 4.1g) is a large figure depicted as a man entwined with Serpens the serpent. Serpens is the only constellation divided into two distinct parts, Serpens Caput, the head of the serpent, separated from Serpens Cauda, the tail, by Ophiuchus. He is identified with Asclepius, the ancient Greek physician, and the grouping with the caduceus, the snake-wrapped staff that is symbolic of modern medicine. Finally, the single constellations, those that generally stand alone, are arranged alphabetically in Table 4.1h.

It is not possible to be completely rigorous when grouping constellations in this manner. Many have a variety of myths told about them, so that families can be ordered in different ways. Hercules is an ancient constellation also called "the Kneeler." It predates the Argo and was originally associated with the defeated Draco. Centaurus was Jason's foster father and tutor and might be included with the Argo group. Scorpius is linked with Orion, having caused his death. The scorpion sets as the hunter rises, so that he need not look upon his killer. Sagitta is said by some to be an arrow shot by Hercules. An interesting major group contains the "wet" constellations of Capricornus, Aquarius, Pisces, Piscis Austrinus, and Delphinus. These are associated with the position of the Sun during a rainy or wet season, though their origins are not clear. And remember, although these constellations may seem natural and definitive to us, they are really only one example of those produced by the myriad cultures of the world: the Arabians, the Chinese, the natives of the Americas, and others all had their own, frequently very different, notions of how the sky is constructed.

4.3 The modern constellations

With the development of modern astronomy after the Copernican revolution, the blank areas of the sky left by the ancients, and particularly the great uncharted regions of the

4.3. Aries, the ram. This depiction of the first sign of the zodiac is taken from Hevelius's *Uranographia* of 1687 (published 1690). The figure is reversed, the stars appearing as they would on a celestial globe. "Majus" is our Triangulum, or "Triangulum Major," to distinguish it from the now defunct "Triangulum Minor," or "Minus" below it. Musca, on Aries' back, has also passed from favor. The ribbon to the left connects the two fish in Pisces to the west of Aries. (University of Illinois Library.)

southern heavens invisible from the Mediterranean world, posed problems. The existence of unnamed areas made it difficult to locate and permanently identify stars, and probably as important, offended the sense of order of the new scientists. It also presented a marvelous opportunity to place new figures in the sky, heretofore unrepresented animals, the devices being developed by a budding technology, and royalty with whom the constellation makers tried (with some success) to curry favor.

The *modern constellations* divide into two subgroups: the ones that passed the test of time and were adopted for general use, and those (many mercifully) that did not. The survivors are listed in Table 4.2 along with their inventors. The dim areas of the northern sky were filled in the seventeenth century principally by the German Johann Hevel, or Hevelius (see Figure 4.3). Those in the south were initially populated around 1600 by the Dutch travelers Pieter Keyser and Frederick de Houtman. Their new figures were made secure when they were included in Johannes Bayer's famous *Uranometria* of 1603 (a northern section of which is illustrated in Figure 4.8). The southern hemisphere was finally completed in the mid-eighteenth century by the Abbé Nicolas de Lacaille, who has been called "the true Columbus of the southern skies."

Several of the modern constellations deserve to be singled out. Grus (the crane) was

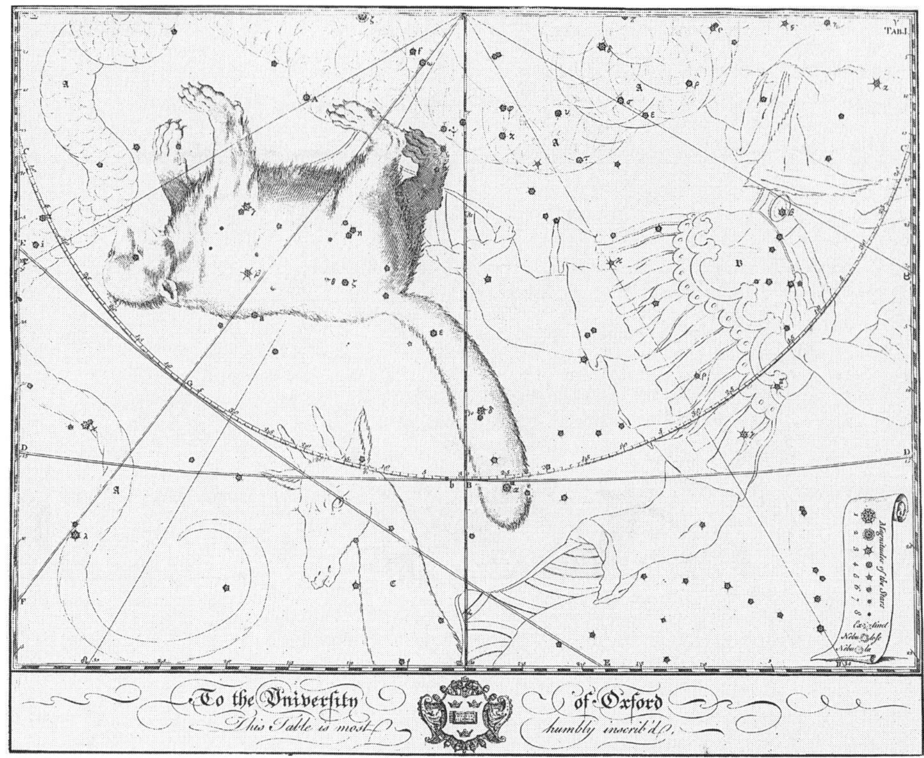

4.4. Ursa Minor. In the *Uranographia Britannica* of 1786, John Bevis used figures created by Jacobo de Gheyn (see Figure 4.8 for another example). Surrounding the little bear are Draco, Cepheus, and the head of the giraffe, Camelopardalis. The ecliptic pole is at the top, with meridians of celestial longitude converging on it, and the celestial pole at lower center, near Polaris. Note the dedication at the bottom. (University of Illinois Library.)

carved from Piscis Austrinus (see Figure 4.6), and the great figure of the south, Crux, the Southern Cross (see Figure 4.7), was at one time the feet of the centaur. Crux is the most famed of the new stellar constructions, and one of the few ever commemorated in music (Richard Rodgers, "Beneath the Southern Cross," from the score to the television production *Victory at Sea*). Coma Berenices (Figure 4.9) is anomalous, not really belonging to the moderns, as it was mentioned as "Ariadne's Hair" by Eratosthenes, the Alexandrian (third century BC) astronomer who first measured the circumference of the Earth (see Section 2.8). This set of dim stars is actually a large nearby cluster that sprawls south of Canes Venatici (see Figure 4.1) below the curve of the Big Dipper's handle. However, because Coma Berenices was not included in the ancient 48 by Ptolemy, it strangely falls among the moderns. The southern Pyxis, the mariner's compass, is nowadays grouped with Argo, against which it nestles, but of course it really does not belong with the ancient vessel. Finally, dim Octans contains the south celestial pole, poorly marked by the weakly glowing Sigma Octantis.

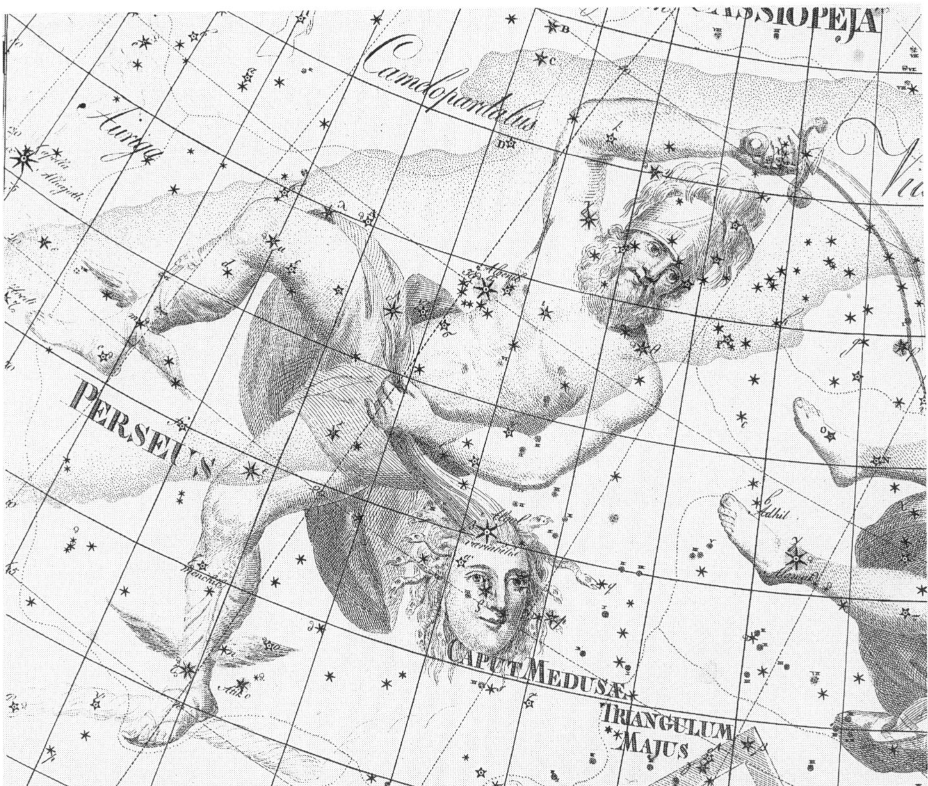

4.5. Perseus, slayer of Cetus (see Figure 4.17). In Bode's *Uranographia* of 1801 he holds the head of the Medusa, personified by the eclipsing binary star Algol, β Persei. Andromeda appears to the right and Triangulum (see Figure 4.3) below him. (University of Illinois Library.)

4.4. The defunct constellations

A large number of modern constellations were ultimately not accepted by either the public or the astronomers. Even erudite sky-watchers are now unaware of most of these, although a few still echo faintly down through our times. A short list, selected to give the flavor of the subject, is given in Table 4.4. Here we find one by the prolific Hevelius, and three (he made more) by Johann Bode, of "Bode's law" fame, whom we will encounter again in Chapter 12. None of his creations survived the passage of time.

One characteristic of the defunct constellations is their appeal to nationalism. It is not surprising that the English chose not to view the memorialization of Frederick the Great, nor that the Germans ignored the great Oak of Charles II. These figures were nevertheless beneficial to their creators: Edmund Halley (who gave his name to Halley's comet) received a master's degree from Cambridge University in return for his tree. None of those honoring people made it to permanent celestial status with the exception of Scutum, the shield that honors the name of John Sobieski, the seventeenth century

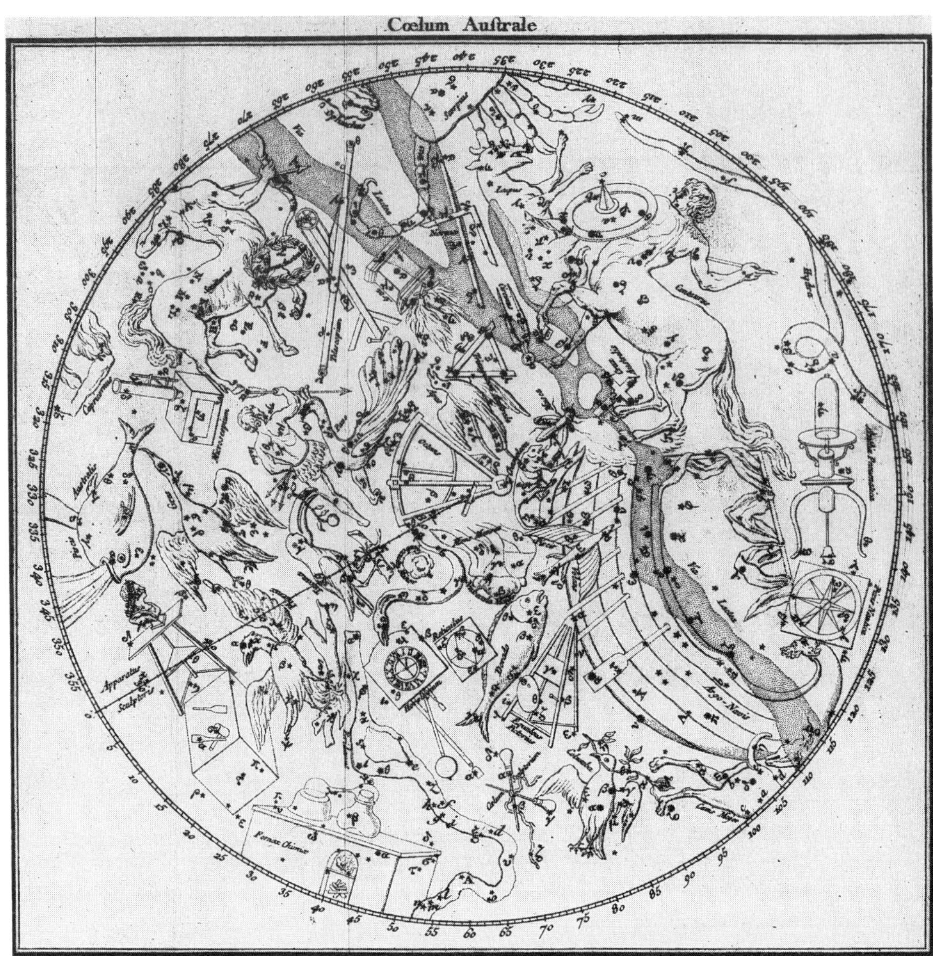

4.6. Lacaille's southern sky. In the *Planisphere contenant les Constellations Célestes*, 1756, he shows his and Bayer's new southern constellations with newly placed Greek letters. Argo is down and to the right of center; Centaurus and Crux are to the upper right. (University of Illinois Library.)

King of Poland, who was widely revered for his defense of Vienna against the Turks.

Some others of these departed figures deserve special attention. Foremost is probably Antinoüs (Figure 4.8), constructed nearly two millennia ago by the Emperor Hadrian from stars just to the south of Altair in the constellation Aquila, the eagle. Even Ptolemy mentioned the constellation (though dismissingly), which memorializes Hadrian's companion, who gave his life in the belief that his remaining years would go to his master. Another that has held some favor is Poniatowski's Bull, a V-shaped figure reminiscent of Taurus's head that is located in northeastern Ophiuchus and honors another great Polish general and king. Finally, Quadrans, near Ursa Major, will live on as the source

4.7. Centaurus and Crux, the Southern Cross. Centaurus is represented here by its two brightest stars, Alpha and Beta, to the left of center (Alpha is at the left). Crux is to the right of center. Even though a modern invention, it is one of the most famous of all constellations. The Coalsack, one of the most prominent of the dark dust clouds of the Milky Way, is down and to the left of the cross. (Akira Fujii.)

of an excellent meteor shower still called the Quadrantids after the direction in which the meteors appear to originate (see Chapter 12).

Some attempts at creating constellation figures went much further than the fill-in-the-blank type of effort. In the early 1600s Julius Schiller simply redrew the entire sky, populating it with figures from Christianity (Figure 4.10). The New Testament was represented in the northern hemisphere, the Old in the southern, and the zodiacal signs were replaced by the Apostles of Christ. Needless to say, none of these survived.

Of course, all constellations are figments of the mind. Antinoüs, Musca Borealis (the northern fly: find the little creature above Aries' back in Figure 4.3), and even Charles' Oak still exist as long as there is anyone to remember them, and they are still there for our perusal. Mere "officializing" is in a sense irrelevant. Try for example, to find Antinoüs. His head is marked by the bright star η Aquilae (see Section 4.7 for an explanation of star names), his right hand by θ, the right knee and left foot by κ and λ. And if you already know Aquila, it will never quite look the same.

4.5. Asterisms and other groups

The Big Dipper, or Plough (Figure 4.11), is among the most prominent of constellations. Yet you will look for it in vain in the lists of Tables 4.1, 4.2, and 4.4. The stars of the Dipper form the tail and hind quarters of Ursa Major, the great bear (Figure

4.8. Aquila and Antinoüs in Bayer's *Uranometria* of 1603. The beautiful figures were drawn by Jacobo de Gheyn. Antinoüs is no longer recognized. The Bayer Greek letters refer to Aquila. The celestial eagle is epitomized by Altair (α) and its two flankers β and γ, which together give the effect of the bird soaring through the sky. The Milky Way runs down the center. The gap between the two branches of the Milky Way is the Great Rift. The star η Aquilae is a Cepheid variable (Section 4.12). (University of Illinois Library.)

4.9. Coma Berenices, Berenice's Hair, a lacy cluster of stars just south of Canes Venatici. It is a wonderful sight in low-power binoculars. (Author's photograph.)

4.1). Many constellations have smaller parts that constitute prominent named figures and that are called *asterisms*. Asterisms can also be formed from the stars of two or more constellations. A short list of important asterisms is given in Table 4.5. Among them are the first learned and best loved of all groups: the Big Dipper, the Little Dipper (Figures 4.1 and 4.4), the Northern Cross (Figure 4.12), and the Great Square of Pegasus.

Three of these configurations are actually physical clusters of stars, and like Coma Berenices (Figure 4.9), are tied together by the force of their own gravity. The most prominent of them is the remarkable Pleiades cluster in northern Taurus (Figure 4.13). To a quick glance it will appear as a fuzzy patch about 1° across that instantly resolves into a nest of closely packed stars. The "seven sisters" were daughters of Atlas who were pursued by the great hunter Orion, and are followed by him yet today as they forever circle the sky. In spite of their mythic number, most people can see but six stars (though the keen-eyed may find as many as eight), a reminder that the constellations only *represent*. Another cluster is the Hyades, also daughters of the prolific Atlas,

Table 4.4. *Some defunct constellations*

Name	Identification	Location	Principal stars[a]	Inventor	Notes
Antinoüs	Hadrian's companion	S Aquila	θ,η,ι,ν,κ,λ Aql	Hadrian	1
Cerberus	guardian of Hell	E Hercules	93,95,96,109 Her	Hevelius	
Frederici Honores	Fredrick (the Great's) honor	Lacerta		Bode	2
Machina Electrica	electric machine	S Cetus		Bode	
Musca Borealis	northern fly	N Aries	33,35,39,41 Ari	Bartschius?	3
Officina Typographica	printing office	NE Canis Major		Bode	
Quadrans	quadrant	Ursa Major		Lalande	4
Robur Carolinium	Charles' (II) Oak	Argo		Halley	
Taurus Poniatovii	Poniatowski's Bull	NE Ophiuchus	66,67,68,70,73 Oph	Puczobut	5
Telescopium Herschelii	Herschel's telescope	Gemini–Auriga	π Gem, ψ Aur	Hell	

[a]See Section 4.7 for a discussion of star names.

Notes

1. An old constellation that does not fit with the others, being formed by the Emperor Hadrian in the second century. It remained popular through the eighteenth century (see Figure 4.8).
2. Attempt to replace Lacerta, which superseded another political figure, the French "Sceptre and Hand of Justice."
3. Popular figure but no longer used.
4. The name still lives as the source of a fine winter meteor shower (Chapter 12) called the Quadrantids.
5. A still-semipopular figure honoring a Polish king, a V-shaped figure easily found in northeastern Ophiuchus.

and half-sisters to the Pleiades. They too reside in Taurus, a V-shaped configuration of stars some 5° or more in extent (see the frontispiece and Figure 11.1). Finally, the Praesepe, or Beehive cluster, is a faint fuzzy patch in Cancer barely visible to the naked eye. While not really an asterism, because the stars form no obvious figure, it is a named group and is included here. It has long carried a relation to Christianity, the cluster representing the manger (*praesepium* in Latin), flanked by two stars called "the asses."

4.10. The three wise men, or kings. They replaced Hercules in Schiller's 1627 biblical version of the celestial sphere, *Coelum Stellatum Christianum*. The figures are reversed as they would be on a celestial globe. St. Sylvester appears in place of Boötes at the far left, and Christ's crown of thorns replaces Corona Borealis. To the right, Lyra reappears as the manger, and Cygnus, at far right, as the cross. (University of Illinois Library.)

Two other asterisms, the Water Jar of Aquarius and the Circlet of Pisces, lie close together in the wet quarter. (These are easy to find on the constellation maps. The circlet represents the head of the western fish, and the water jar – a Y-shaped figure – is just to the west.) Also included in the table are several other well-known star pairs. Scorpio's claws (Libra) have already been mentioned. Two other pairs are drawn from the front bowl stars of the Dippers: one pair points to the north star, the other guards it. Another duo flanks Antares, the luminary of Scorpio, and yet another, Mizar and Alcor (Figure 4.11), are a true physical pair, a test of minimal vision the Arabians called the "horse and rider."

Two famous large figures encompass triplets of constellations. The Winter Triangle (Sirius in Canis Major, Orion's Betelgeuse, and Procyon in Canis Minor) can be seen in Figure 4.2. The Summer Triangle is made from Deneb (see Figure 4.12) in Cygnus, Altair in Aquila (see Figure 4.8), and Vega in the exquisite Lyra (Figure 4.14). The

4.11. The Big Dipper or Plough. This famous figure is seen above the silhouette of the 2.3-meter telescope of the University of Arizona. Mizar, with its companion Alcor, is second in from the end of the Dipper's handle. (Author's photograph.)

Great Diamond is more obscure. It is a northern spring figure formed from Cor Caroli (the bright star in Canes Venatici), Arcturus in Boötes, Denebola in Leo, and Spica in Virgo, the first three of which are included in Figure 4.1. Through the Great Diamond, one can see (telescopically only) great masses of other galaxies.

The existence of these figures again reminds us that the "official" constellation list shows but a small part of the respect and fascination humanity has had for the sky. Considering the cultures of the world, hundreds, even thousands, of constellation figures exist, which reflect our histories, joys, sorrows, and dreams.

4.6. The brightnesses and colors of the stars

Over the entire celestial sphere, the naked eye can detect about 6000 to 8000 individual stars, which range in brightness from those brilliant ones that can be seen in early twilight to those that can be viewed only on an extremely clear, moonless, country night. At any given time, about a third of them are visible because absorption by the Earth's atmosphere removes many faint stars from view as our gaze approaches the horizon (see Section 13.15). The faintest stars are always best seen directly overhead.

By a custom that originated with Hipparchus in the second century BC, the naked-eye stars are divided into six brightness groups called *magnitudes*. The brightest stars are said to be *first magnitude* and the faintest that we perceive are of the sixth. The

Table 4.5. *Asterisms*

Asterism	Constellation	Description[a]
		Groups within constellations
Beehive	Cancer	Praesepe, Crib, Manger, star cluster
Big Dipper	Ursa Major	seven bright stars (α–η UMa)
Circlet	Pisces	head of the western fish (γ, κ, λ, ι, θ, 7 Psc)
False Cross	Vela, Carina	resembles Southern Cross (κ, δ Vel; ε, ι Car)
Hyades	Taurus	star cluster, around Aldebaran
Keystone	Hercules	four stars in northern Hercules (η, π, ε, ξ Her)
Kids	Auriga	triangle of faint stars near Capella (ε, ζ, η Aur)
Little Dipper	Ursa Minor	seven stars, Polaris at end of handle (α–η UMi)
Little Milk Dipper	Sagittarius	five-star dipper (λ, τ, σ, φ, ξ Sgr)
Northern Cross	Cygnus	Deneb at top of cross (α, γ, δ, ε, β Cyg)
Pleiades	Taurus	Seven Sisters, star cluster NW of Aldebaran
Sickle of Leo	Leo	Regulus at end of sickle (α, η, γ, ξ, μ, ε Leo)
Square of Pegasus	Pegasus	central figure of Pegasus (α And; α, β, γ Peg)
Tea Pot	Sagittarius	Little Milk Dipper plus δ, γ, ε Sgr
Urn or Water Jar	Aquarius	four stars in "Y", northern Aquarius (γ, η, ζ, π Aqr)
		Well-known pairs
The Asses	Cancer	γ, δ Cnc; enclose (with η and θ Cnc) the Beehive cluster (Section 4.15)
The Claws	Libra	α, β Lib; claws of Scorpio
Guardians of Pole	Ursa Minor	β, γ UMi
Mizar and Alcor	Ursa Major	ζ, 80 UMa; the horse and rider
The Arteries	Scorpio	τ, σ Sco, flanking Antares
The Pointers	Ursa Major	α, β UMa; point to Polaris
		Large configurations
Great Diamond	—	Cor Caroli, Denebola, Spica, Arcturus
Summer Triangle	—	Deneb, Vega, Altair
Winter Triangle	—	Betelgeuse, Procyon, Sirius

[a]See Section 4.7 for a discussion of star names.

first-magnitude list includes the 21 brightest stars. All but one of the Big Dipper are second magnitude, the pair that flank Altair in Aquila are third, the central stars of the Little Dipper's handle belong to the fourth magnitude, and the faintest star of its bowl (not easy to see from town) is of the fifth. In the southern hemisphere, the Southern Cross displays a nice array, with two of the stars of the first magnitude, and one each representing the second and third. Between the brightest and faintest of the four lies another of the fourth, and in the middle, two that belong to the fifth (look for them all

4.12. Cygnus, the swan. Deneb, the great bird's tail, is the bright star near top left. Albireo (β cygni), his head, lies at bottom right. Gamma Cygni is just up and to the left of center, the swan's wings extending outward from lower left to upper right. The figure also makes the Northern Cross. (Author's photograph.)

in Figure 4.7). The southern pole star, Sigma Octantis, is also a dim fifth-magnitude representative.

The fainter the star, that is, the larger the magnitude number, the more stars there are in the group. Because of the way our eyes detect light, the first-magnitude stars are about 100 times brighter than those of the sixth, and in the nineteenth century, the old magnitude convention was mathematically quantified. Five magnitude divisions were set to correspond exactly to a factor of 100 in brightness. One magnitude interval (from second to third for example) then corresponds to a factor that is the 5th root of 100, or 2.512. (If each magnitude is 2.512 times brighter than the next, 5 magnitudes must be 2.512 multiplied by itself 4 times, or 100.)

The modern magnitude scale was established by adopting a set of faint stars that surround the north celestial pole and calling their average magnitude exactly 6.0. The brightness of any other star can then be found by comparing it to the standards. (Magnitude determination is always relative: other standard stars are in use today.) With this quantification, the very brightest stars become brighter than first magnitude.

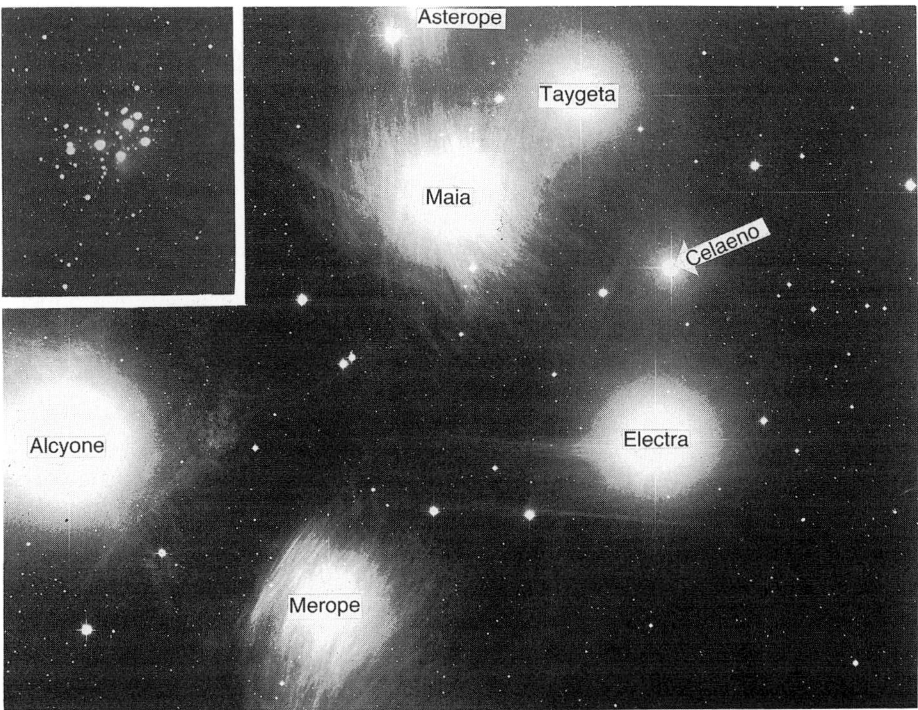

4.13. The Pleiades, the seven daughters of Atlas and Pleione. The inset at upper left shows the whole cluster, in which the unaided eye can see the brightest six, and the larger picture a magnified view of the central region. The sisters are named in the large photo. Atlas and Pleione are the pair to the left of Alcyone in the inset; Atlas is the lower and brighter. The cluster is passing through a cloud of interstellar gas and dust, its delicate wisps illuminated by the reflection of light from the bright stars embedded within. (Large photograph: National Optical Astronomy Observatories; inset: Lowell Observatory.)

There are several of the zeroth magnitude (which is brighter than first), and two go into negative numbers. (These are still generically referred to as "first magnitude.") The magnitude of a star is one of its fundamental observable properties, and astronomers today can measure it to a precision of 0.01 of a division or better. By convention, any star with a magnitude between 1.50 and 2.49 is called second magnitude, between 2.50 and 3.49 third magnitude, and so on. Modern measurement is accomplished by electronic means: a star is imaged by a telescope onto a light-sensitive surface that induces an electric current or charge that is directly proportional to brightness. Measurement of the current or charge then allows the determination of the star's magnitude.

Another basic observable quality of stars is their color. They come in shades from obvious red to somewhat bluish. The first-magnitude zodiacal star Antares, which means "like Mars," is called so because of the reddish color that it shares with that

4.14. Lyra, the lyre or harp. The little constellation consists of the bright star Vega, a parallelogram down and to the right of Vega, and the double star (ε Lyrae) down and to the left. Beta Lyrae, the upper of the two at the right, is an eclipsing variable. (Author's photograph.)

planet. Recall also that sunlight falling through a window is yellowish. Look to see how many hues you can discern.

Physically, the colors reflect surface temperatures. As the electric element in a toaster or oven heats, the color changes from a deep red to orange. If we could increase the temperature further, it would turn yellow, then white, then blue (by which time it would have melted). The sequence of the colors in the rainbow is matched in the metal by increasing the heat. We see the same phenomenon in stars. Red belongs to the coolest stars with temperatures roughly 3000 K (kelvins, K, are centigrade degrees above absolute zero, which is −273 °C). Blue belongs to the hottest at 20 000 K and above. Table 4.6 gives the correspondence between temperature and color and some examples of stars in each category. Note that green, which falls between yellow and blue in the rainbow, is left out. This color is replaced by white; there are no green stars. The one traditional exception is Scorpio's northern claw in Libra (Zubeneschamali, or β Librae: see the next section), which is often touted as green, though most observers have difficulty convincing themselves of that color.

Magnitudes are color sensitive. The human eye is most sensitive to yellow light, and though it can see a variety of colors from red to violet, it is basically a yellow-sensitive

Table 4.6. *Star colors*

Color	Temperature	Examples
red	3 000 K	Betelgeuse, Antares
orange	4 500 K	Arcturus, Aldebaran
yellow	6 000 K	Capella, the Sun
white	10 000 K	Sirius, Vega, Deneb
blue	20 000 K	δ, ε, ζ Orionis (Orion's belt)

detector. Other detectors are sensitive to different colors. For example, an untreated photographic film or plate "sees" blue and violet the best. Therefore, given a hot and cool star of equal total energy output, the hot blue one will appear the brighter on a photograph, the cool one, be it yellow or even red, brighter to the eye. The eye records what are called *visual magnitudes*; the others are often referred to as *photographic* or *blue* magnitudes. Modern measurements are made with electronic devices equipped with filters that mimic the effect of the eye and photographic plate. The professional astronomer uses numerous magnitude systems that cover a variety of colors from ultraviolet to infrared (details of colors in the spectrum may be found in Chapter 13).

The constellation Orion provides a superb example of color effects. Red Betelgeuse and blue-white Rigel appear about the same brightness to the human eye (as they do in Figure 4.2, which was originally made on color film to mimic the human response). However, on older photos made on black and white film with its inferior red sensitivity, Betelgeuse shrinks from first magnitude nearly to third. It is wise to remember that our view of the Universe is quite personal, and depends on our limited abilities of vision.

4.7. The names of the stars

At first sight, the stars present a chaotic sight; how to tell one from another seems a near-insurmountable problem. Some means of systematic identification is necessary, and many have been devised. The constellations are indispensable for naming the naked-eye stars because they section the sky into small areas that are much more easily handled and the visual star patterns arising out of the random scatter provide a superb aid to memory.

Three kinds of star names are commonly used: the *proper name*, the *Greek letter name*, and the *Flamsteed number*; brighter stars commonly carry all three. The proper names of stars and constellations contain within them a simple summary of western history. The concepts are basically Greek and the constellation names we use are the Roman, or Latin, translations. The proper names of the stars, however, are mostly Arabic. During the time that Greek astronomy was nurtured by the Arabic civilizations, the Arabians gave most of the brighter naked-eye stars their names.

Many of these proper names are single words or corruptions of longer phrases descriptive of the star's location within the constellation. For example, the bright blue star in Orion, Rigel, comes from *rijl al-jauza*, the foot of the "central one." This person, *al-jauza*, is not the mighty Orion of today (who would have been *al-jabbar*, the giant), but is a mysterious feminine figure. Some of the Arabic names have no relation at all to the Greek constellation and arise directly from the rich Arabic constellation mythology that is no longer in general use today. An example is the star at the end of the handle of the Big Dipper, Alkaid, which refers to the "chief of mourners," or a "daughter of the bier," having nothing whatever to do with a bear. Still others are derived from numerous European mistranslations of the Arabic language, such as Betelgeuse, which initially meant the "hand of *al-jauza*, the original phrase being misspelled and corrupted into "armpit". In addition, stars were frequently confused and names transferred from one to another.

A few of the proper names we use were carried down directly from Greece and Rome: the brightest star in the sky, Sirius (Figure 4.2), comes from the Greek for "scorching," and the bright star in Leo, Regulus, is the Roman diminutive for "king." A few stars carry true proper names of people, for example Canopus in Carina and the warrior twins Pollux and Castor in Gemini.

Proper names, however, are cumbersome and are hard to memorize. They can also be ambiguous. There are two stars named Algenib, one in Pegasus and one in Perseus, each designating the "side" of the constellation figure, and four variations on "Deneb," which refers to a figure's tail: Deneb itself at the top of the Northern Cross; Denbola, the tail of Leo; Deneb Kaitos, the tail of Cetus the whale; and Deneb Algedi, Capricornus's tail. Two others share the name Alnair, "the bright one."

A more methodical system of nomenclature was introduced in the early seventeenth century by Johannes Bayer. He gave the stars lower-case Greek letter names, organized by constellation, more or less in order of brightness. To the Greek letter we affix the Latin possessive form of the constellation name, given in Tables 4.1 and 4.2. The brightest star in Canis Major, Sirius, is thus α Canis Majoris, and the second brightest star in Taurus, Elnath, is β Tauri. These are conveniently abbreviated to α CMa and β Tau. A list of official constellation abbreviations is given in the tables. There are many exceptions to the brightness rule: β Ori is brighter than α Ori and the seven stars of the Big Dipper are named α through η in order from west to east. Lacaille later extended the system to the stars of the deep southern hemisphere that were beyond the realm of the classical constellations.

The Greek letters now provide the principal means of naming the brighter naked-eye stars. However, there are only 24 letters in the Greek alphabet, an insufficient number to do justice to all the stars that can be seen. Bayer supplemented the Greek letters with lower-case Roman letters, beginning where the Greek left off, and then continuing with upper-case Roman down to the letter Q. These, however, have largely fallen into disuse. The problem of insufficient letters can also be alleviated slightly by affixing numerical superscripts to the Greek letters. For example, a string of stars in western Orion, representing a pelt held aloft by the hunter, are named π^1 through π^6 Ori.

A considerably improved and more consistent system for naming the naked-eye stars was developed in the eighteenth century by the well-known English astronomer and celestial cartographer John Flamsteed and the French astronomer Joseph Lalande. In the companion catalogue to his great star atlas (see Section 4.8), Flamsteed arranged all the fairly prominent stars that could be seen within a given constellation in order of increasing right ascension. Many years later, Lalande applied consecutive numbers to these stars. Although Flamsteed never used them, they are still called "Flamsteed numbers," and are widely employed today. Like the Bayer letters they use the Latin possessive. A well-known example is 61 Cygni, the first star to have its distance measured.

Most of the brighter stars carry all three kinds of appellations: Vega, α Lyrae, is also 3 Lyrae. The modern naming convention is to use the proper names for only the first magnitude stars (Deneb, Regulus, and the like) or for those of special importance (Algol = β Persei, Mizar = ζ Ursae Majoris, Mira = ο Ceti). The Greek letters follow, though these are commonly employed for the first magnitude stars as well. The Flamsteed numbers are then called out for those lacking Greek letters. If a star has no other designation, the old Roman letters will occasionally surface, for example Q Scorpii just southwest of λ in the scorpion's stinger. (Additional Roman letters denote stars that vary in brightness: see Section 4.12.)

These naming systems have several ambiguities and aberrations. Constellation boundaries in the past were often rather haphazard; the current boundaries, squared off by parallels of declination and hour circles of right ascension, were adopted by the world's astronomers only in 1922. There is therefore some confusion caused by overlap and interlocking constellations. The star β Tauri, now formally in Taurus, is shared by the constellation figure outlining Auriga, and is also called γ Aurigae; α Andromedae is one of the corners of the Great Square of Pegasus and historically is shared by that constellation. The Greek letter names in Carina, Puppis, and Vela were distributed over the original constellation Argo: α and β (once called α and β Argus) are in Carina, and γ and δ are in Vela. As a result, α and β Velorum do not exist. Epsilon is again in Carina, and thus the stars of Puppis start with bright (second magnitude) blue ζ.

In addition, some stars that carry the names of one constellation are actually within the formal confines of another: 3 Ari is just across the border in Pisces, 25 Cam is in Auriga, 62 Serpentis is in Aquila, and 31 Crateris is in Corvus. This sort of confusion reaches a pinnacle with regard to the modern constellations, which often use stars borrowed from the ancient. For example α, β, and γ Scuti were (and for that matter still are) 1, 6, and 9 Aquilae. Foremost in this category is Crux, which was taken out of Centaurus from what were originally ε, ζ, ν, and ξ Cen, designations that have since been assigned to other stars. Finally, note that though Serpens is divided in two, it is one constellation, and the letters and numbers are distributed accordingly: α through ε are in the western half (Serpens Caput) while ζ, η, and θ are in the eastern part (Serpens Cauda).

Confusion can be avoided by the use of catalogue systems that are divorced from the constellations. All the naked-eye (and many fainter) stars are listed simply by number in order of increasing right ascension (starting at 0^h) in the Yale *Bright Star Catalogue*. The numbers are prefixed by the designation "HR," which stands for "Harvard Revised" from an earlier incarnation. Thus Arcturus = α Boo = 16 Boo is also HR

5340. The *Bright Star Catalogue* lists a total of 9110 stars, and cross-lists proper names and both the Bayer and Flamsteed designations. Still, the HR numbers are generally used only in the absence of the more common designations. Even though a strict numbering system is more logical, it is impersonal and the HR designations are hard to remember; we still (even professionally) admire and recognize Capella and 30 Cygni rather than HR 1708 and 7730. And so the constellations remain respected thousands of years past their births.

We can perhaps now appreciate the problems faced by the professional astronomer in dealing with the enormous number of observable telescopic stars. The naming systems used, and there are several, employ methods similar to the Yale bright star list. All stars down to magnitude 9 (2.5^3 or 16 times fainter than the unaided eye can see) and many fainter than that have long carried some kind of catalogue number (to be discussed in Section 8.2). These identifications are being extended down to the 19th magnitude to a billion stars.

A list of 144 bright stars, arranged alphabetically by constellation abbreviation, is given in Table 4.7. The table contains all the stars through the second magnitude, plus a number of fainter ones that may be important in outlining a constellation figure, or otherwise have some significance. As extensive as it may appear, it is by no means an exhaustive list of proper star names. The column headings are obvious. The visual magnitudes are given to one-hundredth of a division, as measured by modern electronic devices. Within each constellation, the stars are arranged in order of decreasing brightness. Notice that the Greek letters are frequently out of order.

The third column of Table 4.7 gives the proper name and the last the generally accepted meaning. Several stars have dual names applied during different historical periods. Generally only the most popular is given here, though in a few cases there are double entries. Only a handful have no proper names. Their absence in the southern hemisphere is easily understood, but it is quite odd that bright ones like γ Cas and ε and χ Sco are lacking them. Some of the southern hemisphere stars carry very obvious made-up names (like *Acrux* for α Crucis) designed simply to fill in the blanks largely for the convenience of navigators.

The meanings in the last column are fraught with hazards because of mistranslations, mis-assignments, and lost and confused origins. The problem involves a scholarly pursuit of the first magnitude itself. Here, only a brief summary can be given, taken principally from *A Short Guide to Modern Star Names and Their Derivations* by Paul Kunitzch and Tim Smart (Otto Harrassawitz, Wiesbaden, 1986), supplemented with the classic but old (and often outdated) *Star Names, Their Lore, and Meanings* by R. H. Allen (Dover Publications, New York, 1963; originally published in 1899). The large majority of the names are Arabic; where they are not, the language (*Lat* for Latin, *Gr* for Greek, *Sum* for Sumerian, and *Tur* for Turkish) is placed in parentheses. Do not think that the meanings are actual translations of the word; the name is commonly just an allusion to the original meaning. In the worst cases the meaning is marked by "corrupt" indicating a severe distortion of the original. In the penultimate example, Albireo (β Cygni), the original Greek word was mistranslated, misread, and finally Arabicized into something quite incomprehensible. For more details, consult the volumes cited above.

Table 4.7. *The bright stars*

Star[a]	Magnitude[b]	Proper Name	Meaning[c]
α And	2.06	Alpheratz	horse's shoulder
β And	2.06	Mirach	girdle
γ And	2.19	Almach	cat-like animal
β Aqr	2.91	Sadalsuud	luckiest star
α Aqr	2.96	Sadelmelik	king's lucky star
α Aql*	0.77	Altair	flying eagle
γ Aql	2.72	Tarazed	scale beam
β Aql	3.71	Alshain	?
α Ari	2.00	Hamal	the lamb
β Ari	2.64	Sheratan	with γ, the signs
γ Ari	4.04	Mesarthim	corrupted from Sheratan
α Aur*	0.08	Capella	she-goat (Lat)
β Aur	1.90	Menkalinan	shoulder of rein holder
α Boo*	–0.04	Arcturus	bear watcher (Gr)
η Boo	2.68	Muphrid	solitary star of the lancer
ε Boo	2.59	Izar	girdle
		Mintaka	belt
γ Boo	3.03	Seginus	Boötes (corrupted)
β Boo	3.50	Nekkar	ox driver
δ Cap	2.87	Deneb Algedi	kid's tail
β Cap	3.08	Dabih	slaughterer's lucky star
α^2 Cap	3.57	Algedi	the kid
γ Cap	3.68	Nashira	ref. to "luck" (?)
α Car*	–0.72	Canopus	proper name (Gr)
β Car	1.68	Miaplacidus	ref. to calm (Lat)
ε Car	1.86	Avior	?
ι Car	2.24	Aspidiske	little shield
α Cas	2.23	Sheda	the breast
β Cas	2.27	Caph	the stained hand (ref. to a const.)
γ Cas	2.47v	. . .	. . .
δ Cas	2.68	Ruchbah	the knee
α Cen*	–0.29	Rigil Kentaurus	centaur's foot
β Cen*	0.61	Hadar	proper name
		Agena	knee
θ Cen	2.06	Menkent	shoulder
γ Cen	2.17	Muhlifain	two things causing dispute
ε Cen	2.30	. . .	. . .
η Cen	2.31	. . .	. . .

Table 4.7. (*cont.*)

Star[a]	Magnitude[b]	Proper Name	Meaning[c]
ζ Cen	2.55	Alnair	bright one in the Centaur
α Cep	2.44	Alderamin	right forearm (corrupt)
β Cet	2.04	Deneb Kaitos	sea monster's tail
α Cet	2.53	Menkar	the nostrils
α CMa*	−1.46	Sirius	scorching (Gr)
ε CMa	1.50	Adhara	the virgins
δ CMa	1.84	Wezen	the weight
β CMa	1.98	Mirzam	the announcer (of Sirius)?
η CMa	2.45	Aludra	the virginity
α CMi*	0.38	Procyon	before the dog (Gr)
β CMi	2.90	Gomeisa	little bleary-eyed one (originally denoting α)
α Col	2.64	Phact	ring dove
α CrB	2.23	Alphecca	part circle
		Gemma	jewel (Lat)
γ Crv	2.59	Gienah	raven's wing
α Cru*	1.05	Acrux	Alpha plus Crux
β Cru*	1.25	Mimosa	tropical plant (Lat)
γ Cru	1.62	Gacrux	Gamma plus Crux
δ Cru	2.80	. . .	. . .
α^1CVn	2.90	Cor Caroli	Charles's (II) heart (Lat)
α Cyg*	1.25	Deneb	hen's tail
γ Cyg	2.20	Sadr	hen's breast
ε Cyg	2.46	Gienah	wing
δ Cyg	2.87	. . .	. . .
β Cyg	2.92	Albireo	bird (Gr, corrupt)
γ Dra	2.23	Eltani	the serpent
δ Dra	3.07	Altais	the serpent (corrupt)
α Dra	3.65	Thuban	the serpent (corrupt)
α Eri*	0.46	Achernar	end of the river
β Eri	2.79	Cursa	footstool of Orion
β Gem*	1.14	Pollux	proper name (Lat)
α Gem	1.58	Castor	proper name (Lat)
γ Gem	1.93	Alhena	mark on camel's neck
α Gru	1.74	Al Nair	bright one of fish's tail
β Gru	2.10	. . .	. . .
α Her	3.31	Rasalgethi	kneeler's head
α Hya	1.98	Alphard	the solitary one

Table 4.7. (*cont.*)

Star[a]	Magnitude[b]	Proper Name	Meaning[c]
α Leo*	1.35	Regulus	little king (Lat)
β Leo	2.14	Denebola	lion's tail
γ Leo	2.30	Algieba	the forehead
δ Leo	2.56	Zosma	girdle (Gr)
α Lep	2.58	Arneb	the hare
α^2Lib	2.75	Zubenelgenubi	southern claw of the scorpion
β Lib	2.61	Zubeneschamali	northern claw of the scorpion
α Lup	2.30	. . .	. . .
α Lyr*	0.03	Vega	swooping eagle
γ Lyr	3.24	Sulafat	tortoise (ref. to lyre)
β Lyr	3.45v	Sheliak	harp
α Oph	2.08	Rasalhague	head of serpent bearer
η Oph	2.43	Sabik	preceding one
δ Oph	2.74	Yed Prior	foremost (Lat) hand
β Oph	2.77	Cebalrai	shepherd's dog
λ Oph	3.82	Marfik	the elbow
β Ori*	0.12	Rigel	foot of the central one (now Orion)
α Ori*	0.50v	Betelgeuse	hand of central one (corrupt)
γ Ori	1.64	Bellatrix	female warrior (Lat)
ε Ori	1.70	Alnilam	string of pearls (for δ, ε, and ζ Ori)
κ Ori	2.06	Saiph	giant's sword
ζ Ori	1.91	Alnitak	girdle
δ Ori	2.21	Mintak	belt
α Pav	1.94	Peacock	Pavo (constellation name)
ε Peg	2.39	Enif	the nose
α Peg	2.49	Markab	the shoulder
β Peg	2.42	Scheat	the shin
γ Peg	2.83	Algenib	the side
α Per	1.79	Mirfak	the elbow (of the Pleiades)
		Algenib	the side
β Per	2.12v	Algol	the demon's head
α Phe	2.39	Nair al Zaurak	bright one of the boat
		Ankaa	phoenix
α PsA*	1.16	Fomalhaut	fish's mouth
α Psc	3.94	Alrescha	the cord
ζ Pup	2.25	Naos	the ship (Gr)
ε Sgr	1.85	Kaus Australis	south (Lat) part of bow
σ Sgr	2.02	Nunki	proper name for celestial city

Table 4.7. (*cont.*)

Star[a]	Magnitude[b]	Proper Name	Meaning[c]
ζ Sgr	2.60	Ascella	armpit (Lat)
δ Sgr	2.70	Kaus Media	middle (Lat) of bow
λ Sgr	2.81	Kaus Borealis	north (Lat) part of bow
γ^2 Sgr	2.99	Alnasl	the point
α Sco*	0.96	Antares	like Mars
λ Sco	1.63	Shaula	the stinger
θ Sco	1.87	Sargas	proper name (Sum)
		Girtab	proper name (Sum)
ε Sco	2.29	. . .	. . .
δ Sco	2.32	Dschubba	forehead
κ Sco	2.41	. . .	. . .
β Sco	2.50	Graffias	claws (Lat)
		Acrab	scorpion
υ Sco	2.69	Lesath	nebulous spots
τ Sco	2.82	Al Niyat	with σ, the arteries of the heart (Antares)
α Ser	2.65	Unukalhay	snake's neck
α Tau*	0.85	Aldebran	the follower
β Tau	1.65	Elnath	the butting one
η Tau	2.87	Alcyone	proper name in Pleiades
α TrA	1.92	Atria	Alpha plus TrA
ε UMa	1.77	Alioth	the bull (corrupt)
α UMa	1.79	Dubhe	bear
η UMa	1.86	Alkaid	chief of mourners
		Benetnasch	daughters of the bier (?)
ζ UMa	2.06	Mizar	groin
β UMa	2.37	Merak	loin of the bear
γ UMa	2.44	Phecda	thigh
δ UMa	3.31	Megrez	root of the tail
α UMi	2.02v	Polaris	pole star (Lat)
β UMi	2.08	Kochab	star (?)
γ UMi	3.05	Pherkad	calf
δ UMi	4.34	Yildun	star (Tur)
γ Vel	1.68	Regor	proper name?
δ Vel	1.96	. . .	. . .
λ Vel	2.21	Suhail	?
κ Vel	2.50	Markeb	ship or vehicle

Table 4.7. (*cont.*)

Star[a]	Magnitude[b]	Proper Name	Meaning[c]
α Vir*	0.98	Spica	ear of wheat (Lat)
ε Vir	2.83	Vindemiatrix	female grape gatherer (Lat)
γ Vir	2.91	Porrima	goddess of prophecy (Lat)

[a]First magnitude stars are denoted by asterisks

[b]"v" denotes that the magnitude is variable

[c]Names are of Arabic origin unless otherwise indicated as "Gr" for Greek, "Lat" for Latin, "Sum" for Sumerian, and "Tur" for Turkish. The principal source for the meanings is *A Short Guide to Modern Star Names and Their Derivations*, Paul Kunitzch and Tim Smart, Otto Harrassawitz, Wiesbaden, 1986.

4.8. Star and constellation maps

The sixteenth century ushered in a three-hundred-year period that was a golden age of celestial cartography. The high art of globe-making arose about the same time, but here we will concentrate on maps. Although charts of the stars had been drawn long before, the first accurately plotted flat map was created in 1515 in the form of twin planispheres (Figure 4.15) by Albrecht Dürer. In the hands of this great engraver the celestial map was also turned into a work of art in which the mythological figures of the constellations are beautifully drawn. This tradition of coupling the ancient heroes and heroines of the sky to real and useful atlases was to continue for another pair of centuries. The figures are not just decorative but serve to provide informal constellation boundaries. The Dürer maps are printed backwards as we would see them on a globe, upon which we look from the outside.

A number of the constellations shown earlier are displayed by a succession of charts that illustrate this delightful melding of science and art. Seven are highlighted here. Unquestionably the most famous of all these atlases is the *Uranometria*, created by Johannes Bayer in 1603. The stars of the celestial eagle in Figure 4.8 carry both the Greek and Roman letters. Antinuoüs, held aloft here by Aquila, was not considered as a separate constellation and is figuratively as well as literally just along for the ride. The superb drawings were done by Jacobo de Gheyn for an earlier atlas created by Hugo Grotius. Borrowing of this sort was endemic. These figures are properly oriented as we would see them from Earth. Shortly after Bayer's work, in 1627, Julius Schiller introduced a set of maps without figures and then proceeded to overlay them with those drawn from Christianity (Section 4.4, Figure 4.10). The orientations of the constellations are again backwards, as they would be seen on a globe. Hevelius, that prolific creator of constellations, followed with several works that culminated in one with fine figures in 1690 (Figure 4.3). The stars are again reversed.

Then along came another master, John Flamsteed (see Section 4.7). His great atlas of 1729 was drawn as directly viewed, with right ascension increasing to the left, and included the Bayer letters. The work went through additional editions in 1753 and

(a)

1781. The figures were drawn by Sir James Thornhill and others. Orion and Taurus from the 1729 edition open this book as the frontispiece.

The work of Lacaille, who helped populate the southern skies, appears in Figure 4.6. This production of 1756 is centered on the south celestial pole, and it is interesting to compare it with Dürer's southern planisphere, which shows vast blank areas. Note that the two are also mirror images of one another, Lacaille's drawn as we see the sky from Earth. Here we see the Bayer letters extended south, Crux with its new letters (α–δ) and ε, ζ, ν, and ξ Centauri transferred to other stars.

This minimal survey is completed with two others: an atlas by John Bevis (produced in 1786) and one by the famous Johann Bode. Bevis drew his figures from De Gheyn (or from Bayer), so we see another example of these fine portraits in Figure 4.4. Each page is uniquely dedicated to an institution or an individual. Bode was a great popu-

(b)

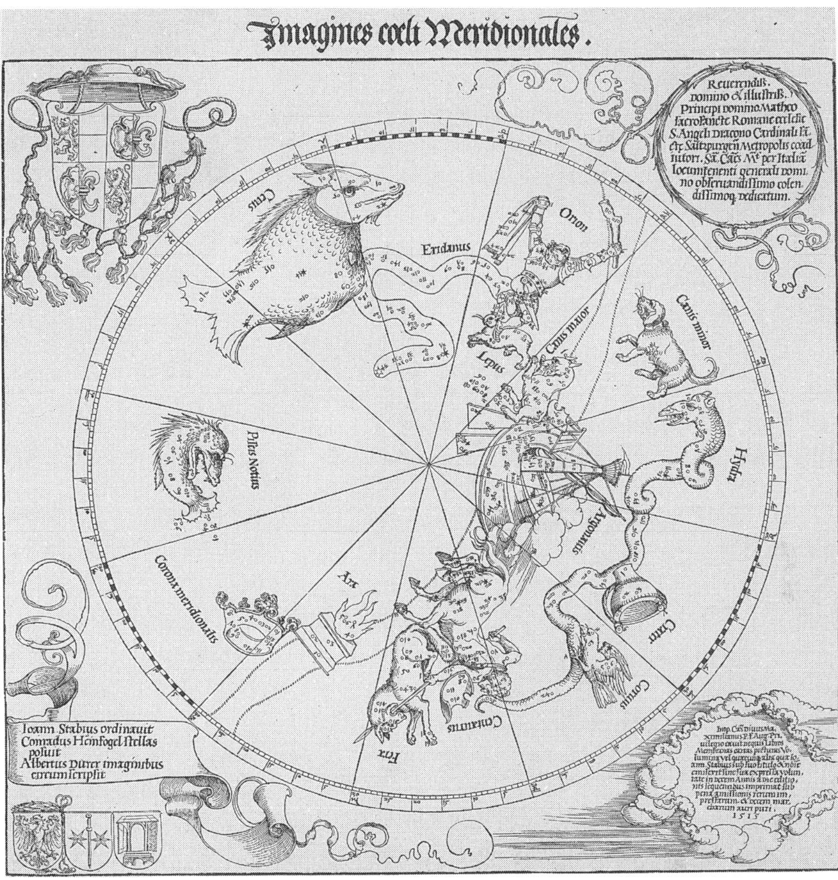

4.15. The planispheres of Albrecht Dürer. These first flat real star maps are a superb combination of art and science. The northern (a) and southern (b) hemispheres are centered on the ecliptic poles, the ecliptic running around the periphery. The figures are backwards, as you would see them on a globe. Note the paucity of constellations in the southern hemisphere, the gap centered roughly on the south celestial pole. (National Gallery of Art.)

larizer. A segment from his 1801 atlas is shown in Figure 4.5. These seven, taken together and compared, give the flavor of an era long gone.

The mythological figures are no longer drawn on our modern star charts. At best, the stars are sometimes connected on simplified maps to show some sort of outline. On the detailed amateur or professional maps now available there would simply be no room, even if we were so artistically inclined. A number of such atlases, described in Section 8.2, are available.

A set of elementary star maps is given in Appendix 2. These show all the 88 currently recognized constellations (though some are indicated by only their brightest stars) and the stars of Table 4.7 plus others that are easily visible and that may be needed to produce a recognizable constellation pattern. Lines that connect the stars to form recognizable figures are included as an aid to memory and identification. The maps also include an overlay grid of right ascension and declination, plus the ecliptic and celestial longitude. Two maps show the polar stars north of +50° declination and south of −50°; the other four are equatorial maps arranged by seasonal visibility. If you are unfamiliar with the skies, start with these maps and move on to the more advanced ones. And by all means, look at the sky through the eyes of the great cartographers presented to you above, and perhaps you will see the figures as they saw them, with Perseus slaying the monster or Antinuoüs forever flying with Aquila.

4.9. Stars and the Sun: the heliacal rising

"Star light, star bright, first star I see tonight . . .;" when do the stars come out, when do they disappear, and what is their relation to the daylight? Except during an eclipse of the Sun, or through the telescope, stars are never visible during daytime hours. The scattering of sunlight by the molecules of the Earth's atmosphere overwhelms the light of even the brightest of them. The first stars are seen shortly after sunset in bright twilight, but there is no unique answer to the above question. Their evening (or pre-dawn) visibility depends on local weather conditions and obviously on the brightness of a particular star. In clear weather, Sirius or Vega will appear within 15 minutes after sundown.

Two extreme cases of stellar visibility merit attention. The chief one of these is the *heliacal rising* of a star, which is its first appearance of the year in the dawn sky as it slips 1° per day to the west. The concept has many historical references, often of mystical significance. A heliacal rising might serve to mark the calendar, since for nearly a century at a time that of a given star occurs on a specific date. The foremost example is the heliacal rising of Sirius, which for the Egyptians of about 2600 BC predicted the rising of the Nile in June. The other extreme, the last glimpse of a star for the year in evening twilight before it is lost to the west, is called its *acronitic setting*. It is not so important, but it does have historical significance: Greek poetry refers to the planting of crops at the heliacal rising of the Pleiades and of harvesting at their acronitic setting.

The heliacal rising of Sirius no longer augurs the rising of the Nile; for us, Sirius is first seen in late August. The date is delayed by about one day every 70 years because of precession, which has been alluded to several times earlier. It causes a change in the direction of the celestial poles, and changes in the visibility of constellations and the locations of the equinoxes and solstices. We will examine this phenomenon in detail in the next chapter.

4.10. Distances

This book is largely concerned with stars as they appear on the celestial sphere, which tacitly implies that they are all the same distance away. It would be remiss, however, to ignore the fact that they are actually at immensely different distances. The problem of the distances to the stars is central to all of astronomy and astrophysics. It began to be solved in 1837 when Friedrich Bessel measured the first *parallax* (to 61 Cygni), the annual angular oscillation caused by the Earth's movement around the Sun (Figure 4.16). Parallaxes are exceedingly small: the largest belongs to α Centauri, and is only 0.76 seconds of arc. (The *parallax* is technically half the total displacement, and is the angular length of the radius of the Earth's orbit as seen from the star.) This value places α Cen four *light-years* (l-y) or 1.3 *parsecs* away. The light-year is the distance a ray of light travels in a year at the speed of 300 000 kilometers (186 000 miles) per second. The parsec, used in professional discussion, is the distance at which the radius of the Earth's orbit subtends an angle of one second of arc. The distance of a star in parsecs is just the reciprocal of the parallax and is equal to 3.26 l-y. Since Bessell's time we have developed a variety of measurement schemes and have determined the distances to thousands of stars. Those in Table 4.7 range from 4 light-years for α Cen to 1500 for Deneb, and beyond.

The apparent brightnesses of the stars are deceiving. The magnitude – more properly the *apparent magnitude*, *m*, that is listed in the table – depends on both the true brightness and the distance. Astronomers rank stars also by *absolute magnitude*, *M*, which is the brightness the star would have at a distance of 10 parsecs, or 32.6 light-years. If we could actually see the sky this way we would be astonished. Our brilliant Sun would shrink to fifth magnitude, Sirius would fade to Vega's brightness and great Deneb would appear at magnitude −7. It would outshine Venus, would throw a visible shadow, and could be seen fairly easily in broad daylight. This discussion, however, opens the door to astrophysics, which is not our subject. Instead, we return to the stars' appearances as we behold them.

4.11. Double stars

Look at Table 4.5 and the stars listed as "well-known pairs." Most of these are widely spaced couples that have some significance to or within a constellation but bear no real physical relation to one another. An exception is the Mizar and Alcor pair, the Arabs' "horse and rider," which represent a true physical double (see Figure 4.11). The stars are moving through space together and slowly revolve about each other, though the separation is so great and the orbital period so long that this motion has never been detected. If we look at Mizar with a telescope, we see something quite remarkable: what appears to be one star is actually two so close together (14 seconds of arc) that the eye sees them as the "single" star we call ζ UMa. If we now examine each of these with a spectrograph – a device that allows us to analyze the component colors of starlight – we can infer that both subdivide again into yet closer doubles. "Mizar" is actually *four* stars, and with Alcor the system contains *five*, all in motion about one another. Other examples include α Librae (magnitudes 2.75 and 5.15, 230 seconds of arc apart) and ε Lyrae (Figure 4.14). A superb eye will see the latter as a close pair of nearly equal brightness (magnitudes 4.68 and 4.50) separated by only 3.5 minutes of arc. A telescope reveals

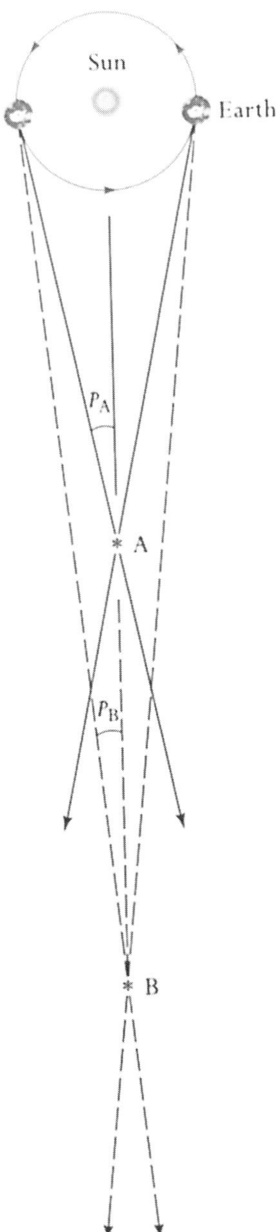

4.16. Stellar parallax. As the Earth orbits the Sun, nearby stars will appear to shift back and forth in position. The parallax is half the total shift angle. The farther the star, the smaller the shift. The distance in parsecs is the inverse of the parallax. (From *Stars*, J. B. Kaler, Scientific American Library, Freeman, NY, 1992.)

a remarkable sight, with each of these split again into components only a few seconds apart: the famous "double-double in Lyra."

Mizar and Alcor vividly illustrate the common binary natures of stars. Over 20% of the apparently single naked-eye stars break into components, or have at least dim companions, as viewed through the telescope, and with more sophisticated astrophysical analysis we find that over half of all stars are involved in double, triple, or more complex multiple systems. Stellar separations range from the above naked-eye pairs to those whose members are literally in contact with each other. A rather large number of the bright stars in Table 4.7 are therefore actually doubles or are even more complicated. Prime examples are α Cru, γ And, β Cyg, γ Leo, γ Vir, α Cen (a triple that includes dim Proxima Centauri), sextuple Castor, and several others. Double stars are extremely important to modern astronomy, as they are used to determine stellar masses, vital to our understanding of stellar evolution (see the discussion of Kepler's laws in Section 11.11).

4.12. Variable stars

The stars seem fixed, constant, and reliable to the casual observer. However, closer examination reveals several unusual characters, stars that do not shine steadily but change in their brightnesses. Some vary by wild extremes. Although the telescopic sky is heavily populated with them, the number of variable stars that can be followed with the unaided eye is relatively small; these, however, can be fascinating to watch.

The most famous variable is probably Algol, β Persei, aptly named the "demon star," which can be seen in the Medusa's head in Figure 4.5. It is normally of magnitude 2.12, but over a period of only a few hours it will suddenly drop to 3.4 and then as quickly recover. It is an *eclipsing binary*, two stars close together whose orbital plane lies almost along the line of sight. One component is bright but small, the other huge and dim. Every 2.9 days the bright one nearly disappears in eclipse behind the big one and the total brightness is cut to one-third normal. It is easy to see the event since α and β Persei are nearly the same brightness outside of eclipse, and consequently the contrast during eclipse is quite noticeable. Another bright example of an eclipsing binary is β Lyrae (see Figure 4.14), which varies between magnitudes 3.3 and 4.2 over a 13-day period.

Eclipsing binaries can take some odd forms, epitomized by two of Capella's "Kids," ζ and ε Aurigae. The first of these displays a 0.2-magnitude depression (which requires skill to detect) every 2.6 years that lasts for 38 days; the second drops by 0.8 magnitudes during an eclipse that persists for an astonishing 714 days every 22 years! The eclipsing body is not even a star, but apparently an orbiting cloud of dust. The next event will take place in 2009.

Eclipsing binaries only appear variable because of the geometry of the binary orbit. There are other classes of single stars that vary all by themselves because of pulsations in radius and surface temperature. The most famous of these *intrinsic variables* is the great star Mira, o Ceti, in the heart of the whale (Figure 4.17: another example of Flamsteed's work). The name alone, which means the "wonderful," or "amazing,"

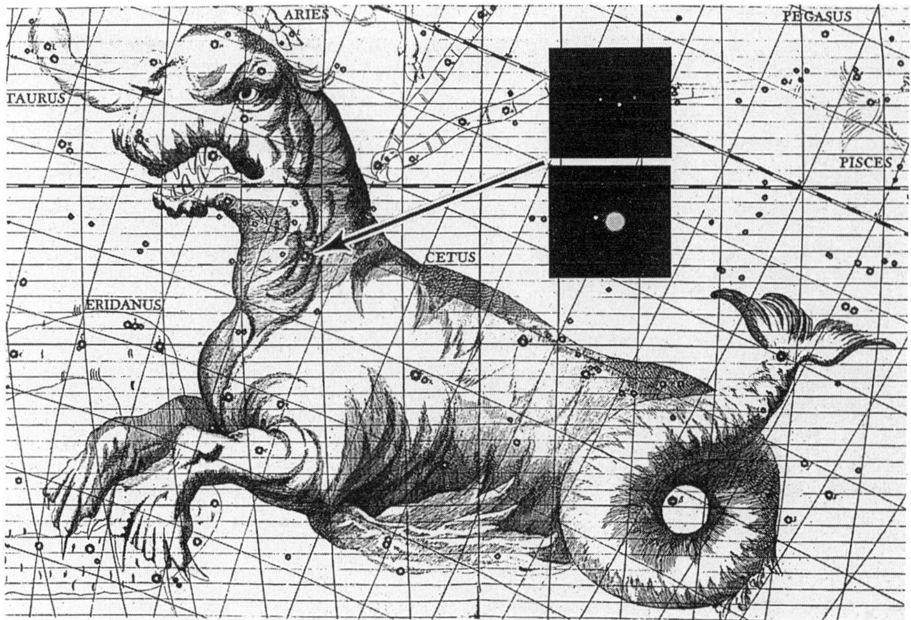

4.17. Cetus and the long-period variable Mira. The arrow points at Mira, seen in the heart of the whale in Flamsteed's 1781 edition of his *Atlas Coelestis*. The inset photos show the star at minimum and maximum about a half-year apart. (Constellation figure: University of Illinois Library; photographs: Lowell Observatory.)

attests to its strange nature. It normally varies between third and ninth magnitude, the latter far below naked-eye visibility, over a period of 330 days, nearly a year. On rare occasions it has brightened even to first magnitude. There are vast numbers of these *Mira*, or *long-period variables*. About a dozen are easily visible at their peaks.

Also accessible to view are the so-called *Cepheid* variables, named after their prototype δ Cephei. This remarkable star swings from 3.9 to 5.1 and back in 5.4 days with the regularity of a clock, steadily pulsing in radius. The variation is easy to follow because of the proximity of neighboring stars ζ and ε Cephei, which are of similar brightness. Other well-known examples are η Aquilae (in Antinuoüs's head: see Figure 4.8), which varies between magnitude 4.1 and 5.4 in 7.2 days, and ζ Geminorum, which goes between magnitudes 3.7 and 4.2 over a 10-day interval.

The brighter variables carry the normal proper, Bayer, and Flamsteed names, as do the other stars. But there are so many – thousands – of variables known that they have their own system of nomenclature. To organize them, Friedrich Argeländer began the use of Roman letters in the late 1800s. Some 300 years earlier Bayer had used Roman letters to supplement the Greek (first lower case then upper) but had not needed any past Q. So Argeländer started with R, which (except for those already named) denotes the first variable to be found in a constellation. However the number discovered rapidly drove the names to Z, whence the sequence was continued with double letters RR. . .RZ, SS. . .SZ, and so on to ZZ (deleting J). Still more variables were discovered

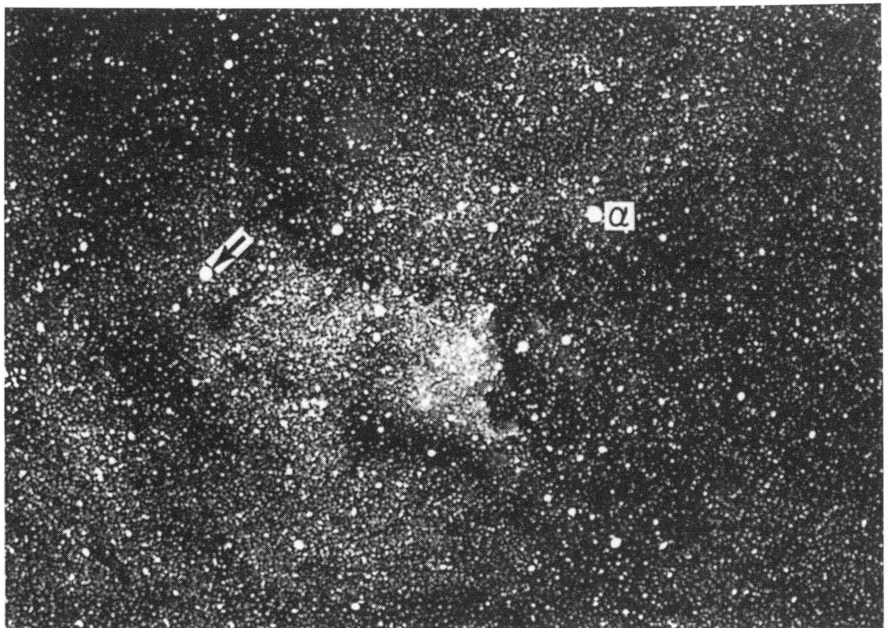

4.18. Nova Cygni 1975 (arrow) near Deneb (α) changed the appearance of the swan for a few weeks. The North America Nebula, a bright cloud of interstellar gas, is at center. (Allan E. Morton.)

so the chain then went to AA. . .AZ, BB. . .BZ, to QZ. That list constitutes 334 letter combinations. At that point, the astronomers gave up and followed QZ with V335 (plus the constellation possessive). Even though the Roman Bayer letters are rarely used, they still call clearly to us over the centuries.

4.13. Exploding stars

The ordinary variables are relatively benign, changing their brightnesses slowly and with some (often remarkable) regularity. There are two classes of variables, however, that are violent and explosive by nature, the *novae* and the *supernovae*. The words alone, "nova" being Latin for "new," reveal the stars' natures: one suddenly appears where none was seen before, and the supernovae are greater, vastly brighter than the novae.

Neither of these are common naked-eye sights. Although several novae erupt every year, none at this writing has been viewable without optical aid since 1976 – a typical wait is a decade or two. Nova Cygni 1975 (Figure 4.18) was a classic example, the "new" star nearly rivaling nearby Deneb. Several others are justly famous, such as Nova Persei 1901 and Nova Herculis 1934. We now know that these exploding stars are caused by matter that is passed from a normal star much like the Sun onto the surface of a highly evolved, nearly dead, close binary companion, one which has shrunk to the size of Earth. When enough matter has accumulated the surface layers explode like a thermonuclear bomb, causing the system to brighten by some dozen magnitudes over only a few days. We then see a brilliant temporary star that slowly decays to normal over a period of months or even years.

The supernovae are brighter yet as the entire star explodes, not just an accumulated layer. They may be either single massive highly evolved stars near death (far larger than the Sun) or odd binary configurations. Naked-eye supernovae are exceedingly rare, none being seen in our Galaxy – the local assembly of 200 billion stars that contains the Sun – since Kepler's Star of 1604. Previous to that, one was observed by Tycho Brahe in 1572. We are long overdue. Tycho's Star and the "guest star" of 1054 recorded by the Chinese were both visible at their peaks in *daylight*. So we await these grand events with considerable anticipation. The best we have seen in the telescopic era is the supernova that appeared in 1987 in the Large Magellanic Cloud (a companion galaxy: see Section 4.17). In spite of its huge distance of 170 000 light-years it reached third magnitude and was easily visible to the unaided eye. With the variables and the novae, however, we again skirt the realm of astrophysics. Yet these odd objects still contribute significantly to the naked-eye sky. The variables and erupters are as much a part of the celestial sphere as are the steady lights of the other stars, and they can be vastly entertaining to watch.

4.14. The Milky Way

How many of us have heard of the Milky Way, and how few in a modern society have really seen it! It comes as almost a shock and a revelation to a city dweller to see this marvelous, amazingly bright band of light spanning the sky (Figure 4.19). As prominent as it is on a dark moonless country night, it takes but a small amount of artificial lighting to extinguish its beauty and subtlety, if not to remove it from view altogether. Once we see it, though, we realize why it is woven into the folklore of the world.

The Milky Way is made of the combined light of billions of stars that inhabit the disk of our Galaxy (Figure 4.20), each separately too faint to see. Since the Sun is set into the galactic disk (about two-thirds of the way from the center to the edge), the Milky Way describes a rough great circle some 10° to 20° wide around the sky. A view through even a small telescope reveals this physical world to us, as it first did to Galileo. However, without a telescope, or any real facts of our true astronomical environment, the Milky Way inspires an instant sense of wonder, and produces sometimes lovely attempts at explaining why the gods put it there. The Greeks gave it the name The Milky Circle. Our "galaxy" comes from the root word in Greek for milk; note our word "lactate," to produce milk. To many peoples it is a celestial river, to some bringing the rain from the seas where it originates; to others it is a pathway of ghosts, the route to heaven.

Not only do the millions of faint stars collect together in this band, but the bright ones do as well. The number of naked-eye stars near the plane of the Milky Way far outstrips that perpendicular to it. Just compare Cygnus and Crux with, say, Ursa Major. In particular, for reasons that involve the way stars are born, the blue ones (like those in Scorpio and Orion) concentrate heavily toward this great disk. This band of blue stars is called *Gould's Belt* after the nineteenth-century astronomer B. A. Gould, and is slightly tilted relative to the Milky Way itself.

There is no physical, or causative, relation between the equator of the Earth and the equator of our flattened Galaxy. Thus the plane of the Milky Way is inclined at an odd angle to the celestial equator of about 63°, which takes its path not far from the celestial poles at extremum. The maximum declinations are over 60° north and south, and consequently, at temperate latitudes, part of the Milky Way is circumpolar, while another part never rises. The Milky Way and its center line, a true great circle called the *galactic equator*, are traced out on the star maps, and pass through such well-known constellations as Cygnus, Sagittarius, Orion, Centaurus, Crux, and Argo.

Like any great circle, the galactic equator has poles. The north galactic pole is located in Coma Berenices at $\alpha = 12^h\ 49^m$, $\delta = +27°24'$, the southern in Sculptor, $\alpha = 00^h\ 49^m$, $\delta = -27°24'$. (These coordinates are for the year 1950. The use of the date is required to account for precession, which will be explained in Chapter 5.) Both are indicated on the star maps. If you can locate these constellations, you can look directly out of the disk of the Galaxy, 90° away from the Milky Way. Upon the galactic equator we can construct a set of galactic coordinates that consist of *galactic longitude* and *galactic latitude* (l and b). Two zero points have been used. The older one, now outdated, employed the intersection of the galactic and celestial equators. The modern one, in which the equator was slightly adjusted, uses the center of the Galaxy in the constellation Sagittarius (see Figure 4.19). For a time during the 1950s, when systems were in transition, they were distinguished by calling the older (l^{I}, b^{I}), and the newer (l^{II}, b^{II}). The superscripts have now been dropped. The systems are included in the summary in Table 3.1. Transformations for any star can easily be made from equatorial to galactic coordinates and back again, and can be estimated from simple diagrams once the poles are located in the alternative system. The system is very useful in problems involving galactic research. Galactic longitude is indicated in the star maps.

Partly because of the Sun's off-center position within the galactic disk and partly because of the distribution of interstellar dust that blocks background starlight, the Milky Way's brightness changes considerably as we look around the galactic circle. It is brightest toward the galactic center in Sagittarius and generally falls off in luminosity in either direction. It remains quite bright, though, from Cassiopeia in the north, south through Cygnus to Sagittarius, and then past the south celestial pole to Carina. It is faint in the other direction between Cassiopeia through Canis Major.

The Milky Way's appearance is anything but smooth, its uniformity badly broken by vast interstellar dust clouds. These dust clouds are very prominent, and some like the *Coalsack* in the Southern Cross (see Figures 4.7 and 4.19), and the *Great Rift* in Cygnus–Aquila–Ophiuchus (see Figure 4.19) are quite famous. The natives of South America even made constellations of them (Figure 4.21). The Coalsack, for example, is a "partridge" to the Incas of Peru and Chile.

All these effects conspire to produce great hourly and seasonal changes in the Milky Way's visibility. From the northern hemisphere we see it best in late summer evenings when Cygnus, Aquila, and Sagittarius dominate the sky. In the winter, the Milky Way

(a)

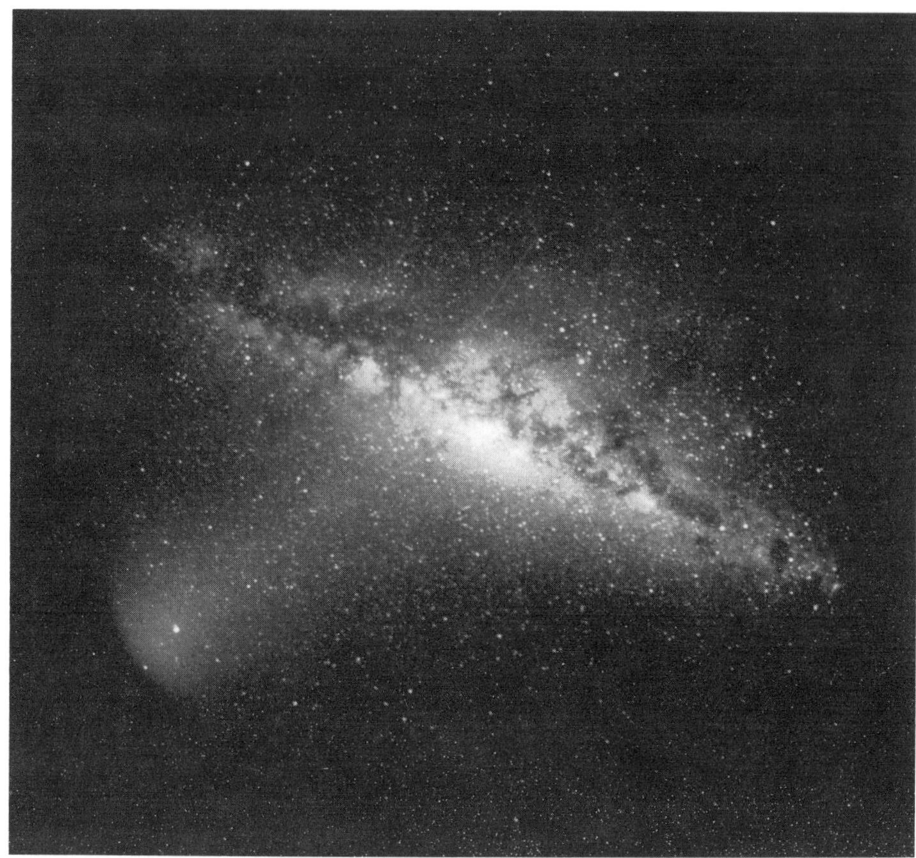

in Orion, Taurus, and Auriga is obscure, and with any light at all – moonlight or streetlamps – is impossible to see. Because of the Milky Way's high inclination to the celestial equator, great stretches of it are never seen by those living in moderate to high latitudes. The full glory of the Milky Way is really the province of the southern hemisphere. There, in South America, Australia, and southern Africa, the galactic center passes overhead, and the now brilliant band can be seen stretching off equally to both sides, an indescribable sight made of millions of stars, clusters, and nebulae, the Galaxy itself, viewed from near its edge.

4.15. Star clusters

Stars not only like to double and triple (and even sextuple) up, but they like to group together in larger families called clusters. Four have already been mentioned (Table 4.5 and Section 4.5) as important parts of constellations: the Pleiades (Figure 4.13), Hyades, Coma Berenices (Figure 4.9), and the Praesepe or Beehive. These are true physical associations in which the stars are bound together under their own gravitational

(b)

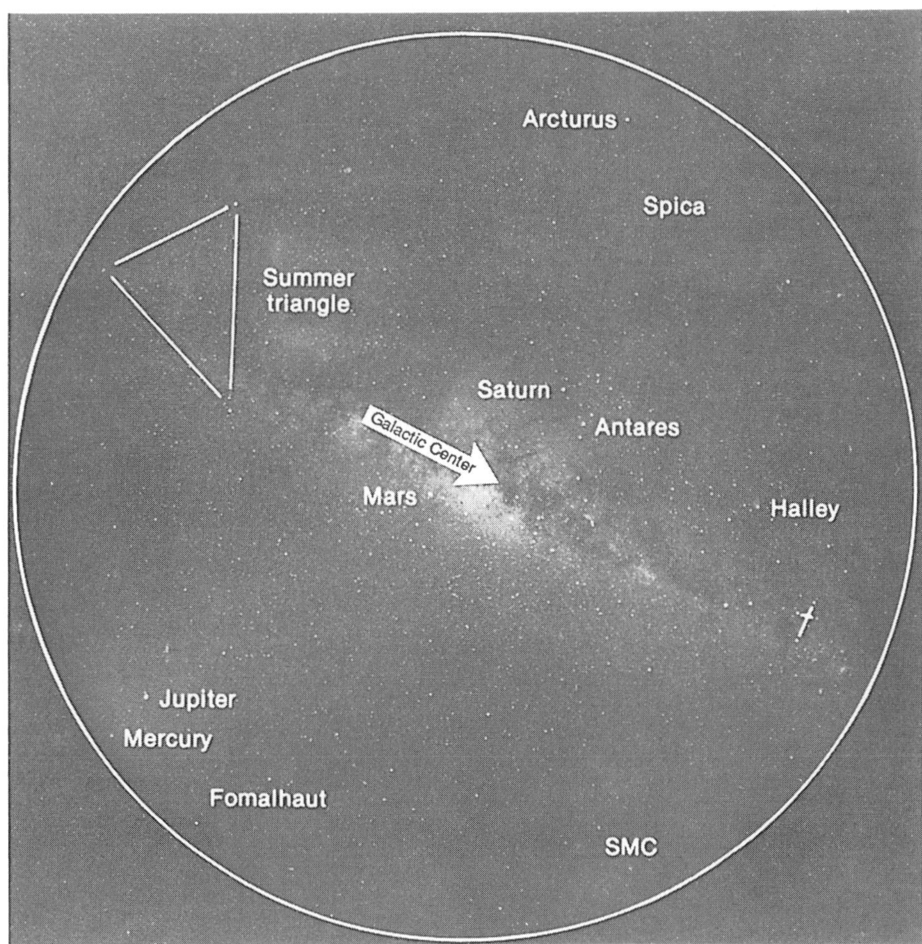

4.19. The Milky Way. (a) A spectacular wide-angle view of the disk of our Galaxy stretches from horizon to horizon. (b) A map shows prominent stars and planets and even Halley's comet. The Milky Way, which features the center of our Galaxy in Sagittarius, runs from the Northern Cross (at the left corner of the Summer Triangle) to the Southern Cross (indicated by a cross at lower right). The dark lane that seems to divide the Milky Way in two is caused by a thin disk of nearby obscuring dust that lies in the spaces between the stars. The northern part of the lane toward upper left is called the Great Rift. The dark cloud called the "Coalsack" is to the left of Crux. The SMC is the Small Magellanic Cloud, a nearby companion galaxy (Section 4.17). The streak across the center was caused by a passing meteor. The glow at the lower left is the zodiacal light, produced by dust in the Solar System reflecting sunlight (Chapter 13). (Dennis di Cicco.)

4.20. A diagram of our Galaxy. It shows the thin disk that contains most of its 200 billion stars, the Sun, interstellar gas and dust, and spiral arms, a common feature of galaxies like ours. We are inside the disk, set off to the side, so that it appears as a band of diffuse starlight (the Milky Way) around us. It is there that new stars are being born. The halo encloses an older, earlier generation of stars that includes the globular clusters and that was born before the disk developed. (From *Stars*, J. B. Kaler, Scientific American Library, Freeman, NY, 1992, art by I. Warpole.)

forces. Stars are commonly born in clusters, from which they later escape as the assemblies slowly dissolve with age. The naked-eye view of these objects is quite limited, as we see only the brightest members. These four collections are all called *open clusters* because of their relatively loose assembly and sometimes *galactic clusters* as they are confined largely to the plane of the Milky Way Galaxy where stars are currently being created out of the interstellar matter.

Several other open clusters are available for naked-eye viewing. In the northern hemisphere, look for the Double Cluster in Perseus, the only known case of a twin cluster, on a line south from γ and δ Cassiopeiae. Because of their brightness the two clusters are also known as h and χ Persei, showing one of the few cases in which a

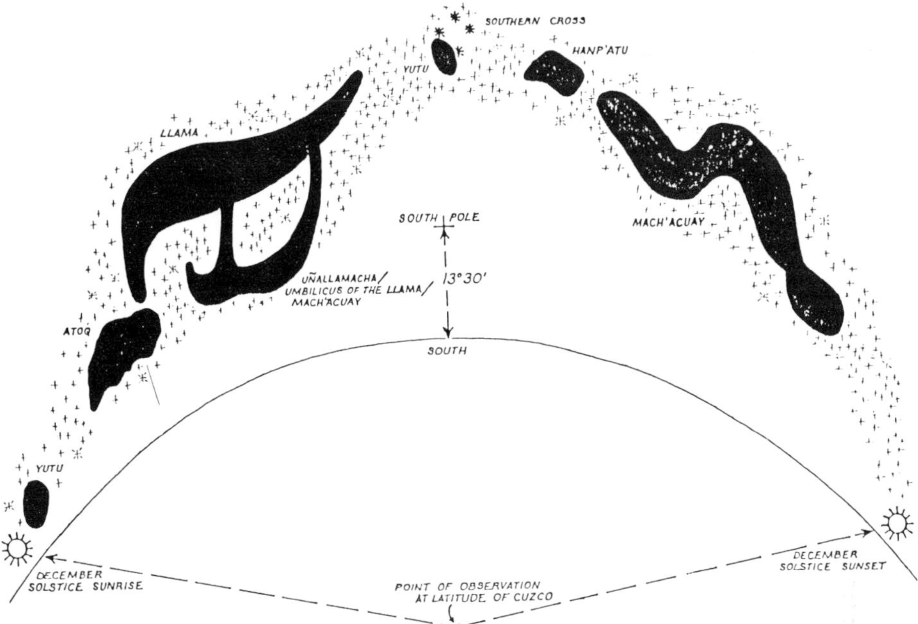

4.21. Dark constellations of the Milky Way, centered on the Coalsack in Crux. These were created by the Incas of South America. "Yutu" is a small game bird, "Mach'acuay" a serpent, "Hanp'atu" a toad, the Llama is obvious, and "Atoq" a fox. The dark features to the left of Crux appear on Figure 4.19; see if you can identify them. Dark constellations are unknown in the northern hemisphere, as the features are much less obvious. (From an article by Gary Urton in the *Proceedings of the American Philosophical Society*.)

lower-case Roman letter is still in use. The best of the rest reside in the southern hemisphere. The most notable include κ Crucis, or the Jewel Box, a bright compact cluster about a degree southeast of β Crucis. Another is Messier 7, an easy naked-eye sight that spans roughly 1° about 5° northeast of the Scorpion's stinger. Four more populate the southern Milky Way and are known only by catalogue numbers: NGC 2451 in southern Puppis, IC 2391 in western Vela, and NGC 2561 and IC 2602 (the Pleiades of the south) respectively in western and southern Carina. All present excellent sights in binoculars or small telescopes. Even a cursory survey of the Milky Way with binoculars will reveal many more.

Rarer to the naked eye, but far more spectacular to the telescope, are the *globular clusters*. These huge assemblies consist of not dozens or hundreds but of thousands or even hundreds of thousands of stars. They are all ancient, being among the first residents of our Galaxy, and distribute themselves around the galactic halo (see Figure 4.20) and concentrating toward the center. Only three can be seen without aid. In the north, a keen-eyed observer can just detect Messier 13 in Hercules' Keystone, between η and ζ Herculis. The best, however, are placed in the south and are bright enough to

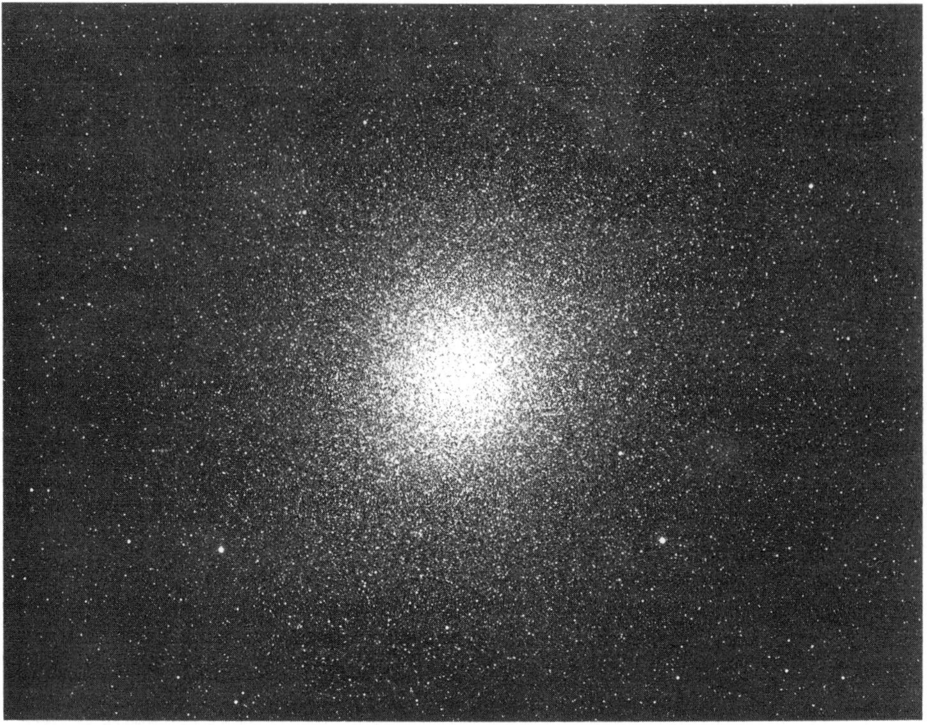

4.22. Omega Centauri. This greatest of all globular clusters appears as an obvious fourth-magnitude patch to the eye and contains nearly a million stars set some 17 000 light-years away. (National Optical Astronomy Observatories.)

carry ordinary star names: ω Centauri (Figure 4.22) and 47 Tucanae. Omega Centauri is a spectacular object easily visible without optical aid, and appears to be the grandest globular in the Galaxy.

The names of these *non-stellar objects* (anything not individual stars though they may be composed of stars) deserve further mention. The Messier appellation is derived from a catalogue of 103 objects compiled by 1784 by the French astronomer Charles Messier and extended by Pierre Mechain to 109. The purpose was to establish a permanent list of these extended patches of light so as to avoid confusion with comets. They are now favorite targets for the amateur observer. Even the Pleiades and the Praesepe have Messier numbers, M 45 and M 44. The first large list was the General Catalogue (GC) of non-stellar objects collected in 1864 by Sir John Herschel, the son of the great Sir William, largely from their own discoveries. This catalogue was followed in 1888 by J. C. E. Dreyer's *New General Catalogue* (NGC), which contains over 7000 objects, and the two *Index Catalogues* (IC) that add over 5000 more.

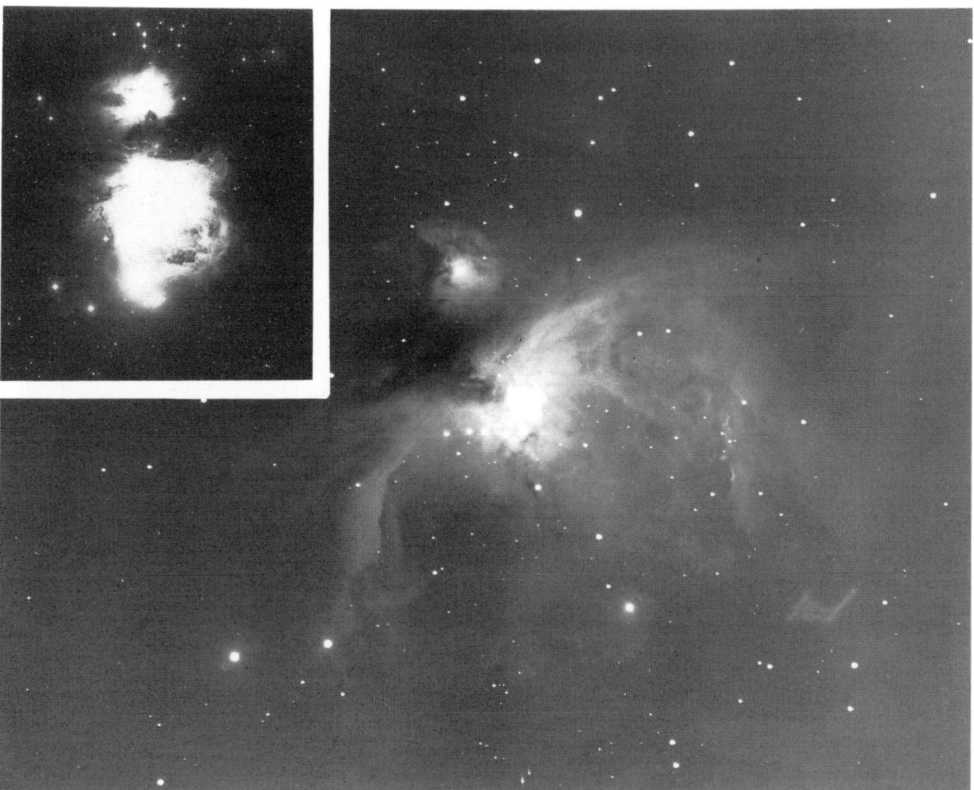

4.23. The great Orion Nebula, a massive cloud of interstellar gas. The main view shows the delicate inner structure. The famous Trapezium, θ^1 Ori (a quadruple star), is at the center of the brightest portion. (The odd-looking feature at the lower right is an internal reflection in the telescope.) The 1/6-scale inset displays the full extent of the object, which is some 20 light-years across. The upper nebula in the inset is NGC 1977, which is out of the field of view of the other picture. (Main photo: University of Illinois Prairie Observatory; inset: Palomar Sky Survey.)

4.16. Nebulae

The word *nebula* in Latin means "cloud" and is applied to glowing clouds of interstellar gas that are the birthplaces of stars and star clusters. The best known is the Orion Nebula (M 42, Figure 4.23), which is centered on θ^1 Orionis in the middle of the Hunter's sword. It is barely visible to the naked eye and stands out beautifully even with low-power binoculars. It fluoresces under the action of ultraviolet radiation from hot embedded stars. A telescopic view reveals extraordinary loops and whirls of gas with a quartet of stars (the Trapezium) at the center.

The unaided eye can also pick out the Lagoon Nebula (M 8) in Sagittarius about 6° north of γ. Again the southern hemisphere provides the showpieces that include the great Eta Carinae Nebula, a specimen some 2° across set within an extraordinary section

4.24. The Large Magellanic Cloud, a companion galaxy to our own. To the eye, it appears as a large fuzzy patch of light. To the telescope it contains vast numbers of stars and nebulae. (Harvard College Observatory.)

of the Milky Way. It can be seen as a bright patch at the far right edge of the Milky Way in Figure 4.19. Eta Carinae itself is not visible without a telescope. This remarkable star (not connected with the luminescence of the nebula) shone at first magnitude in the nineteenth century, and has since faded to invisibility as it has enshrouded itself with a dust cloud of its own manufacture.

4.17. Galaxies

The grandest of all non-stellar objects are *galaxies*, external star systems much like our own. They contain billions of stars apiece and are isolated with vast spaces in between them. Our own provides us with the stars of the nighttime sky and the glory of the Milky Way. A handful are visible to the naked eye. The best are the *Magellanic Clouds*, named after the explorer whose crew first brought news of them back to Europe from the southern hemisphere. They appear as diffuse patches of light, like detached sections of the Milky Way. The *Large Magellanic Cloud* (Figure 4.24) is about 7° across and

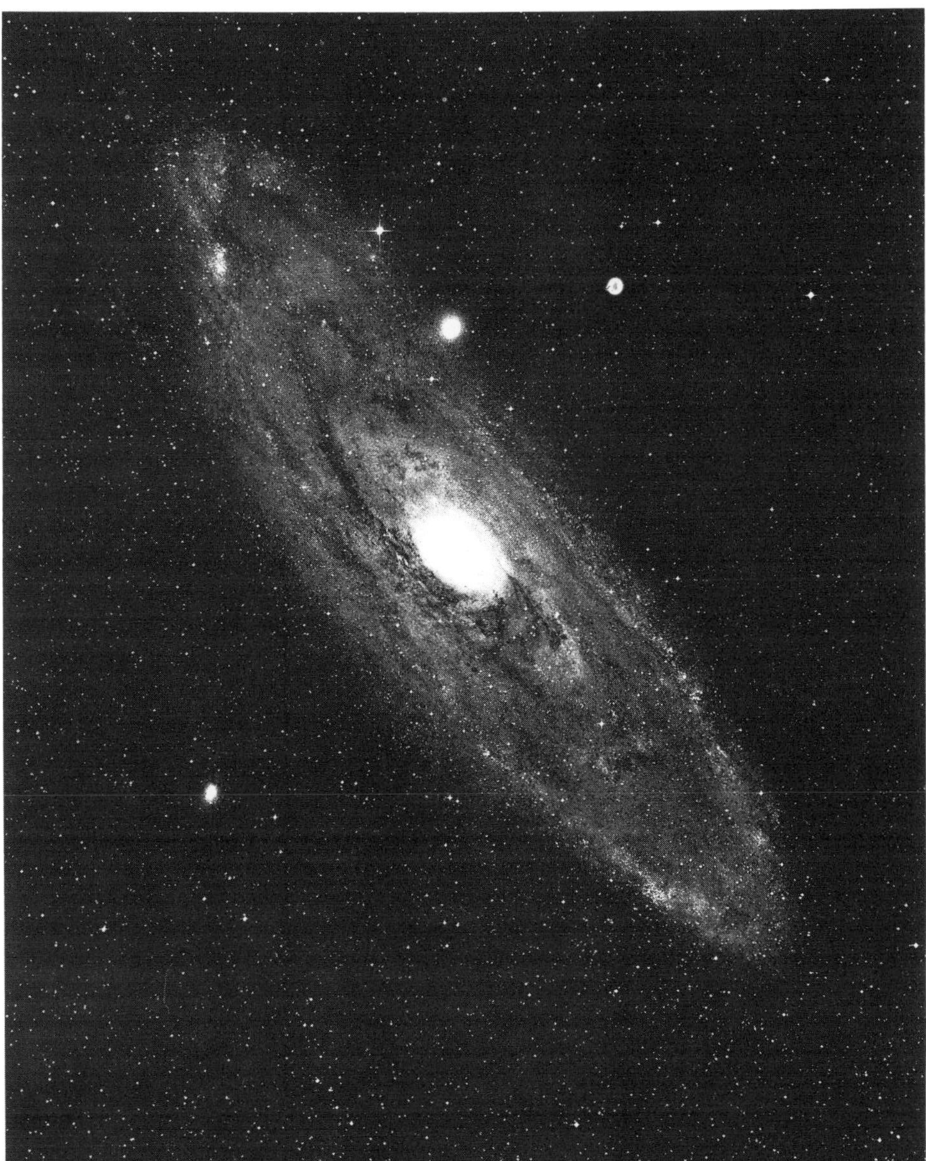

4.25. The nearby galaxy M 31 in Andromeda. Easily visible to the naked eye, it contains half a trillion stars 2.5 million light-years away. It is actually a circular disk like our own but tilted at a 70° angle. Only the bright central portion is seen with the naked eye. (National Optical Astronomy Observatories.)

about double the size of the *Small Magellanic Cloud.* (The Small Cloud can be seen in Figure 4.19.) Both are far south in Dorado and Tucana at $\delta = -70°$ and $-75°$ respectively. They are actually small companions to our own Milky Way Galaxy and are relatively close, only about 150 000 light-years away. As galaxies, they contain all the kinds of objects – clusters, variables, and nebulae – that are found in our own.

The northern hemisphere also displays two more galaxies for our naked eye view. These are farther away but are far grander in scale than the Magellanic Clouds. The principal one is *M 31*, the *Great Nebula in Andromeda*, an elliptical, 2°-long patch of light of about fourth magnitude near μ Andromedae (Figure 4.25). (The term "nebula" was applied to galaxies before it was known that they are resolvable into stars and is used today only for a few of the very brightest.) M 31, 2.5 million light-years away, is the most distant object that can be viewed on the vault of the sky without optical aid.

The last of them – M 33 in Triangulum – is very difficult to see and requires an extremely dark sky. It is about 1½° across and is located about midway between α Tri and β And. It is best to locate it first with binoculars. Then with averted vision (you will see a star better if you do not look directly at it but view slightly to the side), you should be able to pick it out. M 33 is roughly the distance of M 31, but is considerably smaller, i.e. it contains far fewer stars. With even a small telescope, hundreds of galaxies are accessible out to distances approaching a hundred million light-years.

Chapter 5

Precession, nutation, and aberration

We are familiar with the two primary motions of the Earth, axial rotation and orbital revolution, almost as soon as we are old enough to go outdoors and be aware of our environment. They are but two of many. The rest, however, are subtle. Only one is visible to the naked eye, and then only over a lifetime.

5.1 The third motion: precession

The next motion following rotation and revolution – alluded to in previous chapters – is *precession*, or more formally, *precession of the equinoxes*. It is a slow wobble of the Earth's axis about the perpendicular to the ecliptic plane that takes nearly 26 000 years to complete. At first one might wonder why we would bother with so slow a movement, but remember that people have been observing, and in at least some sense recording, the heavens for perhaps 5000 years, which is a significant fraction of the full period. Over historical time the changes wrought by precession are quite evident and must be taken into account when examining old data or documents. With modern positional measuring devices, an astronomer can actually sense the motion from one night to the next.

Precession's effects are profound. They include the change of the pole star, the constellations that are circumpolar from a given location, those that are not seen, the seasonal visibility of the stars, the mismatch between the astronomical constellations and the astrological signs, and even to some extent long-term climate changes on Earth. To the professional astronomer, the effect is of paramount importance in that precession and its attendant irregularities produce variations in the coordinates of stars that complicate the establishment of stellar locations and even affect such workaday tasks as setting a telescope. The phenomenon may not be ignored.

5.2 Polar flattening

The combination of the Earth's rotation and the tilt of its axis are responsible for precession. As a result of its rotation, the Earth is not perfectly spherical. The forces

5.1. The polar flattening of Saturn. The planet is viewed perpendicular to its rotation axis so that the famed rings are edge-on and invisible. The dark band across the middle is their shadow. The high equatorial speed, 20 times that of Earth, leads to an equatorial diameter 10% larger than the polar. Denser Jupiter is flattened by about 6%. (Lick Observatory.)

generated by our spinning planet cause it to bulge slightly at the equator where the rotation speed and apparent outward force is greatest, much as water may be thrown from a spinning wheel. The complementary effect is that the world will appear flattened at the poles, taking on the figure of an *oblate spheroid*. All rotating astronomical bodies are flattened. The rapid rotations of Jupiter and Saturn, with radii some 10 times that of Earth and days of only about 10 hours, cause polar flattenings easily seen through a small telescope (Figure 5.1). More extreme is the disk of our Solar System, which contains the planets. The original cloud that produced the Sun and its family flattened through rotation to form the ecliptic plane. Even the disk of the Milky Way owes itself to the spin of our Galaxy (see Figures 4.19 and 4.20).

The degree of terrestrial flattening is not large. The equatorial radius of the Earth, 6378 kilometers, is only 20 kilometers greater than the polar radius. The difference is enough, however, to cause a weight-driven pendulum clock to run noticeably faster at the pole than at the equator. The force of gravity at the surface of a planet depends on its mass divided by the square of its radius. A clock at a pole is 0.33% closer to the Earth's center than it is at the equator, and the gravity, as well as the acceleration of a falling body, is some 0.67% greater. Consequently the pendulum swings more quickly, and a clock transported from Ecuador to the Arctic will run several minutes fast per

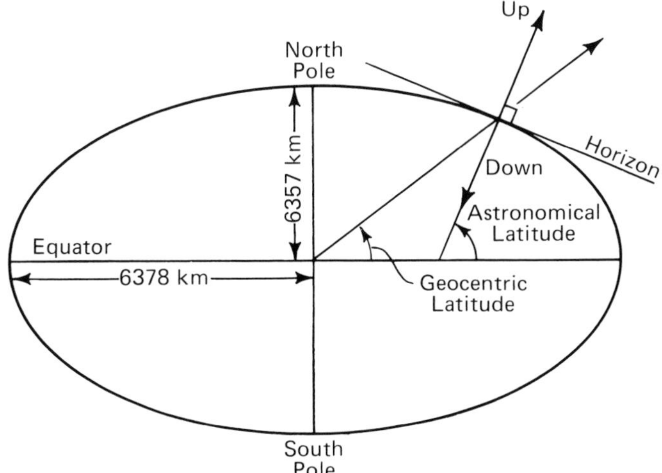

5.2. Astronomical and geocentric latitude. The different definitions are caused by terrestrial flattening. The first is measured relative to the perpendicular to the tangent plane, the latter relative to the line to the Earth's center. The figure is enormously exaggerated for clarity.

day. The rate of such a clock can in fact be used to measure the shape of the Earth. If the average person were to travel from the equator to the pole, he or she would become about one-half kilogram, or one pound, heavier.

The polar flattening of the Earth causes a dichotomy in the definition of latitude. We can define latitude either as the angle measured from the Earth's center (the *geocentric*), or as measured from the perpendicular to the Earth's surface (the *astronomical*), and because of the non-spherical shape of the planet, the two will not be the same (Figure 5.2). The astronomical latitude is the one normally used since it is directly accessible to observation.

5.3 The cause of precession

The bulge of extra mass around the Earth's equator is acted upon gravitationally by both the Moon and the Sun. In Figure 5.3, the Earth is shown at the time of the winter solstice. The Sun is only in the equatorial plane twice *per annum*, at the equinoxes, and the solar gravity over the course of the year applies a net force to the bulge in an attempt to straighten the Earth's axis and reduce the obliquity of the ecliptic. More importantly, the Moon, as we will see in Chapter 9, is always within 5° of the ecliptic plane, so that on the average it acts on the Earth just like the Sun. Because of its proximity to us, it has over twice the effect.

The attempt to pull the Earth's axis to the orbital perpendicular does not succeed, however, because the world is spinning, which imparts great stability. The net effect on our rotating planet is to cause its axis to *precess*, or to move perpendicular to the direction of the force. In Figure 5.3 the north celestial pole moves directly into the

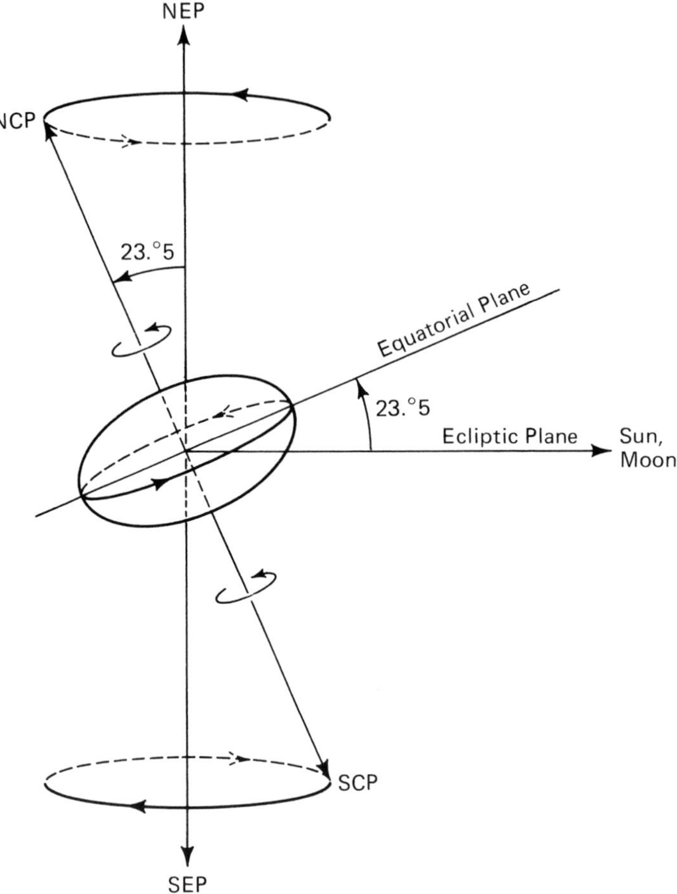

5.3 The precession of the Earth's axis. The combined gravity of the Moon and Sun on the equatorial bulge of the Earth exerts a force that causes the rotational axis to gyrate, or *precess*, about the orbital axis. The celestial poles therefore move around the ecliptic poles in a counterclockwise direction as viewed from Earth. In this drawing, the NCP is moving into the plane of the paper and the SCP out of it. The celestial poles will complete a full circuit in 25 800 years.

page. Since the motion of the Earth's axis is always perpendicular to the force, the axis will gyrate slowly about the orbital perpendicular, clockwise as seen looking down from the NEP and counterclockwise as viewed from Earth, with the obliquity remaining at a fairly constant $23^1/_2°$. This precessional wobble is very much like that of a spinning top whose axis rotates around the perpendicular to the floor. The observed period of precession, 25 725 years, agrees beautifully with that calculated from the science of mechanics. The fact of precession, however, has been known for far longer than its cause. It was discovered by the Greek astronomer Hipparchus of Nicaea in the second

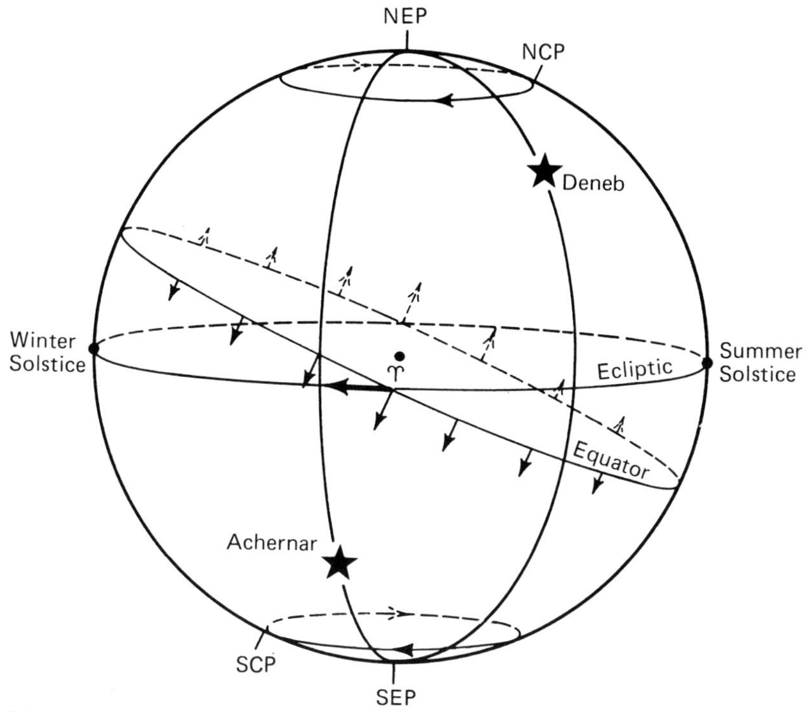

5.4. The motion of the celestial poles and equator resulting from precession. The NCP is coming directly out of the page as it precesses around the NEP; the SCP is going directly into the page. As a consequence, the equator moves downward in the front and upward in the back, shifting the vernal equinox to the left in the drawing, and to the right (west) in the sky. The declinations of stars increase (algebraically) on the front of the sphere, and decrease on the back. The rate of change in declination is a maximum at the equinoctial colure ($\alpha = 0^h$ and 12^h) and is zero at the solstitial colure ($\alpha = 6^h$ and 18^h). Right ascension generally increases except within regions above $\delta = +66\frac{1}{2}°$ for α between 12^h and 24^h and below $\delta = -66\frac{1}{2}°$ for α between $0^h + 12^h$, where it can decrease.

century BC, testimony to its obvious nature. To see how Hipparchus found this motion, we must next look at its effect on the sky.

5.4 Polar motion

The orbital axis of the Earth stays fixed in space, but its rotational axis continuously changes its direction (Figure 5.3). As a consequence, the north and south celestial poles must also change their positions among the stars, moving steadily around the ecliptic poles (Figure 5.4). The north ecliptic pole, NEP, is $23\frac{1}{2}°$ from the NCP in the sky, and once a day the NEP, in Draco, traces out its daily path around the NCP as would any star. Simultaneously, however, the NCP is slowly describing a small circle about the NEP, with counterclockwise motion as seen from Earth, with a period of nearly 26 000 years. If you could watch over a long period of time, the NCP would appear to drift counterclockwise around the NEP through the background stars. In the southern hemisphere we would see the reverse, that is, the SCP would travel *clockwise* around the SEP.

The Earth rotates, but we are not directly aware of it. Instead, we see the reflection of the rotation in the apparent daily motions of the stars. We are also not directly aware of the orbital revolution and see only its reflection in the annual path of the Sun around the ecliptic. Precession behaves similarly. At a given location, the celestial pole will always seem to be in the same place. At a latitude of 30° N, the north celestial pole must always be 30° above the northern horizon. What we would see, could we live long enough, is the reflection of the precessional motion. The stars would appear to move slowly *clockwise* past the NCP, rotating as on a plate around the NEP, and *counterclockwise* past the SCP.

The first thing we might notice on Earth is the change of the stars near the celestial poles, more specifically, the apparent movement of the celestial pole star. The path of the poles through the stars can easily be found by drawing a circle of 23½° radius around the respective ecliptic poles, the NCP appearing to move counterclockwise, and SEP clockwise. These paths are shown on the star maps in Appendix 2. The position of the poles at any time can then be simply determined by sectioning the circle into fractions of the full period. At the moment, the NCP points about 0°.5 away from Polaris, α Ursae Minoris. But for the Greeks of 1000 BC, β UMi, or Kochab, in the bowl of the Little Dipper, was the better pole star, at 6° from the NCP instead of its present 16°, and for the Egyptian civilizations of 2500 BC, α Draconis (Thuban) held that distinctive position.

From Polaris and Ursa Minor, the NCP will advance into Cepheus. Our next north pole star of consequence will be γ Cephei, which will pass within about 3° of the NCP in 2200 years, and in the year 7500 α Cephei will make a fair marker at about the same separation. Roughly 12 000 years from now, after the NCP passes δ Cephei, brilliant Vega will light the way north, as it travels a daily path 5° from the pole. The NCP then moves on to ι and τ Herculis (the latter to be a superb pole star) and finally back to Thuban where we began the historical record. Will the people of the time still recall that distant past? Use the stars to plot out this grand circuit on the nighttime sky.

This polar motion is obvious over only short intervals. Look again at Ursa Minor as depicted in the Bevis atlas of 1818, Figure 4.4. The stars are plotted for the position of the pole in 1744, which is about half as far away from α UMi as it is from δ UMi. Compare with the view on the north polar map of Appendix 2 in which the NCP is practically atop Polaris; it will actually improve as a polar marker into the twenty-first century.

The southern hemisphere is not currently blessed with a vivid pole star. All that is available is dim fifth-magnitude σ Octantis, which resides about 1° from the SCP. Around 1000 BC the Small Magellanic Cloud was less than 2° away. It must have been an impressive sight to its viewers. The SCP is currently within a long gap between decent polar markers, taking some 6000 years to go from β Hydri through Octans and unimpressive Chamaeleon to ω Carinae. It then has a rich path through Argo, past ι Car, γ Vel, L^2 (a variable star) and ν Pup, and in 15000 AD, past η Columbae. Then follows a lengthy period in which it goes through faint figures to α Hyi, whence it finally returns to the Small Magellanic Cloud. Note that stars now seen to mid-northerners – γ Vel, ν Pup, and η Col – will someday be invisible, and Vega, now seen in Australia, will disappear into the north polar regions.

5.5 Precession of the equinoxes

As the celestial poles move through the stars, the celestial equator, which is always 90° from the poles, must move as well. In Figure 5.4, the NCP is at this moment coming out of the page directly at the viewer. Consequently, the celestial equator must be moving downward on the front side of the sphere, and upward at the back. Since the ecliptic remains fixed, the vernal equinox, at the forward intersection, must move left as viewed from outside, or to the right as viewed from Earth. The autumnal equinox obviously moves in the same direction. The motion of the equinoxes must therefore be to the west along the ecliptic. Since the equinoxes take just under 26 000 years to make a full circuit of 360°, they must move at a rate of 50″.4 arc per year. Over the course of an average lifetime, the equinoxes will move just under one degree. The solstices are always in between the equinoxes and must follow along.

Remember that what we see on Earth is the *reflection* of this motion. At a given latitude, the equator, poles and ecliptic will maintain their same configurations relative to the horizon year after year; it is the *stars* that will *appear* to move in the opposite direction. Since the equinoxes and solstices are always on the ecliptic, the whole set of stars on the celestial sphere will thus appear to rotate parallel to that path, the constellations of the zodiac sliding *easterly* past the points that mark the seasons. With 12 zodiacal constellations, the equinoxes and solstices will on the average remain in any one for about 2000 years at a time. Currently, the vernal equinox, the most significant of the four points, since it heralds spring and rebirth, is in Pisces. For the classical civilizations of 2500 years ago, however, whose people devised our system of celestial circles, the vernal equinox was in Aries, which is why Aries has its position of prominence at the top of the astrological forecasts. This equinox is still referred to as "the first point of Aries," and its symbol, ♈, represents the ram itself. For the true ancients of 4000 years ago, the equinox was in Taurus. The great bull, the ancient fertility symbol, is therefore directly connected with the once beginning of Spring. In about the year 2700 the vernal equinox will cross the modern boundary of Aquarius, an event of some astrological significance. Table 5.1 gives the years in which the vernal equinox has resided or will reside within the different astronomical zodiacal constellations.

The three other ecliptic points have their significance as well. Libra, the balance, is opposite Aries, and in ancient times held the autumnal equinox, perhaps an allusion to the balance of days. Fall now begins with the Sun in Virgo. The summer and winter solstices are respectively now in Gemini and Sagittarius. Two and a half millennia ago, however, they were in Cancer and Capricornus, hence our names for the Earth's tropics.

Table 5.1. *The motion of the vernal equinox through the zodiac*

Aries	2000 BC – 100 BC
Taurus	4500 BC – 2000 BC
Gemini	6600 BC – 4500 BC
Cancer	8100 BC – 6600 BC
Leo	10800 BC – 8100 BC
Virgo	AD 12000 – AD 15300
Libra	AD 10300 – AD 12000
Scorpius[1]	AD 8600 – AD 10300
Sagittarius	AD 6300 – AD 8600
Capricornus	AD 4400 – AD 6300
Aquarius	AD 2700 – AD 4400
Pisces	100 BC – AD 2700

[1]Combined with the non-zodiacal constellation Ophiuchus, through which the ecliptic also passes.

The slippage of the equinoxes to the west is responsible for the current difference in position between the astrological "signs" and the astronomical constellations. The 12 signs parcel the ecliptic into equal 30° segments that once approximately overlaid the constellations and took on magical properties associated with the mythological figures. The signs are tied to the vernal equinox. They have been transported westward by one constellation yet still retain their original meanings to the superstitious. Thus the Sun was really in Sagittarius when all the Capricorns were born, and so on. We will look at this weird subject again in Chapter 11.

A variety of observations were included in Hipparchus's discovery and study of precession, one being the actual observed motion of the equinoxes. The right ascensions and celestial longitudes of stars cannot be derived from nighttime observations alone. They require the use of the Sun, which alone defines the vernal equinox, crossing it at a declination of 0°. The celestial longitude and the declination of the Sun are intimately related, as may be quickly seen from the star maps in Appendix 2. Measurement of solar altitude at noon thus immediately gives the solar declination and longitude. To be in the Earth's shadow, the eclipsed Moon must be exactly opposite the Sun in the sky and have a celestial longitude 180° different. Longitudes of stars can therefore be measured relative to the Moon during the eclipse and can be tied to the Sun and the vernal equinox.

Hipparchus found that the star Spica in Virgo had increased its longitude by 2° as compared with similar observations made some 150 years earlier by the Alexandrian astronomers Timocharis and Aristyllus. He thus understood that the vernal equinox must be slipping westward relative to Spica, and derived a precessional rate of 45 seconds of arc per year, quite close to the modern value. It is little wonder that Hipparchus is commonly considered to be the greatest of all ancient astronomers.

5.6 Changes in coordinates

As Hipparchus found, celestial longitudes increase with time as the equinoxes move steadily westward. Since the rotation of the celestial sphere is parallel to the ecliptic – easily seen in Figure 5.4 – the rate of change of stellar longitudes is uniform with time and the celestial latitudes do not change at all. Deneb, for example, must always be the same angular distance from the ecliptic. Precession is supremely simple if we look only at ecliptic coordinates.

The right ascensions of stars must change as well in response to the shifting equinoxes, but since they are measured along the moving celestial equator, the alteration will be more complex, and the declinations must change as well. In Figure 5.4 the equator and the equinox are both moving away from the star Deneb, and both α and δ are increasing. The equator is moving toward Achernar, and its declination is becoming less negative, and is therefore increasing as well (as is its right ascension). A star on the back of this sphere, however, would suffer a *decreasing* declination.

The numerical rates of precession in α and δ are difficult merely to estimate from the figure. Only at the equinoctial colure (the hour circles at $\alpha = 0^h$ and 12^h) and on the celestial equator are the rates simply expressible. In Figure 5.5 we look at a close-up of the vernal equinox formed from the intersecting ecliptic and equator. The drawing appears as seen on the sphere of Figure 5.4, and is reversed compared to the way the circles are seen in the sky. The change in declination is the simpler of the two to follow. In Figure 5.5 we see that the equator must drop by $\Delta\delta$ at the vernal equinox and, from Figure 5.4, that it must climb by a like amount at the autumnal equinox (Δ means "change in"). From the rate at which the NCP moves around the NEP (Figure 5.4) we can calculate that the downward rate of movement of the equator is $20''.0$ of arc per year ($\Delta\delta_o$) at ♈, so that at 0^h and 12^h declinations δ must respectively increase and decrease by that amount annually. At 6^h and 18^h, however, the equator is simply rotating about a point, and the annual change in declination, $\Delta\delta$, is zero. The value of $\Delta\delta$ must therefore drop continuously away from the vernal equinox and go through a minimum, the largest negative value, $-\Delta\delta_0$, at 12^h.

Because the celestial pole moves from 23½° on one side of the ecliptic pole to 23½° on the other side, all stars will show variations of twice that amount, or 47°, in their declinations over a precessional cycle. Polaris now has a declination of 89°, but 13 000 years from now the celestial pole will be on the opposite side of the ecliptic pole. Polaris's declination will then be 89° −47° or 42°, and it will pass nearly through the zenith in temperate northern latitudes. Declination changes were well known to Hipparchus. By comparing coordinates from measurements made some 200 years apart he concluded that they increased by as much as 1° on one side of the sky and decreased by the equivalent on the other, demonstrating that the precessional motion is parallel to the ecliptic.

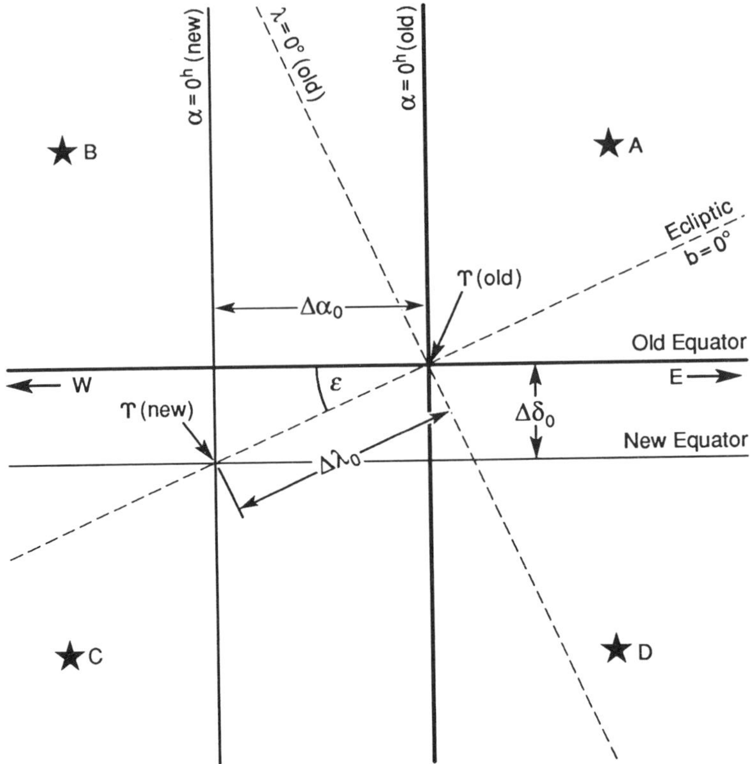

5.5. Precessional rates at the vernal equinox. The ecliptic and equator are represented as they would be on a globe (as in Figure 5.4), so that the Sun moves up and to the right. The equator moves downward by $\Delta\delta_0$ in a year (20 seconds of arc) causing the vernal equinox to move 50″.3 ($\Delta\lambda_0$) along the ecliptic to the left, or westward. Since the equator intersects at a 23½° angle (ϵ), the annual shift in right ascension ($\Delta\alpha$) must be smaller, 46″.1 or $3^s.1$ per year. This rate of declination change is valid only at the equinoctial colure (increasing declination at 0^h, decreasing at 12^h) and goes to zero at the solstitial colure. The equator is not displaced exactly parallel to itself, but tilts down and to the left as the NCP moves out of the paper in Figure 5.4. The new hour circles therefore tilt as well, so that the right ascension of stars *A* and *C* increase faster than $\Delta\alpha_0 = 3^s.1/\text{yr}$, and those of stars *B* and *D* increase more slowly.

The annual drop in the equator in Figure 5.5 through $\Delta\delta_o$ causes the intersection, the equinox, to be transported annually through an angle $\Delta\lambda_o$. This movement carries the equinoctial colure westward by $\Delta\alpha_o$, which since the ecliptic is tilted (by ϵ) relative to the equator, must be less than $\Delta\lambda_o$. From simple trigonometry (see Appendix 3) $\Delta\alpha_o$ is $\Delta\lambda_o$ times the cosine of the obliquity, $\varepsilon = 23° 26'$, which is 0.9175, so that in one year $\Delta\alpha_o = 0.9175 \times 50''.3 = 46.1$ seconds of arc, or 3.1 seconds of time.

For any star exactly on the equator ($\delta = 0°$ for any α) or any star exactly at $\alpha = 0^h$ or 12^h (for any δ), α will be incremented in a year by $\Delta\alpha_o$. The problem is that Figure 5.5 is only an approximation drawn on a flat map under the premise that the NCP moves uniformly downward from the top of the page, and that the equinoctial colure and hour circles stay parallel to their original positions, which they do not. As we see from Figure 5.4, the NCP is always moving toward the equinox, and consequently, the new equator actually tilts down and to the left, which causes the colure and hour circles to slant as well. As a consequence the precessional rates in right ascension will be larger or smaller than $\Delta\alpha_o$ depending upon the coordinates. In Figure 5.5, for example, stars *A* and *C* have rates greater than $\Delta\alpha_o$, and *B* and *D* less. For stars with positive declination, $\Delta\alpha$ first increases above $\Delta\alpha_o$ up to 6^h right ascension, decreases back to it at 12^h, then drops below $\Delta\alpha_o$ to 18^h, and back once again to $\Delta\alpha_o$ at 0^h. The effect is exaggerated with increasing δ, such that $\Delta\alpha$ can go negative between 12^h and 24^h north of 66½° N declination. The changes in $\Delta\alpha$ are reversed in the southern hemisphere, that is, $\Delta\alpha$ is smaller than $\Delta\alpha_o$ between 0^h and 12^h (and can be negative south of −66.5° declination), and larger between 12^h and 24^h. The positions of all stars are continuously changing, and consequently the precessional rates for each star are changing as well. Every star in the sky moves through an entire 360° of right ascension every 25 800 years, so the rate of precession must average out to $3^s.4$/yr. The precessional increments that take place over a 50-year period (a standard interval) are given in Table 5.2.

In the earlier development of coordinates in Chapter 3, it looked as if we had finally converged upon a solid stable system, the equatorial right ascension and declination, which were unchanging since the reference circles moved with the stars. Now, however, we find that α and δ are anything *but* fixed; the changes are just slow on a day-to-day basis. The system may seem needlessly complex, but in fact it is not. There are no stable reference frames (other than one fixed to the distant galaxies), and besides, the rates of change are accurately calculable with the aid of spherical trigonometry. A set of measurements of α and δ (or of α) are correct only for a particular instant. Consequently, it is necessary to assign to any catalogue of stellar positions (α,δ) an *epoch*, or a time for which the coordinates are valid. Star catalogues may be given for 1950.0 or 2000.0, where the decimal ".0" signifies the beginning of the year. If we know α and δ for a given epoch, we can always calculate what they should be for any other time since the variations produced by precession are very well known.

Although these changes may at first appear very small, precession is extremely important to anyone using a telescope (see Section 3.17). After only a few years an observer would be off by minutes of arc, and miss a sought-for star entirely. Most astronomers have been embarrassed at one time or another by wasting valuable telescope time vainly searching for an object that was not there, the instrument set in the wrong direction as a result of an incorrect epoch. Before the advent of computers, astronomers had to calculate the coordinates for the observing night from the catalogued positions by long equations or extensive tables. Modern computer-controlled telescopes, however, make the task of dealing with precession almost trivial. The computer contains a program that can solve the trigonometric equations that govern precession. Along with the catalogued right ascension and declination, the astronomer need only enter the catalogued α, δ, and the epoch, and the coordinates for "tonight" are instantly available. The hour angle is then computed from a built-in sidereal clock, and the telescope automatically driven to the proper position.

Table 5.2. *Precession over a 50-year interval*[a]

R.A.	Prec. in Dec.	Precession in Right Ascension δ =+85°	+80°	+75°	+70°	+60°	+50°	+40°	+30°	+20°	+10°	0°	-10°	-20°	-30°	-40°	-50°	-60°	-70°	-75°	-80°	-85°	Prec. in Dec.	R.A.
h m	′	m	m	m	m	m	m	m	m	m	m	m	m	m	m	m	m	m	m	m	m	m	′	h m
0 00	+16.7	+ 2.56	+2.56	+2.56	+2.56	+2.56	+2.56	+2.56	+2.56	+2.56	+2.56	+2.56	+2.56	+2.56	+2.56	+2.56	+2.56	+2.56	+2.56	+2.56	+2.56	+ 2.56	-16.7	12 00
0 30	+16.6	4.22	3.38	3.10	2.96	2.81	2.73	2.68	2.64	2.61	2.59	2.56	2.53	2.51	2.48	2.44	2.39	2.31	2.16	2.02	1.82	+ 0.90	-16.6	11 30
1 00	+16.1	5.85	4.19	3.64	3.36	3.06	2.90	2.80	2.73	2.67	2.61	2.56	2.51	2.45	2.39	2.32	2.22	2.06	1.77	1.48	0.93	- 0.73	-16.1	11 00
1 30	+15.4	7.43	4.98	4.15	3.73	3.30	3.07	2.92	2.81	2.72	2.64	2.56	2.49	2.40	2.31	2.20	2.05	1.82	1.39	0.97	+0.14	- 2.31	-15.4	10 30
2 00	+14.5	8.92	5.72	4.64	4.09	3.52	3.22	3.03	2.88	2.76	2.66	2.56	2.46	2.36	2.24	2.09	1.90	1.60	1.03	0.46	-0.80	- 3.80	-14.5	10 00
2 30	+13.2	10.31	6.40	5.09	4.42	3.73	3.37	3.13	2.95	2.81	2.68	2.56	2.44	2.31	2.17	1.99	1.75	1.39	0.70	+0.03	-1.28	- 5.19	-13.2	9 30
3 00	+11.8	11.56	7.02	5.50	4.73	3.92	3.50	3.22	3.02	2.85	2.70	2.56	2.42	2.27	2.11	1.90	1.62	1.20	0.40	-0.38	-1.90	- 6.44	-11.8	9 00
3 30	+10.2	12.66	7.57	5.86	4.99	4.09	3.61	3.30	3.07	2.88	2.72	2.56	2.40	2.24	2.05	1.81	1.51	1.03	+0.13	-0.74	-2.45	- 7.54	-10.2	8 30
4 00	+ 8.3	13.58	8.03	6.16	5.21	4.23	3.71	3.37	3.12	2.91	2.73	2.56	2.39	2.21	2.00	1.75	1.41	0.89	-0.09	-1.04	-2.91	- 8.46	- 8.3	8 00
4 30	+ 6.4	14.32	8.40	6.40	5.39	4.34	3.79	3.42	3.16	2.93	2.74	2.56	2.38	2.19	1.97	1.70	1.33	0.78	-0.27	-1.28	-3.27	- 9.20	- 6.4	7 30
5 00	+ 4.3	14.85	8.66	6.58	5.52	4.42	3.84	3.46	3.18	2.95	2.75	2.56	2.37	2.17	1.94	1.66	1.28	0.70	-0.40	-1.45	-3.54	- 9.73	- 4.3	7 00
5 30	+ 2.2	15.18	8.82	6.68	5.60	4.47	3.88	3.49	3.20	2.96	2.75	2.56	2.37	2.16	1.92	1.63	1.25	0.65	-0.47	-1.56	-3.70	-10.06	- 2.2	6 30
6 00	+ 0.0	15.29	8.88	6.72	5.62	4.49	3.89	3.50	3.20	2.97	2.76	2.56	2.36	2.16	1.92	1.62	1.23	0.63	-0.50	-1.60	-3.75	-10.17	0.0	6 00
12 00	-16.7	2.56	+2.56	+2.56	+2.56	+2.56	+2.56	+2.56	+2.56	+2.56	+2.56	+2.56	+2.56	+2.56	+2.56	+2.56	+2.56	2.56	+2.56	+2.56	+2.56	+ 2.56	+16.7	24 00
12 30	-16.6	+ 0.90	1.82	2.02	2.16	2.31	2.39	2.44	2.48	2.51	2.53	2.56	2.59	2.61	2.64	2.68	2.73	2.81	2.96	3.10	3.38	4.22	+16.6	23 30
13 00	-16.1	- 0.73	0.93	1.48	1.77	2.06	2.22	2.32	2.39	2.45	2.51	2.56	2.61	2.67	2.73	2.80	2.90	3.06	3.36	3.64	4.19	5.85	+16.1	23 00
13 30	-15.4	- 2.31	+0.14	0.97	1.39	1.82	2.05	2.20	2.31	2.40	2.49	2.56	2.64	2.72	2.81	2.92	3.07	3.30	3.73	4.15	4.98	7.43	+15.4	22 30
14 00	-14.5	- 3.80	-0.60	0.46	1.03	1.60	1.90	2.09	2.24	2.36	2.46	2.56	2.66	2.76	2.88	3.03	3.22	3.52	4.09	4.64	5.72	8.92	+14.5	22 00
14 30	-13.2	- 5.19	-1.28	+0.03	0.70	1.39	1.75	1.99	2.17	2.31	2.44	2.56	2.68	2.81	2.95	3.13	3.37	3.73	4.42	5.09	6.40	10.31	+13.2	21 30
15 00	-11.8	- 6.44	-1.90	-0.38	0.40	1.20	1.62	1.90	2.11	2.27	2.42	2.56	2.70	2.85	3.02	3.22	3.50	3.92	4.73	5.50	7.02	11.56	+11.8	21 00
15 30	-10.2	- 7.54	-2.45	-0.74	+0.13	1.03	1.51	1.81	2.05	2.24	2.40	2.56	2.72	2.88	3.07	3.30	3.61	4.09	4.99	5.86	7.57	12.66	+10.2	20 30
16 00	- 8.3	- 8.46	-2.91	-1.04	-0.09	0.89	1.41	1.75	2.00	2.21	2.39	2.56	2.73	2.91	3.12	3.37	3.71	4.23	5.21	6.16	8.03	13.58	+ 8.3	20 00
16 30	- 6.4	- 9.20	-3.27	-1.28	-0.27	0.78	1.33	1.70	1.97	2.19	2.38	2.56	2.74	2.93	3.16	3.42	3.79	4.34	5.39	6.40	8.40	14.32	+ 6.4	19 30
17 00	- 4.3	- 9.73	-3.54	-1.45	-0.40	0.70	1.28	1.66	1.94	2.17	2.37	2.56	2.75	2.95	3.18	3.46	3.84	4.42	5.52	6.58	8.66	14.85	+ 4.3	19 00
17 30	- 2.2	-10.06	-3.70	-1.56	-0.47	0.65	1.25	1.63	1.92	2.16	2.37	2.56	2.75	2.96	3.20	3.49	3.88	4.47	5.60	6.68	8.82	15.18	+ 2.2	18 30
18 00	- 0.0	-10.17	-3.75	-1.60	-0.50	+0.63	+1.23	+1.62	+1.92	+2.16	+2.36	+2.56	+2.76	+2.97	+3.20	+3.50	+3.89	+4.49	+5.62	+6.72	+8.88	+15.29	+ 0.0	18 00

[a]Adapted from The Observer's Handbook, an annual publication of the Royal Astronomical Society of Canada.

The Bayer names of the stars are unaffected by precession, however. The constellation right ascensions and declinations of the constellation boundaries (Section 4.7) were fixed for the epoch of 1875 and precess along with the stars.

5.7 The visibility of the constellations

The next effect of precession (and of course all these phenomena are intimately related) is a change in the visibility of stars and constellations for an observer at a given latitude. A star will alter its declination and move north and south over a range of 47° during the precessional cycle. The bright star Achernar, at 57° south declination, does not now rise at New York (latitude 40° N) where the southern horizon limit is 50° S. But note in Figure 5.4 that the north celestial pole and the celestial equator are moving toward Achernar, whose declination is now increasing (i.e. becoming less negative). Once again, at New York, as everywhere else, the latitude is not affected by precession, which means that the celestial equator appears fixed; the southern declination limit is always (90° – latitude). The stars *reflect* the precessional motion, and *appear* to move north and south. By about the year 3400 Achernar will be at declination −50°, and will be barely visible to future New Yorkers. By about AD 9000 it will reach its maximum declination of −35°, be visible from London and Winnipeg, and then reverse its trek and appear to move south. Alternatively, Deneb is now at $\delta = 45°$ N and can be seen low in the north from southern Chile. In 8000 years it will be a rather poor pole star with a declination of 84° N and people south of northern Peru will only be able to read about it.

Think of the concept from a different point of view. As the equinoxes and solstices slide westward, the zodiacal constellations appear to move continuously eastward, as if they were on a moving belt. The winter solstice, at a declination of 23½° S, is now in northern Sagittarius. From New York, we can see only 26 or so degrees farther south of Sagittarius when it transits the meridian; only Corona Australis and a little bit of Telescopium is visible. Sagittarius is now just starting its northerly motion; in 13 000 years, it will have moved halfway around the ecliptic to the position of the summer solstice at declination 23½° *north*. Instead of being low in the sky in the northern summer, it will be high in the sky in northern winter. As it transits, an observer in New York will then view some 70° south of Sagittarius, 23½° (from northern Sagittarius, where the winter solstice now resides, to the equator) plus another 50° (to the visible horizon). The dim stars of Octans, now surrounding the SCP, will actually be seen. It is intriguing to realize that 6000 years ago the Southern Cross was visible to the natives of North America.

Precession allows the people at a given location (were they to live long enough) to witness a vast extent of sky over the precessional period. Those living in the tropics, between 23½° N and 23½° S latitude, would be able to see the entire celestial sphere over the 26 000-year interval. People living north of the tropic of Cancer would see most of the sky except for a particular patch of radius (latitude −23½°) surrounding the south ecliptic pole and those south of the tropic of Cancer could see all except the stars within that angle of the NEP.

The effect of the polar wobble is detectable in the tilts of formal constellation boundaries. These were originally set along hour circles and parallels of declination for the epoch of 1875. An atlas constructed for the year 2000 clearly shows a slight rotation away from parallelism. The constellation Hydra provides another example of change. Some 3500 years ago it lay (perhaps deliberately) along the celestial equator. Look at it in Appendix 2 to see how much it now cants. The reality of precession can also be seen in the distribution of the ancient 48 constellations. Their southern boundaries are not concentric with the present pole but center near the Small Magellanic Cloud, which was near the pole between 2000 and 2400 BC. It can therefore be surmised that these constellations were invented somewhere near this time period.

5.8 The length of the year

How do you now define the length of the year? To measure the period of the Earth as it moves about the Sun, or the apparent period of the Sun's passage along the ecliptic, we need an external reference. There are two of them, one defined by the sphere of stars, the other by the vernal equinox, and they are moving relative to each other.

Precession leads to dual definitions for the length of the year. Because we want the seasons to synchronize with the calendar, the *calendar* or *tropical year* is defined as the interval between successive passages of the vernal equinox by the Sun, a length of 365.2422. . . days. However, with the vernal equinox moving westward, the tropical year must be slightly shorter than the true or *sidereal* year, the interval between successive passages of the Sun across a given star. The latter is 365.2564. . . days long, 20 minutes longer than the tropical year. That is, it takes the Sun 20 minutes to move through the 50″ annual westward displacement of the vernal equinox. Twenty minutes is an impressive length of time, and helps make the point that precession is not just a long-period event noticeable only over centuries, but is an annual affair, one of considerable short-term significance.

The sidereal year does not enter directly into civil timekeeping. However, it is the true year, the one that would be measured by an outside observer. As such, it is the basis for any calculations involving the motions of the other bodies in the Solar System. And, once again, enter Hipparchus. His actual discovery of precession derives from his observation that the year of the seasons was slightly less than 365¼ days, smaller than the Babylonian year – based on the stars – which was a bit greater than 365¼ days: a remarkable achievement more than 2000 years old.

5.9 Nutation

The complexities of the Earth's motion do not yet end. The precessional path of the celestial pole through the stars is not a perfect circle. It wobbles back and forth and varies in speed as it moves, and after the nearly 26 000-year passage it does not quite return to its starting point. For the interpretation of naked-eye observations, the perfect circle is generally adequate, but the other motions are conceptually interesting and their consideration is mandatory for detailed telescopic observations.

The deviation from the smooth circular path, the *nutation*, results from the variation of the gravitational forces that act on the Earth's equatorial bulge. The principal cause is the 5° tilt of the lunar orbit to the ecliptic plane (which we will discuss further in Chapter 9). The Moon dominates precession, so look at its motion alone. The Moon will actually cause the NCP to precess not about the NEP but around the pole of the lunar orbit, which is displaced from the NEP by 5°9′. However, the lunar orbit *itself* precesses around the NEP to the west with a period of 18.6 years. Consequently, the average center of precession will be the NEP, with the oscillatory action of the Moon's orbit producing an 18.6 year wobble in the NCP's path. The result is that the true NCP can be as much as 17″ ahead or behind the smoothly moving average NCP, and as much as 9″ off in the perpendicular direction. The effect of the 9″ deviation toward and away from the NEP also produces a periodic variance in the obliquity of the ecliptic.

There are a variety of other effects. First, the Moon and Sun are continuously moving back and forth across the terrestrial rotation bulge. The force acting to produce precession is at a maximum when these two bodies are at the solstices, and are zero when they are at the equinoxes and in line with the Earth's bulge and center. These variations produce tiny wobbles that have periods of half a year, half a month, a full year, and a full month (where the "month" here means the lunar orbital period). The eccentricities of the terrestrial and lunar orbits (their deviations from perfect circles) also cause variations in the gravitational forces that induce additional wobbles. The entire picture is extraordinarily complex and mathematically challenging: there are a total of over 100 recognized variations!

5.10 Planetary precession

Everything that we at first regard as fixed and unchanging seems upon closer analysis to move, true even of the ecliptic itself. The Moon and Sun are not the only bodies that influence the Earth gravitationally. So do the other planets. Because of their distances and (compared with the Sun) their low masses, their effect is small, but it is measurable. As seen from other planets, the Earth is tiny, almost point-like, and consequently the planets cannot exert a pull on the bulge nor affect the NCP. Instead, their combined gravities slowly change the position of the Earth's orbit, or that of the ecliptic plane, in space. The motion of the ecliptic plane produces a precession of the equinoxes analogous to (but opposite and only 0.2% of) that caused by the motion of the equator. We must now distinguish between the effects caused by the Moon and the Sun – referred to as the *luni-solar precession* – from those produced by the planets, the *planetary precession*. The luni-solar precession rate (smoothing out the nutation) is 50.4 seconds per year westward. The eastward planetary precession of 0″.12/year is then subtracted from the luni-solar precession to produce the *general precession* rate of 50″.3/year, which is the actual observed value and the one (intellectually – he could not have observed to that precision) appropriate to Hipparchus's great discovery and to Table 5.2.

A main point of interest is that the planetary influences on the Earth's orbital plane cause a slow change in the obliquity of the ecliptic, which varies between 24½° and 21½° over a 41 000 year interval. At present, the obliquity is decreasing at a rate of 0″.5 per year independently of nutational wobbles. Although this figure may seem insignificant to us, over a long period it could be noticeable to the unaided eye. In 1000 BC, the obliquity was closer to 24° rather than the 23½° now observed. Thus at the season of the summer solstice, the Sun rose and set slightly farther to the north than it does now. An attempt to use this effect to date Stonehenge, the ancient monument on the southern plains of England that is aligned to the extreme point of sunrise, has actually been made, but the precision of construction is apparently inadequate to produce a reliable result.

Over the full precessional period, these slow changes cause the path of the celestial pole not to close on itself. At the next passage of the pole to Polaris, it will miss by some 3°, instead of the present 0°.5. The astronomer quickly learns that nothing is truly stationary. There are no fixed reference frames; all we can do is recognize those that change very slowly, and evaluate the natures of their motions.

As their most profound effect, the precession, the change in the obliquity, and a slow periodic change in the shape of the Earth's orbit (collectively called the *Milankovitch cycles*) may together trigger long-term climate changes on Earth, and may be responsible for the ice ages that periodically glaciate much of the terrestrial surface.

5.11 Aberration of starlight

As if all these variations, of increasing levels of subtlety, were not enough, there is one more to consider, the *aberration of starlight*, a result of the orbital (and daily) motion of the Earth coupled to the finite velocity of light. Think of standing in the rain on a windless day with the raindrops falling straight down from overhead. Now move forward. Your motion combines with that of the falling water, which now, as it hits your face, seems to come from a point shifted ahead of the zenith in the direction of your motion. The faster you go the more the apparent direction shifts into your face, and if you drive rapidly in a car, the drops will seem to fly at you almost parallel to the ground (Figure 5.6a).

The same thing happens with light. Look at a star at the north ecliptic pole, in the direction perpendicular to the plane of the Earth's orbit and to the direction of our planet's motion (Figure 5.6b). Since the Earth is moving, the star will not actually be seen at the ecliptic pole, but will be shifted ahead of it. The angle will depend on the ratio of the Earth's orbital velocity, v (29.8 kilometers per second on the average, found by dividing the circumference of the orbit by the number of seconds in a year), to the speed of light, c (2.998×10^5 kilometers per second). Simple trigonometry gives the angle of displacement, 20.5 seconds of arc. It is a small effect, well below the capability of the naked eye to behold, but still comparable to the maximum displacement caused by nutation.

Because the Earth has a nearly circular path, a star at the NEP will trace out a small annual circle of radius 20″.5 about it, always shifting in the forward direction. Stars on

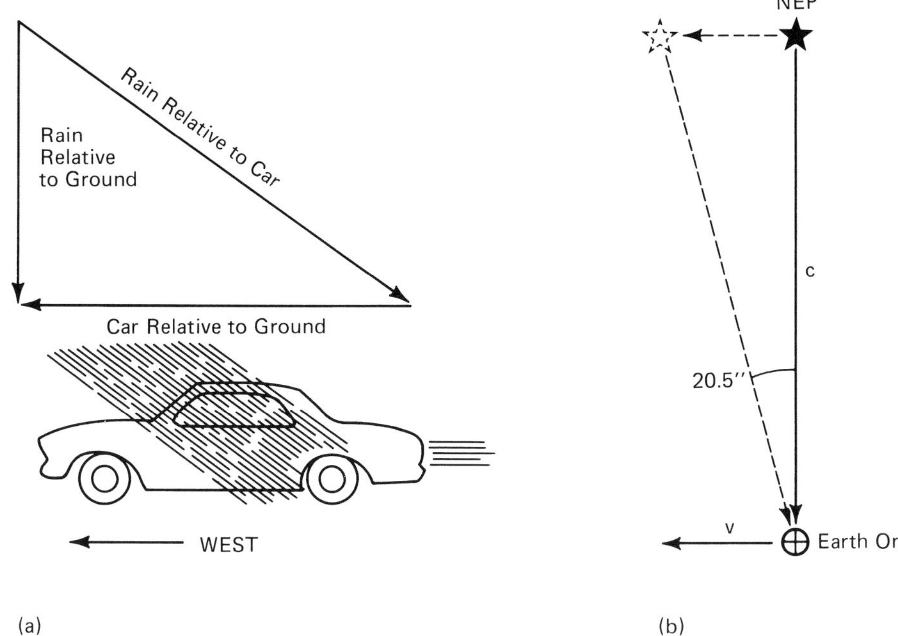

5.6. The aberration of starlight. Rain falling vertically onto a moving car (a) will appear to be coming from the forward direction. Light falling from "overhead" from the north ecliptic pole (b) will be shifted forward, so that it seems to come from an angle 20″.5 in front (relative to the Earth's motion) of the NEP. ((a) from *Principles of Astronomy*, S. P. Wyatt, Allyn and Bacon, Boston, 1964.)

the ecliptic will move back and forth along a line 41″ long. Those in between will trace out ellipses. The only stars that display no aberration will be those that lie on the ecliptic ($\beta = 0°$) directly in the Earth's path, which have longitudes of $\lambda_{\odot}$ (the solar longitude) plus or minus 90° (Figure 5.7). The maximum aberration must therefore take place for stars along the direction to the Sun, at $\lambda = \lambda_{\odot}$ and $\lambda = \lambda_{\odot} + 180°$ (for example, star *A* in Figure 5.7).

Aberration is symmetrical about the ecliptic and must be translated into displacements in α and δ through the application of spherical trigonometry. Even though annual aberration is not readily visible, the angular displacement is huge compared to our ability to measure stellar positions and must be taken into account. It was discovered in 1727 by James Bradley during his unsuccessful attempt to find the annual parallaxes of stars (the minute shifts caused by looking at stars from different directions as the Earth orbits: see Section 4.10). The aberration of starlight is considered to be a principal proof that our world actually does revolve about the Sun.

Even terrestrial rotation produces an aberrational effect. The Earth's rotation speed at the equator is 0.46 kilometers per second. When stars are on the meridian directed perpendicularly to the Earth's motion, they are shifted eastward by 0.19 seconds of arc: small, but still detectable. Daily aberration is zero for stars rising

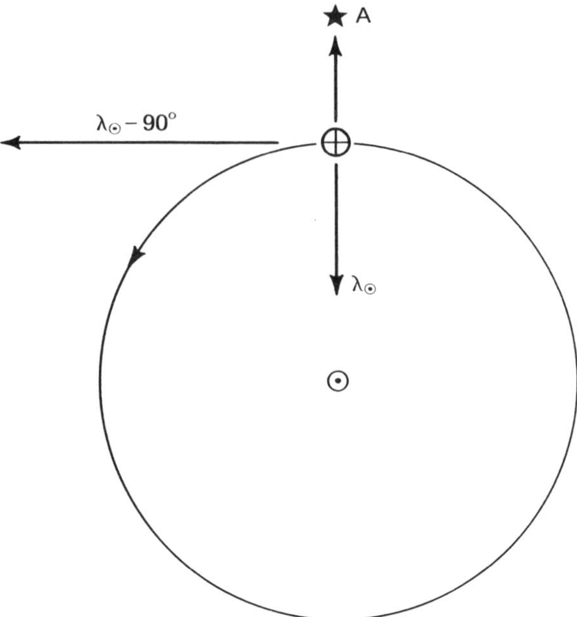

5.7. Aberration on the ecliptic. Stars *A* and *B* are on the ecliptic. Star *A*, perpendicular to the Earth's motion, will show the full aberrational shift. Star *B*, at a celestial longitude 90° less than the solar longitude ($\lambda_\odot$), will show no shift at all.

east and setting west, and diminishes toward higher latitudes as the apparent rotational speed drops.

Nothing seems stable; the stars are constantly shifting. We will encounter even more motions in Chapters 6 and 8. Yet remarkably, we know about these motions, and more importantly can correct for them, which restores us to some sense of serenity.

Chapter 6

Time

We accept the clock with little thought. If it runs down or stops as a result of a power failure we merely turn on the radio to get the correct time to reset it. But how do the radio stations know the time? Who tells them? How are the world's clock's set? Modern nations make use of a synchronized clock and timing system, one so precise that it can be used to help hurl spacecraft to distant planets to arrive on time after journeys of millions of kilometers at speeds of thousands of kilometers per second. Timekeeping to better than a millionth of a second, however, requires that numerous subtle effects be taken into account and increasingly complex definitions of what we actually mean by the correct time.

6.1. Local apparent solar time

We begin the tale with the simplest kind of time. Because we live our daily lives according to the Sun, the most obvious way to keep time is with the visible Sun itself. Solar time was defined in Section 3.7 as the hour angle of the Sun plus 12 hours. This kind of time is based on the real, observed Sun, the body the astronomers call the *apparent sun* and, as pointed out before, it is therefore called *apparent solar time*.

For the moment ignore the Earth's rotation, that is, in your mind, bring it to a halt so that the Sun stops in its daily path. If you walk to the east through a specific arc of longitude, the stars and Sun would appear to shift to the west in hour angle by the same amount. As you move east, then the time gets later by the amount you have moved in longitude (the time in Los Angeles is earlier than that in New York). As a specific example, if you walk 15 seconds of arc in longitude to the east (only about 460 meters at the equator; less elsewhere), the time will increase by one full second (see Table 2.1), and if you walk a degree, the time changes by 4 minutes. Conversely, if you move west, the Sun moves east, and the time gets earlier. Longitude differences and time differences are therefore seen to be equivalent. Time is clearly a *local* measurement, and the time told by the hour angle of the apparent sun is properly called *local apparent solar time*, LAST. The first rule in time measurement is thus

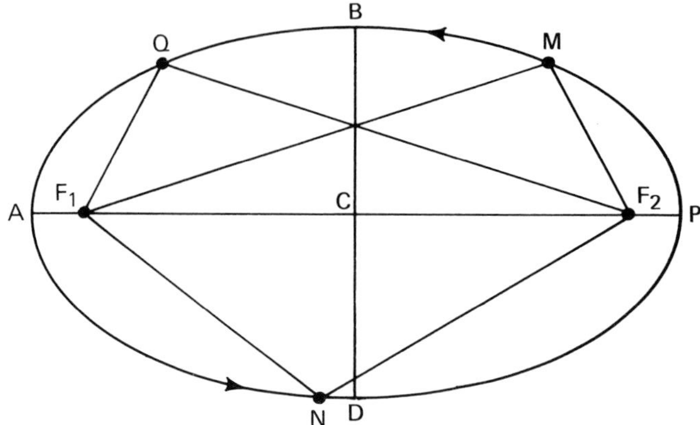

6.1. The ellipse. The sum of distances from any location on the curve to the two foci (F_1 and F_2) is the same ($MF_1 + MF_2 = NF_1 + NF_2 = QF_1 + QF_2$ etc.). The size of the ellipse is given by the semimajor axis $CP = CA = a$, the eccentricity by $e = CF_1/CP = CF_2/CA$. If the ellipse is a planetary orbit, the Sun is at one focus and the other is empty. If the Sun is at F_2 then P is perihelion, the point of closest approach, and A is aphelion, the point of greatest separation.

$$\boxed{\text{LAST} = \text{hour angle of the apparent sun} + 12 \text{ hours}} \quad .$$

LAST is not, however, the time we use in daily life.

6.2. The Earth's orbit

Anyone who has used a sundial knows that it rarely reads the "correct" watch time and may be as much as an hour in "error." In fact, the sundial is absolutely accurate; it is just that the observer's watch does not keep local apparent solar time. There are two reasons. The first involves the orbit of the Earth. Like all planetary orbits, it is elliptical, a fact noted by Johannes Kepler in 1609 (and part of a statement known as *Kepler's first law of planetary motion*, which will be discussed in more detail in Chapter 11).

An ellipse is defined as a curve for which the sum of the distances from all points on the curve to two interior points, each called a *focus*, is constant (Figure 6.1). F_1 and F_2 are the foci, and points Q, M, and N are three randomly selected locations on the curve. The distance $QF_1 + QF_2$ equals $MF_1 + MF_2 = NF_1 + NF_2$, and so on. You can easily draw an ellipse by putting two tacks in a board and connecting them with a loose string. Pull the string taut with a pencil placed, for example, at point Q in Figure 6.1, with the string stretched from F_1 to Q and Q to F_2. Then move the pencil, keeping the string tight and you will trace out an ellipse.

Ellipses are characterized by a pair of numbers that express size and shape. The longest line that can be drawn within an ellipse, which in Figure 6.1 extends between points A and P through the foci and the center C, is called the *major axis*. The *minor axis* is perpendicular to it, and goes between B and D through the center. The size of

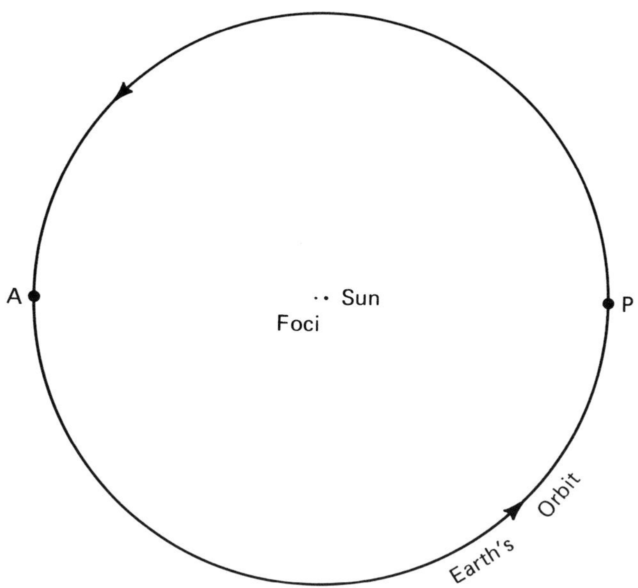

6.2. The Earth's orbit drawn to scale. The two dots at the center are the two foci. The center of the orbit is exactly between them and unmarked. The Sun, drawn to scale, is arbitrarily placed at the right-hand one. Looking down from the north ecliptic pole, the Earth orbits counterclockwise, and P is perihelion and A aphelion. The eccentricity is only 0.0167, so the ellipse is very close to being a circle.

the ellipse is given by the *semimajor axis*, a, which is one-half the length of the major axis. The shape, or degree of flattening, is given by the *eccentricity*, e, the ratio of the distance from the center to a focus (CF_1 or CF_2) divided by a. The eccentricity of the ellipse in Figure 6.1 is 0.83. If the two foci are brought together at the center, the ellipse reduces to a circle, CF_1 is zero, and $e = 0$. If the distance CF_1 is increased, it comes ever closer in dimension to a, and the eccentricity, e, approaches 1. All ellipses have eccentricities between 0 and 1.

In all planetary orbits, the Sun is at one of the foci of the ellipse (the other part of Kepler's first law). The other is an unimportant empty point in space. The astronomical unit (Section 3.1), the distance to the Sun (149.6979 million kilometers), is the semimajor axis of the terrestrial orbit and is the *average* solar distance. The eccentricity of our orbit is (fortunately) quite low, only 0.0167. The foci and the Sun are therefore only 2.50 million kilometers from the orbital center, less than twice the solar diameter (Figure 6.2). The minimum distance occurs where the Earth crosses the major axis at P, a point called *perihelion* (from the Greek meaning "closest to the Sun"). The distance then must be a minus the distance between the center and a focus, or (149.6 – 2.5) million kilometers = 147.1 million kilometers (or 91.4 million miles). The maximum distance, when the Sun is at A in Figure 6.2, at *aphelion* (meaning "farthest from the Sun"), must then be (149.6 + 2.5) million kilometers or 152.1 million kilometers (94.5 million miles). The Earth passes perihelion about January 2, and aphelion about July

4. As pointed out in Section 3.10, the variation has no significant effect on the severity of the seasons.

Of much greater importance is that the changing distance gives rise to a variation in the Earth's orbital velocity (described by *Kepler's second law*, also to be discussed in Chapter 11). The greatest speed (30.4 kilometers per second) occurs at perihelion in January and the smallest (29.4 kilometers per second) at aphelion in July. As a consequence, the apparent angular rate of the Sun as it moves along the ecliptic is also variable. From Section 3.13, the Sun moves at a speed of 0°.9856 per solar day. This value is an average: the actual angular speed varies from 1°.002/day in January to 0°.9701/day in July.

The most obvious result of the irregular angular motion of the Sun is the *inequality of the seasons*. It takes the Sun a shorter period of time to go from the vernal equinox to the autumnal than from the autumnal to the vernal. Northern hemisphere spring and summer (each 93 days) are longer than autumn and winter (90 and 89 days respectively). More subtly, the velocity variation also causes a 2% variation in the degree of the aberration of light over the course of the year.

6.3 Local mean solar time and the equation of time

Since the apparent sun moves at a variable rate along the ecliptic, it cannot be an accurate clock. If you were to try to synchronize a mechanical clock to the apparent sun, you would continuously have to change its rate to follow the speeding, then slowing, Sun. When the Sun moves the fastest to the east on the ecliptic the day is slightly prolonged. Thus the solar day (the interval between successive meridian passages) is longer in January than July; your clock would have to be made to run slower in January to fill the longer day. Since there are a constant number of seconds in a day ($24^h \times 60^m \times 60^s = 86\,400$), the length of the second would therefore have to be longer in January than in July.

Even if the Earth's orbit were a perfect circle and the apparent sun did move at a constant rate it would still be unsuitable. The reason is that the hour angles, from which time is determined, are measured on the celestial equator, whereas the apparent Sun moves on the ecliptic, which is pitched at an angle to the equator because of the tilt of the Earth's axis. The hour circle through the Sun moves faster when the Sun is near the solstices than when it is near the equinoxes. The result again is a day and second of inconstant length. Thus the simple notion of solar time turns out to be rather complex.

An irregularly moving clock with a variable-length second is not tolerable in a world in which it may be necessary to keep time to one-millionth of a second. The apparent sun is no longer used in timekeeping. To resolve the problem of an inconstant day, astronomers and navigators invented an imaginary *mean* (average) *sun* that moves along the celestial equator at a constant rate. The mean sun represents the average position of the apparent sun's hour circle. The mean sun is used to define *local mean solar time*, LMST, which, in parallel with LAST, is

$$\boxed{\text{LMST} = \text{hour angle of the mean sun} + 12 \text{ hours}}$$

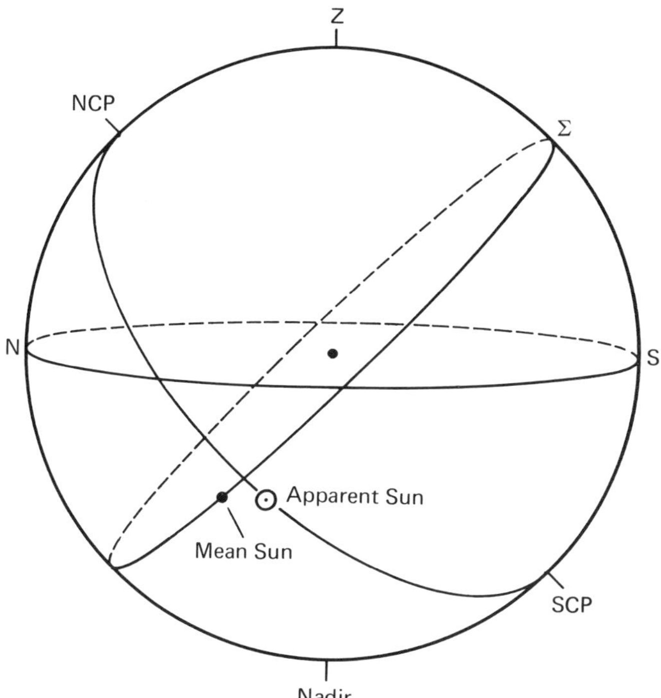

6.3. Apparent and mean solar time. The local apparent solar time is the hour angle of the apparent sun plus 12 hours. The hour angle is 3 hours, so that the LAST is 15^h. The solid dot on the equator represents the mean sun, which is ahead of the apparent sun for this diagram set in mid-February. The local mean solar time is the hour angle of the mean sun $+12^h$. The mean sun's hour angle is about $3^h\ 14^m$, so that the LMST is $15^h\ 14^m$, and the equation of time is -14^m, as negative as it gets.

Local mean solar time has a second of constant length and keeps an average pace with local apparent solar time, that is, the mean sun moving at its uniform pace completes a tropical year in the same interval as the apparent sun moving at its variable rate.

If we could see both the apparent and mean suns in the sky we would see one getting ahead of the other in hour angle, then falling behind (Figure 6.3). The difference between local apparent solar time and local mean solar time depends on the difference in hour angle between the apparent and mean suns and is expressed by a quantity called the *equation of time*, *E*,

$$\boxed{E = \text{apparent solar time} - \text{mean solar time}} \quad .$$

To derive the equation of time we must look at each of its sources individually (Figure 6.4). First, correct for the eccentricity of the Earth's orbit by adopting a fictitious *intermediate sun* that moves uniformly along the ecliptic and define a *local intermediate solar time*, LIST, as the hour angle of the intermediate sun plus 12 hours. The intermediate and the apparent suns start at the same position when the Earth is at

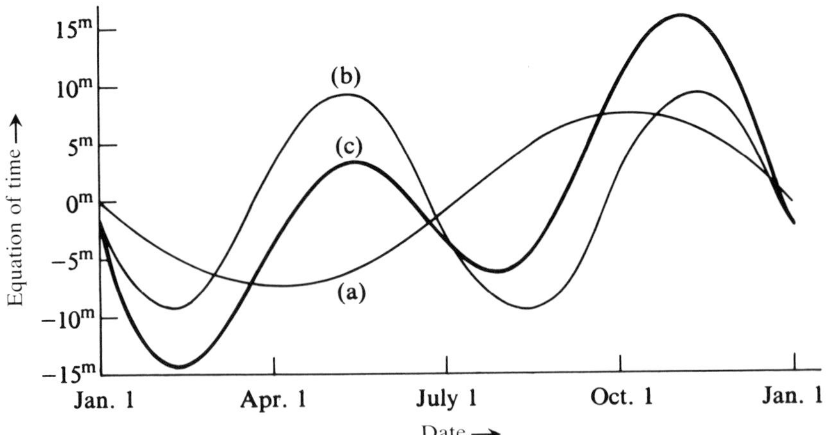

6.4. The equation of time. Curve (a) gives LAST − LIST, where LIST is the local intermediate solar time told by the "intermediate sun", a fictitious point that averages the motion of the apparent sun on the ecliptic. Curve (b) gives LIST − LMST, and curve (c) the sum of the two, LAST − LMST = E, the complete equation of time. A more accurately readable graph is given in A1.4.

perihelion near January 2: the two suns then give the same time. Since the Earth is now moving fastest in its orbit, the apparent Sun immediately moves east of the uniformly moving intermediate sun. The time told by the intermediate sun will then be later than that told by the apparent sun and (LAST − LIST) will be negative: see curve (a) in Figure 6.4. From about January 2 to July 4 (at aphelion), the Earth steadily moves more slowly. The intermediate sun will reach a maximum of about 10 minutes east of the true Sun in April, and then it will begin to lag, falling back to merge with the intermediate sun on July 4. Then it will fall behind, or west, of the intermediate sun, and until the next January 2, the curve will be positive.

Next, correct for the obliquity of the ecliptic. Place the intermediate and mean suns together as the intermediate sun crosses the vernal equinox on March 21. The time told by the two are then the same. Since the intermediate sun now moves at an angle to the equator, its hour circle will fall behind, or west, of that through the mean sun, and (LIST − LMST) will be positive: see curve (b) in Figure 6.4. As the summer solstice approaches, the ecliptic becomes both more parallel to the celestial equator and closer to the celestial pole. The motion of the hour circle increases, and by June 22 the intermediate sun will have the same right ascension as the mean sun; the two times are again the same. Following the solstice, the intermediate sun moves past and east of the mean sun, and until September 23, the curve becomes negative. Analogously, it is positive between the autumnal equinox and the winter solstice, and negative between the winter solstice and the vernal equinox. The complete equation of time, E, shown in Figure 6.4 by the heavy curve labeled (c), can be expressed as (LAST − LIST) + (LIST − LMST), or as curve (a) plus curve (b). It is also shown with greater scale for easier reading in Appendix 1 as Figure A1.4.

The equation of time reaches a maximum of 16 minutes in early November, when the apparent sun crosses the meridian 16 minutes before noon local time. This effect might be visibly masked by the observer's longitudinal displacement from the standard meridian; however the consequences of the equation of time are easily seen in sunset and sunrise. The shortest day of the year must be December 22, the day of the winter solstice. But the date of the earliest sunset occurs two weeks before, on December 7, since the true Sun in December is west of the mean sun, and consequently sets earlier. Similarly, latest sunrise occurs two weeks after the solstice, about January 7, when E has swung negative, and the true is east of the mean sun. In the summer, the effect is not so noticeable as the equation of time is nearer zero; latest sunset occurs about June 27 and earliest sunrise about June 18.

People commonly encounter the equation of time in a strangely distorted figure 8, the *analemma* (Figure 6.5), frequently seen on terrestrial globes. The analemma is the equation of time plotted against the Sun's declination. Note that since the equation of time is not zero at the solstices or equinoxes, the analemma is skewed somewhat and does not line up precisely with the vertical axis. The analemma is also reversed from the usual way of presenting a graph: the x axis is positive to the left. This presentation displays the position of the apparent sun in hour angle relative to that of the mean sun as viewed from outside the celestial sphere. It shows the location of the apparent sun relative to the mean sun whether ahead or behind. The reality of the analemma and the equation of time is vividly seen in a masterful multiple exposure of the Sun presented in Figure 6.6. A stationary camera photographed the Sun at the same mean solar time about every 8 days for a whole year. Here we see the Sun move north and south, move ahead and fall behind in a wonderful display of natural mathematics.

With E known, you can find apparent solar time from mean solar time (apparent = mean + E) or mean solar time from apparent solar time (mean = apparent − E) where attention must be paid to the algebraic sign of E. For example, say a sundial reads $11^h\ 16^m$ on July 3. From Figure A1.4, $E = -4^m$. The mean solar time is then $11^h\ 16^m - (-4^m) = 11^h\ 16^m + 4^m = 11^h\ 20^m$.

6.4 Standard and Universal Time

The second reason for the difference between sundial and watch (that is, civil) time involves longitude. A century or more ago, everyone kept local times on their clocks. Travel was slow and it did not much matter if someone's clock to the east of you was somewhat later. However, rapid communication and travel made local times cumbersome. Imagine taking a train trip and finding a different time at each station. Worse, imagine trying to construct a train schedule for different towns. The problem was addressed and solved by the American educator Charles F. Dowd. In 1883, the railroads and most of the United States adopted his invention of *standard time*. (Europe adopted it somewhat later.) In his scheme, the world is divided into 24 *time zones*, each centered on a *standard meridian* starting at Greenwich. Standard time therefore is also commonly known as *zone time*. In principle, every location within 7°.5 of the standard meridian will keep the local mean solar time of that meridian. Then, when a person passes from

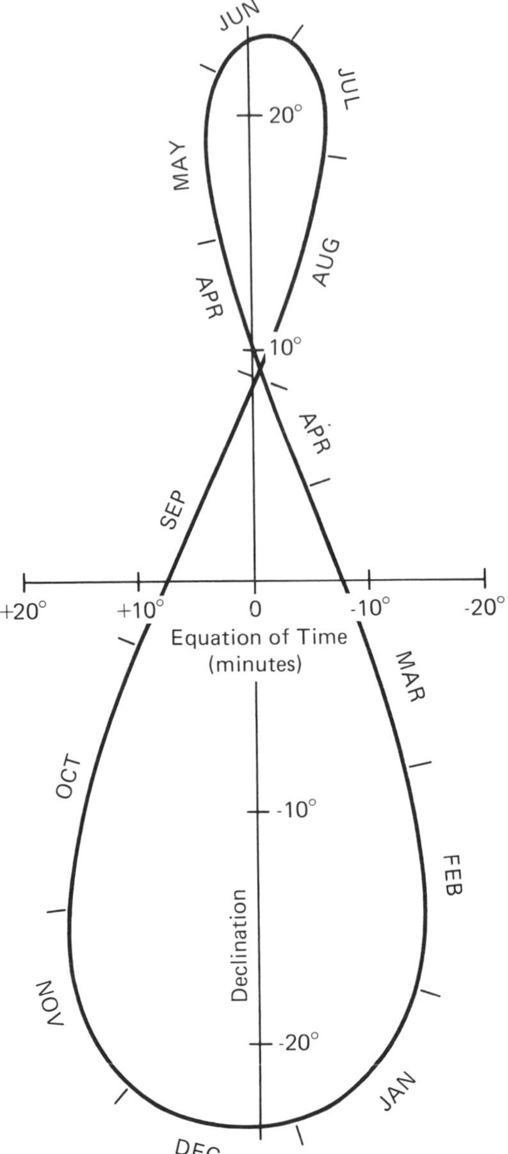

6.5. The analemma, a graph of the equation of time (on the horizontal axis) against solar declination (on the vertical). The value of the equation of time increases to the left. The figure, placed on a globe, shows when the real Sun is ahead of or behind the mean sun.

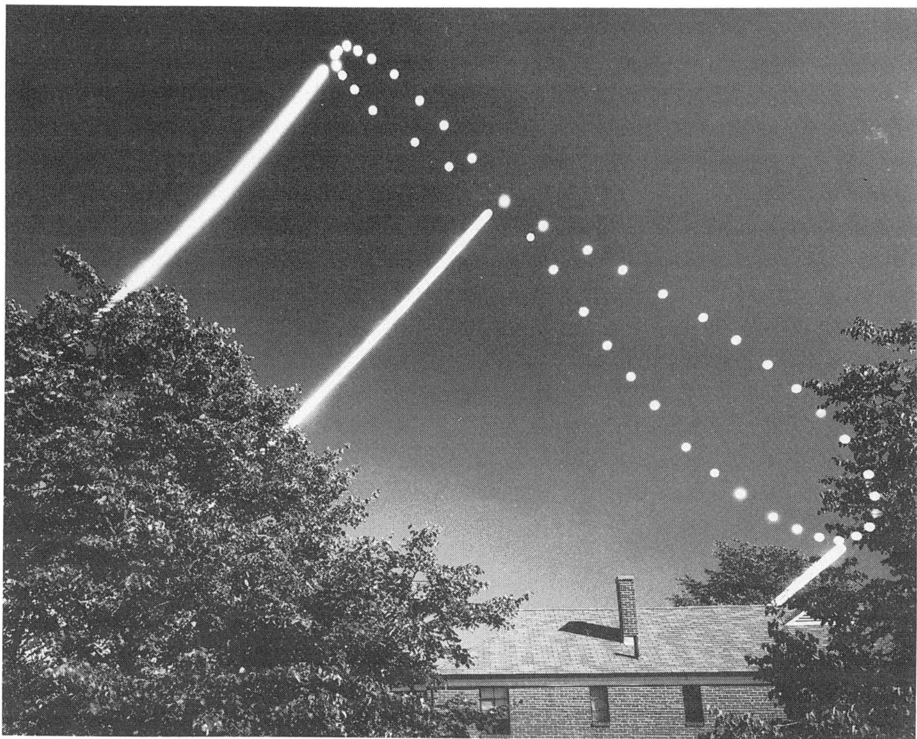

6.6. A natural solar analemma. The Sun was photographed in multiple exposures at the same time about every 8 days, allowing us to see the relation between the equation of time and declination as it really is. The solar photos were of very short duration. One standard exposure served to record the background. (Dennis DiCicco.)

one time zone to the next, only the hour changes, not the minute or the second. In the United States, for example, everyone within 7°.5 of the 75th meridian (75° W longitude) should keep *Eastern Standard Time*, and in Europe, everyone within 7°.5 of 15° E longitude should keep *Central European Time*. An incomplete list of zone names is given in Table 6.1. In practice, however, time zone boundaries (Figure 6.7) often zig-zag wildly to conform to political divisions. Moreover, there are a few odd zones in which the legal time is on the half (or even three-quarter) hour. You can find your standard time algebraically adding $(\lambda - \lambda_{\text{standard}}) \times 4^{\text{m}}$ to your LMST if you are in the western hemisphere and by subtracting it if you are in the eastern. As an example, assume you are at longitude 87°45′ W and that the local mean solar time is $13^{\text{h}}\ 22^{\text{m}}$. You are 2°15′ east of the 90th meridian, Central Standard Time in the United States. An arc of 2°15′ corresponds to 9 minutes of time (see Table 2.1). Your standard time is thus $13^{\text{h}}\ 22^{\text{m}} - 9^{\text{m}} = 13^{\text{h}}\ 13^{\text{m}}$ CST. To find local mean solar time from standard time, reverse the procedure.

It is convenient to pick one time zone as a world-wide standard, most logically the one at 0° longitude. Local mean solar time at 0° longitude is called *Universal Time*, or

Table 6.1. *An incomplete list of time zones*

Standard Meridian	Zone Description[a]	Name
15° E	-1^h	Central European Time
0°	0^h	Greenwich Time = UT
60° W	4^h	Atlantic Standard Time
75° W	5^h	Eastern Standard Time
90° W	6^h	Central Standard Time
105° W	7^h	Mountain Standard Time
120° W	8^h	Pacific Standard Time
150° W	10^h	Alaska–Hawaii Standard Time

[a]Add to standard time to obtain UT

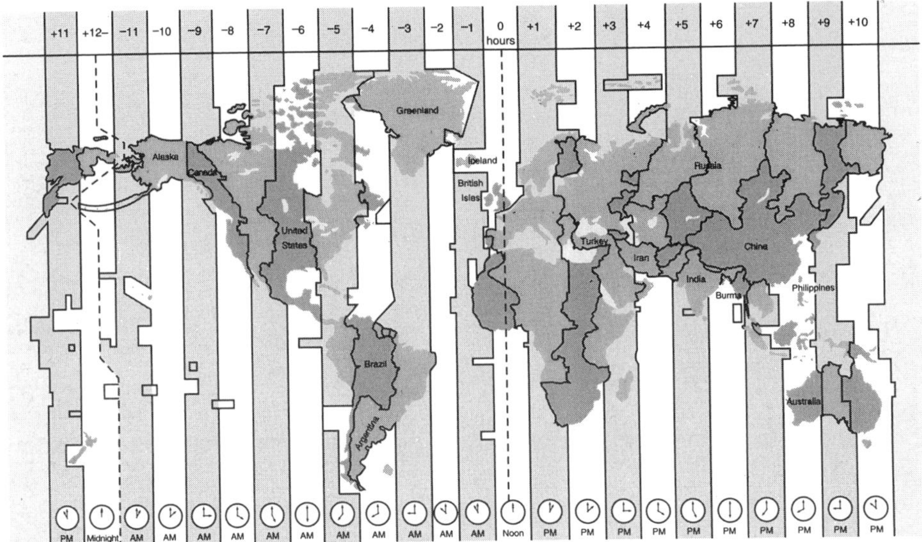

6.7. Time zones of the world, which on the average are spaced 15° or one hour apart. The international date line is on or near longitude 180°. (From *Astronomy!*, 3rd ed., J. B. Kaler, HarperCollins, New York, 1994.)

UT. Since 0° longitude is defined by the meridian that passes through a telescope at the observatory at Greenwich, England (Figure 1.10), UT used to be called *Greenwich Mean Solar Time*, *Greenwich Mean Time* (GMT), or just *Greenwich Time*, and the terms are still in common use in England. There are a variety of sub-definitions of UT that differ from one another by fractions of a second and that will be examined in below.

UT differs from any given standard time by an integral number of hours, called the *zone description*, equal to the longitude of that time zone's central meridian divided by 15°. For example, to find UT from Pacific Standard Time in the United States and

Table 6.2. *Solar right ascension from the* Astronomical Almanac

Date	Apparent right ascension	
	h m s	
Feb. 15	21 52 47.68	
		233.67
Feb. 16	21 56 41.35	
		232.92
Feb. 17	22 00 34.27	

Canada, which is centered on longitude 120° W, add $^{120}/_{15}$ = 8 hours; to find UT from Central European Time, appropriate to 15° E, subtract 1 hour (15° E/15°).

Finally, consolidate all the definitions and corrections. To find standard time from a sundial (local apparent solar time), algebraically subtract the equation of time and then add or subtract the difference in longitude (in time units) between your location and your standard meridian. Then add (or subtract) the zone description to find UT. The sundial in your back yard in Washington, D.C., longitude 77° W, reads $15^h\ 22^m$ on October 1, when the equation of time (Figure A1.4) is $+10^m$. The local mean solar time is therefore $15^h\ 12^m$ and the Eastern Standard Time (appropriate to the 75th meridian) is 8 minutes later, or $15^h\ 20^m$, 3:20 PM. The Universal Time is 75°/5 = 5^h later or 20^h 20^m.

6.5 The *Astronomical Almanac*

A wide variety of information pertaining to events in the sky is available in the *Astronomical Almanac*, published coöperatively by the United States and British governments. Universal Time is the basis for the various tables. You can, for example, find the exact right ascension, declination, and celestial longitude of the Sun for the beginning of each day of the year at Greenwich, UT = 0 hours (or at "ephemeris time" = 0 hours, where the distinction will be made later.)

To find the Sun's right ascension for any given standard time at your location, calculate the UT and interpolate in the table. As an example of how to execute an interpolation, assume you want to know the right ascension of the Sun on February 15 of a particular year at $10^h\ 52^m$ Central Standard Time (LMST at 90° W). Table 6.2, taken from an anonymous year of the *Astronomical Almanac*, gives $\alpha_\odot$ on February 15, 16, and 17 at 0^h UT. The zone description of CST is 6 hours, so your Central Standard Time corresponds to a Universal Time of $16^h\ 52^m = 16.87^h$ = 0.7028 of a day. The decimal date for your time is then Feb. 15.7028. At 0^h UT February 15, the right ascension of the Sun was $21^h\ 52^m\ 47.68^s$. Between February 15 and February 16, the Sun moved in right ascension by 233.67 seconds of time, so from February 15 at 0^h to your time it moved by (233.67 × 0.7078) = 165.39 seconds of arc = $2^m\ 45^s.39$. Add this value to $\alpha_\odot$ at 0^h on February 15 and you find $\alpha_\odot$ at $10^h\ 52^h$ CST to be $21^h\ 55^m\ 33^s.07$.

If you had wanted to know the solar right ascension at 20^h CST, the UT would be 26^h or 02^h *on the next day*, and you would have to interpolate between right ascensions for 0^h UT on February 16 and February 17.

The equation of time is provided for each day of the year in the *Astronomical Almanac* as the "Ephemeris Transit Time of the Apparent Sun," or the UT when the Sun is on the Greenwich meridian. E is then $12^h\ 00^m$ minus the listed value. Additional tables give information on precession and nutation and other information to be described later.

6.6 The international date line

If strictly followed, the above definitions of time lead to an ambiguity in the date. Assume it is noon in New York on September 23. Ignore the rotation of the Earth and the normal advance in time and mentally travel east. It will be 17 hours, or 5 PM, in London, and 20 hours in Moscow. As you cross Mongolia you pass midnight and *enter a new day*. A little farther on, it is now 02 hours in Tokyo on September 24. If we continue back to New York in this simple way, we would arrive back at noon, but now it is both September 23 and September 24 at the same time. Somewhere on your mental travel around the globe you must return a full day, so that when you come back to New York it is still September 23. By world-wide agreement an *international date line* is established at or near 180° longitude, opposite the prime meridian. It follows the 180° meridian, except for an eastward jog through the Bering Sea around Siberia, a westward shift around the Aleutians, and another shift to the east between Fiji and the Samoan Islands (see Figure 6.7). As western Aleuts and eastern Siberians face one another over a small strip of water it is one full day later for the Siberians.

There is no astronomical reason to pick 180° longitude. Any arbitrary north–south line would serve as long as everyone agreed to use it. The 180° meridian is a good choice because not only does it satisfy an aesthetic symmetry, but it passes almost entirely across water in the middle of the Pacific Ocean, so that confusion to civil life and timekeeping is minimized.

6.7 The calendar

A complete system of time measurement also involves keeping track of the days, and where we are in the year, by the *calendar*. Calendars fall into two broad categories, solar and lunar. In both cases, the calendar-maker is faced with the problem of incommensurability. The periods of the Earth's rotation and the Earth's and Moon's orbital revolutions are all independent, and no one value goes into another an even number of times. The problem is especially serious in the construction of a lunar calendar since all three periods must somehow be reconciled. The detailed examination of lunar calendars is deferred to Chapter 9.

The story of our modern system, the *Gregorian calendar*, begins in ancient Italy about 1000 BC with a calendar that had a year of 10 months and 304 days. The calendar started in March, which explains the meaning of our modern September as the "seventh month," October as the "eighth month" and so on. Two months (January and February) were added about 700 BC to raise the number of days to 355. This *Roman calendar* was synchronized with the Moon, which takes 354 days to go through 12 sets of phases, one for each month. About 450 BC the months were set into their current order and the start of the year moved to January.

Such a system, with 355 days, can obviously maintain no correlation with the seasons. In 46 BC the Roman calendar was reformed by the astronomer Sosigenes under the order of Julius Caesar. The *Julian calendar* is synchronized only with the Sun. An ordinary year has 365 days and every fourth year is a *leap year* of 366 days. The average number of calendar days in a year is therefore 365.25, satisfyingly close to the actual number of days in a tropical year, 365.2422. The date of the solar vernal equinox passage will move forward for 3 years (from March 20 in our modern calendar to March 21) and then in a leap year will snap back to March 20. Sosigenes began the Julian calendar by extending 46 BC to 445 days, placing the first day of spring on March 25.

The Julian calendar served the western world for over 1600 years. However, over a long period of time, the small error in the number of days in the Julian calendar will accumulate. Since 365.25 is slightly greater than 365.2422, the date of solar passage across the vernal equinox will on the average slowly move backward, or earlier, in the calendar. By 1582, the equinox had moved up 13 days, so that the equinox passage occurred on March 12. In that year, Pope Gregory XIII ordered the calendar to be reformed once again. To straighten matters out, Aloysius Lilius and Christopher Clavius slightly modified the Julian calendar, successfully proposing that leap years be skipped in century years not divisible by 400. Thus 1900 was not a leap year, but 2000 will be. This procedure slightly shortens the average calendar year to 365.2425 days, which results in an error of only one day in 3000 years. At the same time, 10 days were dropped from the calendar so that spring began on March 21 to coincide with the Julian calendar of the year 325, the time of the Council of Nicaea, at which the date of Easter was decided. As a result, October 4 1582 was followed by October 15.

The Gregorian calendar was adopted by all the Catholic countries, but England (and its colony, now the United States) did not recognize it for nearly 200 years, not until 1752. By that time, the equinox had moved up yet another day, and 11 days had to be dropped to make the conversion. Thus September 2, 1752 was followed by September 14, 1752. The jump caused considerable confusion for many years (as well as rioting and discontent at the time), and we frequently find "os" (old style) and "ns" (new style) appended to dates of the period.

However, the civil calendar can be inconvenient for astronomers, who often need to know the number of days that have elapsed between events. It is not a simple matter to subtract, for example, January 21, 1962 from July 7, 1968. The problem gave rise to the *Julian Day Calendar*, in which the days are numbered consecutively from a far-distant epoch. The Julian Day number, JD, (or the *Julian Date*) is the number of days that have elapsed since January 1, 4713 BC of the Julian calendar. Julian Day numbers begin not at midnight but at *noon*, as did the astronomical civil day prior to 1925. They can be found in the *Astronomical Almanac*, a page of which is reproduced as Table 6.3.

Table 6.3. *A Sample of Universal and Sidereal Times from the Nautical Almanac*

Date 0^h UT	Julian Date	G. SIDEREAL TIME (GHA of the Equinox) Apparent	Mean	Equation of Equinoxes at 0^h UT	GSD at 0^h GMST	UT at 0^h GMST (Greenwich Transit of the Mean Equinox)
	244	h m s	s	s	**245**	h m s
Jan. 0	**7160·5**	6 35 33·6021	33·5392	+0·0630	**3862·0**	Jan. 0 17 21 35·3542
1	**7161·5**	6 39 30·1642	30·0945	·0696	**3863·0**	1 17 17 39·4448
2	**7162·5**	6 43 26·7281	26·6499	·0782	**3864·0**	2 17 13 43·5353
3	**7163·5**	6 47 23·2930	23·2053	·0877	**3865·0**	3 17 09 47·6258
4	**7164·5**	6 51 19·8575	19·7606	·0969	**3866·0**	4 17 05 51·7164
5	**7165·5**	6 55 16·4207	16·3160	+0·1047	**3867·0**	5 17 01 55·8069
6	**7166·5**	6 59 12·9818	12·8714	·1104	**3868·0**	6 16 57 59·8974
7	**7167·5**	7 03 09·5403	09·4267	·1135	**3869·0**	7 16 54 03·9879
8	**7168·5**	7 07 06·0961	05·9821	·1140	**3870·0**	8 16 50 08·0785
9	**7169·5**	7 11 02·6497	02·5375	·1123	**3871·0**	9 16 46 12·1690
10	**7170·5**	7 14 59·2018	59·0928	+0·1089	**3872·0**	10 16 42 16·2595
11	**7171·5**	7 18 55·7531	55·6482	·1049	**3873·0**	11 16 38 20·3501
12	**7172·5**	7 22 52·3050	52·2036	·1014	**3874·0**	12 16 34 24·4406
13	**7173·5**	7 26 48·8585	48·7589	·0996	**3875·0**	13 16 30 28·5311
14	**7174·5**	7 30 45·4149	45·3143	·1006	**3876·0**	14 16 26 32·6217
15	**7175·5**	7 34 41·9749	41·8697	+0·1052	**3877·0**	15 16 22 36·7122
16	**7176·5**	7 38 38·5388	38·4250	·1137	**3878·0**	16 16 18 40·8027
17	**7177·5**	7 42 35·1058	34·9804	·1254	**3879·0**	17 16 14 44·8933
18	**7178·5**	7 46 31·6741	31·5358	·1383	**3880·0**	18 16 10 48·9838
19	**7179·5**	7 50 28·2412	28·0911	·1500	**3881·0**	19 16 06 53·0743
20	**7180·5**	7 54 24·8047	24·6465	+0·1582	**3882·0**	20 16 02 57·1649
21	**7181·5**	7 58 21·3633	21·2019	·1614	**3883·0**	21 15 59 01·2554
22	**7182·5**	8 02 17·9173	17·7572	·1600	**3884·0**	22 15 55 05·3459
23	**7183·5**	8 06 14·4682	14·3126	·1556	**3885·0**	23 15 51 09·4365
24	**7184·5**	8 10 11·0184	10·8680	·1505	**3886·0**	24 15 47 13·5270
25	**7185·5**	8 14 07·5698	07·4233	+0·1465	**3887·0**	25 15 43 17·6175
26	**7186·5**	8 18 04·1238	03·9787	·1451	**3888·0**	26 15 39 21·7080
27	**7187·5**	8 22 00·6807	00·5341	·1466	**3889·0**	27 15 35 25·7986
28	**7188·5**	8 25 57·2403	57·0894	·1509	**3890·0**	28 15 31 29·8891
29	**7189·5**	8 29 53·8019	53·6448	·1571	**3891·0**	29 15 27 33·9796
30	**7190·5**	8 33 50·3646	50·2002	+0·1644	**3892·0**	30 15 23 38·0702
31	**7191·5**	8 37 46·9271	46·7555	·1715	**3893·0**	31 15 19 42·1607
Feb. 1	**7192·5**	8 41 43·4884	43·3109	·1775	**3894·0**	Feb. 1 15 15 46·2512
2	**7193·5**	8 45 40·0478	39·8663	·1815	**3895·0**	2 15 11 50·3418
3	**7194·5**	8 49 36·6046	36·4216	·1829	**3896·0**	3 15 07 54·4323
4	**7195·5**	8 53 33·1587	32·9770	+0·1817	**3897·0**	4 15 03 58·5228
5	**7196·5**	8 57 29·7104	29·5324	·1781	**3898·0**	5 15 00 02·6134
6	**7197·5**	9 01 26·2603	26·0877	·1726	**3899·0**	6 14 56 06·7039
7	**7198·5**	9 05 22·8092	22·6431	·1661	**3900·0**	7 14 52 10·7944
8	**7199·5**	9 09 19·3583	19·1985	·1598	**3901·0**	8 14 48 14·8850
9	**7200·5**	9 13 15·9087	15·7538	+0·1548	**3902·0**	9 14 44 18·9755
10	**7201·5**	9 17 12·4614	12·3092	·1521	**3903·0**	10 14 40 23·0660
11	**7202·5**	9 21 09·0172	08·8646	·1526	**3904·0**	11 14 36 27·1565
12	**7203·5**	9 25 05·5766	05·4199	·1567	**3905·0**	12 14 32 31·2471
13	**7204·5**	9 29 02·1392	01·9753	·1639	**3906·0**	13 14 28 35·3376
14	**7205·5**	9 32 58·7039	58·5307	+0·1732	**3907·0**	14 14 24 39·4281
15	**7206·5**	9 36 55·2687	55·0861	+0·1826	**3908·0**	15 14 20 43·5187

Note that at the beginning of each civil day the Julian day is halfway over, since it started the previous noon. The JD beginning on January 1, 1990 was 2447892.

6.8 Sidereal time

Sidereal time was defined in Section 3.14 as the hour angle of the vernal equinox or the right ascension of the meridian (see Figures 3.19 and 3.20). Like solar time, sidereal time depends on the observer's longitude and is correctly called *local sidereal time*, LST. It is based on the sidereal day, the interval between successive passages of a star across the meridian. The solar day is $3^m\ 56^s.56$ longer than the sidereal day (Section 3.13). As a result, a sidereal clock will gain $3^m\ 56^s.56$ per day on a solar clock and a full day over the course of the year. There are therefore 366¼ sidereal days in the tropical year. Sidereal time allows the location of stars and other celestial bodies, as it is always equal to a body's hour angle plus its right ascension (see Section 3.16). It is a necessity in the setting of a telescope (see Section 3.17).

Unlike solar time, there is no standardization of sidereal time. Standard sidereal times could certainly be defined, but there is no need or use for them. The single exception is *Greenwich Sidereal Time* (GST), the sidereal time at 0° longitude, which, like Universal Time, is a world-wide standard. Neither are their calendar dates (with sidereal months and days) in sidereal time. The extra day would necessitate a different calendar, with the "sidereal dates" changing at odd times relative to the standard solar date. None are needed: we simply use the ordinary civil calendar instead. Astronomers do, however, keep a running tab on sidereal days as the *Greenwich Sidereal Date* (GSD), which is the number of sidereal days that have elapsed since the start of the Julian Day calendar on January 1, 4713 BC The difference between GSD and JD is obviously the number of years since 4713 BC.

Section 3.14 demonstrated that sidereal and solar time are the same at the time of the autumnal equinox. However, no distinction was then made between local apparent solar time and local mean solar time. To compare sidereal and mean solar time, we must incorporate the equation of time. Local apparent solar time and LST are identical at the instant of the solar passage across the autumnal equinox, when the Sun is exactly opposite the vernal equinox (Figure 3.18). Then the hour angle of the vernal equinox is the same as the hour angle of the apparent sun $+12^h$. However, on September 23, the equation of time is $+7^m$ (Figures 6.4 and A1.4), so LMST is behind LAST, or the mean sun is 7^m or $1°.75$ of right ascension east of the apparent sun. Since the mean sun moves $0°.985\ 65$/day, it must have passed the autumnal equinox 1.77 days *before* the moment of the autumnal equinox, at which time LMST had been equal to LST. So LMST and LST will be equivalent about September 21, not September 23. In parallel, LST and LAST will be 12^h apart on March 20, the usual date of the vernal equinox passage, when E is -7^m. Consequently, the mean sun is $1°.75$ *west* of the apparent sun, and similarly, LMST and LST differ by 12^h exactly 2 days later, on March 22.

The exact times of equality actually depend on the precise moment of equinoctial passage and the equation of time for a given year. To find sidereal time from mean solar time exactly, take the number of days (with fractions) past the instant of the passage of the autumnal equinox by the mean sun, multiply by 3^{m} $56^{s}.56$, and add to solar time; reverse the procedure to find LMST from LST.

Because of precession and nutation (see Chapter 5), the vernal equinox has a rather complicated motion through the stars, which leads to two slightly different definitions for sidereal time. What we have considered so far is called *mean sidereal time*. It is defined for the mean, or average, equinox, for which only the smooth westwardly general precession term is included. Mean sidereal time averages out the irregularities introduced by nutation and has a sidereal second of constant length. We can, however, also define an *apparent sidereal time* based on the irregularly moving apparent or actual equinox, in which the effect of nutation is taken into account. The second in this system clearly has a variable length. The difference between the two, apparent minus mean sidereal time, is called the *equation of the equinoxes* and is a measure of the nutational effect. Since the Moon can push the true NCP 17″ ahead or behind the mean NCP (see Section 5.9), the equation can amount to a bit more plus or minus one second of time, with a major periodic of variation of 18.6 years. A sidereal clock always keeps mean sidereal time; apparent sidereal time has only a limited technical use. The equation of the equinoxes is also listed in the *Astronomical Almanac* (and shown in the example in Table 6.3).

Unfortunately, the abbreviation for Greenwich Mean Sidereal Time or Local Mean Sidereal Time is the same as for Greenwich Mean (or Local) Solar Time. The ambiguity is avoided by referring to Greenwich Mean Solar Time as UT. GMST then refers to sidereal time. LMST alone, however, means solar time and GST and LST always express mean sidereal time. The use of apparent sidereal time must be expressly stated.

Precession produces a minor inconsistency in the definitions of the sidereal year. In Section 6.8 we defined two kinds of year, the tropical year defined by the Sun's passage across the vernal equinox, and the sidereal year defined by the Sun's passage of a particular star. The tropical year is 20 minutes shorter because the vernal equinox is moving westward. The sidereal day, however, is inconsistently defined by the successive daily passage of the equinox across the meridian. In one day, the equinox will move westward by $50''.2/(15 \times 365.2442) = 0^{s}.009$ in right ascension, so the *real* sidereal day, the one with respect to the stars themselves, is 9 milliseconds (thousandths of a second) longer than the sidereal day of the definition.

Local sidereal time for any given civil date and time is most easily found from the *Astronomical Almanac*. The second and third columns of Table 6.3 respectively give the apparent and mean Greenwich Sidereal Times for zero hours UT for every day of the year. Add your zone description to your standard time (subtract it for the eastern hemisphere) to find UT. Add the UT to the sidereal time at 0^{h} UT for the appropriate Greenwich date. Then add the amount by which the sidereal clock has gained over the solar during that period (interpolating from the table). Finally, subtract your longitude (add it for the eastern hemisphere).

As an example, assume you are at longitude 77°32′16″ W on January 15 and that the Eastern Standard Time (that for longitude 75° W) is 09^{h} 36^{m} $03^{s}.0$. What is the local mean sidereal time? Since EST's zone description is 5^h west of Greenwich, the UT is January 15 at $14^h\ 36^m\ 03^s.0$. From January 15 to 16 the sidereal clock gains $3^m\ 56^s.56$ over the solar, and in $14^h\ 36^m\ 03^s.0 = 0.60837$ days gains $2^m\ 23^s.92$. On January 15 at 0^h the Greenwich Mean Sidereal Time was $7^h\ 34^m\ 41^s.87$. Therefore, Greenwich Mean Sidereal Time is

$$\begin{array}{r} 7^h\ 34^m\ 41^s.87 \\ +\ 14^h\ 36^m\ 36^s.03 \\ \underline{+\ 2^m\ 23^s.92} \end{array}$$

$$= 21^h\ 72^m\ 101^s.82 = 21^h\ 73^m\ 41^s.82 = 22^h\ 13^m\ 41^s.82.$$

At your longitude of 77°32′16″ your time in any system is $5^h\ 16^m\ 01^s.07$ earlier than it is at Greenwich (Table 2.1). Therefore, your LST is

$$\begin{array}{r} 22^h\ 13^m\ 41^s.82 \\ \underline{-\ 05^h\ 16^m\ 01^s.07} \end{array} = \begin{array}{r} 21^h\ 73^m\ 41^s.82 \\ \underline{-\ 05^h\ 16^m\ 01^s.07} \\ 16^h\ 57^m\ 40^s.75. \end{array}$$

The last column of Table 6.3 gives the reverse information, "UT at 0^h GMST" (Greenwich Mean Sidereal Time). To find local mean solar time (and standard time) from local sidereal time, reverse the above procedure. Add your longitude to your LST (subtract it if you are in the eastern hemisphere) to obtain Greenwich Mean Sidereal Time. Add it to the listed "UT at 0^h GMST" and *subtract* the interpolated amount by which the sidereal clock gains over the solar during the sidereal day ($3^m\ 55^s.91$). Then subtract your longitude for LMST or your zone description for standard time (add them for the eastern hemisphere).

6.9 The determination of time

It would at first seem logical to determine solar time – standard or Universal – directly from the solar hour angle. All one needs to do is find local apparent solar time, subtract the equation of time, E, and correct for longitude. However, the apparent sun is a bright extended disk with a variable angular diameter owing to the eccentricity of the Earth's orbit, and it is not a simple matter to locate its exact center to measure its hour angle or to time its meridian transit. In contrast, the stars are points and their positions are far easier to determine. Although solar time ultimately does depend on careful measurements of the position of the Sun, needed to locate the equinoxes, in daily practice solar time is determined by first finding the sidereal time and then by applying the known difference between the sidereal and mean solar clocks.

Local sidereal time is both the hour angle of the vernal equinox and the right ascension of the celestial meridian (see Section 3.16). Accurate time can be determined by noting the meridian passages of stars with precisely known right ascensions, which can be done with a specialized kind of telescope called a *meridian transit circle* (Figure 6.8). The meridian circle is free to rotate only along an east-west axis and can consequently point only at the celestial meridian. The axis is affixed to a pair of highly accurate

6.8. A meridian transit circle. The telescope moves only on one axis, north and south along the celestial meridian, and can determine either local sidereal time from right ascension or right ascension from local sidereal time. Declinations of stars are found by measuring their altitudes at meridian transit and knowledge of the latitude. (United States Naval Observatory.)

graduated setting circles that allow measurement of the altitudes of celestial objects. Observations made through the telescope itself serve to define precisely the celestial meridian and the visible pole. The meridian is identified as the location where stars have their maximum altitudes; the determination of the instrument's latitude and the elevation of the pole are discussed in Chapter 8.

In older times, the observatory's sidereal clock was electrically connected to a pen that traced a line on paper affixed to a rotating metal drum. At the start of each second, the pen would jump and produce a tick mark on the line. The astronomer waited for a star with known right ascension to pass a central cross hair in the field of view of the transit circle, and when it was centered would trip a telegraph key that also placed a mark on the drum. The positioning of the stellar mark relative to the time ticks served to find the correction to the clock. It is not actually necessary to reset the clock. The astronomer only keeps track of the changing correction that is applied to find the correct time.

In the 1930s the meridian circle began to be replaced for time determination by the inherently more stable *photographic zenith tube* or PZT (Figure 6.9), a vertical telescope fixed in place to observe stars at the zenith. The zenith is precisely defined by a pool of liquid mercury that acts as a flat horizontal mirror. Photographs of stars were timed

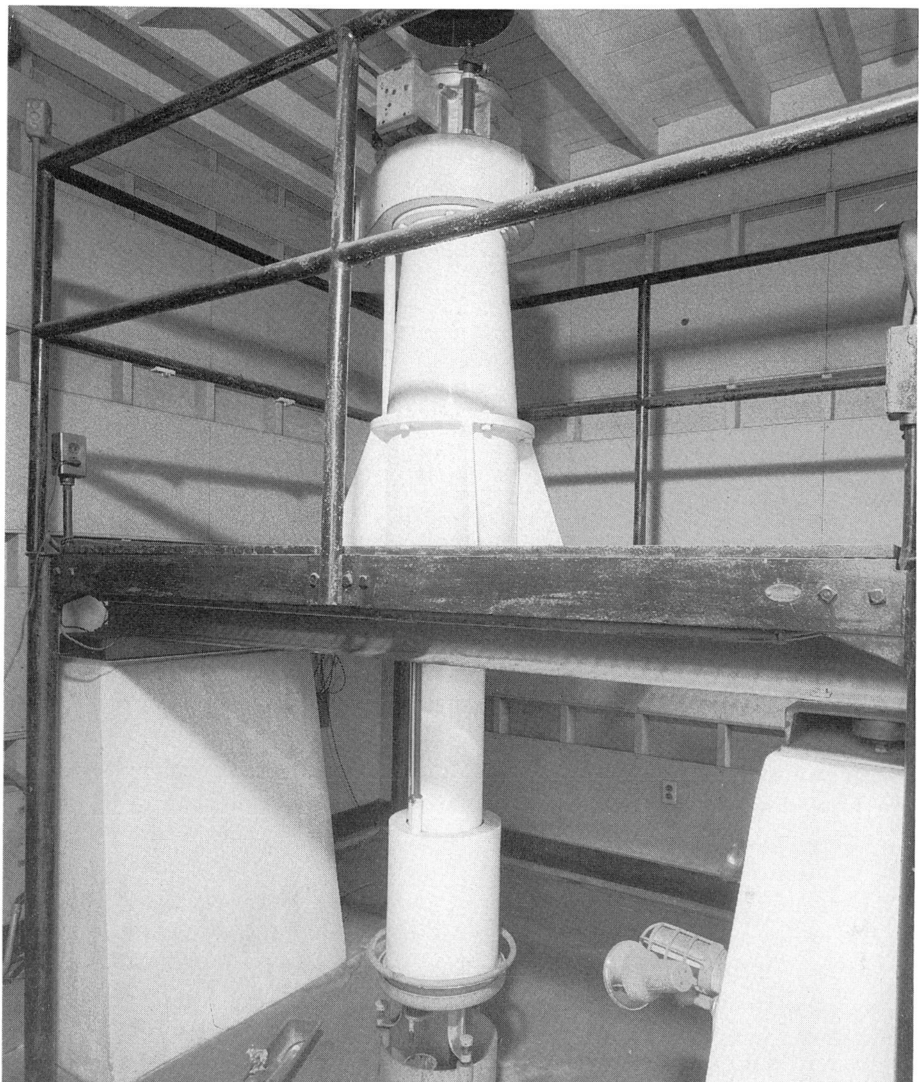

6.9. The photographic zenith tube. The telescope points at the zenith and is used to determine sidereal time from the right ascensions of stars. The horizon and zenith are defined by a liquid mirror made of mercury that is near the ring at the bottom. (United States Naval Observatory.)

with the observatory's sidereal clock. Their images were measured against the known position of the meridian on the photographic plate and the correction to the clock was then determined.

Sidereal time was determined in the United States from 1935 until the mid-1980s by PZTs in Washington, D.C. and Florida under the control of the United States Naval Observatory. The Navy has historically been charged by the United States

Table 6.4. *Positions of some prominent stars*

Star	α (1980)[a]	δ (1980)
Achernar (α Eri)	$1^h\ 37^m$	$-57°20'$
Aldebaran (α Tau)	$4^h\ 35^m$	$+16°28'$
Sirius (α CMa)	$6^h\ 44^m$	$-16°41'$
Regulus (α Leo)	$10^h\ 07^m$	$+12°04'$
Spica (α Vir)	$13^h\ 24^m$	$-11°03'$
Antares (α Sco)	$16^h\ 28^m$	$-26°23'$
Vega (α Lyr)	$18^h\ 36^m$	$+38°46'$
Deneb (α Cyg)	$20^h\ 41^m$	$+45°12'$
Fomalhaut (α PsA)	$22^h\ 57^m$	$-29°44'$

[a]Precession is unimportant for naked-eye observations.

Government with timekeeping duties because of the need for precise time in the practice of navigation; in Great Britain the responsible agency is the National Physical Laboratory. All major countries have similar timekeeping services whose data are correlated and combined into an agreed-upon world-wide standard system. The PZT is capable of sub-millisecond (thousandth of a second) precision. Even more precise modern methods will be addressed in Section 6.11. The principle, however, will stay the same.

Once local sidereal time has been found, it is a simple matter to compute local mean solar time, since the difference between the two clocks is a function of the interval between the time of observation and the instant of the passage of the mean sun across autumnal equinox. Standard or Universal Time is then calculated from the longitude of the observing telescope at the rate of 4^m per degree, or 15° per hour. The actual location of the vernal or autumnal equinox is determined by observations of the solar declination, which must be 0° when it crosses the celestial equator. The Sun's declination can be found by measuring its daily altitude (or zenith distance) at meridian transit with a transit circle and a knowledge of the exact latitude.

To see how this scheme works, try to make your own estimate of standard time by personal observation. First, learn the right ascensions of a few prominent stars distributed about the sky such that one is always visible from your latitude (listed in Table 6.4), or locate the approximate positions of the equinoxes and solstices from the star maps in Appendix 2. Find the star or equinox in the sky and mentally draw the hour circle through it to the celestial equator. Next, estimate the hour angle to the intersection of the hour circle and equator. (If you measure east, the hour angle is negative.) Add the hour angle to the right ascension and you have your local sidereal time. From the date, you know approximately the advance of the sidereal clock over the solar clock, at 2 hours per month, 4 minutes per day past about September 21. Subtract this time difference from your LST to obtain your local mean solar time. Finally, make the longitude correction to your standard meridian to get standard time, adding 4^m for each degree you are west of your standard meridian, subtracting 4^m for each degree east.

The procedure is identical to that used by the professional astronomer, the only difference being that of precision. With practice, a naked-eye observer can estimate the standard time to an accuracy of about 10 minutes, whereas that with professional instruments is about a ten-thousandth of a second.

6.10 Clocks and the Earth's rotation

The purpose of a clock is to keep track of the Earth's rotation. Neither the clock nor the Earth is a perfect mechanism: they both contain irregularities and the clock must periodically be set (or corrected) by observations of the stars and Sun. These irregularities ultimately limit our ability to measure right ascension. Great effort has been expended in the search for the perfect clock so that at least this source of error can be eliminated, and we have very nearly succeeded.

Through the nineteenth century and until the middle of the twentieth the pendulum clock (Figure 6.10) was standard. The clock works on the principle that the period of a pendulum's swing is independent of the amplitude, or size, of the swing: the amplitude depends only on the length of the pendulum and the force of gravity. Each swing of the pendulum releases a spring-loaded ratchet in the clock mechanism, which drives the hands. If left alone, the pendulum would eventually stop because of frictional forces, so the clock must contain a weight-driven or electrically operated mechanism that periodically pushes the pendulum to keep it swinging. Pendulum clocks are dependent upon external conditions. As we first saw in Section 5.2, the clock rate depends on the local force of gravity. A clock made in Chicago will run slower in Mexico City where, because of our planet's oblateness, it is farther from the Earth's center and the gravitational pull is smaller. This effect is not a problem if the clock is kept in one location. Much more serious is the effect of local temperature, which will cause the pendulum to expand or contract, and the clock to run slower or faster. Elaborate temperature-compensating clocks were built in the first half of this century and kept in sealed, controlled rooms to attain maximum possible accuracy. Even so, the best that could be done was a precision of about one part in 10^7, about a second per year. ("Precision" in such a clock does not mean that it has to "read" the exact time. What is important is that the *rate*, the degree that it gains or loses in an interval of time be known. That way we can always correct the clock time to the true time.)

The keeping of time has been revolutionized by modern technology. By the middle of this century, the quartz crystal clock was in operation. The basis for any clock is a cyclic motion. For the pendulum, we simply substitute the natural vibration of quartz, which can control the oscillation in an electric current, and which can then be used to count the flow of time. The frequency of vibration for a given crystal is quite constant and is largely independent of local conditions. These clocks can keep time to about one part in 10^{10}, or about a millisecond (one thousandth of a second) per year.

The final stage in the evolution of the modern clock came in the 1950s with the inven-

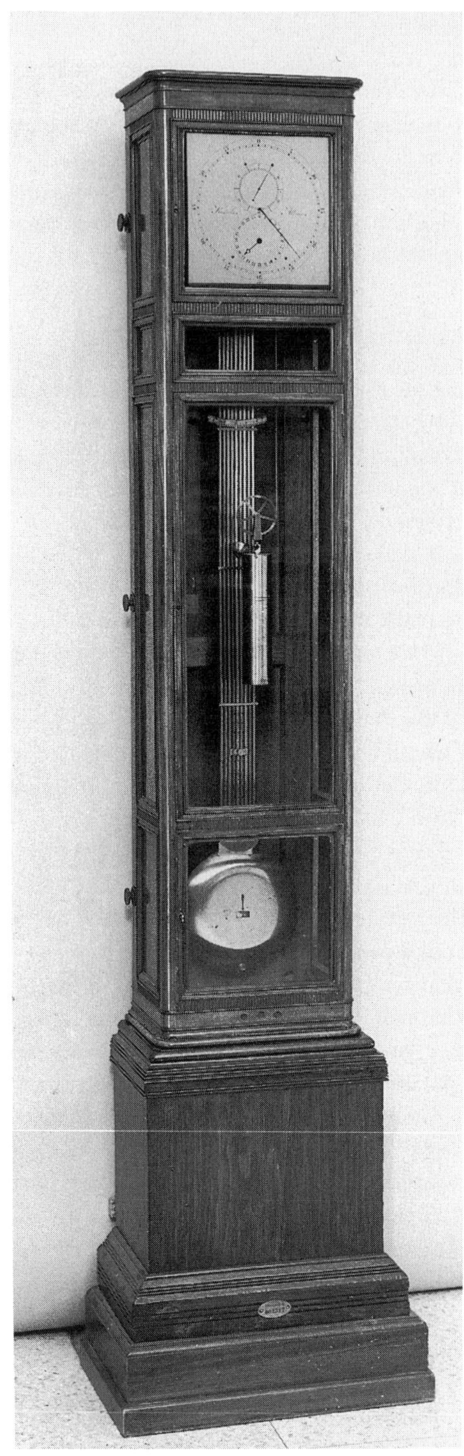

6.10. A pendulum clock. This old master clock of the United States Naval Observatory was driven by the Earth's gravity, which controls the pendulum's period. (United States Naval Observatory.)

tion of the *atomic clock*. This device takes its rate from the vibrations of atoms themselves; the earlier ones worked on a particular oscillation of the cesium atom, which now actually defines the length of the second, and the modern ones work on hydrogen. The atomic clock may be the ultimate attainable device; a set of them can keep time to nearly one part in 10^{15}, or about one second in over a million years! Indeed, the clocks are so good that they *become* the standards, as there is no way to check them.

For most purposes we think of the Earth as a steady rotator, but there are indeed irregularities and over time they become quite noticeable. It has long been known that the Earth is slowing down and the length of the day increasing at a rate of about 0.002 seconds per century because of drag induced by lunar and solar tides (the gravitational stretching of the Earth and its waters by the Moon and Sun, to be described in Chapter 10). Superimposed upon the steady lengthening of the day are several short-term irregularities that cause the Earth alternately to speed up a bit and then slow down (Figure 6.11a). The principal terms have periods of one year and six months, but there are many others, including one a decade long. There are a variety of causes that include tides raised in the Earth's atmosphere and solid body, interaction between the Earth's fluid core and solid mantle, and rebounding of the Earth after the recession of heavy glacial ice.

Another kind of irregularity – the *polar motion* – has been known since the 1890s (Figure 6.11b). The Earth actually moves slightly about its rotation axis, which causes the north and south poles to shift on the planet's surface over an area about 15 meters across. The two principal periods are a year and 14 months (the latter called the *Chandler wobble* after its nineteenth-century discoverer Seth C. Chandler). The first is caused largely by changes in the distribution of atmospheric and surface water masses, and the second is a free wobble induced by the off-center distribution of the Earth's mass. The effect produces minute changes in latitude and longitude, or in the observer's celestial meridian. Errors in the early clocks masked these irregularities, but with the modern precise atomic devices they can be studied in considerable detail.

The above variations cause serious difficulties with the definitions of time. For civil purposes, we want a clock to be synchronized with the passage of the Sun across the sky, from which comes our basic solar time. But such a clock must run slower every year, and if we want to keep track of the periodic variations, it must occasionally speed up and slow down at an irregular rate. Over the years, the length of the second in solar (or sidereal) time must increase. This type of timekeeping can be very unsatisfactory. For example, if we want to be able to predict the positions of the Earth or other planets in their orbits, we need a time system in which the length of the second is absolutely constant. As a crude analogy, think of trying to time a horse race with a watch that was geared to change its rate in an unpredictable manner; you cannot chart the course of planets with such a watch either. The resolution of the problem has produced a variety of time systems that attempt to satisfy all the needs.

6.11 Modern timekeeping

So far we have defined three different timekeeping systems: sidereal time, apparent solar time, and mean solar time. Universal Time (UT), which embodies the concept of

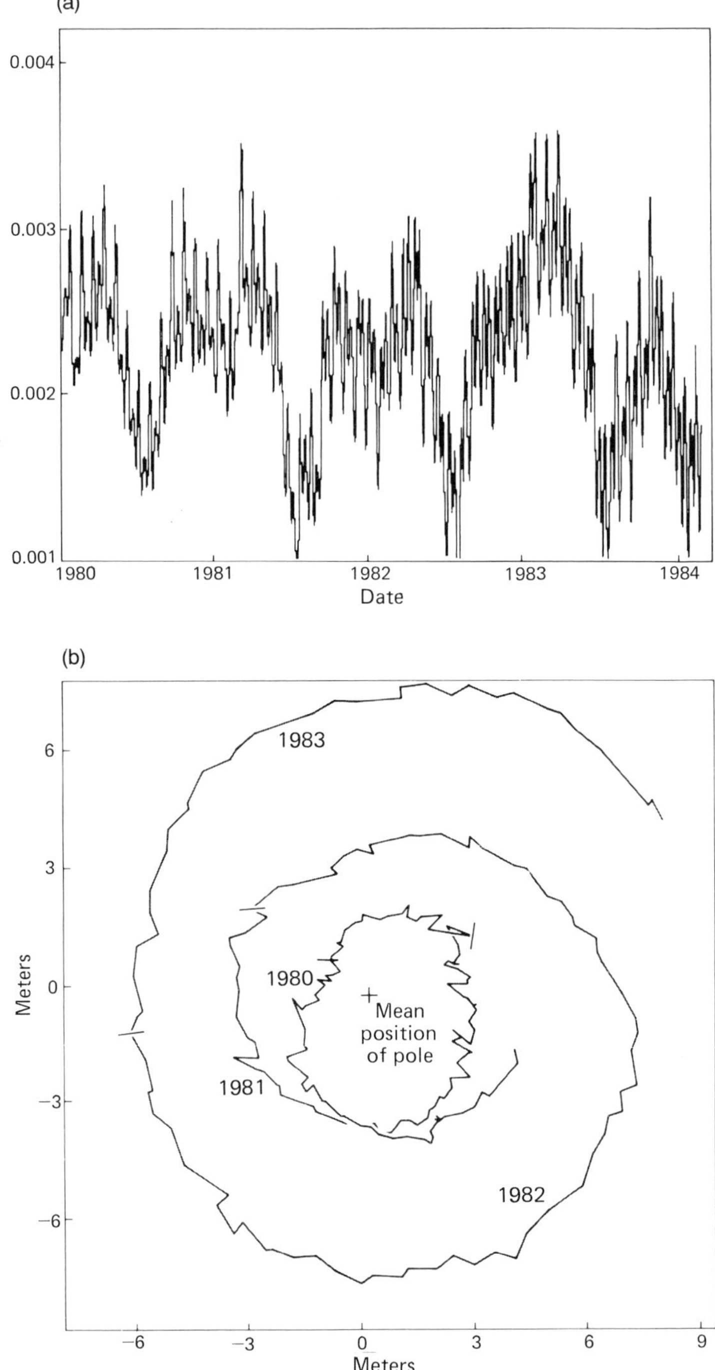

6.11. Irregularities in the Earth's rotation. (a) The graph shows the change in the length of the day from 1980 through 1984. (b) The plot shows the movement of the north pole about its mean position. ((a): Jet Propulsion Laboratory, M. Eubanks; (b): Lageos.)

standard time, was defined as local mean solar time at Greenwich. Actually, there are technical differences between Greenwich Mean Solar Time and UT that result from slightly different definitions of the mean sun, but they need not concern us here: we will consider the two as identical.

The search for a useful general time system that can reconcile an ideal uniform time with the variable rotation of the Earth has led to a number of different definitions, including four subdivisions of UT. UT as measured at the observatory with a meridian circle or PZT is called UT0. UT0 is the "natural system" that contains all the variables of the Earth's rotation, including polar motion. The length of the second in UT0 is erratically variable. From observations made at stations spotted around the globe that observe or measure latitude, the effect of the polar motion can be measured and removed from UT0, which results in a system called UT1. UT1 is the "astronomer's time," the system most applicable to positional measurements made at the observatory (Chapter 8). But UT1 contains the effects of the Earth's variable rotation rate and is unsuitable for general civil purposes as well as for the prediction of planetary positions.

An early attempt to solve the problem generated UT2. This system is UT1 from which seasonal terms are removed. UT2 can be precisely corrected after the fact, when the true behavior of the Earth has been measured against an atomic time standard. UT2 served well for many years as a civil time standard. It still contains the secular slowing of the Earth and it is in synchronism with the Earth's long-term behavior, the length of the second in UT2 slowly and steadily lengthening. But because instantaneous UT2 relies upon the *prediction* of the Earth's behavior, which may not be truly accurate, it is not entirely satisfactory. Its use has been discontinued in favor of a system described below. But it is important, as it is likely to encountered in historical and other contexts and because it is used in defining the starting points, or epochs, of other systems.

For some 300 years, since Newton, we have understood that we must have a uniform time system against which to measure the passage of events in the Universe. With the relatively inaccurate pendulum clocks (even as sophisticated as they became), and with primitive calculation procedures that make it nearly impossible to compute a planet's orbital position very far into the future or the past, the rotating Earth served admirably as a uniform clock. But with modern accurate atomic clocks and the huge capacity of modern computers, it does so no longer.

The first general attempt at uniform time was a system called *Ephemeris Time*, ET, adopted for general use in 1955. ET is based not upon the rotation of the Earth, but upon the Earth's orbital revolution, and philosophically upon the gravitational theory of the Solar System in general. The length of the second is defined as a fraction ($^1/_{31\,556\,925.9747}$) of the tropical year 1900. By convention, ET is set to equal UT2 at the beginning of January 1900 (called the *epoch* of the system). Since the UT clock is gradually slowing down because of the decrease in the rotation speed of the Earth, ET is gradually pulling ahead of UT1 or UT2 by about half a second per year; by 1990 the difference was about 57 seconds. Even though the day lengthens by only 0.0015 seconds per century, the effect on clocks running at even slightly different rates is cumulative, so that ET − UT advances quite noticeably.

We set the UT clock by the positions of the stars in the sky relative to the meridian; analogously, we set the ET clock by comparing the observed positions of the Sun, planets, and particularly the Moon, with theory. The use of ET as the independent, or time, variable in predicting planetary positions presents some serious difficulties. First, accurate ET (represented by the difference ET − UT) cannot be known until the various celestial bodies have been carefully observed and the data reduced, a process that takes a few years. Second, it is a self-defined system: it is used to chart the course of planets but is also defined from them. Errors in the gravitational theory of the Earth and Moon, which clearly exist, will surface as errors in the ET clock.

In response to these problems, a new system called *Atomic Time*, or AT, was developed. AT is based upon the vibrational frequencies of atoms (specifically a spectrum line of the cesium-133 isotope at 9 192 631 770 cycles per second), and as such it is independent of the motions of the planetary bodies. A system of AT called A.1 went into use in 1967, with an epoch, the moment when A.1 = UT2, of January 1, 1958. In the early 1970s the *International Atomic Time System* (IAT or TAI), which differs only slightly in epoch from A.1, also came into use. Since ET was $32^s.18$ ahead of UT1 for January 1, 1958, ET = TAI + $32^s.18$. In principle, TAI is the ultimate constant terrestrial time standard, but even it is not without its problems. The measurement of the frequency of the cesium vibration is affected very slightly by the Earth's gravity and magnetic field, so that errors will indeed accumulate even in atomic time.

Finally, all clocks and all time standards are affected by the precepts of *general relativity*, the theory of motion and gravity formulated by Albert Einstein. In any moving system, time itself becomes distorted, and in fact, there is no such thing as a universal absolute time standard. The system that takes this problem into account is termed *dynamical time*. There are two kinds: *Terrestrial Dynamical Time* (TDT) is but an extension of ET in which we base the length of the second upon the atomic standard instead of the Earth's orbital period. The offset from TAI is the same as it is for ET and is defined as TDT = TAI + $32^s.184$. The other kind, *Barycentric Dynamical Time* (TDB), relativistically corrects TDT for the Earth's orbital velocity, and reduces the time standard to the center of mass (the barycenter) of the Solar System. TDB is the time variable needed for the calculation of planetary positions, and TDT is the standard used to mark their locations as viewed from Earth. They differ by at most 2 milliseconds.

The discussion of star positions in Section 5.6 referred them to an epoch, the beginning of a particular year. Prior to 1984 the astronomical year, or the *Besselian year*, began when the Sun passed a right ascension of $18^h\ 40^m$. Because there are a non-integral number of days in a year, this moment moved around relative to the calendar year and could fall on December 30 or December 31 (oftentimes called "January 0"). After 1984, the epoch of star catalogues was taken to be January 1, 12^h TDB. The modern epoch for star positions is called J2000, and is set at that time for the year 2000. Not only has the old standard Besselian year been dropped from "official" use, but so has the tropical year. For simplicity, long intervals are now reckoned in *Julian centuries* of 36 525 days (the name derived from the Roman calendar: see Section 6.7).

The study of time determination can be confusing: there are so many different systems. Time measurement is a continuously evolving and changing subject. As our techniques and our clocks improve, we must develop new and better systems, each of which rests on one that was defined earlier. No one of the systems discussed above really dominates. Different problems require the use of different kinds of time. Nevertheless, we must find some system to agree upon for use in general worldwide affairs. The current solution is a hybrid structure called *Coordinated Universal Time* (UTC). In this system, the length of the second is defined by AT: we no longer have to worry about the second becoming longer and changing the rate of the clock. The epoch, however, or the beginning of the day, is defined according to UT1. Now, from above, it is apparent that since the Earth is slowing down, UTC will gradually creep ahead of UT1. The difference between the two is kept below 0.9 seconds by adding an extra whole second – a *leap second* – to UTC to slow the UTC clock and redefine the epoch. The leap second may be introduced as necessary, preferably on December 31 or June 30. Summary definitions of the kinds of time we have examined are presented in Table 6.5.

The precision of the definitions and the accuracy of the clocks is being matched by great advances in the technology of the observations. The PZT – still useful for understanding the concepts involved – has been replaced by a far more sophisticated system, which consists of a network of radio telescopes linked into what is called a *very long baseline interferometer* (VLBI). The precision to which we can locate a body in the sky depends on the diameter of the telescope (a subject to be explored in Chapter 13) as well as on the smearing effects of the Earth's atmosphere, which are not evident at radio wavelengths. Observations of distant quasars – point-like radio sources in the outer reaches of the Universe – are made simultaneously but independently at each telescope and synchronized by atomic clocks. Analysis of the data by computer gives the effect of our having observed with a huge instrument equal in size to the intercontinental spacing between the instruments. The result is precision in the range of a ten-thousandth of a second of arc, allowing us to observe the Earth's rotation (and UT1) to amazing accuracy. The analysis of the data also yields the exact distances between the telescopes. With telescopes in Europe and in North America we can, over a long period of time, measure the rate of continental drift and watch the United States and England move away from one another at a centimeter or so per year! Laser ranging to artificial satellites and the Moon also give information regarding the Earth's rotation. A pair of organizations, the International Earth Rotation Service (IERS) and the International Bureau of Weights and Measures (BIPM) collect and analyze the data. The first of these is responsible for data on the rotation of the Earth and astronomical time, the second for international scientific and civil time systems. Astronomers have advanced a remarkable distance from the time that the visual observer tapped a telegraph key in the dark to put a mark on a mechanically rotating cylinder.

6.12 Time services and time signals

All major countries have their national time services, and each has its own manner of operation. In the United States, precise time and frequency are maintained by the

Table 6.5. *Time systems*

System	Meaning	Definition
LAST	local apparent solar time[a]	Hour angle of the true Sun + 12h
LMST	local mean solar time[a]	Hour angle of the mean sun + 12h
GMST[b]	Greenwich Mean Solar Time	LMST at Greenwich
UT	Universal Time	General term for GMST, UT2, or UTC, depending on context
LST	local sidereal time	Hour angle of the vernal equinox
	apparent sidereal time[c]	Based on nutating equinox
	mean sidereal time[c]	Based on mean precessing equinox
GST	Greenwich Sidereal Time	LST at Greenwich
UT0	. . .	UT as directly observed
UT1	. . .	UT0 with polar motion removed
UT2[d]	. . .	UT1 with periodic annual and semiannual terms in Earth's rotation removed
ET	ephemeris time	"Constant" time defined by earth's orbital revolution
AT,A.1,TAI[e]	atomic time	Constant time defined by vibration of cesium atom; A.1 differs slightly from TAI
TDT	terrestrial dynamical time	ET with atomic time second, TDT = TAI + $32^s.184$
TDB	barycentric dynamical time	TDT reduced to the center of mass of the Solar System
UTC	Coordinated Universal Time	Atomic time synchronized to UT1 to within $0^s.9$ by adding leap seconds

[a]The difference LAST − LMST = equation of time
[b]No longer an "official" name but still in public use
[c]The difference apparent-mean sidereal time = equation of the equinoxes
[d]No longer used
[e]International Standard

National Institute for Science and Technology (NIST, formerly the *National Bureau of Standards*, NBS), which uses a battery of several atomic clocks in Boulder, Colorado. Universal Time is both maintained and determined directly from astronomical observations by the Precise Time Service Department of the United States Naval Observatory in Washington, D.C. The data are analyzed to find the difference between the constant time standard and that based on the rotating Earth. NIST and the Naval

Table 6.6. *Time signal stations*[a]

Station	Country	Frequency (MHz)	Comments
WWV WWVH	United States	2.5, 5, 10, 15, 20[b]	voice announcement every minute
CHU	Canada	3.3, 7.335, 14.670	voice announcement every minute
LOL1	Argentina	5, 10, 15	transmits 1 hour at 11^h, 14^h, 17^h, 20^h, 23^h; voice announcement every 5 minutes
IBF	Italy	5	15^m preceding 7^h, 9^h to 18^h
YVTO	Venezuela	6.10	announcement every minute

[a]All information subject to change
[b]Not for WWVH

Observatory synchronize their clocks, which are capable of keeping time that is accurate in the nanosecond range, that is, they are accurate to a *billionth of a second.* Most countries cooperate to synchronize their time systems through the BIPM to define the international standard (TAI). The IERS then determines when it is necessary to add a leap second to UTC. Various publications are available from the different agencies giving data on precise time, the differences between TAI, UTC, and UT1, and other information.

Precise time is disseminated throughout the world via telephone, radio, and satellite. Most major countries have time services that broadcast the time on specialized civilian radio stations. The military have their own, which are also often accessible to the general public. In the United States, time signals are broadcast on radio station WWV (Table 6.6), operated by the NIST, in Fort Collins, Colorado (and by WWVH in Hawaii), which transmits on several frequencies. WWV provides a number of services. It is a clear-channel radio-frequency standard; for example, when you tune to it at 10 megahertz (megacycles per second) you know that you are precisely on 10.000 megahertz. In the background, you hear a tone that is an audio-frequency standard, which alternates between 500 hertz and 1000 hertz. Superimposed on the tone is a clock tick, which consists of several cycles of an audio tone; the beginning of the tick announces the start of the second. Coded into the tick is information that allows a user to convert from UTC, the broadcast time, to UT1, with an accuracy of ±0.1 seconds. Canada operates a similar system through station CHU. In Great Britain, the National Physical Laboratory (NPL) near London disseminates time and frequency via a very low frequency (VLF) transmitter, whose frequency is below the standard broadcast band, and for civil purposes by the BBC. International time signals are also available from the Navstar Global Positioning System (GPS). Table 6.6 lists the six stations (from dozens) that give voice announcements of the time. Complete lists are usually available from national time services.

In using any system for precise timing, it is important to be aware of the "propagation time" for the radio station. Because the speed of light (and, of course, radio) is finite, 300 000 km/s in a vacuum, there is a short delay between the transmission and the reception of the signal that depends on the intervening distance. (A properly situated listener, that is, one for whom the propagation times are different, can hear WWV and WWVH separately as two closely spaced clock ticks.) If one knows the distance, and thus the delay, a correction can be made to the received signal. However, there is an inherent limitation. A radio signal in the short-wave bands, which include all the WWV frequencies, travels by bouncing (actually refracting: see Chapter 7) off the Earth's ionosphere. The height of this layer above the Earth depends on the degree of solar heating and solar activity. Consequently, the propagation interval varies slightly in an unpredictable manner. To overcome this problem, one must go either to surface telephone transmission, with known delay time, which is available for WWV and from the Naval Observatory, or to a very low frequency (VLF) station (most are military). The VLF signal travels along the ground and the propagation time can be accurately established. The most precise radio time signals available are those of the GPS, from which microsecond (10^{-6} s) times can be derived.

Now we have the time of day, both in simple and precise forms. In the next two chapters, we will use it.

Chapter 7

Sunrise and sunset

Among the most important astronomical calculations that deal with human affairs are those of sunrise and sunset, the moments that turn the night to day and the day to dark. Ideal sunset and sunrise azimuths and times were introduced in Section 3.11. We are now in a position to extend the discussion, to examine the exact horizon positions and standard times of these two instants, wherein we take into account a variety of distorting details.

7.1. The solar disk

The Sun is not a point, as was assumed in Chapter 3, but a circular disk with an angular diameter of 32 minutes of arc. The edge of its disk is called its *limb*, the top part the upper limb, the bottom the lower limb. Sunrise and sunset respectively occur with the first and last appearances of the upper limb. Such a definition will make sunrise and sunset take place closer to the visible celestial pole than they would if we assumed that the Sun were a point (Figure 7.1). In northern latitudes the azimuth of sunset is increased and that of sunrise is decreased. In southern latitudes the azimuth changes are reversed.

At the Earth's equator, daily paths of celestial objects are perpendicular to the horizon, and the azimuths of sunset and sunrise show no change as a result of the solar diameter, but as the observer proceeds north or south, the line of the daily paths flatten out near the horizon and the *difference* between the setting (or rising) azimuths of the solar center and limb increases. In middle temperate latitudes, the change in azimuth is about a quarter of a degree and it increases greatly toward the poles.

The Arctic Circle was earlier defined as the latitude at which the Sun is circumpolar on June 21, 66°33′ N. That definition, however, again assumes the Sun to be a point, not an extended object. Barring refraction, which will be fully discussed in the next section, at the Arctic Circle on June 21 you would see the upper half of the Sun and would have to move southward by one-half the solar diameter (16 minutes of arc), to a latitude of 66°17′, to reach the actual southern limit of circumpolarity. The formal

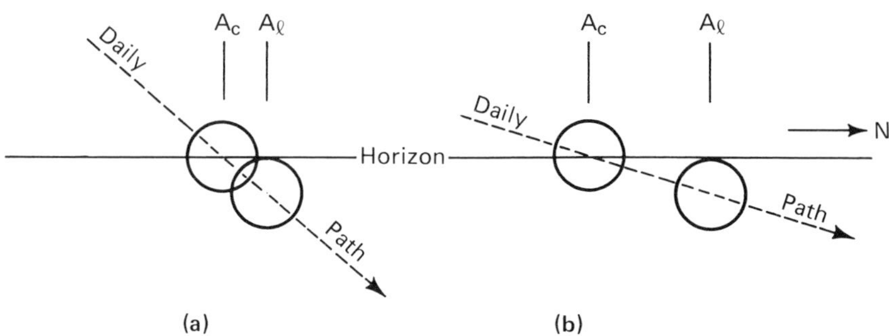

7.1. The effect of the solar diameter on sunset. (a) shows sunset along a daily path from a northern middle latitude. If sunset is defined by the appearance of the upper limb (A_l, where A is azimuth) rather than from the center (A_c), it will occur farther along the horizon to the north point. The time of sunset will also be delayed. (b) shows sunset at a higher northern latitude. Since the daily path in (b) is flatter relative to the horizon, the azimuth difference between the setting of the center and the upper limb, $A_l - A_c$, will be increased, as will the time delay.

Arctic Circle, however, still remains at +66°33′. Alternatively, on December 22 when the Sun is not supposed to rise, you would see the upper half above the horizon at noon, and would have to travel 16′ north, to a latitude of 66°49′, before the Sun was not visible at all during the day.

Not only are the azimuths of sunrise and sunset changed by our using the upper limb of the Sun as the reference point, but so is the length of the day and the time of sunrise and sunset. It takes the Sun 24 hours to move 360° along its daily path, or 4 minutes to move a degree. At the Earth's equator, where the Sun rises perpendicular to the horizon, the Sun's upper limb makes its appearance ($^{16'}/_{60'}$) × 4 minutes, or $1^m\ 4^s$ earlier than the center. Sunset is delayed by the same amount, and the length of the day so defined is $12^h\ 2^m\ 8^s$, instead of 12 hours. The effect increases with latitude, as the Sun's daily path at sunrise and sunset becomes more nearly parallel to the horizon (Figure 7.1). As you travel north, the Sun must move farther along its daily path between the setting of the solar center and the setting of the solar limb. For example, at the equinox and at a latitude of 45°, the day is $12^h\ 2^m\ 44^s$ long.

7.2. Refraction

Even more important than the solar diameter is the effect produced by *refraction*. Refraction is the bending of light by its passage through a substance. Refraction effects are seen everywhere: it is the physical principle that underlies the operation of a camera lens and the appearance of a rainbow, and it causes the distortions that appear in a fish aquarium. Figure 7.2 shows how it works. As a light ray passes from a rarefied medium, air for example, into a denser substance, perhaps glass or water, it changes its direction so that it goes more toward the perpendicular: in Figure 7.2a, the angle r (the *angle of refraction*) is smaller than i (the *angle of incidence*). The greater i, the greater the change

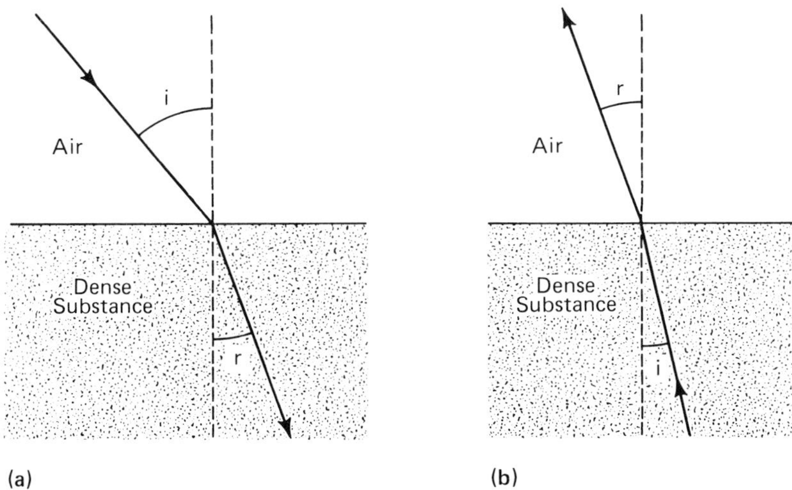

7.2. Refraction of light. A light ray bends toward the perpendicular as it enters a denser substance (a), and bends away if it moves in the other direction, and enters a less dense substance, (b). The light path is independent of direction. The greater the angle of incidence (i), the greater the difference ($i - r$) or ($r - i$).

of direction, i minus r. As a general rule, the denser the substance, the greater will be the directional change for any value of i. If the light ray moves in the opposite direction (Figure 7.2b), the effect is reversed. A ray passing out of a dense substance changes direction so that it moves *away* from the perpendicular. The path itself is independent of the direction in which the light ray is moving. The change in direction of the origin of the light will cause the source to appear to be shifted in its position. And since the change in direction, ($i - r$) in Figure 7.2a, or ($r - i$) in Figure 7.2b, is proportional to the angle at which light enters or leaves the substance, objects will appear distorted; the bottom of a flat swimming pool, for example, will seem to curve.

We live within a refracting substance, the Earth's atmosphere. It may seem to have an insignificant density to us, but it is vastly more dense than the near vacuum of interplanetary space through which sunlight must first pass to reach the ground. As light enters the atmosphere it begins to change direction toward the perpendicular to the atmosphere's surface (Figure 7.3). The atmosphere becomes continuously denser downward, and as a consequence, the directional change continuously increases. A person standing on the Earth's surface therefore perceives the Sun or a star displaced in a direction higher toward the zenith than it truly is.

Atmospheric refraction is not great enough to cause any immediately noticeable distortion in the appearance of the sky. Constellation patterns, for example, do not sensibly change their shapes as they rise. Nevertheless, the effect is large enough that it was known to Ptolemy. For altitudes greater than 45°, refraction elevates an astronomical image by less than 1 minute of arc, which is an angle smaller than can be perceived with the unaided human eye. But at 10° altitude, the lofting effect is nearly 6 minutes

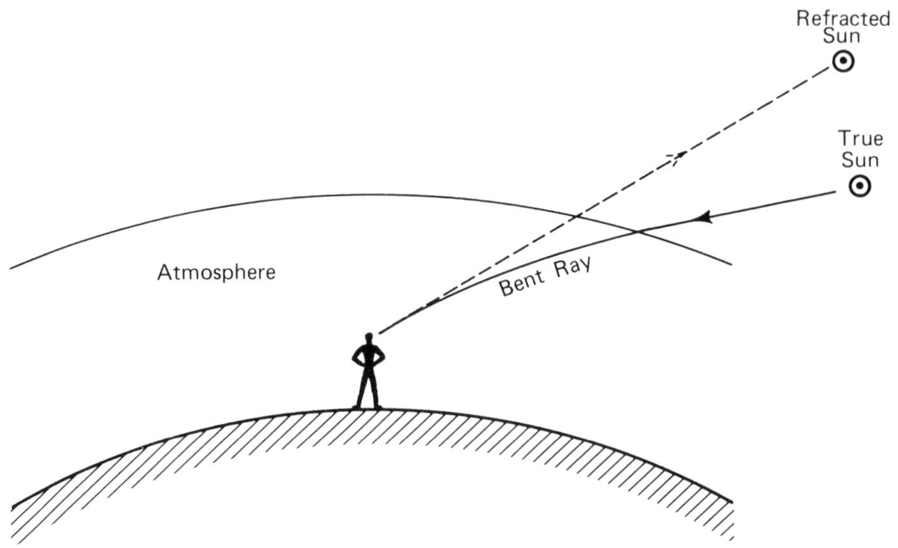

7.3. Atmospheric refraction. As sunlight enters the Earth's atmosphere, it bends gradually, though only slightly, toward the perpendicular. A person on the ground always sees the Sun or a star in a false position, displaced upward toward the zenith. The effect is grossly exaggerated here.

of arc. Near the horizon, where sunlight enters the Earth's atmosphere almost parallel to the ground, the amount of refraction becomes quite large. The exact increase in the solar altitude depends on weather conditions, specifically barometric pressure and temperature, but on the average, the setting or rising Sun is lofted by some 35 minutes of arc. This number is greater than the solar diameter, which means that when we see the lower limb touch the horizon at sunset, the true Sun (that which we would see if we had no atmosphere) has already set!

Refraction has about twice the effect on the rising and setting azimuths of the Sun, and on the extension of daylight hours, as the solar diameter. The solar radius (or *semidiameter*) extends the Sun toward the zenith at sunset by 16′, and refraction by 35′. Thus at visible upper-limb sunset, or sunrise, the center of the true Sun has an altitude of −16′ −35 ′ = −51′, close to a full degree. On June 21 we can see a circumpolar Sun 51′ south of the arctic circle, or at a latitude of 66°33′ − 51′ = 65°42′ N; and on December 22, we must be north of the arctic circle by the same 51′, or at 67°24′ N for the Sun not to rise at all. The Sun is actually seen at the north pole about two days before it passes the vernal equinox. The effects are obviously the same in the antarctic, just six months out of phase.

7.3. The location and time of sunrise and sunset

The combined effects of refraction and the solar diameter are shown in Figures A1.5 and A1.6 in Appendix 1. Figure A1.5 gives the difference in sunset (or sunrise) azimuths (ΔA) between the refracted upper limb and the unrefracted center for various declinations of the Sun. The displacements are always toward the visible celestial pole. In the northern hemisphere, sunset azimuth is increased (ΔA is positive) and sunrise azimuth is decreased (ΔA is negative). The signs of ΔA will be reversed for an observer in the southern hemisphere. In Figure A1.5, ΔA is given for every 10° of solar declination plus the solstice Sun at plus and minus 23°27′. The displacements are not quite symmetrical, that is, they are a bit different for plus and minus any value of non-zero latitude. Use the solid lines if declination and latitude have the same sign and the dashed ones if one is plus and the other minus. The range in ΔA is so great that the curves are broken into two parts. The left-hand set of curves is to be read along the left axis. This set slowly increases from zero to about 1° at a latitude of 45°. At this point, ΔA increases drastically, so the right-hand set is read along the right axis, where the scale is changed.

The observed azimuth of sunrise of sunset is then that found from Figure A1.1, for which the Sun is assumed to be an unrefracted point, plus (or minus) the displacement given in Figure A1.5. A pair of examples show their use. First, what is the actual azimuth of sunrise for 30° N latitude on May 11, when the Sun has a declination of +20°? From Figure A1.1, the unrefracted azimuth for the solar center is 66°45′ (the figure cannot be read more accurately than that). From Figure A1.4, ΔA is −32′ (remember, sun*rise*), so the Sun will come up at A = 66°15′ (where the 32′ is rounded to 30′). Actually, there is little difference at ϕ = 30°. But now what is the azimuth of sunset on that date at 60° south latitude? From Figure A1.1, A(rise) = 47°, so a(set) = 360° − 47° = 313°. Then from A1.5, ΔA = −1°58′ (call it −2°). The correct azimuth must be 313° − 2° = 311°, a quite noticeable change.

Figure A1.6 shows the analogous delay in sunset, or the advance in sunrise, again for every 10° of solar declination and the solstices. Solid and dashed lines mean the same as in Figure A1.5. These numbers are added to the hour angle of the unrefracted setting solar center, subtracted from that of the unrefracted rising solar center, or subtracted from the (negative) eastern HA of the unrefracted rising solar center. The result is the hour angle of the unrefracted actual solar limb at the time that the Sun is *seen* to set (or rise). Sunset delay accelerates markedly with latitude, so Figure A1.6 is again broken into two parts. The left-hand curves that run up to latitudes of 45° to 55° go with the left-hand scale and the right-hand curves go with the right-hand scale. Note the large values of the delays at high latitudes as the Sun slowly edges below the horizon on a nearly flat daily path.

The duration of daylight is double the actual solar hour angle at sunset, or twice the sum of the hour angle from Figure A1.2 plus the delay from Figure A1.5. Or, expressed somewhat differently, the duration of daylight is extended beyond the value given by Figure A1.2 alone (twice that hour angle) by twice the sunset delay time from Figure A1.6. The two examples above will serve here as well. From Figure A1.2 on May 11 at 30° N latitude the hour angle of the setting Sun is $6^h\ 48^m$, which would yield a daylight duration of $13^h\ 36^m$. From Figure A1.6, the delay in sunset is $4^m\ 17^s$, which extends daylight hours by $8^m\ 34^s$, to the nearest minute, 9^m. The period of daylight, from the first-to-last glimpse of the upper limb of the Sun is thus $13^h\ 45^m$. On that date at 60° S, the hour angle of sunset from Figure A1.2 is $3^h\ 24^m$, for a daylight duration of $6^h\ 48^m$. The extension is $9^m\ 20^s$ times two, or $18^m\ 40^s$. Daylight is consequently about $7^h\ 07^m$ long.

Calculation of the actual time of sunrise or sunset is slightly more complicated. Figure A1.2 gives the hour angle of sunrise or sunset as a function of latitude and the date through solar declination (Figure A1.3), from which we readily find the local apparent solar time by adding 12^h. We then apply the equation of time in Figure A1.4 to determine the local mean solar time as in Section 6.3. Finally, we must take into account the observer's longitude to find the standard time. See the summary at the end of Section 6.4.

The moment of sunrise or sunset in standard time will therefore depend on the observer's longitude. If you are west of the standard meridian, you will see the Sun go down later in standard time than will someone east of it. The effect can be quite large for locales that for political reasons adopt a standard time that is technically outside their own actual time zone.

In working real problems for specific dates, note that the declination of the Sun changes continuously during the day, so in calculating daylight durations you should use the declination for midday. To calculate the times of sunrise or sunset, you should use the solar declination at those moments. Moreover, Figure A1.3 assumes that the Sun passes the vernal equinox at 0^h March 21 and that the year has exactly 365 days. However, the year actually contains 365.2422. . . days. As a result, the Sun will pass the equinox at a different time every year. Figure A1.3 can then be as much as one day off and the solar declination in error by as much as 24 minutes of arc. Given the precision to which the graphs can be read, however, the effect is not that important.

7.4 Summary: finding sunset and sunrise

In summary, to find the standard time of sunset (or sunrise) at any time of the year and at any place on Earth:

1. Find the declination of the Sun from Figure A1.3.
2. Find the solar hour angle of sunrise or sunset (defined for the unrefracted solar center) for your latitude and the above declination from Figure A1.2. Consider the hour angle negative for sunrise. Then add 12^h to find local apparent solar time.
3. From Figure A1.6 add the delay for sunset (or subtract the advance for sunrise) for your latitude to account for refraction and the solar semidiameter.
4. Subtract the equation of time found from Figure A1.4. Be sure to use the correct algebraic sign: 18 hours – (–4 min) = $18^h + 4^m = 18^h\ 04^m$.
5. Add 4 minutes times the difference in degrees between your longitude and standard longitude if you are west of the standard meridian and in the western hemisphere and subtract it if you are east of the standard meridian; reverse the signs of the longitude correction if you are in the eastern hemisphere. Convert between longitude in degrees of arc and time units via Table 2.1.

Examine more examples. First, what is the time of sunset at Dallas, Texas, on May 9? From a map, Dallas has a longitude of 96°45′ W and a latitude of 33° N (these figures are rounded off). The declination of the Sun on May 9 (Figure A1.3) is +17°10′. The hour angle of sunset from Figure A1.2 is $6^h\ 45^m$; the time of setting of the unrefracted solar center would thus be $18^h\ 45^m$. The delay from Figure A1.5 is $4^m\ 20^s$ so the true hour angle is $6^h\ 49^m\ 20^s$, and the local apparent solar time of sunset is $18^h\ 49^m\ 20^s$ (carry the seconds for the moment to avoid a possible round-off error). The equation of time from Figure A1.6 is $+3^m\ 40^s$, so local mean solar time is $18^h\ 45^m\ 40^s$. The longitude places Dallas 6°45′ west of the 90° meridian used for its time zone of US Central Standard Time, which converts to 27 minutes of time. The standard time is therefore $18^h\ 46^m + 27^m = 19^h\ 13^m$ (or as used in the US, 7:13 PM).

Next, what is the time of sunrise at Leeds, England, 1.5° W, 54° N, on June 21, when $\delta_\odot = 23°.5$ N? The hour angle of the unrefracted solar center is $-8^h\ 27^m$. The advance is $7^m\ 50^s$, which is negative (i.e. the hour angle of true sunrise must be more negative). The local apparent solar time of sunrise is consequently $-8^h\ 27^m - 8^m + 12^h = 3^h\ 25^m$. On June 21, the equation of time is -2^m, which renders sunrise at $03^h\ 27^m$. The longitude correction to Greenwich ($\lambda = 0°$) is $+6^m$, so the Sun will pop up in Leeds on the first day of summer at $03^h\ 33^m$ Greenwich Mean Time.

The precision of this graphical method is limited by the accuracy of Figures A1.3 and A1.4 in the determination of solar declination and the equation of time. This source of error, however, is not important compared to the problem of interpolating between the curves in the other graphs. Although the graphical method is generally useful to a precision of a minute or so, and is highly instructive (really its purpose here), it is more accurate and certainly simpler to look up the times of sunrise and sunset in an annual almanac such as the *Astronomical Almanac* (see Section 6.5). The volume gives the local times of sunrise and sunset at Greenwich for a given year every 4 days for a variety of latitudes (see Table A1.2), and the user must interpolate for latitude and for the local date. To find the Greenwich date needed for the table, simply add to your local date (expressed in decimal fractions of a day) your longitude divided by 360° if you are in the western hemisphere, or subtract it if you are in the eastern. The resulting interpolations from the table yield your local times of sunset and sunrise; then correct to standard times as described in rule 5 above.

The *Astronomical Almanac* provides local times to the nearest minute, and to maintain that precision, one should use the volume for the current year. A single universal table for sunrise and sunset cannot be given because the year does not contain an integral number of days (see Section 6.7) and the Sun will not set at exactly the same time on the same date in successive years. But Table A1.2, taken from the *Astronomical Almanac* for a year midway between leap years, is generally usable without much loss of accuracy. Below latitude 50°, the largest error is only 1 minute; above 50°, 2 minutes. Accuracy much better than 1 minute is rarely needed, and is actually difficult to achieve because of variances in refraction caused by local weather conditions. (The *Almanac* uses 50 minutes of arc for the combined elevating effects of semi-diameter and refraction as opposed to the 51′ used in this volume. The difference is not significant.)

Use Table A1.2 to find sunset and sunrise for the above examples. May 9 falls between the entries for May 6 and 10. At 30° N, sunset on these dates takes place at $18^h\ 40^m$ and $18^h\ 43^m$ respectively, and at 35° N at $18^h\ 49^m$ and $18^h\ 52^m$. The difference over 4 days is 3 minutes, so over 3 days will be 2 minutes (rounding to the nearest minute), and sunset on May 9 will be at $18^h\ 42^m$ and $18^h\ 51^m$ for the two latitudes. At 33°, sunset will then take place at $18^h\ 47^m$ local time, only 1 minute different from the LMST figured from the graphs. (Remember that you still must make the longitude correction.)

On June 21 at 54° latitude, Table A1.2 shows that sunrise will fall between $3^h\ 27^m$ and $3^h\ 28^m$ local time, the same as found above. The agreement between the two methods is very satisfying. In spite of the seeming complexities, the determination of sunset and sunrise turns out to be surprisingly easy.

7.5 The visible horizon

So far, we have assumed that we can see the formal, or astronomical, horizon, which divides the sky into two exactly visible hemispheres. That, however, is rarely the case. The visible horizon, where the sky seems to meet the land, may be formed by hills that lie well above the astronomical horizon. In calculating local sunset times, azimuths, and visibilities, local topography must be taken into account. Only when you are on a still sea with your eye at water level is the visible horizon equivalent to the astronomical horizon.

With regard to local land forms, there can be no general rules. In the prairies of the United States, the visible horizon is indeed very close to the astronomical; in the Andes, the observer has a serious problem. A 1 kilometer high mountain 10 kilometers away elevates the horizon by 5°43′ and, for a vertical sunset at the equator, will advance the time of sunset by $22^m\ 50^s$. The effects on sunset(rise) azimuths and times can be addressed only by the application of trigonometry to the known local elevations. If contour maps are available, the problem is in fact not very difficult; the methods of solution can be derived from Appendix 3. Listed sunset and sunrise times always ignore this problem and refer to the astronomical horizon.

The problem of the sea horizon is much simpler because of the sharpness and uniformity of the distant watery surface. If you are elevated in any way above the ocean, you will be looking below the astronomical horizon because of the curvature of the Earth (Figure 7.4). Therefore, you can see the Sun and Moon when they have negative altitudes, or zenith distances greater than 90°. In Figure 7.4, the astronomical horizon is defined, as always, by the plane tangent to the Earth's mean or average sphere at the observer, which must be perpendicular to the line to the Earth's center. The person in Figure 7.4 sees below the formal horizon because of the mountain's elevation. The angle formed by the two lines, or the displacement between the astronomical and visible horizons, is called the *dip* of the horizon. From trigonometry, we find that the dip in minutes of arc is equal to 1.811 times the square root of the elevation in meters or, by coincidence, very nearly equal to the square root of elevation in feet:

$$\text{Dip}' = 1.811\ \sqrt{\text{elevation (meters)}} = \sqrt{\text{elevation (ft)}}.$$

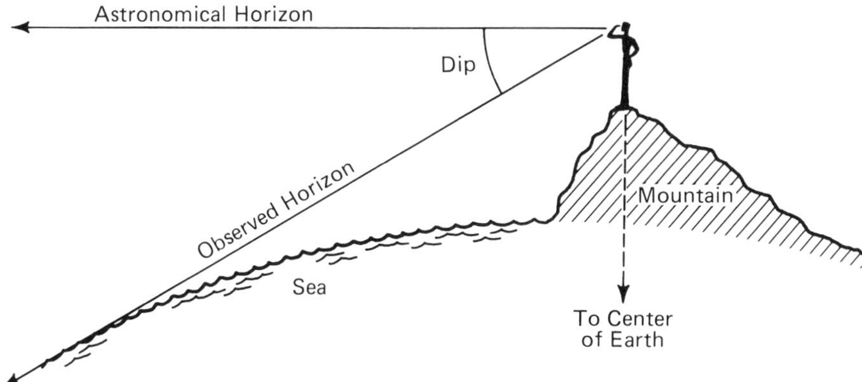

7.4. Horizon dip. The observer, on a mountain overlooking the sea, sees the true horizon below the formal, astronomical horizon.

The actual azimuths and times of sunset and sunrise for the observed horizon must be calculated individually for each separate case. A graphical representation becomes too complex, since the phenomena of the visible sunset now depend on three variables: latitude, solar declination, and dip. The effect of horizon dip can be quite significant and noticeable. Even for a 2-meter tall person on the shore, the dip is over 2′ and the delay in a vertical sunset (zero degrees latitude) is 10 seconds. At an elevation of 800 meters, the dip increases to 51′, equal to the combined effects of refraction and solar diameter.

7.6 Twilight

The sky does not instantly dim as soon as the Sun goes down and we can easily see outdoors well in advance of sunrise. *Twilight* is the interval between sunrise or sunset and the time when the sky is truly dark. It is caused by sunlight caught by the upper atmosphere, the glow illuminating the ground (Figure 7.5). It is intimately related to

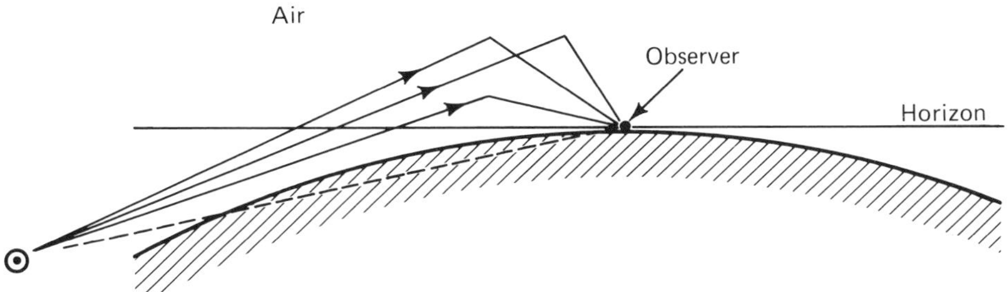

7.5. Twilight. The Sun is below the horizon for the observer, but its light can still reach the air high above, which then illuminates the ground. The diagram shows the end of nautical twilight, with the Sun's altitude at −12°. It does not get perfectly dark on the ground until the Sun is 18° below the horizon.

the concept of horizon dip. The pilot of an airliner flying at 10 000 meters has a horizon dip of 1.8 $\sqrt{10\,000}$ minutes of arc = 180′, or 3°. For a vertical sunset at the equator, the pilot will see the Sun for $3° \times 4^m$ = 12 minutes after it has set for an observer on the Earth; the person watching the airplane then sees the glint of sunlight from the fuselage 12 minutes after sunset and also sees the illuminated surrounding air. At a latitude of 45°, where daily paths make a large angle with the horizon, the Sun (if it is on the celestial equator) will take longer, 17 minutes, to achieve an altitude of −3°. To determine when twilight is over, we figure the dip for the highest layer of atmosphere that will relay significant light to the ground, and for our latitude compute how long it will take the Sun to drop that far below the true horizon.

There are three definitions of twilight. *Civil twilight* ends (or begins) when the center of the true unrefracted Sun is 6° below the horizon ($h = -6°$, $z = 96°$); at that time it is considered too dark for usual outdoor activities. *Nautical twilight* ends with the Sun 12° below the horizon ($h = -12°$, $z = 102°$); then it is too dark to see the sea horizon from shipboard. Finally, at 18° below the horizon ($h = -18°$, $z = 108°$), *astronomical twilight* ends. Now, the sky is as dark as it will get for the night.

The duration of twilight depends on how long it takes the Sun to achieve the specific angle below the horizon. This interval in turn is dependent upon the latitude and the solar declination. At one extreme is the equator. Here the Sun sets vertically, and the interval of astronomical twilight varies in length from $1^h\,09^m$ at the time of the equinoxes to $1^h\,15^m$ at the time of the solstices. Civil twilight lasts only one-third as long; full darkness descends suddenly at the equator and in the tropics in general, always a surprise to visitors from temperate latitudes.

At the other extreme is the pole. The Sun is not actually seen to rise above the horizon at the north pole until about 2 days before it encounters the vernal equinox (see Section 7.2). However, astronomical twilight begins when the Sun is at declination −18°, which happens about January 30. We then would see a steadily brightening twilight all through February and into March, until the Sun rises about March 19. Then, after sunset about September 25, we would live in a deepening twilight until it is fully dark about November 14, when the Sun again reaches declination −18°. At the north pole we would see 6 months of full daylight, about 2½ months of full night (November 14 to January 30), and 3½ months of varying levels of twilight.

The Sun's upper limb is seen as circumpolar at 65°42′ N on June 21 (51′ south of the Arctic Circle because of the combined effect of the finite solar disk and refraction). However, at latitude 48°33′ N, the Sun passes 18° under the north point of the horizon on June 21 and astronomical twilight lasts all night. The closer you live to the arctic circle, the longer will be the time around June 21 when you will experience continuous twilight. Above the arctic circle, although the Sun may not rise around the time of the winter solstice, there may be a considerable twilight period around noon, the duration depending on latitude and solar declination.

The latitudes above which twilight lasts all night for a certain date can be found from Figure A1.3, which gives the solar declination. The NCP is $\phi°$ (the latitude) above the horizon, and the Sun must be 18° below the horizon for there to be twilight. Therefore, for the sky to be light at all the Sun must be $\phi° + 18°$ from the pole. Its declination must therefore be $90° - (\phi° + 18°)$. Thus the limiting latitude for which twilight is observed at midnight is $\phi°$ N $= 72° - \delta_{\odot}$. If in the southern hemisphere, $\phi°$ S $= -(72° + \delta_{\odot}) = -72° - \delta_{\odot}$. Similar calculations can be made for civil and nautical twilight, for which one would just substitute 6° and 12° for the 18° in the relation.

Figure A1.7 shows the length or duration of astronomical twilight for different declinations and latitudes, calculated by spherical trigonometry from the astronomical triangle. The durations of nautical and civil twilights can be found by respectively taking two-thirds and one-third the value found for astronomical twilight. To find the standard time of the ending of twilight simply add the duration of twilight and the standard time of sunset found in Sections 7.3 and 7.4; or to determine the beginning of twilight, subtract this duration from the standard time of sunrise. In Figure A1.7 the solid lines refer to declinations with the same signs to the latitudes (northern declinations for the northern hemisphere) and the dashed lines refer to declinations with the opposite signs to the latitudes (southern latitudes for northern declinations and vice versa). Note the way in which the curves for negative declinations reverse themselves. For a given latitude, say 65° N, the shortest twilight is for a solar declination of −10°. (The actual minimum at this latitude is for $\delta = -8°$, for which the curve is not shown.) The difference is not very important, and it is preferable to keep the graph simple. At higher latitudes, it is better to use Table A1.3 discussed below.

Before that, though, refer to the two examples used in the last Section. At 33° N latitude (Dallas), on May 9, when the Sun is at 17° N declination, astronomical twilight lasts for $1^h\ 33^m$; adding that figure to the standard time already derived yields $20^h\ 19^m$ local mean solar time or $20^h\ 46^m$ central standard time. At 54° N (Leeds) at the time of summer solstice passage by the Sun, astronomical twilight lasts all night.

Table A1.3 gives the time for the beginning and ending of astronomical twilight from the *Astronomical Almanac*, again for every four days for different latitudes for a year midway between leap years. Look up the example of Dallas, $\phi = 33°$, on May 9. Interpolation in the table as before gives a local mean solar time of $20^h\ 21^m$, sufficiently close to the $20^h\ 19^m$ found from the graphs. These tables should be used for near-polar latitudes.

All these calculations presume that we know where we are. That information is determined by applying our concept of time and the clock to the all the bodies of the sky, not just to the Sun.

Chapter 8

Positions in the sky and on Earth

The determination of time requires that we know exactly where the stars are in the sky, that we have their exact right ascensions. Once right ascensions and declinations of celestial bodies are known we can use them and the time to tell precisely where we are on Earth. Knowledge of celestial coordinates is also vital for the progress of astronomy and astrophysics, since they allow us to chart stellar movements and to understand the construction and evolution of our Galaxy.

8.1 Astrometry

Like all other fields of endeavor, astronomy is divided into a wide variety of specialties. One of the oldest, *astrometry*, involves finding the locations of stars in the sky, their right ascensions and declinations, that is, their positions relative to the celestial equator and the vernal equinox. It is a demanding subject that requires extreme care, a thorough understanding of the complex motions of the Earth and stars, and extensive calculation.

Stellar coordinates are traditionally determined with meridian transit circles (or related instruments) like the one shown in Figure 6.8. The principles involved are simple. As described in Section 6.9, the transit circle is constructed so as to point only to the celestial meridian, which can be defined by a set of crosshairs seen through the instrument. We can then observe a star the instant it is seen to cross the meridian. Graduated circles allow the determination of the altitude of the star above the horizon, h. With the latitude known and the relations developed in Section 2.6, we can then find the star's declination. For right ascension, we reverse the procedure of the previous section. Instead of using the stars to find the local sidereal time, LST, we use the LST to find right ascension, α. Since LST is by definition the instantaneous right ascension of the meridian, we determine the right ascension of a star by reading the LST from the observatory clock at the moment of meridian transit. No instrument can be constructed perfectly. A variety of self-correcting procedures are applied to the transit circle to determine its "error," the displacement of its crosshairs from the true meridian. Appropriate corrections are then applied to produce the final accurate positions. The astronomer no longer looks through the telescope to determine either time or coordinates. The procedures are wrought electronically and automatically. Nevertheless, the principles remain the same. With the best instrumentation and automatic measurement techniques, the coordinates of the few thousand brightest "fundamental" stars can be measured to a precision of $0''.01$ or better.

The procedure seems quite straightforward. But there is an intriguing dilemma. Sidereal time is used to find right ascensions and right ascensions are needed to find sidereal time: the astronomer appears to be trapped in a circular argument. The problem is resolved by referring finally to the Sun. Since the position of the ecliptic is defined in the sky, the right ascension of the Sun is a mathematical function of its declination, a concept used since before the time of Hipparchus (Section 5.5). A measurement of $\delta_\odot$, which is determined by the solar altitude and which is made independently of right ascension, gives $\alpha_\odot$. We therefore find the exact positions of the equinoxes by locating the Sun at 0° declination. With $\alpha_\odot$ known, we determine the right ascensions of a few of the brightest stars (or planets) by observing them in the daytime with the transit circle, timing their meridian passages with a sidereal clock that was "set" (or corrected) by the meridian passage of the Sun. The right ascension of any star can now be found at night by measuring the time interval of meridian transit between it and one of these bright daytime stars.

Even though the complexity of the procedure is increased, it still seems relatively straightforward: to find δ we use the measured meridian altitude and the known latitude; to find α we use the rotating Earth to measure the angle along the celestial equator between stars or between a star and the vernal equinox. In practice, however, there are a number of complications that make it a lengthy procedure and limit the attainable accuracy. One of the worst difficulties involves atmospheric refraction (Section 7.2). All stars appear higher in the sky than they actually are by amounts that depend on their altitudes. The astronomer must then make a correction to the altitude measured with the transit circle before calculating the declination. Great pains are taken to evaluate the degree of refraction, which depends strongly on local conditions. Refraction is important not only in declination measurements, but for right ascension as well, since the declination of the Sun is used to establish the position of the vernal equinox. The Earth's irregular rotation and the polar motion (Section 6.10 and Figure 6.11) cause continuous variations in the observer's latitude and in the sidereal time that are reflected into stellar coordinates. These are largely eliminated by applying the known secular (steady) rotational decay of the Earth and by averaging many observations.

Additional, and very serious, problems are caused by the precessional and nutational motions of the Earth and by the motions of the stars themselves. The coordinates of the several thousand bright stars cannot be measured all at once, but require several years at the telescope. During the course of the observing program, the stars move with respect to the Sun and with respect to one another as they pursue their paths around the Galaxy. This *proper motion* (Figure 8.1) is not visible to the naked eye over a human lifetime, but over a period of several thousand years it will produce a distortion of our familiar constellation patterns (Figure 8.2). Proper motions can be detected for the nearer stars telescopically over a period of a few years or decades, and are important to

(a)

(b)

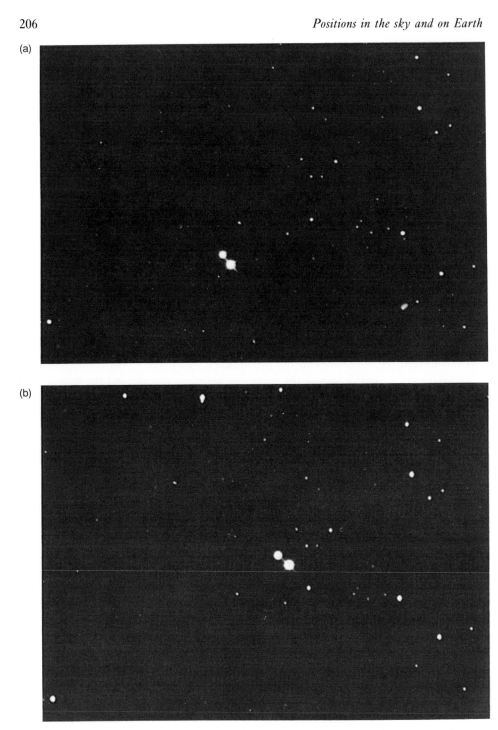

8.1. The proper motion of 61 Cygni. This binary star moved nearly three minutes of arc between 1916 (a) and 1948 (b). The star was the first candidate for parallax measurement because its rapid motion implied proximity. (Lowell Observatory.)

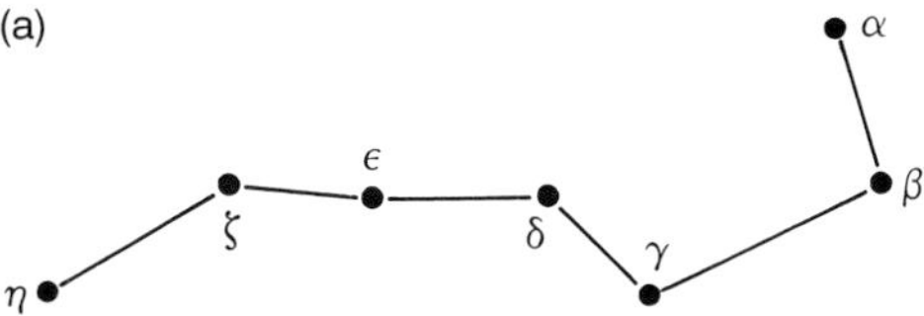

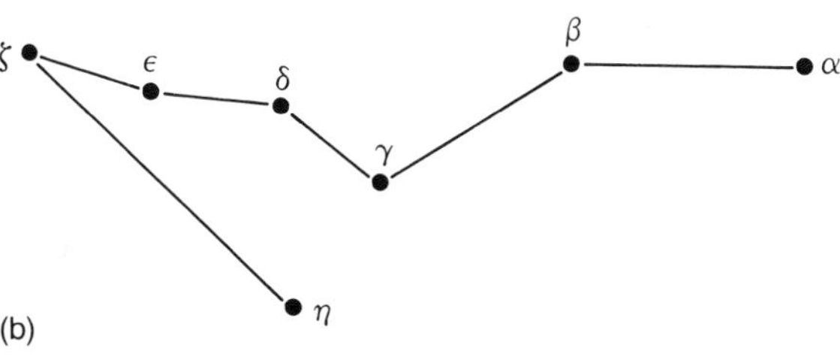

8.2. The changing shape of the Big Dipper, or Plough. (a) shows the present shape. In 250 000 years it will be distorted into the figure seen in (b). Only the middle five stars retain anything like their original configuration as they move east at about the same rate.

the astronomer for what they can tell about the gravitational forces and the dynamics of our Galaxy. The motions show up as changes in α and δ that are independent of precession. They are largely random in nature, that is, there is little relation between the proper motion of one star and that of another. Coupled to this movement is the star's *parallax*, the annual shift in the star's position caused by the orbital motion of the Earth (Section 4.10 and Figure 4.16), which is easily removed from the steady effects of proper motion. The reality of proper motion is vividly revealed by the star ρ Aquilae: in 1992 it moved across the border into the constellation Delphinus.

We are trying to measure the positions of moving objects – stars A and B – from a shifting platform. We determine the coordinates of star A first. By the time star B is observed, both the Sun (and Earth) and A have moved. We know where star B was at time b, and A at time a, but we do not know their relative positions at the same instant. The problem is resolved by observing each star twice.

Say we have a program to catalogue and measure the positions and proper motions of N stars. Each is observed once, the whole operation taking a few years. We correct each observation for precession, nutation, and aberration, that is, we determine the locations of the stars in a corrected coordinate system appropriate to a specific moment, the epoch (Section 5.6) of the catalogue. We have now observed where each star was at a particular time, the time of observation, where the position has been corrected for precession, but we do not yet know where they all were at the *same* time. After a few years, or even decades, we repeat the whole program, observing all N stars again. By comparing the two sets of data, we can see how the stars have moved, and we simultaneously determine (for each star) both its position (α and δ) at the chosen epoch and its proper motion. We next test all the derived proper motions to see whether or not they are truly random. If the rates of precession that were used contain errors (which at some level they most certainly do), the proper motions should show systematic shifts that should reflect them. That is, all the stars in one part of the sky might appear to have an average motion in one direction. We then adjust and improve the precession constants so as to make the proper motions truly random again. In effect, we take all the available data and solve simultaneously for right ascensions, declinations, proper motions, and precession (with its attendant variations) so as to obtain the most self-consistent results. However, it is clear to us that some stars really *do* have systematic motions that are related to the dynamics of the Galaxy. These motions become confused with precession, limiting our ability to separate the effects without additional data.

Astronomers define three kinds of star positions that have different corrections for precession, nutation, and aberration. The *mean* coordinates (or *mean place*) of a star are determined relative to the mean position of the equator and equinox for a specific date, that is, nutation (see Section 5.9) is smoothed out and aberration (Section 5.11) ignored. The *true place* takes nutation into account, that is, the coordinates are determined relative to the equator and equinox as shifted by nutation. In the *apparent place* both nutation and the effects of aberration are included.

8.2 Star catalogues and atlases

Stars are organized and their coordinates and other data are listed in a great variety of catalogues, starting with that of Aristyllus and Timocharis in the third century BC, which formed the basis for Hipparchus's discovery of precession (see Section 5.5). Modern star catalogues fall into three broad kinds: precision catalogues, which contain highly accurate positions of a relatively small number of stars; survey catalogues that contain coordinates of lower accuracy for a much larger number; and specialty catalogues that are concerned more with other data than they are with coordinates. There are also those that fall between.

Precision catalogues incorporating proper motions first appeared in the nineteenth century. Each new effort results in an improvement in the proper motions and in the mathematical constants of precession. The usual procedure is to observe the few thousand brightest stars with the meridian circle to establish a *fundamental catalogue*, and then to measure tens of thousands more by photography relative to the fundamental stars. The most important of these volumes is the FK5 (*Fundamental Katalog* number 5), a compilation of the precise coordinates of 1535 bright stars. Every few decades this fundamental catalogue is updated with a new series of observations: the FK3 was made standard in 1935, the FK4 in the 1950s, and the FK5 in 1988. The stellar positions are given for every 10 days throughout the year in an annually issued volume called *Apparent Places of Fundamental Stars*. Another catalogue, one of great historical importance, is the *Boss*, or *Albany*, *General Catalogue of 33,342 Stars for the Epoch 1950*, otherwise known as the GC. The results of several efforts are compiled in the *Smithsonian Astrophysical Observatory Catalogue* (SAO Catalogue), which gives mean coordinates and proper motions of about 250 000 stars, all reduced to the epoch 1950.0. The *Astronomical Almanac* (see Section 6.5) gives mean places precessed to the middle of the year for about 2000 stars, taken largely from the FK5 or, for fainter stars, precessed from the SAO.

One of the most important of all survey catalogues is the *Bonner Durchmusterung* (the famous BD, the *Bonn Survey*, epoch 1855), which covers the northern hemisphere and part of the southern to a declination of −23°. It and its southern hemisphere companion, the *Cordoba Durchmusterung* (CD, epoch 1875), include nearly three-quarters of a million stars, the less accurate positions serving mostly for stellar organization and nomenclature for stars well below naked-eye vision. The BD and CD divide the sky into one-degree-wide bands. Within each band, stars are numbered consecutively from 0^h right ascension. Thus Vega = α Lyrae = 3 Lyrae is also BD +38°3238 as well as SAO 067174; Canopus is α Carinae and CD −52°914. The BD and CD terms are commonly combined into DM for "Durchmusterung," so Vega is also called DM +38°3238. Nearly 150 years of precession, however, have moved a number of stars out of their declination bands. As an example, the star 49 Arietis is BD +25°477 (or DM +25°477), but its epoch 2000 declination is +26°04′. The original designation still applies, however, in spite of precession.

There are numerous other specialized catalogues. The HR numbers, for example, are used to identify the naked eye stars (see Section 4.7). The most extensive of the others is the *Henry Draper Catalogue* (HD), which orders a quarter of a million stars according to increasing right ascension, and which provides basic spectroscopic data.

The companion to and the visual analogue of the star catalogue is the star map or atlas. Older maps that plot naked eye stars together with constellation figures were encountered in Section 4.8 and an elementary atlas is presented in Appendix 2. Star atlases come in two basic varieties: the map that is plotted by hand or computer from a catalogue and the photographic atlas. The stars of the BD and CD catalogues are hand-plotted in three folio-sized volumes: the original BD from the NCP to a declination of −2°, a BD southern extension to a declination of −23°, and the CD, which plots stars from −21° to the SCP. The stars of the SAO catalogue are computer-plotted in a quarto-sized set of 152 loose charts (Figure 8.3 inset).

A photographic atlas consists of carefully matched and exposed images of the sky, and can show literally billions of stars as faint as the 20th magnitude (see Section 4.6). The most famous of these atlases is the *Palomar Observatory Sky Survey* (Figure 8.3), made with the 48-inch Palomar Schmidt camera (see Section 13.10) and completed in the mid-1950s. It consists of nearly 1000 pairs of photographs, each about 6° square, covering the sky from the NCP to a declination of 33° S, and made in blue and red light (the photos themselves are in black and white, just taken through colored filters). A 12° southern extension (the "Whiteoak fields"), taken in red light only, was added some years later. The Scientific Research Council of Great Britain and the European Southern Observatory have jointly extended this fundamental photographic reference atlas into the southern hemisphere with similar pairs of photographic plates that cover

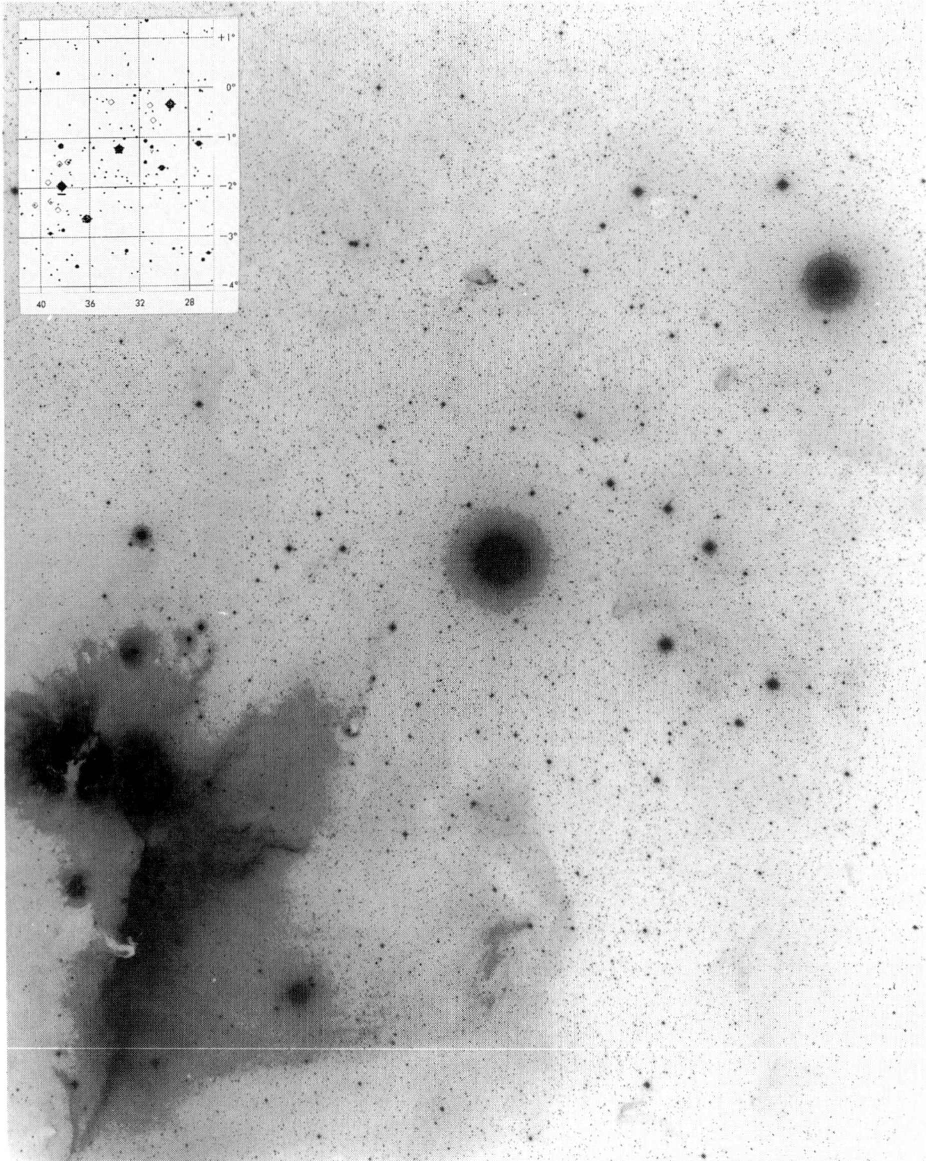

8.3. Star atlases. The photograph shows a 2°.3 × 3°.5 segment of the blue print of the Palomar Observatory Sky Survey centered on Orion's belt and reproduced to nearly original scale. Astronomers commonly use such negative prints. The two bright stars at center and to the right are ε and δ Ori respectively. Numerous nebulae are seen, including the famed dark Horsehead Nebula at lower left, the upper part of which nearly obscures ζ Ori. The faintest stars are near magnitude 18. The inset at upper left shows that section of the sky from the Smithsonian Astrophysical Observatory (SAO) atlas, also nearly to scale. The declination and the minutes of right ascension (5^h) are indicated. Try to match the stars. (Photograph: National Geographic–Palomar Sky Survey; SAO atlas: Smithsonian Astrophysical Observatory.)

the sky from 20° S declination to the SCP. The northern hemisphere is again being photographed from Palomar, but now in three colors: ultraviolet, blue, and red. This version records fainter stars and, by comparison with the original atlas, also allows the detection of the motions of a huge number of nearby stars.

The photographic atlas allows the reversal of the older procedures of plotting maps from catalogues. The Palomar Survey forms the basis for the *Minnesota Catalogue*, which contains data on a *billion* stellar images. In the 1980s, the Palomar Schmidt telescope and its southern hemisphere sister were used to take a (relatively) quick survey of the entire sky for the compilation of the *Guide Star Catalogue* of the Hubble Space Telescope (see Section 13.12). It contains highly accurate positions of 15 million stars and numerous non-stellar objects, and is readily accessible on optical disks at relatively low cost.

These photographic atlases, however, are expensive (the last printing of the old Palomar Sky Survey cost $6000, and the new one $60 000) and are beyond the means of the average person. Numerous other atlases and maps, however, are available to both the casual observer and the active amateur. The best known are probably *Norton's Star Atlas*, the out-of-print *Skalnate Pleso Atlas of the Heavens*, and W. Tirion's *Atlas 2000* (see the Bibliography). The latter two plot stars brighter than magnitude 7.75 and 8.0 respectively on several individual maps, and include a wealth of non-stellar objects. As the sizes and quality of amateur telescopes increase, so do the quality and complexity of their atlases, so that now the professional and amateur are beginning to cross over and merge. An example is the *Uranometria 2000* by W. Tirion, B. Rappaport, and G. Lovi, which records stars to tenth magnitude (with a huge variety of non-stellar objects). The photographic *Atlas Stellarum* by Hans Vehrenberg, which covers the sky to a declination of 25° S, is also readily available (see the Bibliography).

8.3 Celestial navigation

Now we look from the sky back to the Earth. The stars are critically important to human affairs. Without them, we would not know where we are, and exploration of our planet would have been considerably more difficult and dangerous. The horizon coordinates of a star, altitude and azimuth (h and A), the equatorial coordinates, right ascension and declination (α and δ), the time of day, and the terrestrial coordinates, latitude and longitude (ϕ and λ), are all interdependent. If we know any three of these four sets of data we can find the fourth. In Section 8.1, we used the time, altitude and azimuth, and the latitude and longitude to find α and δ (stars are observed on the meridian, with the azimuth precisely known to be 180°). In Sections 6.10 and 6.11, we used the terrestrial, equatorial, and horizon coordinates to find the time. In this section, we will look at another permutation, and use the horizon and equatorial coordinates, plus the time, to find latitude and longitude.

This discussion has already been well anticipated: recall the dictum referred to several times that the altitude of the visible celestial pole equals the latitude of the observer, and the discussion in Section 6.1 regarding the equality of time and longitude differences. Early navigators in fact used primitive instruments to measure the angular elevation of the pole above the horizon. Today, we can simply estimate the altitude of Polaris, and have our latitude to well within one degree. Centuries and millennia ago, Polaris was not near the pole because of precession; nevertheless, it was not particularly difficult to locate the point of zero rotation in the northern sky. The measurement of the navigator's latitude, or north-south position, is just that easy.

The determination of the east–west, or longitudinal, position provides a dramatic contrast. To find longitude, the navigator measures time differences. In principle, the mariner finds the local mean solar time from the Sun (correcting for the equation of time) at, say, London, and sets a clock. He or she then sails to New York, taking the clock along. At New York, the LMST is again measured from the Sun's hour angle. A comparison of the local time with the clock that reads London time yields the longitude. This procedure means that the navigator must have a clock that can keep precise time for as long as it takes to make the journey, which in the early days of exploration meant many months.

Two methods became available in the mid-eighteenth century, the first ingeniously using the Moon as a clock. By around 1750, the theory of the motion of the Moon, as calculated by the German astronomer Tobias Mayer, was sufficiently good to tell time by the lunar position relative to the stars. A mariner anywhere in the world could then know the time at London by consulting a previously prepared lunar *ephemeris*, a table of lunar positions (either right ascension and declination or celestial latitude and longitude) listed against time. The accuracy of the procedure is limited, however. Our satellite moves slowly, and small errors in position yield large ones in time and hence in longitude. What was required was a true, reliable mechanical clock. The need was sufficiently great that the British Crown (George III) put up a prize of £30 000 for its invention. The award was ultimately won by John Harrison for his shipboard chronometer (Figure 8.4), the final version being tested about 1765. The two methods competed for a time, but the latter clearly won out because of the steadily improving accuracy of mechanical clocks. The effect of the "discovery of longitude" as it is called is obvious: look at an old map of the world, or of North or South America, drawn before 1700. The maps are invariably highly distorted. But careful examination shows the north–south proportions to be accurate; the distortion is east-west, reflecting the lack of good clocks.

In practice, latitude and longitude are determined simultaneously. There is a unique place on Earth where any particular star has a given altitude and azimuth at a given time. Any motion of the observer will change one or both of them. Consequently, a measurement of altitude and azimuth (with, of course, the correct time and the equatorial coordinates of the star) is sufficient to determine the observer's latitude and longitude, ϕ and λ. Mathematically, we have two unknowns, ϕ and λ, and two equations, those linking ϕ and λ to altitude and azimuth, h and A.

The measurement of altitude is quick, simple, and accurate. The azimuth of a celestial body, however, is not so easy to find, as one must first accurately establish north or south. During the day, north or south can be found by tediously observing the solar culmination to find the direction of maximum solar altitude. We then simultaneously find h and A (the latter being 0° or 180°) for the Sun, allowing calculation of ϕ and λ.

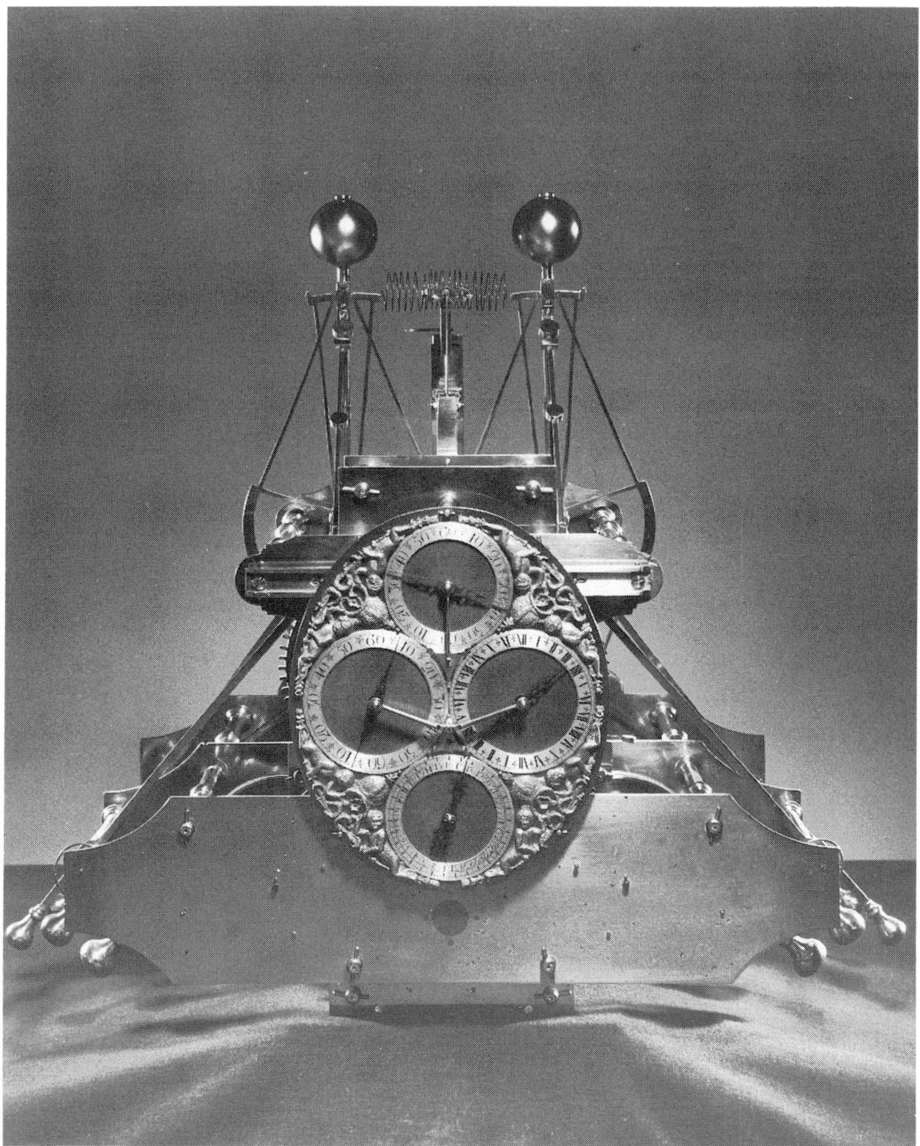

8.4. An early Harrison chronometer. It made precise navigation possible. (National Maritime Museum, London.)

This case is the simplest: since $\delta_\odot$ is known on a given day, altitude defines latitude and meridian transit defines apparent solar noon, from which longitude is easily derived. During the night, for a star, the evaluation of accurate azimuth at sea is nearly impossible. The problem is easily overcome not by measuring the altitude and azimuth of one body, but by measuring only the altitudes of two or more bodies. The mathemat-

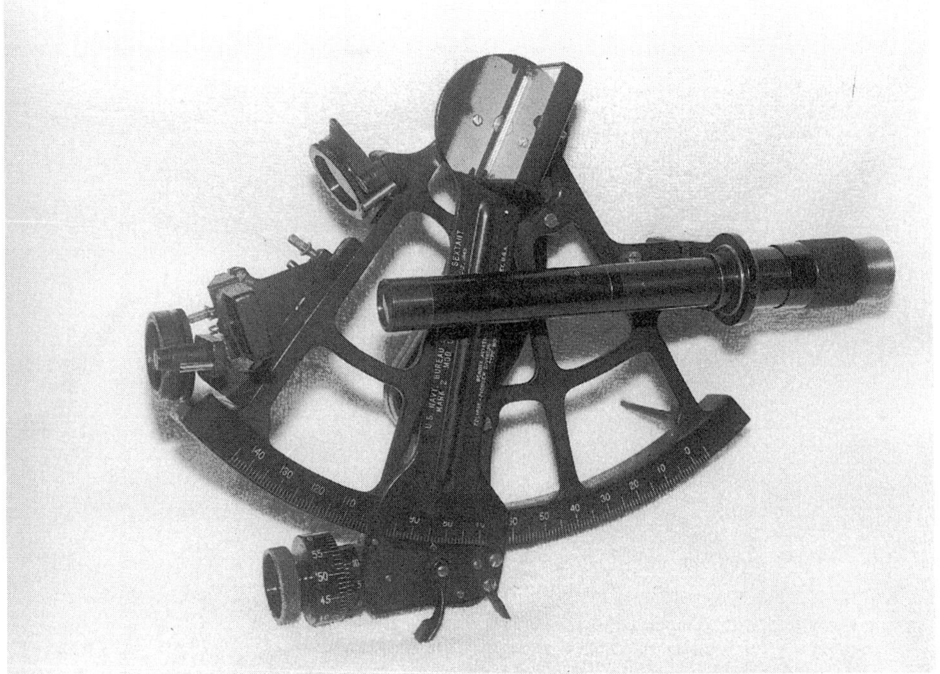

8.5. A sextant. This US Navy Mark II sextant is of the kind used in World War II. Compare it with the diagram in Figure 8.6. A solar filter is just down and to the left of the upper mirror. (Department of Astronomy, University of Illinois; author's photograph.)

ical requirement of two equations in two unknowns is satisfied, and we can still determine ϕ and λ.

At sea, altitudes of stars are measured with the hand-held *sextant* (Figure 8.5), so called because it is based on a sixth of an arc of the circle (as an *octant* is based on an eighth of the circle). It is a venerable instrument. Modern ones look remarkably similar to the one used by the great explorer Captain James Cook, the first navigator to use the new chronometer. The device (diagramed in Figure 8.6) consists of a telescope through which the navigator simultaneously looks at and through a half-silvered mirror. A second mirror reflects starlight to the half-silvered mirror and then through the telescope. The navigator sights a star (or a planet) at twilight when the horizon is visible, and rotates the upper mirror until the star appears to touch the horizon. The altitude of the star can then be read from the graduated section of the circle (because of the design, an angle of up to 120° can actually be read). The observed altitudes are subsequently corrected for atmospheric refraction (Section 7.2) and horizon dip (Section 7.5).

During the day, the navigator can use the Sun (observed with a filter that can be rotated into the field of view) and the Moon. The observer sets the limb of either body on the sea horizon. Additional corrections must be made for the solar or lunar radius (or semidiameter, see Sections 7.1 and 7.4) and for the lunar *parallax*. The parallax in

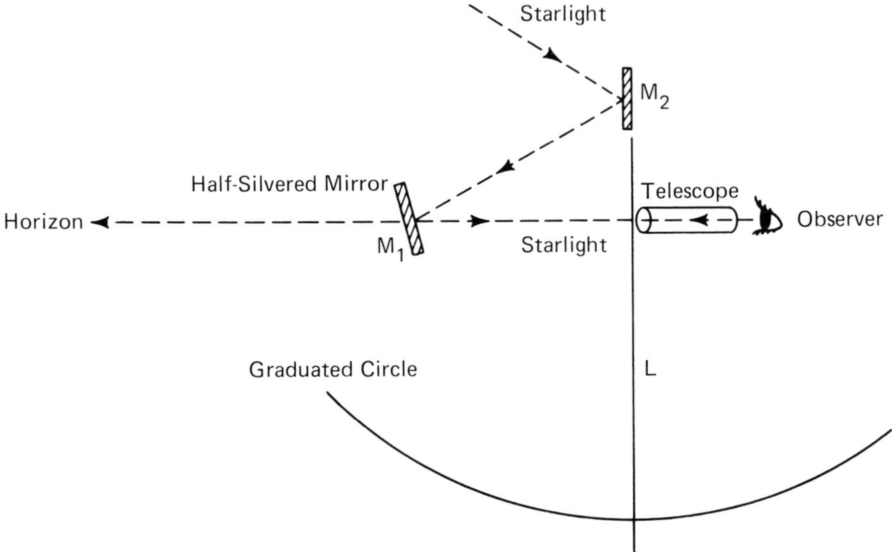

8.6. The principle of the sextant. The observer sees the superimposed image of the star and horizon through the telescope and the half-silvered mirror M_1. By rotating mirror M_2 with the arm L, the navigator makes the image of the star appear on the horizon. The altitude of the star is then easily read on the graduated circle.

this instance is an effect of the finite size of the Earth, which shifts the observer's position relative to the Moon, and hence changes the altitude that would be observed were the Earth a point (see Section 9.8).

A navigator uses each star observed to establish a *line of position*, or *Sumner line*, named after Captain Thomas Sumner, who accidentally discovered the concept in the early nineteenth century. Listen to the tale direct from his ship's log:

> Having sailed from Charleston, S.C., November 25th, 1837, bound for Greenock [Scotland], a series of heavy gales from the westward promised a quick passage; after passing the Azores the wind prevailed from the southward, with thick weather; after passing longitude 21° W no observation was had until near the land, but soundings were had not far, as was supposed, from the bank. The weather was now more boisterous and very thick, and the wind still southerly; arriving about midnight, December 17th within 40 miles, by dead reckoning, of Tuskar light, the wind hauled SE true, making the Irish coast a lee shore; the ship was then kept close to the wind and several tacks made to preserve her position as nearly as possible until daylight, when, nothing being in sight, she was kept on ENE under short sail with heavy gales. At about 10 a.m. an altitude of the Sun was observed and the chronometer time noted; but, having run so far without observation, it was plain the latitude by dead reckoning was liable to error and could not be entirely relied upon.
>
> The longitude by chronometer was determined, using this uncertain latitude, and

> it was found to be 15′ E of the position by dead reckoning; a second latitude was then assumed 10′ north of that by dead reckoning, and toward the danger, giving a position 27 miles ENE of the former position; a third latitude was assumed 10′ farther north, and still toward the danger, giving a third position ENE of the second 27 miles. Upon plotting these three positions on the chart, they were seen to be in a straight line, and this line passed through Smalls light.
>
> It then at once appeared that the observed altitude must have happened at all of the three points and at Smalls light and at the ship at the same instant.
>
> Taken from Bowditch's "Practical Navigator."

He did not know exactly where he was but knew he must be somewhere on that line, which he had only to follow to reach safety.

Consider star A observed to be at a given altitude h_A and zenith distance $z_A = 90° - h_A$, and examine Figure 8.7. There will be a small circle a of radius z_A, centered on the substellar point S (the point on Earth directly under the star, where the star appears at the zenith), for which the star will have that particular zenith distance z_A. Having observed the star, the navigator knows that he or she must be somewhere on that small circle. A second star, B, is observed that places the observer somewhere on a second circle b. The ship must then be at one of the two intersections. These two points are usually so far apart there is no trouble discriminating between them. As a general rule a third star is always observed. The intersection of the three circles pinpoints the correct location.

In actual practice, the navigator calculates and plots on a chart a small section of each circle approximated as a straight line, as in Figure 8.8. The lines are actually the tangents to the circles of position, and as we can see in Figure 8.7, the tangent at any point must be perpendicular to the direction to the substellar point. A line of position in Figure 8.8 must therefore be at right angles to the direction of the star's azimuth and displaced from the substellar point by an arc equal to the observed zenith distance. Because of the unavoidable observing errors, the lines derived for the three stars will not pass through a point, but will define a triangle, or *circle of confusion*, within which the navigator and ship must lie. An experienced navigator can locate the ship within a few nautical miles, well within the area bounded by the observed horizon.

8.4 Precise terrestrial position

Sea travel, without identifiable landmarks such as rivers or mountains, provided the first real need for celestial navigation. Land maps could be drawn by eye, and from the ancient traveler's point of view, distortions were not terribly important. But in a modern world we need to know correct relative locations and boundaries. Errors in a map are unacceptable to an airplane pilot and highly accurate ϕ and λ must be available at the telescope for astrometric purposes. Accurate meridians and parallels must also be available even to define land ownership. On a moving, rolling ship, observation is made with a hand-held instrument with a resulting error of a few kilometers. On land,

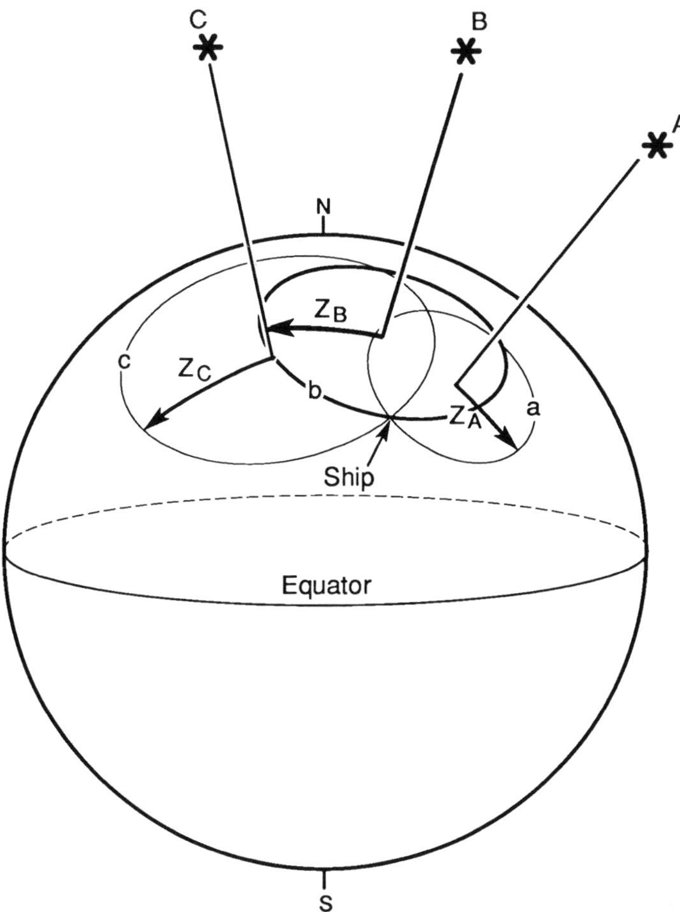

8.7. Circles of position. The lines from the stars A, B, and C intersect the Earth at their substellar points, the positions at which the stars appear in the zenith. The small circles a, b, c on the Earth are the loci for which the stars A, B, C, have constant zenith distances z_A, z_B, z_C, respectively. All circles are on the front of the sphere; that for star B is heavily drawn to distinguish it easily from the others. The position where the circles all cross is the observer's location.

we can apply these same principles, but now with a fixed device capable of measuring altitudes with very high precision.

An example of such an instrument is the pendulum astrolabe, which consists of a telescope that is constrained to point to an altitude of 60°. The astrolabe can be fixed to the ground and be precisely leveled, unlike the sextant. The observer times the passages of many stars – perhaps 10 or 20 – as they cross the $h = 60°$ almucantar on their daily paths. Each one yields a line of position, and the latitude and longitude can be accurately fixed by a statistical analysis of the circle of confusion. Coordinates of a

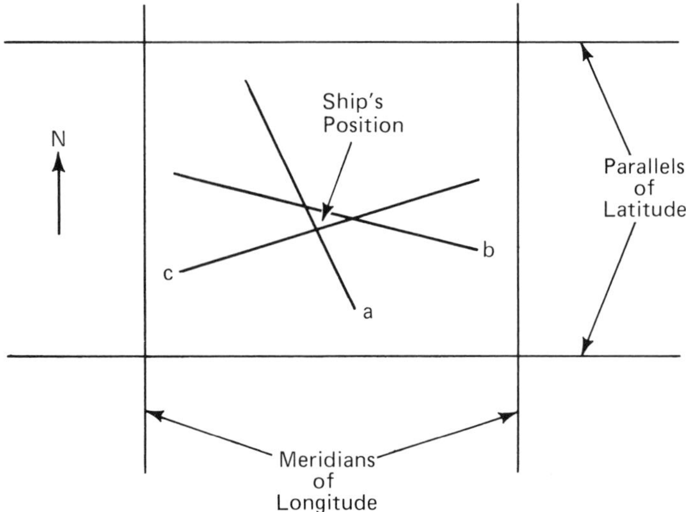

8.8. Lines of position, or Sumner lines. Straight-line approximations to the three circles in Figure 8.7 are plotted and greatly expanded. The ship is within the triangle bounded by the lines. A circle inscribed into the triangle is the *circle of confusion* and is typically only a few kilometers across.

fraction of a second of arc (which corresponds to about 100 meters) or better can easily be defined in this way. The use of separated radio telescopes working as an interferometer can determine relative coordinates (see Section 6.11) to within a few centimeters, a mere ten-thousandth of a second of arc.

At the observatory, precise latitude, necessary for the astrometric determinations of declination, can be measured without prior knowledge of stellar declination by observing the zenith distances (corrected for refraction) of both the upper and lower transits of circumpolar stars. From Section 2.6 and Table 2.2, we see that the zenith distance of a star north of the zenith and above the pole (upper transit, z_u, in the northern hemisphere) is

$$z_u = \delta - \phi,$$

and that the zenith distance for a star below the pole and above the horizon (lower transit, z_l) is

$$z_l = 180° - \phi - \delta.$$

If we add z_u and z_l, δ cancels, and we find

$$z_u + z_l = 180° - 2\phi,$$

or

$$\phi = 90° - {}^1\!/_2(z_u + z_l).$$

If declinations are known, we can also observe two stars (called *Talcott pairs*) that are very close to and just north and south of the zenith. To the south, the zenith distance

$$z_S = \phi - \delta_S,$$

or

$$\phi = \delta_S + z_S;$$

to the north,

$$z_N = \delta_N - \phi$$

or

$$\phi = \delta_N - z_N.$$

Adding the two expressions for ϕ,

$$2\phi = \delta_S + \delta_N + z_S - z_N,$$

or

$$\phi = {}^1\!/\!_2(\delta_S + \delta_N) + {}^1\!/\!_2(z_S - z_N),$$

which practically eliminates refraction effects.

The longitude at the observatory can be determined by finding the local time through transits of the Sun or stars and by comparing the local time with broadcast UT. Longitude can be found without an exact knowledge of stellar right ascensions; all we need is the *relative* time difference between two observatories, which can be found by comparing the transit times of the same star.

8.5. Satellite navigation

The major problem of celestial navigation is the weather. It may be cloudy for several days, during which time a vessel can accumulate a large and potentially dangerous positional error. For both sea and land navigation, the stars have largely been supplanted by satellite navigation.

The *Global Positioning System* (GPS) consists of a "constellation" of 24 Earth satellites. Each carries atomic clocks that are synchronized by radio signals from NIST. A receiver picks up time signals from the satellites that are above the horizon, and from the time delays can fix an absolute position to within about 10 meters and can give relative location to even better precision.

But sophisticated systems have a way of failing at the most inopportune moments. The sky always remains as a reliable backup system in case of radio or electrical failure. And ultimately, even the sophisticated electronic gear is dependent on celestial observations through absolute time determination and measurement of the slowing of the Earth's rotation. As good as electronic navigation gets – and the technology will rapidly advance – we can always return to the perpetual stars.

We now know the time and we know where we are. We next turn our eyes again to the sky to the other naked-eye bodies of the Solar System, the Moon and the planets, to explore their motions and other related phenomena.

Chapter 9

The Moon

Aside from the Sun, the Moon (Figure 9.1) has had a more profound influence on humanity than any other celestial body. It illuminates the night, largely controls the ocean tides, and affects the growth rates of some tidal marine animals. Our lives and languages testify to its significance. Ancient calendars are based upon the motion of the Moon, explaining the wandering dates of many religious holidays within our civil calendar. A major time unit, the *month* is derived from the period of the Moon about the Earth, as is the name of a day of the week: "Monday," "Montag," "Lundi" (from the Latin "luna" for "moon.") The folklore surrounding it is nearly endless: the orientation of the crescent's cusps foretell rain; crops must be planted at certain phases; the full Moon can bring on insanity ("lunacy"); and so on.

The Moon is physically unusual. Aside from Charon, the planet Pluto's moon, it is bigger relative to its parent body than any satellite in the Solar System, and with Earth forms something of a double planet (Figure 9.2). Most satellites of the Solar System orbit in the equatorial planes of their planets, testifying to their simultaneous origins. But the Moon orbits very nearly in the ecliptic plane, suggesting that it was created as part of the Solar System at large and not strictly as a by-product of the development (4.6 billion years ago) of our Earth; it seems to have been assembled from the debris created in a giant collision between the primitive Earth and a Mars-sized body in solar orbit. By coincidence, the angular diameter of the Moon is the same as that of the Sun, one-half degree, a fact responsible for spectacular solar eclipses. Even casual observation shows that the Moon keeps one face pointed toward us, the result of gravitational tides raised upon it by our planet.

9.1 Distance and size

Except for passing meteoroids – debris from space that sometimes hits us, to be examined in Chapter 12 – and artificial satellites, the Moon is the closest celestial body to the Earth. From the angular size of the Earth's shadow during a lunar eclipse, in the second century BC Hipparchus estimated its distance to be 59 times the radius of the

9.1. The rising Moon. Its size and brightness have profoundly influenced human affairs. (United States Navy.)

Earth (the actual value is 60.3). Combination of this ratio with Eratosthenes' measure of the Earth's circumference (see Section 2.8) then readily gives the actual distance. Modern measures initially used the Moon's parallax (see Section 4.10), in this context the angular position of the Moon as seen from two places on Earth (to be discussed in detail in Section 9.8). We now readily find the distance to the Moon to within a meter or so by bouncing laser beams off mirrors left by astronauts. The average distance is 384 400 km. When coupled to the lunar angular diameter, we find a physical diameter of 3476 km, about a quarter that of the Earth.

9.2 The lunar phases

The most obvious single feature of the Moon is its continuous change of shape, or *phase*. At times it is all but invisible, while at others it casts enough light to enable us walk around easily at night. Our satellite shines by the reflected glow of sunlight, and as it moves around our Earth we see varying degrees of its daylight and night-darkened sides. Its phase depends upon its spatial relation to the Sun and the Earth, as does its apparent position in the sky. Thus the Moon's phase intimately correlates with its time of visibility in the sky and with times of moonrise and moonset. The crescent always closely follows sunset or precedes sunrise, and the full Moon rises at sunset.

Figure 9.3 shows the Moon orbiting the Earth with the Sun shining far off to the right. For now, we ignore the orbital motion of our planet about the Sun and the deviations of the lunar orbit from circularity. The cycle of the phases takes 29½ days

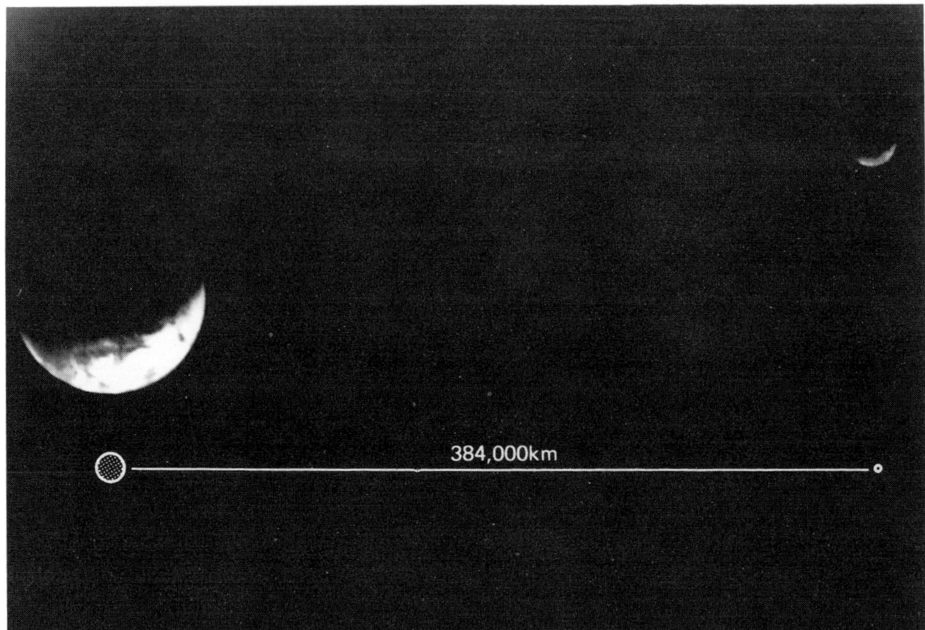

9.2. The Earth-Moon system. Our double planet is photographed from a distance of 11 million kilometers by *Voyager 1* on its way to Jupiter and Saturn. Because the Moon is much darker than the Earth, its brightness is enhanced in this picture for visibility. A true scale drawing of the Earth and Moon is shown below the image; the bodies in the photo appear closer because of foreshortening. (NASA/JPL.)

to complete, and is the interval between successive *conjunctions* with the Sun. The geometry of lunar and planetary angles (or aspects) has a rich vocabulary. *Conjunction* means that two celestial bodies have the same celestial longitudes (or in some contexts right ascensions); if they are 180° apart, they are said to be in *opposition* to one another. We will encounter these and several other terms when we discuss the planets in Chapter 11.

Assume first that the Moon is in conjunction with the Sun, when it is at position 1 in Figure 9.3, directly between the Sun and the Earth. The Sun illuminates the right-hand hemisphere only; we on Earth are left looking at the nighttime or dark side, and the Moon is invisible to us. This *new Moon* is the start of the phase cycle. Since the Moon is seen in the direction of the Sun, the two travel nearly together through the day, the new Moon rising at sunrise and setting at sunset.

Note here two cautionary facts. First, do not confuse the *dark*, or *nighttime* side of the Moon with the *far side*. The far, or back, side is never seen from the Earth since our satellite always keeps one face pointed toward us. The far side witnesses as much sunlight as the *near side*, that visible to us from Earth. At new Moon, the far side is in full daylight, whereas all of the near side is in darkness. Second, if the Moon were exactly between the Earth and the Sun, it would cover the face of the Sun and we

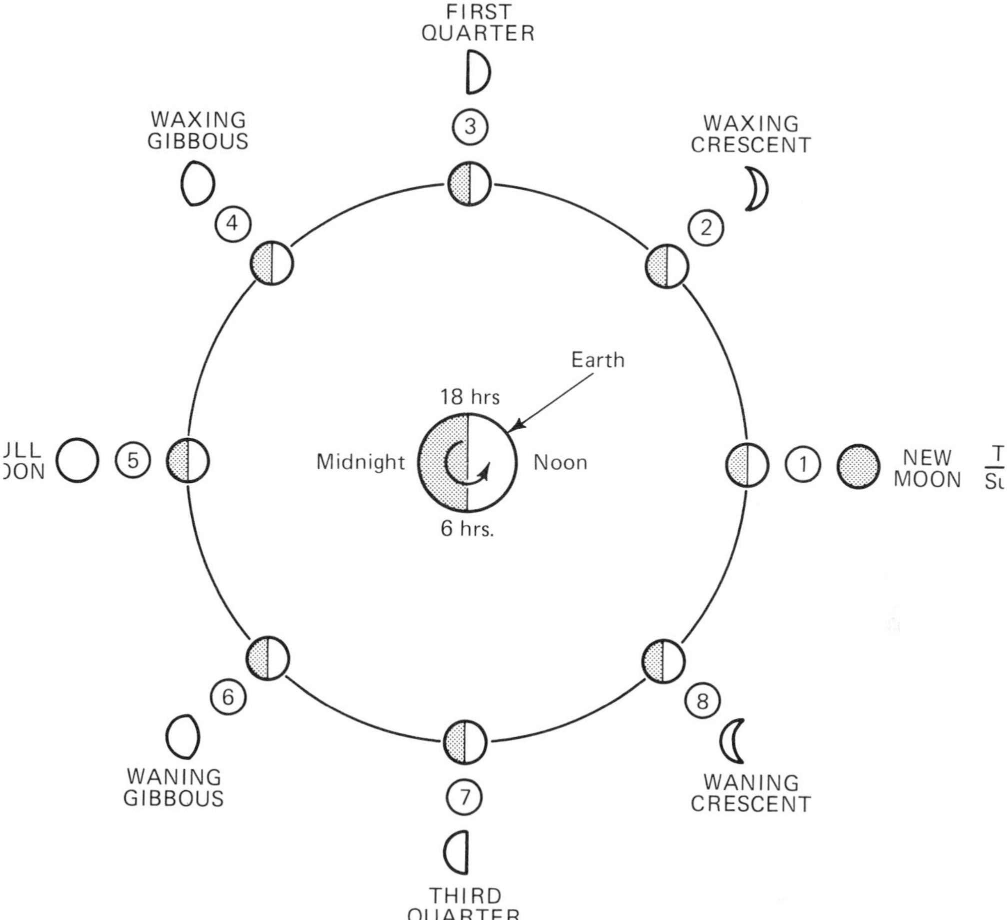

9.3. The phases of the Moon. The Earth, rotating counterclockwise, is at the center. The shaded areas denote the nighttime sides of the Earth and Moon, those opposite the Sun, which is far off the page to the right. The apparent lunar shape as viewed from Earth depends upon the relative amount of the daylight side that can be seen. Since phase depends upon the angle between the Moon and Sun, it must correlate with moonrise and moonset.

would witness a solar *eclipse*. Occasionally such eclipses take place, but remember from Section 5.9 that the lunar orbit, although close to the ecliptic plane, is inclined to it by 5°, and usually the Moon passes somewhat north or south of the Sun. *Conjunction* means that two bodies have same celestial longitudes, but the latitudes need not be the same.

The Moon orbits the Earth counterclockwise as viewed from the north, the same direction as the Earth's rotation and revolution. After just a bit over one week, it will move a quarter of the way around, and will be at position 3 in Figure 9.3, called the *first quarter* (Figure 9.4). Our companion is now said to be "7 days old," the *age* of the Moon here meaning the number of days since the new phase. As we look at the Moon from Earth, we will now see half of it in daylight, the other half in night (half of the far side is in daylight as well), and the Moon will have the visible shape shown above the position number. This phase is popularly called a "half-moon." The *terminator* is the line dividing day from night: a person standing at this position on the Moon would be seeing sunrise. A line bisecting the terminator, the line drawn perpendicular to it at the center, points directly at the Sun at all lunar phases. You can verify this statement by looking at the Moon in the daytime sky.

Now think of yourself on the rotating Earth. The first quarter Moon is 90° east of the Sun, and it must cross your meridian at 18 hours local time, on the average at sunset; it thus rises on the average at noon and sets at midnight. Consequently this phase is easily visible in the eastern sky in the afternoon and in the western sky after sunset. The Moon closely follows the ecliptic, and the vagaries of moonrise and moonset are the same as those described in Chapter 7 for the Sun, complicated by the lunar orbital inclination of 5°. For example, if first quarter occurs on March 21, the Moon must be near the summer solstice, and in the northern hemisphere it must rise *before* noon. We will examine the details of moonrise and moonset in Sections 9.11 and 9.12. For now, consider only mean times; averaged out over the year, the first quarter rises at noon.

Between new and first quarter we will look mostly at the Moon's dark side, and will see only a portion of daylight. At all locations between 1 and 3, for example at position 2, the Moon appears to us as a *crescent* (Figure 9.5). The points, or horns, of the crescent are called the *cusps*. This phase between 1 and 3 is called the *waxing* crescent (related through Anglo-Saxon to the German *wachsen*, to grow), since the area of visible daylight grows larger with time. The waxing crescent is less than 90° east of the Sun, and it must therefore rise before noon and set before midnight, crossing the observer's meridian during the morning.

The Moon first becomes visible as a thin, narrow crescent about one day after new in western twilight, setting just behind the Sun. This first appearance of the crescent is in fact sometimes called the "new Moon," and its sighting begins the new month in a lunar calendar. Technically, even at the true new phase, the Moon appears as a crescent north or south of the Sun unless it exactly covers the Sun and we see an eclipse. But since it is at conjunction, the Moon must always be less than 5° from the Sun and is never visible because of the brightness of the blue sky.

Past first quarter, we will see more sunlit face than nighttime side, and now the invisible nighttime portion is the crescent. Between points 3 and 5 (for example point 4) in Figure 9.3, we see the out-of-round figure of the *gibbous* Moon growing toward full, or the *waxing gibbous* Moon (Figure 9.6). Rising progressively later as it increases its easterly angle from the Sun, the waxing gibbous Moon rises in the afternoon, is seen in the east at sunset, and sets between midnight and dawn.

At point 5 the Moon becomes *full* (Figure 9.7). It is opposite, or in opposition to, the Sun, and we see the entire sunlit face. The far side, that which we never see from Earth, is now in darkness and the terminator is coincident with the apparent edge, or

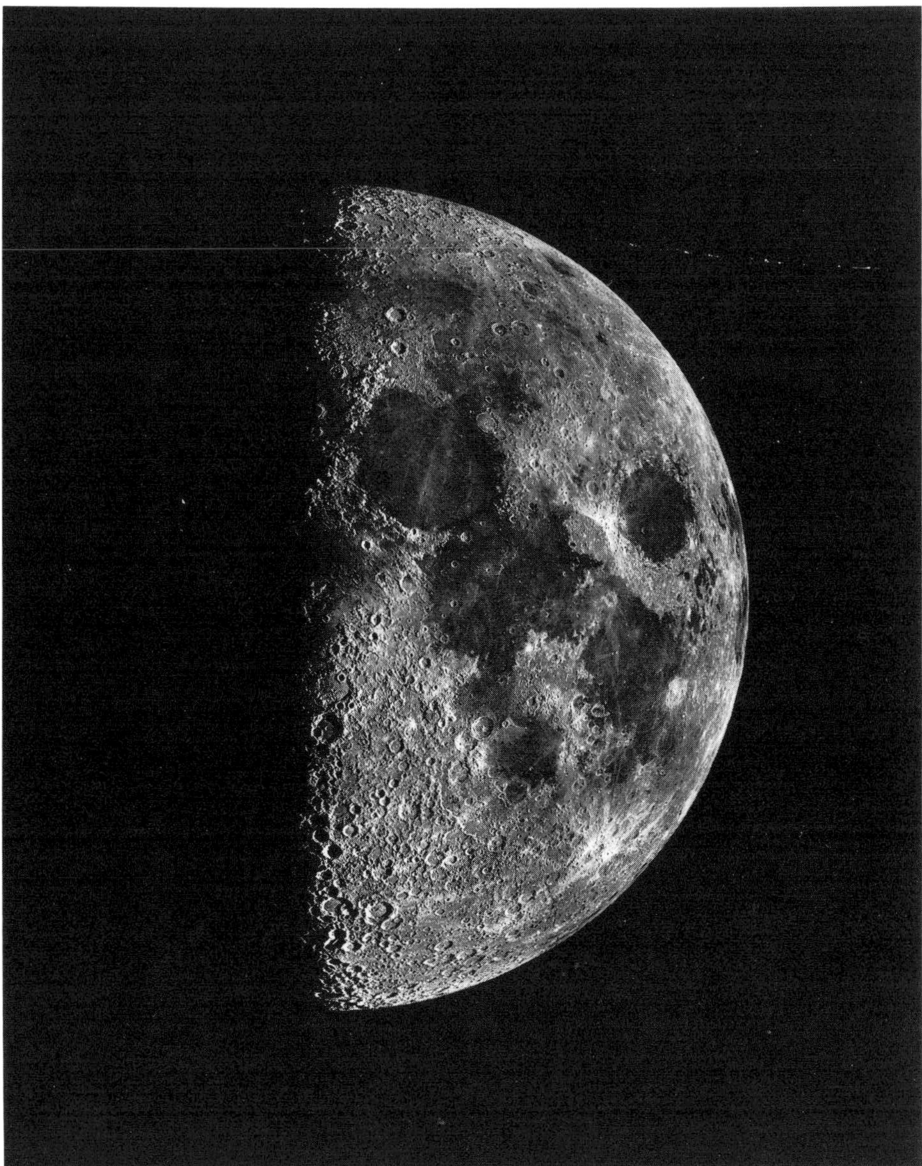

9.4. The first quarter Moon. The Moon has orbited 90° counterclockwise from the Sun and is at position 3 in Figure 9.3. The terminator, which divides day from night, runs down the middle. Sunrise is creeping to the left. Several maria (dark, lava-filled basins) are visible as are myriad craters. (Lick Observatory.)

9.5. The crescent Moon (near position 2 in Figure 9.3) only 4 days after new. The terminator now appears curved, and a different set of craters stands out in stark relief. (Lick Observatory.)

limb, of the Moon. Since the observer is now on a line between the Moon and the Sun, the full Moon must rise at sunset or set at sunrise. Unlike the crescent and gibbous phases, which are phases of duration and grow and shrink with time, the new, full and quarter Moons are phases of the instant. Quarter separates the gibbous from

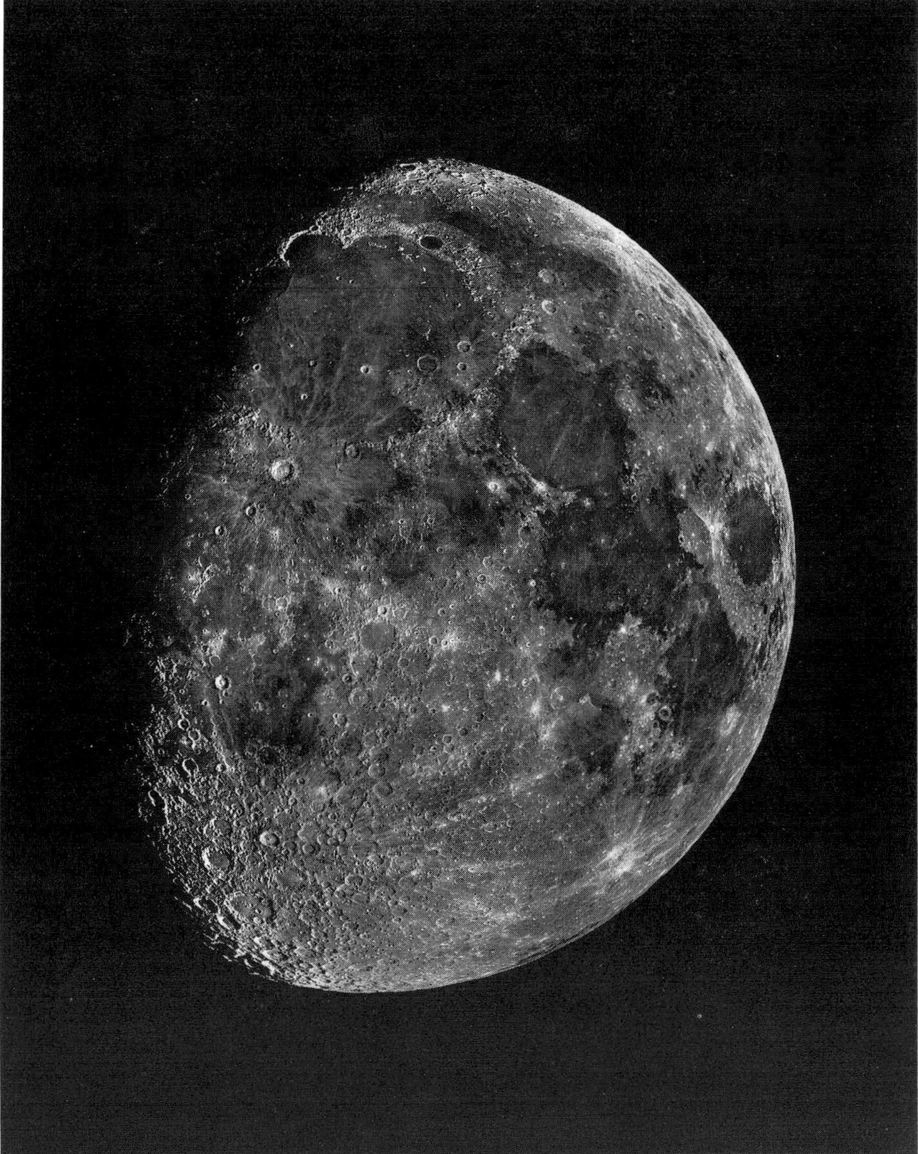

9.6. The waxing gibbous Moon, 10 days old (near position 4 in Figure 9.3). The terminator now begins to approach the Moon's eastern limb. Large areas of maria are now visible. (Lick Observatory.)

the crescent; quarter and full occur at a given moment that can be timed to the second, and then they are over, although to the eye the phase seems to be kept for the night.

If we ignore refraction and semidiameter, the Moon rises at sunset only if it is *exactly* full at that time. By sunrise, it will have moved several degrees past full, and will set

9.7. Full Moon. The entire lunar face is lit, and the terminator has disappeared. The maria, which form the "man in the Moon" and other fanciful figures, stand out clearly. Craters are hard to see because the shadows have disappeared. The rays – splashes that emanate from young craters – are now seen at their best. (Lick Observatory.)

after sunrise. The Earth's shadow, obviously, must point directly away from the Sun. If the Moon passes through the shadow, we will witness a lunar eclipse, which can only occur at full phase. Usually, however, the full Moon misses the shadow, passing north or south of it. Technically at this moment the Moon is then still in a gibbous phase, although the eye cannot sense any distortion from the perfect circle.

After full, the phases will just repeat themselves backwards. Between points 5 and 7 we see the *waning*, or diminishing gibbous Moon (Figure 9.8), which rises at night before midnight, and sets the following day before noon. The waning gibbous is a common pleasant sight in the western morning sky as we begin work for the day. At point 7 we reach *third quarter* (Figure 9.9), which rises at midnight and sets at noon. Third quarter occurs at the three-quarter point in the Moon's monthly journey. The "second quarter," a term never used, would be the full Moon. At first quarter, we saw the western half of the Moon's visible face (western as viewed from Earth); at third we see the eastern half. The third quarter Moon is the reverse of the first quarter and points in the opposite direction; the waning and waxing gibbous and crescent Moons have similar mutual relationships.

Finally, the waning crescent (Figure 9.10) makes its last appearance in the predawn hours, and after 29½ days (29^{d} 12^{h} 44^{m}), the Moon will once again be new, and the whole process repeats itself. This interval is called the *synodic period* of the Moon, or the *synodic month* (from the Greek *synodos*, or "meeting," referring to the Moon's conjunction with the Sun). By coincidence, 235 synodic periods equals 19 Julian years of 365.25 days to within 2 hours. Thus the lunar phases repeat themselves on the same dates (or within 1 day depending upon the intervening number of leap years) every 19 years, a period called the *Metonic cycle* (after the fifth century BC Athenian astronomer Meton, even though knowledge of the cycle is really much older).

An observer on the Moon (Figure 9.11) sees the Earth pass through the same set of phases, except in reverse: the Earth's phase is always opposite to the lunar phase seen from Earth. When the Moon is near new for us, the lunar resident sees the Earth near full (see Figure 9.3).

The apparent brightness of the Moon changes considerably over the period of its phases. At crescent, it casts very little light on the Earth, while at full it reaches a magnitude of −12, bright enough nearly to allow one to read. Although at full phase we see twice as much sunlit area as at first quarter, the full Moon is about 8 times as bright. At the quarters, lunar relief (hills and crater walls to be described below) casts visible shadows, especially near the terminator where the Sun is just rising or setting. The shadows cannot be seen with the naked eye, but their combined effect reduces the amount of reflected sunlight. At full, when we look at the Moon with the Sun directly behind us, the shadows are not seen, and we receive the full effect of the reflected sunlight. More importantly, the lunar surface reflects (a better word is "scatters," since the light is reflected in all directions) more efficiently directly back at the observer, less so to the side, as it must do at the quarters. Oddly, first quarter is nearly twice as bright as third. The dark areas of the Moon are not uniformly distributed, and more of the Moon is covered by them at third.

Recall how light it can be outdoors under a full Moon; because of the Earth's greater

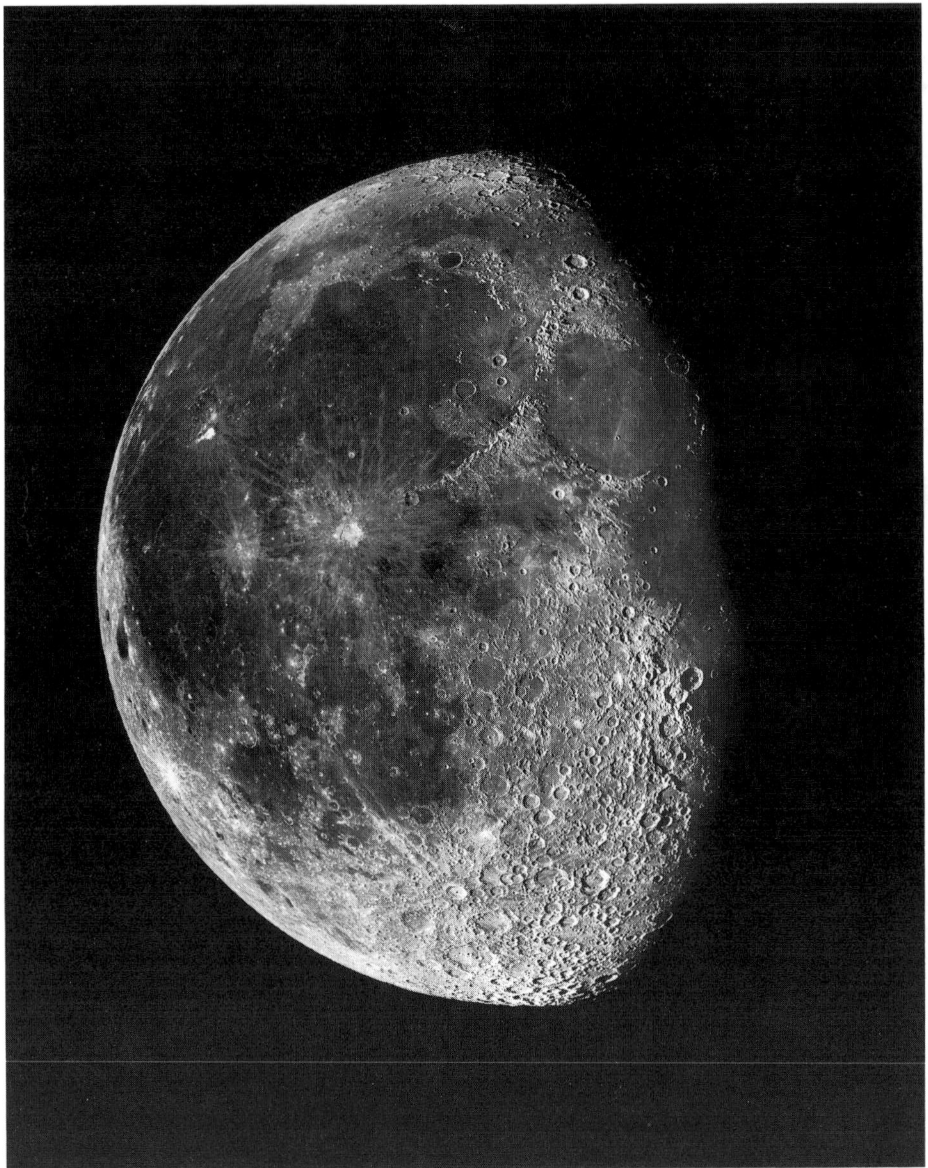

9.8. The waning gibbous Moon, 20 days old (near point 6 in Figure 9.3). The crescent of Figure 9.5 is now in darkness as the terminator rolls on to the lunar west (the observer's east). (Lick Observatory.)

size and the cloud and water cover that cause high reflectivity, the full Earth as seen from the Moon is some 65 times brighter. The effect of this *earthlight* can easily be seen during the crescent lunar phases. In deep twilight, we see the full outline of the Moon (Figure 9.12), where the area in night is lit by the brilliant Earth. The phenom-

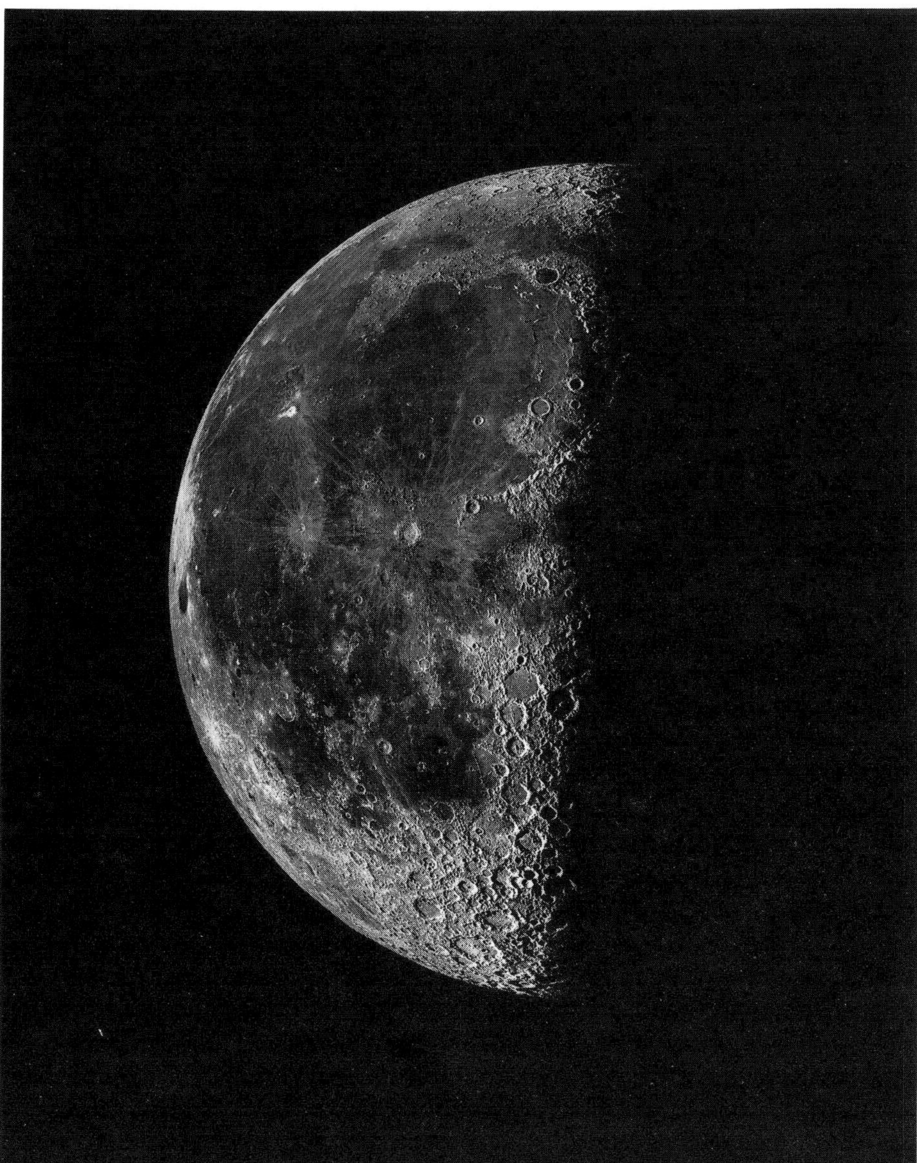

9.9. Third quarter (7 in Figure 9.3). Now the terminator moves steadily to the left, and shortly the Moon will be in its waning crescent phase. (Lick Observatory.)

enon is sometimes called "the old Moon in the new Moon's arms." By the time of the quarters, the Earth is also in a quarter phase and is no longer as bright, and the earthlight becomes invisible against the contrast of the brighter daylight side.

The brightness of the Moon is a serious problem for the astronomer. At full, it makes the sky so bright that the faint fifth and sixth magnitude stars cannot be seen

9.10. The waning crescent (near 8 in Figure 9.7), here 26 days old, only 3.5 days before new. It rises as a silver sliver just before dawn. (Lick Observatory.)

with the naked eye, and weak images ordinarily observed or photographed through the telescope are washed out. Observatory schedules are traditionally divided into "dark runs," from third quarter to first, when faint objects are studied, and "Moon runs," from first to third, when only the brighter things can be examined. The intervals around the quarters are sometimes called "gray time."

9.11. The waning gibbous Earth. The photograph was taken by an *Apollo* astronaut in orbit around the Moon, the bleak lunar landscape seen below. (NASA.)

The progression of the phases is a source of continuing fascination for the dedicated sky-watcher. It is one of those astronomical events that may first spark our interest in the heavens, and was of great and often mystical significance to our ancestors. Watch them all and correlate them with their times of visibility.

9.3 Synodic and sidereal periods

Several kinds of period can be defined for the Moon, depending upon the reference point chosen. During the synodic month, the Sun moves approximately $^1/_{12}$ of the way around the ecliptic. By the time the Moon catches up with the Sun, the Moon must have moved more than one full revolution with respect to the stars. The true orbital period of the Moon, that with respect to the stars, or the *sidereal period*, must be shorter than the synodic period.

At *A* in Figure 9.13 we see the Moon in conjunction with both the Sun and a distant star. The star is off the page far to the right, effectively at infinity. At *B*, 27.3 days later, the Moon has orbited the Earth and has returned to conjunction with the star, but because of the Earth's motion, the Sun is east of the star, and the Moon is seen in the waning crescent phase. Since the Earth has moved about $^1/_{12}$ of its circuit, the Moon must travel an additional $^1/_{12}$ of *its* orbit of 27 days in order to come to new, which takes about another 2 days. The Earth, of course, moves a bit during this short last interval, but that is not shown in the figure. The exact sidereal period is

9.12. Earthlight. As the crescent Moon rises in morning twilight, we see the entire lunar outline, with the nighttime portion lit by the brilliant gibbous Earth. The bright "star" to the left is the planet Mercury. (Rick Olson.)

$27^d\ 7^h\ 43^m$. As a consequence, the Moon moves at an average rate of 13°.2 per day, or about 0°.55 (roughly its own diameter) in an hour.

The sidereal period is the one to be used in any gravitational calculations involving the Earth and Moon. The synodic period is dependent on both the orbital characteristics of the Moon *and* the Earth. If the sidereal period were to remain constant, but if we could push the Earth closer to the Sun to shorten the year and to make the Sun move faster on the ecliptic, the synodic month would lengthen quite independently of the gravitational forces between the Earth and Moon. We will encounter sidereal and synodic periods again in our discussion of the planets two chapters hence.

9.4 Visible aspects of the Moon

The Moon displays a number of features and aspects visible to the naked eye, some so obvious that they are noticed in childhood, others fairly subtle. As we watch the orbiting Moon move through its phases, we easily see that it is not a smooth, uniform body, but that its surface is heavily mottled with dark areas prominently visible in the sequence of photos displayed in Figures 9.4 through 9.10. These permanent features are best seen at full Moon, and make the popular face of the "man in the Moon." There is a great deal of folklore associated with the dark markings. In addition to the face, others see a lady seated in a chair, some a dog, and yet others a rabbit, whatever

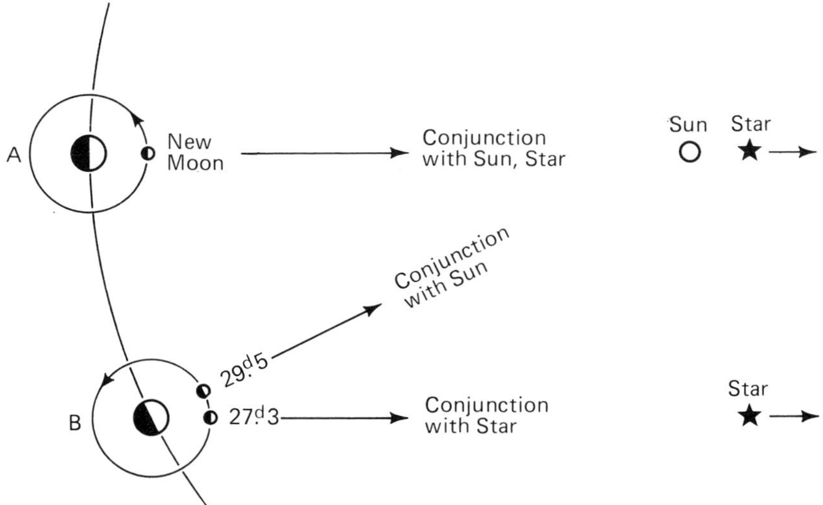

9.13. Sidereal and synodic periods of the Moon. At *A* the Moon is new and we see it lined up with the Sun and a distant star, which on this scale may be several kilometers to the right. In 27.3 days the Earth moves to *B*, the Moon orbits the Earth once, and again is in line with the star, but because of the Earth's motion, the Sun is now east of the star and the Moon will not be new for another 2.2 days. The motion of the Earth in the 2.2-day period is ignored in the diagram.

the human imagination can conjure. The dark areas are called *maria*, Latin for "seas," and at one time they were thought to be lunar oceans, their names – the Sea of Showers, the Ocean of Storms, the Bay of Rainbows – again testifying to the flight of the mind (Table 9.1). In reality, they are great basins made by awesome collisions between the Moon and giant interplanetary bodies (Figure 9.14).

The bright areas, the so-called lunar highlands, are older than the maria and were formed in the first half-billion years of lunar history by vast numbers of impacts that cratered and crushed the surface (Figure 9.15), a time called the heavy bombardment, the impacting bodies the debris of the newly formed Solar System (they will be discussed in Chapter 12). Craters piled on craters, as older generations were wiped out by newer. Cratering of this sort is commonly seen on the surfaces of planetary bodies, vividly demonstrating the violence of the early Solar System. (The Earth was bombarded as well, but the craters were erased by erosion and tectonic processes associated with continental drift. Nevertheless, over 200 terrestrial impact craters are recognized.)

A crater is formed when a body moving at high speed, some tens of kilometers per second, is suddenly stopped. The energy of motion is transformed into heat, which vaporizes all or part of the body and some of the lunar surface, and which sends powerful shock waves into the rock below. When the compressed surface suddenly rebounds, huge quantities of rock are excavated, forming a hole about ten times the size of the impacting body and raising a crater wall. If the impact was large enough, rebound can form a central mountain peak. The ejected rock is sprayed in all directions,

Table 9.1. *Lunar maria and related features*

Latin	English	Location[a]
Mare Crisium	Sea of Crises	West
Mare Frigoris	Sea of Cold	North
Mare Foecunditatis	Sea of Fecundity	West
Mare Humorum	Sea of Humors	Southeast
Mare Imbrium	Sea of Showers	North
Mare Nectaris	Sea of Nectar	West
Mare Nubium	Sea of Clouds	East central
Mare Serenitatis	Sea of Serenity	North central
Mare Tranquilitatis	Sea of Tranquility	West central
Mare Vaporum	Sea of Vapors	Central
Oceanus Procellarum	Ocean of Storms	East
Lacus Mortis	Lake of Death	W Mare Frigoris
Lacus Somniorum	Lake of Dreams	S Mare Serenitatis
Palus Nebularum	Marsh of Clouds	W Mare Imbrium
Palus Putredini	Marsh of Decay	W Mare Imbrium
Palus Somnii	Marsh of Dream	NW Mare Tranquilitatis
Sinus Iridum	Bay of Rainbows	NE Mare Imbrium
Sinus Medii	Central Bay	Central
Sinus Roris	Bay of Dew	E Mare Frigoris

[a]As viewed from Earth; consult Figure 9.16

spreading rubble (an *ejecta blanket*) around the crater and creating long radial features or *rays* that consist of secondary craters made by falling debris.

The largest impacting bodies, those 50 or 100 km across, produced huge basins, most of which were obscured by the cratering of the heavy bombardment. As the amount of falling debris lessened toward the end of the heavy bombardment about four billion years ago, the Moon was struck by several huge bodies that produced the basins we now see and that wiped out the cratered highlands that once existed there. The "mountain ranges" on the Moon are merely the walls of the great basins thrown up by the impacts. There is no continental drift on the Moon (there are no continents), and consequently no mountain ranges like those on Earth. These impacts were so powerful they cracked the underlying rock deeply enough to allow lava to flow upward from the then-hot lunar interior (it has since cooled and solidified). The lava filled portions of the basins in a series of flows that ran with some vigor for about a billion years then slowly tapered off, leaving smooth, dark, mare floors. Since then, the impact rate has dropped dramatically, so volcanic mare surfaces are only lightly cratered.

The largest of the craters is just barely too small to be seen with the naked eye, but several can be observed with binoculars. A small telescope shows hundreds. The craters and other detail on the Moon are best seen near the terminator, where the structures

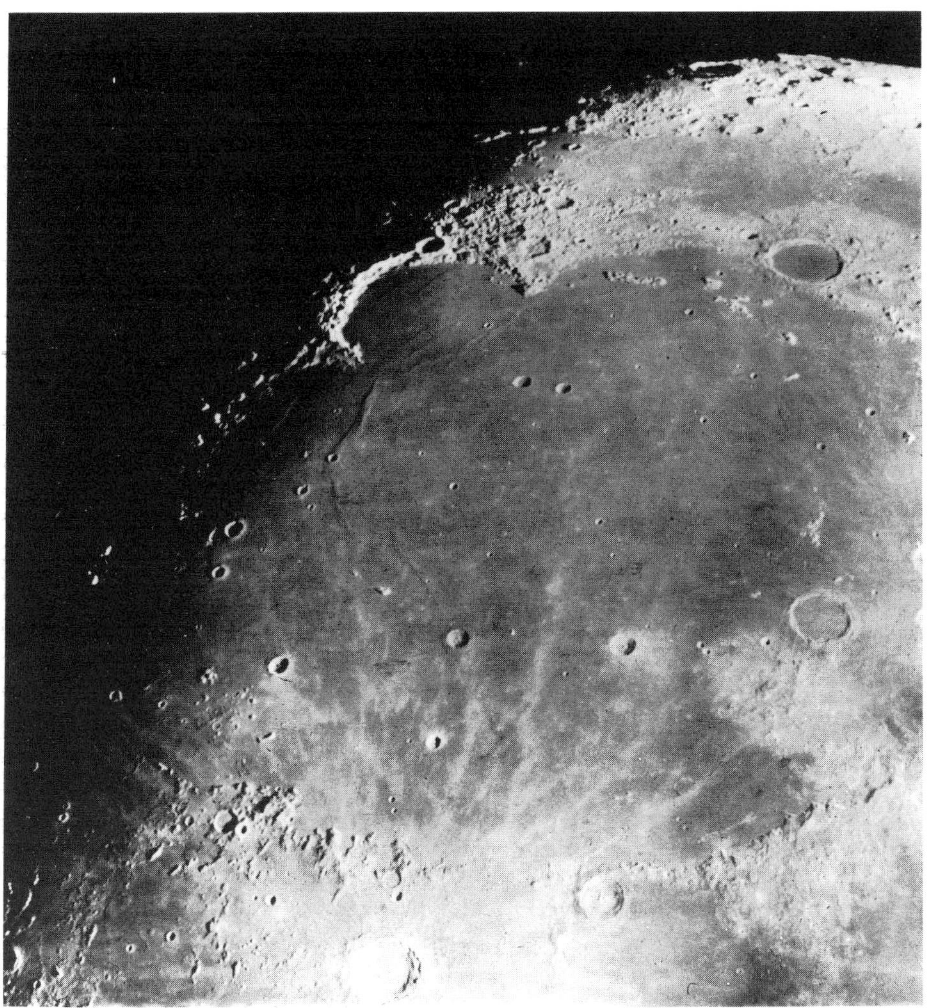

9.14. Mare Imbrium, the Sea of Showers. This classic view of one of the great lunar basins shows a flat lava plain pocked by a few craters. The mare is 3.5 billion years old and from its smoothness we see that there has been little activity since that time. The mare is bordered by the lunar Alps, which are not mountains as we know them, but are the walls of the immense crater that actually forms the basin. The bight of Sinus Iridum is at upper center. The bright rayed crater at the bottom is Copernicus and the dark-floored crater at upper right is Plato. A close-up of Copernicus is seen in Figure 9.17. (Yerkes Observatory, University of Chicago.)

throw contrasting shadows (note Figures 9.4 through 9.6, 9.8 through 9.10, 9.14, and 9.15). The lunar rays are best seen at full Moon (Figure 9.7), those from the large crater Tycho making the Moon look something like a naval orange. The rays slowly disappear under the action of continuing bombardment, so they mark the youngest craters. Tycho appears to be only about 100 million years old.

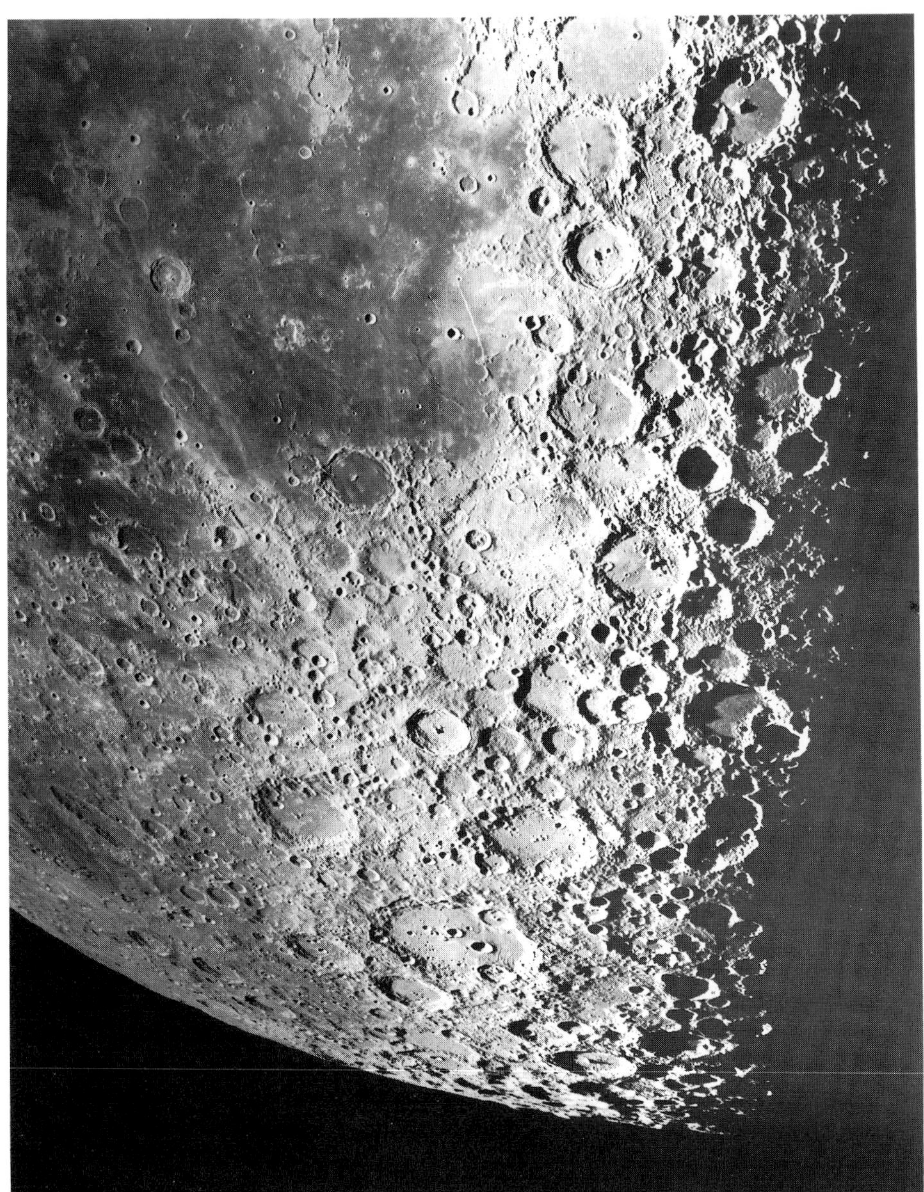

9.15. The cratered southern uplands of the Moon. The jumble of ancient craters that were created shortly after lunar formation was not obliterated by the later impacts that formed the maria, and they bear silent witness to the awful violence of the early Solar System prior to 3.8 billion years ago. The southern part of Oceanus Procellarum appears at the top. The huge crater near the bottom is Clavius. (Yerkes Observatory photograph, University of Chicago.)

The principal lunar features are identified in Figure 9.16. Use the map to locate the various craters and maria on the above photographs. However, it is far more enjoyable to go outside on a clear night and locate them yourself. Start with the naked eye and work up through binoculars to a telescope if one is available. However, note that astronomical telescopes invert the image, up for down, left for right. The map and all the photographs reproduced here have north at the top, appropriate to the naked-eye view from the northern hemisphere. For a telescopic view (or one with the naked eye from the southern hemisphere) rotate the map and photographs 180°. Vastly more features have been discovered and named as a result of detailed terrestrial views and the United States and Soviet space programs. The Moon has been mapped by close orbiters and examined by landers and directly by astronauts, all of which have provided stunning views of a real world (Figure 9.17).

Notice that the lunar features are always visible, that they are never obscured by clouds or haze as are terrestrial features viewed from space. With neither air nor water, the Moon has no weather, nor any erosion other than the occasional impact to wear away its surface. Over the lunar period, the dark markings do not even change their apparent positions; the Moon keeps one face perpetually pointed at Earth. The features on the back, or far side, of the Moon were not seen by us until the Soviet and United States space programs sent cameras into orbit behind it. (The far side is as heavily cratered as the near side, and contains basins, but few maria. The lunar crust is thinner on the near side, possibly as a result of the impact that made the 3400-km-wide Procellarum basin, which made it easier for mare lavas to flow.)

Recall (see Section 6.10) that our planet is slowing down because of the tidal action of the Moon and Sun (we will examine tidal phenomena in Chapter 10). Because the Moon is so much less massive than the Earth, the tides, which cause a gravitational distortion of the Moon, have slowed the lunar rotation to its final state, in which it is said to be rotating *synchronously* with its orbital revolution. As a result, the Moon has zero rotation with respect to the Earth, and keeps one hemisphere on the terrestrial side at all times. But because of its orbital motion, the satellite rotates once every revolution with respect to the Sun and stars. At a given point on the Moon, an observer would see a sunrise or a sunset every synodic period of 29.5 days (the lunar "solar day"), and would see a given star rise or set every sidereal period of 27.3 days (the lunar "sidereal day").

Sunrise and sunset on the Moon are actually rather easy to see with a small telescope. From night to night, or even over the course of a few hours, we can watch the sunrise terminator creep over the lunar landscape. Sometimes we will even see mountain peaks on the dark side of the terminator catching the first rays of the morning Sun – or the last of the evening light as night descends. Notice the sharpness of the daylight line. With no air, there is no twilight (although on the Moon it takes the Sun 29 times longer to set as it does on Earth: nearly an hour at minimum). Synchronous rotation is commonly seen when two bodies orbit close to each other.

9.5 The Moon's orbit

The Moon has played a key role in our understanding of the nature of gravity, the pervasive force that holds you to the ground, keeps the Moon as Earth's companion,

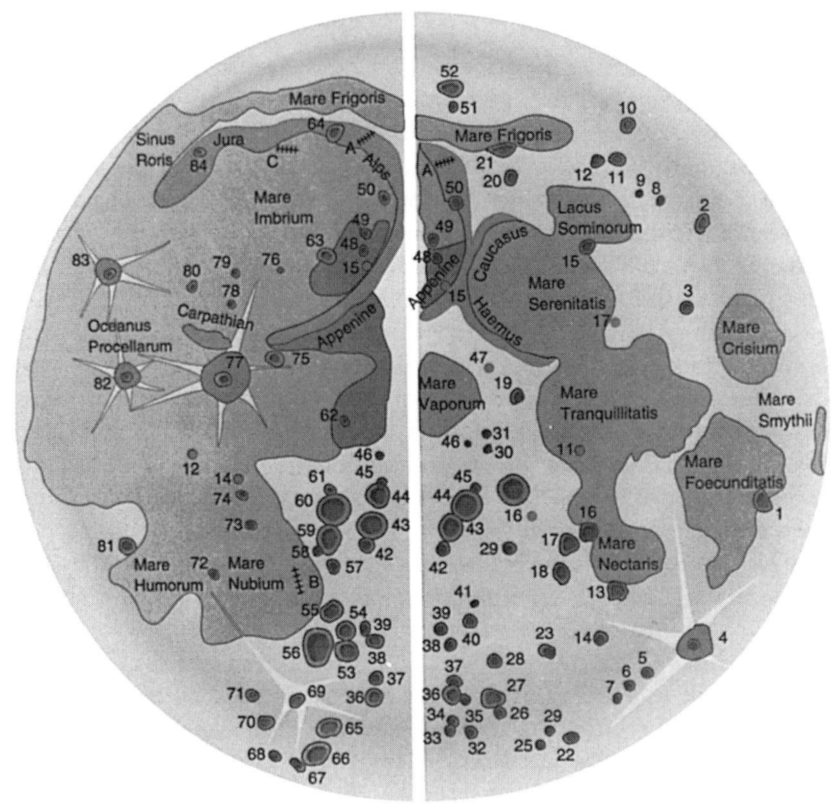

Key to Figure 9.16

1. Langrenus	22. Vlacq	43. Albategnius	64. Plato
2. Geminus	23. Zagut	44. Hipparchus	65. Maginus
3. Macrobius	24. Pitiscus	45. Horrocks	66. Clavius
4 Stevinus	25. Nearchus	46. Rhaeticus	67. Blancanus
5. Rheita	26. Barocius	47. Manilius	68. Scheiner
6. Metius	27. Maurolycus	48. Autolycus	69. Tycho
7. Fabricius	28. Gemma Fricius	49. Aristillus	70. Longomontanus
8. Franklin	29. Abulfeda	50. Cassini	71. Wilhelm
9. Cepheus	30. Godin	51. Archytas	72. Bullialdus
10. Endymion	31. Agrippa	52. Bond	73. Guericke
11. Atlas	32. Cuvier	53. Walter	74. Parry
12. Hercules	33. Heraclitus	54. Regiomantanus	75. Eratosthenes
13. Fracastorius	34. Licetus	55. Purbach	76. Timocharis
14. Piccolomini	35. Faraday	56. Lexell	77. Copernicus
15. Posidonius	36. Stöfler	57. Arzachel	78. Pytheas
16. Theophilus	37. Fernelius	58. Alpatragius	79. Lambert
17. Cyrillus	38. Aliacensis	59. Alphonsus	80. Euler
18. Catharina	39. Werner	60. Ptolemaus	81. Gassendi
19. Julius Caesar	40. Apianus	61. Herschel	82. Kepler
20. Eudoxus	41. Bohenenberger	62. Pallas	83. Archimedes
21. Aristoteles	42. Parrot	63. Archimedes	84. Mairan

9.16. A map of the Moon identifying prominent maria and craters. (From *Astronomy!*, J. B. Kaler, HarperCollins, New York, 1994.)

and ties the Solar System together. Isaac Newton is commonly said to have "discovered" gravity, a phenomenon discovered by anyone who has ever jumped in the air. His discovery was far more brilliant: he found how gravity works. By comparing the rate at which a body falls to the ground with the rate at which the Moon is deflected from straight-line motion, Newton demonstrated that the gravity that makes you fall is the same force responsible for the lunar orbit and that the Moon is in fact a falling body. Our satellite quite literally is falling toward, and fortunately for us, *around* the Earth. By analogy, Newton then realized that gravity must also be the force that keeps the planets in orbit around the Sun.

The nature of gravity will be examined further in Chapter 11 when we discuss the planets. Here we look only at the basic nature of an orbit. Many learn as children that the "pull of gravity is balanced by the outward centrifugal force of orbital revolution, and consequently the Moon, or the Earth, stays in orbit about its central body." That explanation is wrong. There is only one force involved, gravity, and a body stays in orbit because it has a motion perpendicular to the direction of that force. Pick up a rock and let it go; it drops to the ground. Now throw the rock on a level with the Earth's surface. As it moves away, it drops to Earth in a curved arc, *at the same rate as before*, that is, the horizontal motion has no effect on the speed with which the rock falls.

However, the Earth's surface is also curved. In your imagination, hurl the rock at such great horizontal speed that the Earth's curvature causes the ground to fall away at the same rate at which the rock drops. The rock cannot catch up with the Earth, and you have launched the body into orbit. The rock and the Moon are perpetually falling toward the Earth, but because of their horizontal motions they cannot ever reach it. The rock will eventually be slowed by friction with the Earth's atmosphere, and will finally come to rest. But there are no forces to stop the motion of the Moon, so it will orbit forever. (The gravity of the Sun disturbs the orbit but does not slow the lunar motion.)

We commonly say that the Moon orbits the Earth, or that the Earth orbits the Sun. But gravity is a mutual affair, and just as the Moon falls toward the Earth, our planet falls toward the Moon. In reality, the two bodies simultaneously orbit a point on a line between them called the *center of mass*, or *barycenter*, as in Figure 9.18. The center of mass of two orbiting bodies is located so that the ratio of the semimajor axes is inversely poroportional to the ratio of the masses, or $M_{\mathrm{Earth}}/M_{\mathrm{Moon}} = a_{\mathrm{Moon}}/a_{\mathrm{Earth}}$, where M stands for mass and a for distance to the barycenter (or the semimajor axis of an elliptical orbit that has the barycenter at the focus). Consequently, the mass of one body times its distance to the center of mass equals the mass of the other times *its* distance, or $M_{\mathrm{Earth}} \times a_{\mathrm{Earth}} = M_{\mathrm{Moon}} \times a_{\mathrm{Moon}}$. If two bodies with equal mass orbit each other, the center of mass is halfway between them. Since the Earth is 80 times more massive than the Moon, the barycenter is $1/80$ the distance from the Earth to the Moon, or 5000 kilometers from the terrestrial center. The Earth orbits this point once every lunar month.

A two-body orbit is always an ellipse (see Section 6.2). (A circular orbit, which is unlikely, is a limiting case of an ellipse.) In Figure 9.18, and for any orbital pair, both

(a)

(b)

(c)

9.17. (a) The crater Copernicus as viewed by *Lunar Orbiter*. Compare this view with that in Figure 9.14. (b) Mount Hadley and roving astronauts from the *Apollo 17* mission to Mare Serenitatis. (c) A bootprint left by an astronaut. It will be there for aeons. (NASA.)

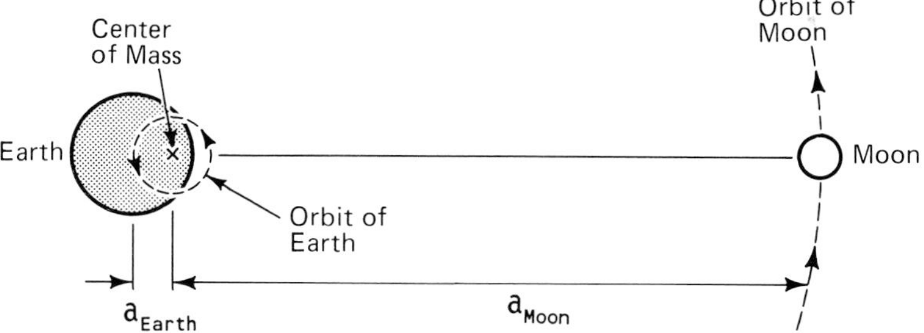

9.18. The center of mass. The Earth and Moon both orbit their mutual center of mass, which is 5000 kilometers from the Earth's center and inside its solid body, once a month. The Moon is actually 10 times farther away on this scale than is shown in the figure. The semimajor axes of the lunar and terrestrial orbits about the center of mass are a_{Moon} and a_{Earth}. The center of mass is defined as that position for which $M_{Earth}/M_{Moon} = a_{Moon}/a_{Earth}$, or for which $M_{Earth}a_{Earth} = M_{Moon}a_{Moon}$, where M stands for mass. The sum of the two semimajor axes is equal to the semimajor axis of the lunar orbit relative to the center of the Earth, $a_{Earth} + a_{Moon} = a$, which is the average distance between the two bodies.

bodies have elliptical orbits of different sizes, but of the same shapes or eccentricities, about the center of mass, which is at a focus of each ellipse. However, the orbit of one body *relative to the center of the other* is also an ellipse, and mathematically we can still speak of the Moon as "orbiting the Earth."

The general problem of orbits is mathematically difficult and exacting. We have thought here of the Earth and Moon as two bodies undergoing a mutual gravitational attraction. A two-body problem in itself is simple to solve; the complications arise from the influence of other gravitational sources in the neighborhood. The Sun, for example, has a strong effect on the Moon and the Moon's orbit, and so we must consider *three* bodies in the affair. It is common in astronomy first to look at an orbital problem as if it were strictly a two-body problem and then to examine how a third or fourth might distort, or *perturb*, the orbit. We will look at the simple two-body case in the next section, and then examine the effects of the complications in the succeeding one.

9.6 The elements of an orbit

An elliptical orbit and the position of the orbiting body are uniquely defined by seven numbers or constants, called the *elements* of the orbit. The first two define its size and shape. They were discussed in Section 6.2 and will be reviewed here. In Figure 9.19, the line through the foci, the longest line that can be drawn within the ellipse, is the *major axis*; the *minor axis* is perpendicular to it at the center. Half the length of the major axis, the *semimajor axis*, or a, specifies the ellipse's size. The sum of the semi-major axes of both bodies relative to the center of mass (for example $a_{\mathrm{Moon}} + a_{\mathrm{Earth}}$) equals the semi-major axis of the orbit of the smaller body relative to the center of the major body (as is evident from Figure 9.18). This summed value is also the average distance between the two bodies; in the case of the Moon, relative to the center of the Earth, it is 384 400 km. To be physically correct, the Earth–Moon barycenter should be at the focus of the elliptical orbit. But since we want to know the location of the Moon relative to the observer, it simplifies things to place the center of the *Earth* at the focus, and to consider the mathematical ellipse about that point. (That is, we force the center of the mathematical coordinate system to be at the center of the Earth.)

The shape of the orbit is defined by its *eccentricity*, e, defined as the distance from a focus to the center divided by a. For a circle, the foci merge at the center, and e is zero. As the orbit becomes more elongated, and as the foci become farther apart for a given minor axis, the ellipse becomes thinner and e approaches 1. The eccentricity of the Earth's orbit about the Sun is only 0.017, so the orbit is nearly circular. That of the Moon about the Earth is much greater, 0.055.

Two points on an elliptical orbit are special: the closest to, and the most distant from, the focus. Their names depend on whatever body is near the focus. In the case of the lunar orbit they are called *perigee* and *apogee* where the "peri" and "apo" prefixes come from Greek, meaning "closest to" and "farthest from" and "gee" is from the Greek "Ge" meaning "Mother Earth." For a planet going around the Sun, the points are *perihelion* and *aphelion* (see Section 6.2), from the Greek "helios" for Sun; for two stars going around one another they are *periastron* and *apostron*, and for a star orbiting

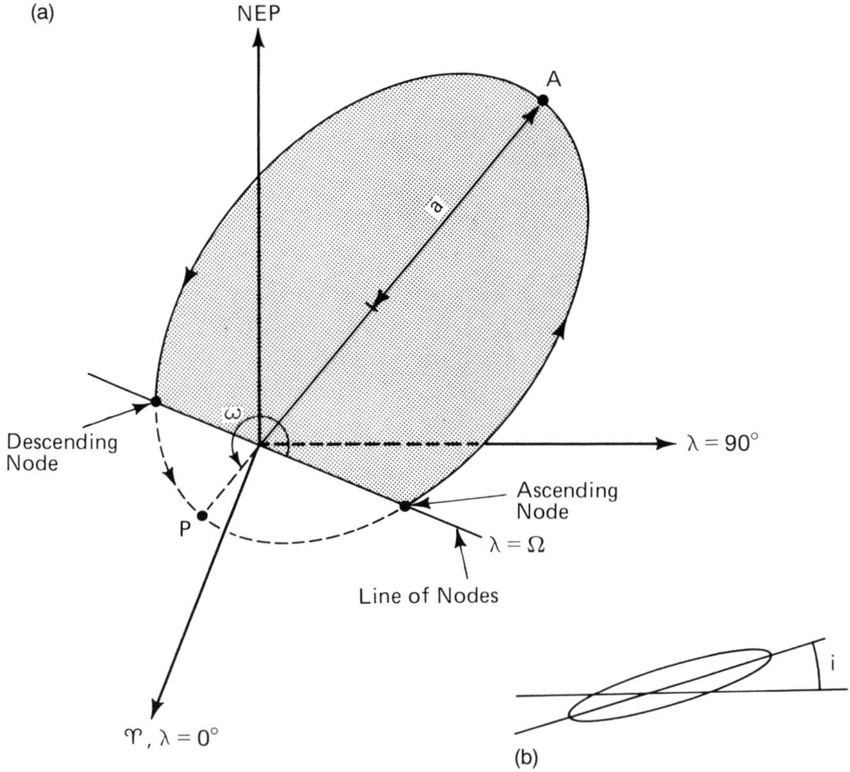

9.19. The elements, or description, of an orbit. (a) In this example of the Moon's orbit, we mathematically consider the center of the Earth to be at the focus and at the zero point of the coordinate system; λ is the celestial longitude. The ecliptic plane is defined by the direction to the vernal equinox, ♈, and the line to 90°. The Moon's orbit, of semimajor axis a, and eccentricity e (greatly exaggerated), is inclined to the ecliptic plane by inclination i as shown in (b). The Moon must cross the ecliptic twice each revolution at the *nodes* of the orbit. The orientation of the orbit with respect to the vernal equinox is given by the longitude of the ascending node, Ω, where the Moon crosses to the northern side of the terrestrial orbit. P is perigee, A is apogee. The orientation of the lunar orbit within its plane is given by ω, the argument of perigee.

the center of the galaxy they are *perigalacticum* and *apogalacticum*. The line connecting the two points and extending outward in both directions, coincident within the ellipse with the major axis, is called the *line of apsides*. From the definition of e, the perigee distance of the Moon must be equal to the $a(1 - e)$, and the apogee distance to $a(1 + e)$, so that they are respectively 363 000 and 406 000 kilometers.

The effect of the lunar orbital eccentricity is quite noticeable. It produces a 12% change in the lunar angular diameter, easily seen in Figure 9.20. (The distance and the angular diameter are tabulated for every day of the year in the *Astronomical Almanac*.) Moreover, as a consequence of the Earth's elliptical orbit (see Section 6.2), the Earth

9.20. The change in the Moon's angular diameter. The left-hand photo (a), taken at apogee, shows the lunar disk to be 12% smaller than the right-hand picture at perigee (b). (From *An Introduction to the Study of the Moon*, Z. Kopal, Reidel, Dordrecht, 1966. Reprinted by permission of Kluwer Academic Publishers.)

moves at a variable rate along its orbit. Because of the Moon's greater orbital eccentricity, our satellite suffers a concomitantly greater variation. The average easterly rate through the stars is 32′.9 per hour, but it varies from 31′.1 per hour at apogee to 35′.4 per hour at perigee. As a consequence, the intervals between the quarter phases will not be identical. If new Moon takes place near perigee, the interval from that phase to first quarter will be shorter than that between first quarter and full. The difference can amount to over a full day.

The next three orbital elements define the orientation of the orbit in space, and are also illustrated in Figure 9.19. The *inclination* of an orbit, i, is the orbit's tilt relative to the ecliptic plane. The lunar inclination of 5°09′ is responsible for the principal term in the nutation of the Earth's axis (see Section 5.9). For the observer on Earth the Moon follows the ecliptic reasonably well and goes through its phases within the constellations of the zodiac, but it can be found as much as 5° north or south of that circle, crossing it twice a month.

The points at which the Moon (or for that matter, a planet) crosses the ecliptic plane are called the *nodes* of the orbit, and the line connecting them is the *line of nodes*. At the *ascending node*, the Moon crosses the ecliptic from south to north and at the *descending node* from north to south. The orientation of the Moon's orbital plane relative to the Earth's solar orbit and the vernal equinox is expressed by giving the celestial longitude of the ascending node, Ω, measured from the vernal equinox counterclockwise as viewed from above. The nodes can easily be found with the naked eye simply by watching the Moon move against the background of the zodiacal constellations.

Now we must orient the ellipse within its own plane. The position of the orbit is expressed by the *argument of perigee* (or perihelion for a planet), ω, the angle measured counterclockwise in the body's orbital plane from the ascending node to perigee (or in the case of a planetary orbit, to perihelion). Sometimes the position of this point is measured counterclockwise directly from the vernal equinox, defining the *longitude of perigee* (or perihelion), ω', which is the sum of ω and Ω. Note that ω' is not a true angle, since it is measured in two different planes. It is especially useful in the case of the lunar orbit, for which ω and Ω change rapidly with time.

The next of the elements is the (sidereal) period of the orbit, P, the time it takes the Moon to make a full swing around the Earth, or the time between successive perigee passes. For a given semimajor axis, the period is a measure of the gravitational force between the two bodies. As the force goes up, so does the falling speed, and so must the velocity of the body in order that an orbit actually be maintained. Therefore, the period goes down.

The above six elements fully define the elliptical orbit with respect to the Earth's orbit, but they do not tell anything about where the Moon may be on that orbit. To locate the lunar (or a planetary) position, we must specify a seventh element, the *epoch*, T. The epoch is usually expressed as the time of some perigee (or perihelion) passage. With that and the period, we would then always know where to find the Moon (or a planet). The elements of the lunar orbit, as well as other lunar data, are summarized in Table 9.2.

Properly, it is the Earth–Moon barycenter that revolves around the Sun on an elliptical orbit and defines the ecliptic plane. We could easily place the barycenter at the focus in Figure 9.19, which slightly redefines the ellipse and its elements; a, for example, would be a few thousand kilometers less (see Figure 9.18). The Earth revolves about the barycenter with a set of orbital elements that equal, or complement, those of the barycentrically defined lunar orbit. The semimajor axis, a_{Earth} depends on a and the ratio of masses (Figure 9.18); e, i, and P must be the same. The Earth will move alternately north and south of the true ecliptic plane with a direction opposite to that taken by the Moon, so that when the Moon is seen at the ascending node, someone at the barycenter would see the Earth at the *descending* node, 180° away. To maintain a

Table 9.2. *The Moon's orbit*

Synodic period: $29^d\ 12^h\ 44^m\ 3^s = 29^d.5305883$
Sidereal period: $27^d\ 7^h\ 43^m\ 12^s = 27^d.32166150$
  Mean daily motion = 13°11′/day
  Mean hourly motion = 32′56″/hour
Metonic cycle: 235 synodic months = 19 Julian years (which average 365.25 days)

Mean elements
  semimajor axis, a = 384 400 km
  eccentricity, e = 0.0549 (range 0.0432 - 0.0666)
    Mean perigee = 363 300 km; mean apogee = 405 500 km
  inclination, i = 5°09′ (range 4°59′ –5°18′)
  period, $P = 27^d\ 7^h\ 43^m\ 12^s$ = sidereal period above
For January 1, 1994:
  longitude of the ascending node, Ω = 241°.0927
  argument of perigee, ω = 197°.5971
  longitude of perigee, ω' = 318°.1199
For 1994:
  new Moon: January 11, $23^h\ 10^m$ UT
  epoch, passage of perigee: T = January 06, 01^h UT
  passage of ascending node: January 8

Perturbations
  Advance of perigee (line of apsides): 8.8505 years
    anomalistic month = $27^d\ 13^h\ 18^m\ 37^s = 27^d.5545503$
  Regression of nodes: 18.6133 years
    draconic month: $27^d\ 5^h\ 5^m\ 36^s = 27^d.212220$
    eclipse year (Section 10.7): $346^d.62$
  Evection: amplitude = 1°16′ in longitude, period = $31^d.807$
  Variation: amplitude = 40′ in longitude, period = synodic
  Annual Inequality: amplitude = 11′, period = eclipse year (346^d)
  Evection in latitude: amplitude = 9′ in inclination and latitude period = eclipse year (346^d)

fixed ratio of distances to the barycenter, the Earth and Moon must be closest to or farthest from the barycenter simultaneously, so T will be the same. Since the Earth and Moon are 180° apart at their points of closest approach, so must be the respective values of ω.

All the planets have orbital elements as well. These are defined with the Sun at the focus of the ellipse and will be examined in Chapter 10. All that is necessary in defining elements for any two-body system is to establish a reference plane analogous to the ecliptic plane. For example, we can establish the elements of the Sun's orbit about the center of the Galaxy by using the plane defined by the average position of the Milky Way, or more properly, the galactic equator (see Section 4.14).

9.7 Perturbations of the orbit

A perfectly elliptical orbit is possible only for two isolated bodies, each of which act upon the other as if their masses were concentrated into perfect points (that is, as if they were spheres with symmetric distributions of matter). Since every body in the Solar System acts upon every other body to some degree, real elliptical orbits are impossible. The deviations from a true two-body orbit are called the *perturbations* of the orbit. The concept of orbital elements is still used, but now they must be allowed to change. We define the elements at any given moment to be those that the orbit would have were the outside gravitational forces suddenly to vanish (sometimes called the *osculating orbit*). A short time later, however, these forces have produced deviations from this calculated elliptical path, and we must then compute a new set of elements. We treat perturbations by defining the elements at a certain epoch and by establishing their rates of change with time.

There are two kinds of perturbations. In *periodic* perturbations, the elements oscillate about a mean value; in *secular* perturbations they change continuously in one direction. Variations in a, e, i, and P are generally of a periodic nature. Sometimes the period of the variation is short, a few days to a few years, in which case the values presented in a list of elements are the means, as given earlier (and in Table 9.2) for the Moon. If the period is hundreds or thousands of years, the values will be the current ones for the year, or even for the century of the table. Longer periodic changes may be treated as secular simply for convenience. True secular perturbations principally affect the orientation angles Ω and ω, and the epoch, T, which are subject to superimposed periodic variations as well.

The orbit of the Moon is awesomely complex. There are hundreds of known periodic and secular perturbations of the orbital elements. Some are so large they are easily visible to the naked eye and have been known since antiquity. The main culprit is the Sun; the Earth's equatorial bulge and the planets produce some additional smaller effects. At its new phase, the Moon is closer to the Sun than is the Earth, and is subject to greater solar gravity, while at full it is subject to less. The variable solar gravity as the Moon revolves severely distorts the orbit from an ellipse.

Two secular perturbations produced by the external forces, the *regression of the nodes* and the *advance of the line of apsides*, are the most widely recognized. The former causes the line of nodes to rotate clockwise, so that the nodes move to the west, and Ω steadily decreases. A full rotation takes a period of 18.61 years. Thus Ω decreases by 360°/18.61 = 19°.34 per year. The 18.6-year interval is the principal period of nutation (see Section 5.9). The regression of the nodes is another way of expressing the precession of the Moon's orbital pole around the ecliptic pole. The phenomenon is easily noted with the naked eye. At the same time, the line of apsides rotates counterclockwise relative either to the ascending node or the vernal equinox. Since Ω changes so rapidly, it is preferable to refer perigee to the equinox through the longitude of perigee, ω', which runs through a full cycle of 360° in only 8.85 years (a rate of 41° per year). These effects cause variations in, for example, the azimuth of the rising Moon, and they are very important in the study of eclipses.

Two additional lunar periods result from these perturbations. The *anomalistic period* or month is the interval between successive perigee passages. Its value is 27.5546 days, longer than the 27.3217 day sidereal period because the perigee point moves ahead by about 3° per month and the Moon must move for an addition 5½ hours (at 0°.55 per hour) to catch up with it. The *draconic* (or *draconitic*) month is the interval between successive passages of the ascending node. Since the nodes move backwards by 1½° per sidereal month, the Moon will catch up with the node 2.6 hours early, for a draconic period of 27.2122 days. The perigee and nodal points (or ω and Ω) are also subject to significant periodic variations.

The other important perturbations are all periodic. The most obvious of them is the *evection*, a variation in lunar orbital eccentricity that can displace the Moon in celestial longitude by 1°16′ (over two diameters). It has a period of 31.807 days, and will alter the distances of perigee and apogee. The effect of solar gravity is to stretch out the Moon's orbit along the line directed to the Sun. If the major axis of the Moon's orbit points at the Sun, the orbit will become more oblate and the eccentricity will increase. If the orbital axis is perpendicular to the direction to the Sun, e will be decreased. Quantitatively, e varies between 0.0432 and 0.0666 and is reflected in changing values of perigee and apogee. The evection is so large it can easily be seen with the naked eye and was known to Hipparchus in the second century BC.

The *variation*, here a name for a specific perturbation, is also the result of variable solar gravity as the Moon revolves about the Earth, continuously changing its distance from the Sun. It produces a change of up to 40 minutes of arc in longitude, greater than the lunar diameter, in a synodic period. It was first noted about the year 1000.

The *annual inequality* comes from the eccentricity of the *Earth's* orbit, which causes a change in the distance to the Sun, and consequently in the perturbing gravitational force. The Moon's longitude can vary by as much as 11′ because of it. The period is the *anomalistic year*, which is the interval between successive passages of perihelion by the Earth. If all these changes act in concert, the Moon can be displaced forward or backward along its orbit relative to a prediction made on the basis of a perfect ellipse by as much as 2°, four lunar diameters.

The last of the big perturbations takes place in the inclination. The average value is 5°09′. When the Moon is at the ascending node its celestial latitude is obviously 0°. Then, one-quarter of an anomalistic month later, it reaches a maximum latitude equal to the inclination, followed by passage of the descending node and an average minimum latitude of -5°09′, returning again to the ascending node after a full anomalistic month. The perturbation in inclination causes i to vary between 5°18′ and 4°59′ with a period equal to the *solar eclipse year* of 346.62 days, the interval between successive solar passages over a node, the maximum value occurring when the Sun is at a node. The maximum celestial latitude seen during any given month therefore depends upon the time of year, and this perturbation is consequently sometimes called the *evection in latitude*.

The orbital properties of the Moon, including the mean elements and some of their superimposed perturbations, are summarized in Table 9.2. The mean orbital elements are given for the year 1994. The time of new Moon, the passages of perigee, and the ascending node are given for January of that year; succeeding (or preceding) dates can be found by applying the synodic, anomalistic, or draconic months respectively. The longitudes of the ascending node and perigee can similarly be found for any date from the values given for January 1 and the periods of regression or advance. In performing such calculations, remember to use the tropical year of 365.2422 days and to count leap years. As an example, from January 1, 1980 to January 1, 1982 the interval is $366^d + 365^d = 731^d = {}^{731}/_{365.2422}$ years = 2.0014 years. In that time, the node regresses $({}^{2.0014}/_{18.6133}) \times 360° = 38°.7093$. Remember also that the 8.85-year period in the advance of the line of apsides refers to the vernal equinox and ω', not ω, where the latter is referred to the regressing node.

The construction of a complete lunar theory is an extraordinarily difficult task, and over the centuries has occupied some of the best scientists and mathematicians. Once the elements and their perturbations are known, however, we can always calculate exactly where the Moon should be at a given time. That is, we can translate the elements and their variations into a lunar ephemeris, a table of celestial latitude and longitude, and hence into right ascension and declination, as a function of time (given in the *Astronomical Almanac*). The last word has not yet been said, however. Observations are still being made, and our knowledge of the orbit continues to improve.

9.8 The parallax of the Moon

The above discussion supposes that we are viewing the Moon from the center of the Earth. But we are over 6000 kilometers from that point, and thus the Moon will not generally appear to us to be in the calculated position. We see an effect rather like three-dimensional vision. In observing the Moon we must always take this angular displacement or *geocentric parallax* into account. We encountered this term in Section 4.10, where it pertained to the displacements of stars as seen from opposite sides of the Earth's *orbit*. The geocentric parallax refers to the Earth's *radius*.

Look first at the Moon when it is on the horizon for the observer in Figure 9.21. If we move from the center of the Earth to the surface, the Moon will appear to drop by an angle p. Since our hypothetical observer is located on a terrestrial radius perpendicular to the direction to the Moon, this quantity is the maximum possible shift, and is called the *horizontal parallax*. Its value depends upon the radius of the Earth and the distance to the Moon, and averages $0°.950 = 56'58''$, about two lunar angular diameters! Since the lunar orbit is elliptical, the horizontal parallax actually varies from $54'$ at extreme apogee to $1°02'$ at extreme perigee; it is tabulated in the *Astronomical Almanac*.

The actual geocentric parallax at the time of observation – unless the Moon is just on the horizon – must always be less than the horizontal parallax. If the Moon is in the zenith, so that the observer is on the line between it and the center of the Earth, the parallax vanishes. At an intermediate lunar altitude, the parallax has a value between p and zero that can be calculated from simple trigonometry. The lunar parallax always

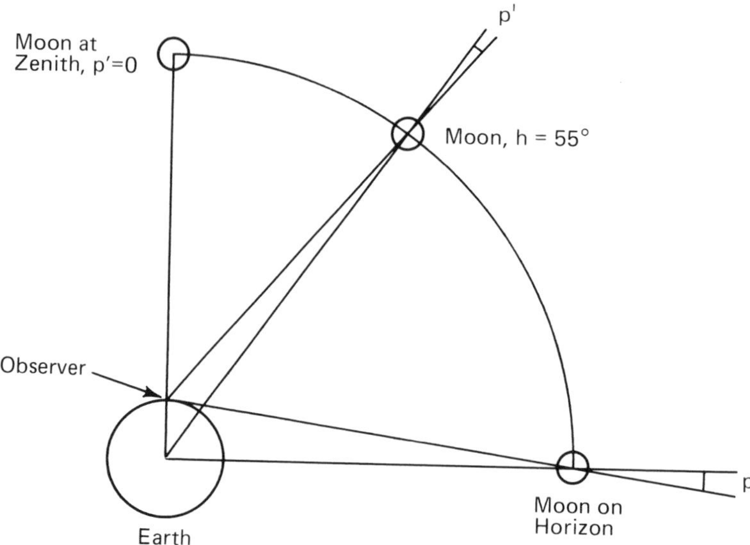

9.21. The geocentric parallax of the Moon. Angle p is the maximum displacement, called the *horizontal parallax*, and occurs at moonrise and moonset. The angle p' is the parallax when the Moon is high in the sky, and clearly is less than p. Parallax disappears when the Moon is at the zenith.

displaces the Moon downward, away from the zenith and it changes continuously as the Moon moves across the sky on its daily path: it is a maximum of p at moonrise, a minimum at meridian transit, and then increases again to maximum at moonset. A graph of parallax as a function of altitude is given in Figure A1.8 for three values of the horizontal parallax, appropriate to mean apogee, mean distance, and mean perigee. (Because of the perturbations, the actual perigee and apogee distances vary by a percent or so.) The parallax at any time from perigee to apogee can be found to within naked-eye accuracy by interpolation.

The Moon's geocentric parallax is important in a number of contexts. The *observed* right ascension and declination will never (except when the Moon is at the zenith) be exactly the same as given in an ephemeris; they must be corrected continuously as a function of the lunar altitude. Since the Moon's apparent position is depressed perpendicular to the horizon, the effects are opposite those of refraction. Failure to take geocentric parallax into account when navigating by day with the Moon and Sun (see Section 8.3) can result in a positional error approaching one degree, the order of 100 kilometers, which can be highly deleterious to the health of the ship's crew.

All nearby celestial bodies must have some geocentric parallax (we can ignore the stars for which only the parallax produced by the Earth's orbital motion has any significance). But the value depends directly on distance, and none are as close to us as the Moon. Since the Sun is 400 times farther than the Moon, the parallax is 400 times less, or only 8.8 seconds of arc, negligible for a naked-eye observer. Even Venus

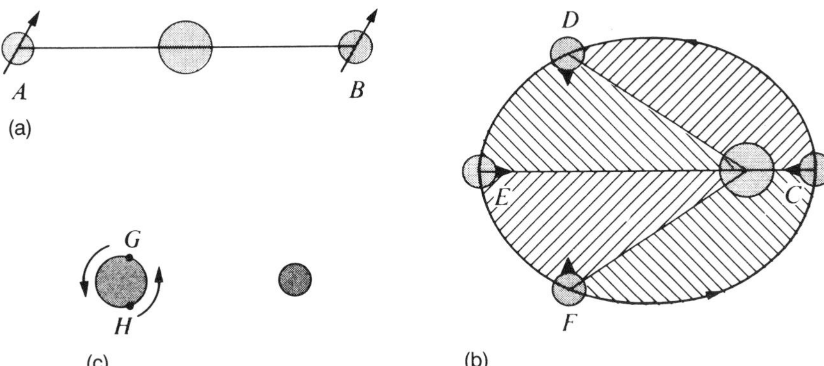

9.22. Librations of the Moon. (a) Libration in latitude. The 6°.5 tilt of the Moon's axis allows us to see first over one pole (*A*), then over the other (*B*). (b) Libration in longitude. The lunar rotation speed is constant but the orbital speed is not, which causes the Moon (whose orbital eccentricity is greatly exaggerated) to appear to oscillate east and west by nearly 8°. At *C* and *E* a feature is centered, but at *D* we see a bit around the western limb, and at *E* somewhat around the eastern. (c) Diurnal libration. The finite size of rotating Earth (on the left) allows a further 1° glimpse around the far side; at *G* we can see around the eastern limb and at *H* around the western. (From *Principles of Astronomy*, 3rd ed., S. P. Wyatt, Allyn and Bacon, Boston, 1977.)

at closest approach to the Earth has a parallax of only half a minute of arc, still less than naked-eye resolution.

An additional parallax is introduced by the motion of the Earth about the barycenter. The radius of this motion is but 500 kilometers, only ¾ the Earth's radius, so that the effect on positions of the Sun and planets is negligible for our purposes, but must still be taken into account for precise orbital determination. Barycentric parallax can be, and was, used to determine the distances of the nearby planets.

9.9 Librations of the Moon

The Moon keeps one face toward the Earth as it rotates in synchronism with its orbital revolution. We commonly say that we can see and study just one-half of the lunar surface from the Earth, the near side, while the back half, the far side, is visible only to orbiting spacecraft. However, there is a set of three phenomena collectively called the *librations* of the Moon, discovered by Galileo, that gives us an angular peek around the far side.

The first is the *libration in latitude* (Figure 9.22a). Just as the Earth's rotational axis is tilted to its orbital plane, the obliquity of the ecliptic, so is the Moon's axis tipped with regard to its terrestrial orbit. The Moon's "obliquity" is much smaller, only 6°.5; but over the month, the axis first tips toward us (position *A*), and then away (position *B*), allowing us a 6°.5 glimpse of the far side, first over the north pole, then over the south.

Next, and largest, is the *libration in longitude* (Figure 9.22b). The Moon rotates on

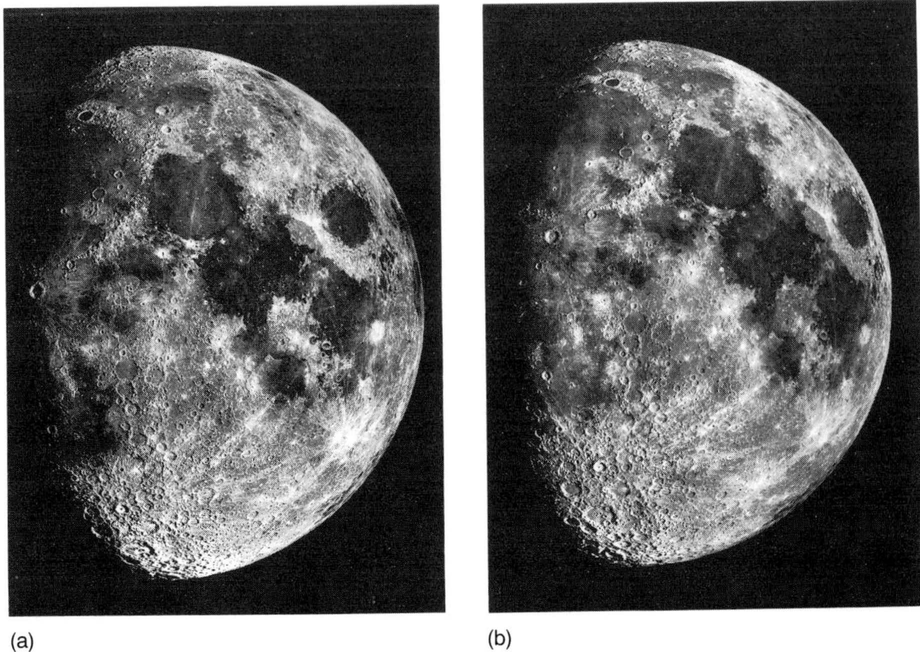

(a) (b)

9.23. Observed librations. The photo in (a) gives a peek over the north pole and around the western limb relative to the picture in (b). (Lick Observatory.)

its axis at a nearly constant rate, as does the Earth, but since its orbit is roughly elliptical, it does not (as we have seen) *revolve* at a constant speed. At point *C*, perigee, a feature is seen at the center of the lunar disk. A quarter of a sidereal period later, at point *D*, the feature has rotated 90°, but because the Moon is moving faster near perihelion, it has gone more than a quarter of the way around the orbit, allowing us to see 7°.75 around the western (as viewed from Earth) limb. Then at apogee, point *E*, the Moon has rotated and revolved halfway, and the feature is re-centered. At *F*, we can similarly look by the same amount around the eastern limb.

Finally, the smallest, the *diurnal libration*, is related to the lunar parallax (Figure 9.22c). We are displaced from the center of the Earth, so that at moonrise, when parallax is greatest, we see about 1° around the western limb. At moonset, we see 1° around the eastern. The effect is a little like craning your head to one side in order to see around a tree.

The observed effect is quite noticeable (Figure 9.23). When all these librations are summed, we find that we can see 59% of the lunar surface from the Earth. Some 41% is always visible (ignoring the phases), 41% is ever-hidden, and 18% swings in and out of view.

9.10 Occultations

The Moon is our nearest neighbor, and as it orbits the Earth it occasionally appears to pass over and cover up more distant bodies. Such an occurrence is called a lunar *occultation*, from the word *occult*, meaning "hidden." Since the Moon moves its own angular diameter in just under an hour, an occultation may last for up to that period of time. The most noted occultation is the covering of the Sun by the Moon, called in that instance an *eclipse of the Sun*, since we are acutely aware of the shadow phenomena involved. We will look at eclipses in Chapter 10, and here briefly examine only the simpler "hidings" of stars and planets.

Every year a variety of publications, including the *Astronomical Almanac*, present a list of predictions of occultations. Those of the first magnitude stars and planets can be seen with the naked eye (the brightness of the Moon precludes fainter stars from being visible), and many more can be seen with binoculars or telescopes. They are fun to watch, especially during the first two quarters when the leading edge of the Moon is in darkness. In that case we see a planet fade away or a tiny star suddenly wink out as it disappears behind the invisible lunar outline. The disappearance is called *ingress*, and the reappearance an hour or less later, *egress*.

The duration of an occultation depends upon the path of the Moon with relation to the position of the star. The average angular diameter of the Moon is $31'05''$ and the average angular motion is $32'56''$ per hour (32.9 seconds of arc per minute of time, $0''.55$/second), so that the average maximum duration, for a central occultation, is 56^{m} 40^{s}. It matters little whether the Moon is at perigee or apogee, since both the angular diameter and orbital velocity are proportionately smaller and larger. If ingress occurs near the upper or lower limb, the occultation will last a shorter period of time. Of special interest is the *grazing occultation*, in which the edge of the Moon just touches the star; with a telescope, you can sometimes see the star blink in and out as mountains and valleys on the lunar limb alternately hide and expose it.

Occultations are astronomically important. Since we have accurate positions of the stars, the exact time of an occultation helps to locate the Moon precisely so that its orbit can be improved. The suddenness with which stars disappear behind the Moon's leading edge testifies to their tiny angular diameters: as far as our eyes can discern, stars are true points. But with careful electronic measurements, an astronomer can actually determine the time it takes for one of the angularly larger stars to disappear, and with the Moon's known rate of motion can compute the star's angular diameter. If we know the star's distance, its physical diameter in kilometers can be found. Even the angularly largest stars are only a few hundredths of a second of arc across. Antares, for example, has an angular diameter of about $0''.02$ and disappears behind the lunar disk (if the occultation is centered) in only about 0.03 seconds. Because of the regression of the lunar nodes, every star within an angle of $5°34'$ (the maximum orbital inclination plus the lunar semidiameter) of the ecliptic must be occulted at some time or other.

Lunar occultations also have revealed numerous double stars that are too close together to separate by eye at the telescope. The occulted star winks out in steps as first one component and then the other is covered.

9.11 Moonrise and moonset: basic concepts

For a given latitude, the time and azimuth of sunrise depend only on the date. The time and azimuth of moonrise are functions of both the date and the lunar phase. We

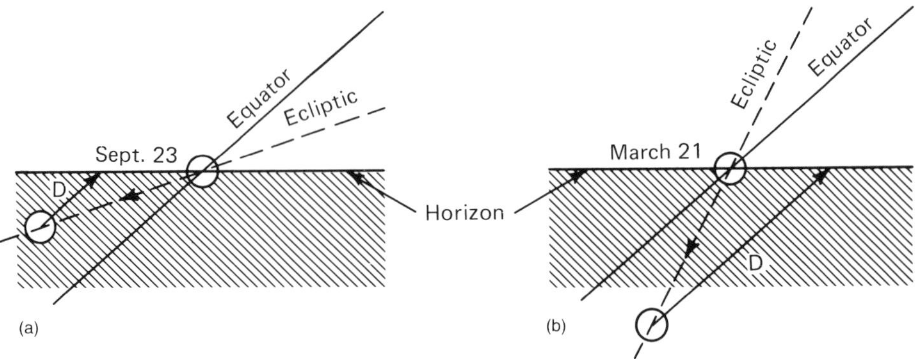

9.24. The harvest Moon. In September (a) the daily delay in moonrise after full Moon, *D*, is much less than it is in March (b) because of the difference in angles between the ecliptic and the horizon.

examined the Moon's general behavior with phase in Section 9.2; now we will look at the problem in more detail. Moonrise and moonset are complicated by a variety of factors. The inclination of the lunar orbit to the ecliptic plane and the regression of the nodes make the Moon's declination more difficult to determine than the Sun's. The orbital eccentricity and the perturbations, particularly the evection and the variation, complicate the calculation of the Moon's celestial longitude, the angle between it and the Sun, and the hour angle for a given time of day. We will first look at the average behavior by assuming the Moon to be in a circular orbit in the ecliptic plane. Then we will examine the principal effects of the true orbit.

The Moon advances along the ecliptic ahead of the Sun at a rate of 12°.19/day, 360° divided by the synodic period. (It moves 13°.18/day against the *stellar* background: the difference of a 0°.99/day, of course, reflects the Sun's daily motion along its path.) Therefore, on the average, the Moon must rise or set $(12°.19/15°) \times 60 = 49$ minutes later every day. But since the Moon is moving either south or north along the ecliptic, the delay in moonrise from one day to the next can vary considerably, depending on the angle that the ecliptic makes with the horizon at that time.

For example, in Figure 9.24, look at the rising of the full Moon in September as compared with that in March, from a latitude of 40° N. If the full Moon occurs on September 23, it will be at the vernal equinox and both will rise at sunset. With the vernal equinox on the horizon, the winter solstice will be at upper transit, and the ecliptic will approach the horizon at a low angle (see Figures 3.5 and 3.6a). The next night, the Moon is 12° farther along the ecliptic at sunset, but because of the low angle of that path, it is not very far below the horizon. The Moon must rise along its daily path (parallel to the equator); the interval to the horizon, *D*, is short, only about half an hour. That is, with full moonrise at 18^h on September 23, moonrise on September 24 will be about $18^h\ 30^m$. In March, however, when the autumnal equinox rises at sunset, the ecliptic makes a much steeper angle with the horizon (see Figure 3.6c). The daily delay, *D*, is much longer, about 70 minutes. Thus in September, we get a great deal of moonlight in the early evening around full phase, with the full or nearly full Moon rising only half an hour later each night. In March the Moon disappears very quickly from the evening sky after full phase passes. The September full Moon (or the one nearest the time of the equinox) gives us an ideal time to work outdoors after sunset, and the phenomenon is called, appropriately, the *harvest Moon*. The situation is not very much different a month later, when the full phase is called the *hunter's Moon* (see Table 10.3).

The calculation of the exact position of the Moon at any time is more complex than is possible to do here, but we can make some simplifications that will allow us to figure an approximate rising azimuth and time by following the procedures outlined earlier. We assume that the Moon has a circular orbit in the ecliptic plane and that there are no perturbations. In a loose sense, we are dealing with an average lunar position. Since the error we will make can be as large as several degrees, we assume that the Earth's orbit is circular as well.

By coincidence, the parallax of the Moon closely offsets the combined effects of refraction and semidiameter. The latter two loft the upper limb relative to the true center by 51′, whereas the horizontal parallax depresses the Moon by a bit more, 57′. We see the Sun rise when it has a true altitude of −51′, but we do not see moonrise until its true altitude (relative to the Earth's center) is $-51' + 57' = 6'$. Thus the azimuth of moonrise (moonset) relative to that given by Figure A1.1 is shifted in the direction *opposite* the shift for the Sun (see Chapter 7), and moonrise is *delayed* relative to Figure A1.2 rather than advanced. The net azimuth and time changes (including parallax, refraction and semidiameter) can be read from Figures A1.4 and A1.5 at $^1/_{10}$ the values given for the Sun. These changes are so small relative to the errors inherent in our approximation that they can be ignored, and the azimuth and hour angle of moonrise or set can be read directly from Figures A1.1 and A1.2.

To proceed further, we need to estimate the Moon's right ascension (to find the rising time from the hour angle) and the declination, which requires that we know the celestial longitude at the time of moonrise. The longitude of the apparent Sun (where we ignore the effect of orbital eccentricity) is just the number of days past the vernal equinox times $^{360}/_{365} = 0.986$. Since the inclination, eccentricity, and perturbations affect the Moon by more than 1°, a tenth of a day is adequate. To find the lunar longitude we add to the solar longitude the product of the number of days past new (at moonrise or set) times the gain rate of 12°.2/day. We then read the right ascension and declination of the Moon from Figure A1.9, which gives the α and δ of the ecliptic as a function of celestial longitude, λ. The azimuth is found immediately from Figure A1.1. The *sidereal* time of moonrise is the hour angle from Figure A1.2 plus the right ascension, α (see Section 3.16). With our known date and Table A1.1, we can convert sidereal time to local solar time (remember that the sidereal clock is ahead of the solar by 3^{m} $56^{s}.56$ for each day past September 21: see Section 6.8). Finally, from the terrestrial longitude, we find the standard time (Section 6.4).

As an example, estimate moonrise on November 25, 1980 for longitude 77° W, latitude 33° N. Since the Moon moves so rapidly, we must first guess a rough time of moonrise so that we know the interval past new Moon to about a tenth of a day. From the *Astronomical Almanac*, the new Moon occurred on November 7 at 21^h UT, or about 16^h local time, or November 7.7. Eighteen days later, the Moon has moved through $^{18}/_{29.5} = 0.6$ of its orbit, so that it will rise roughly $0.6 \times 24^h = 15^h$ after sunrise. From Table A1.2, sunrise takes place at about $6^h.6$, and thus the Moon rises at about 22^h. We then need the lunar longitude on November 25 at 22^h = November 25.9, 18.2 days past new. From Table A1.1, November 26 (the closest whole date) is 250 days past March 21, so that the longitude of the Sun is $0.985 \times 250 = 246°$. At 12°.2 per day the Moon has moved $18.2 \times 12.2 = 222°$ from the Sun, so that its longitude is 246° + 222° = 468°. We subtract 360°, leaving us with λ(Moon) = 108°. From Figure A1.9, α(Moon) = $07^h\ 17^m$ and δ(Moon) = +22°10′, so that from Figure A1.1 the rising azimuth is about 64°, and from Figure A1.2 the rising hour angle is about $-07^h\ 00^m$ or $17^h\ 00^m$. The sidereal time, which is hour angle plus right ascension, is then $00^h\ 17^m$. To the nearest whole day, November 25.9 is 66 days past September 21, so that the sidereal clock is 66 days × $3^m\ 56^s.56$ per day = 260^m = $04^h\ 20^m$ ahead of the mean solar clock. Subtracting $04^h\ 20^m$ from $00^h\ 17^m$ = $24^h\ 17^m$ = $23^h\ 77^m$ gives us moonrise at $19^h\ 57^m$ local mean solar time. We are at 77° W longitude, so that the standard time of moonrise is (77° − 75°) × 4^m = 8^m later, or $20^h\ 05^m$ (8:05 PM) Eastern Standard Time, to the northeast, at an azimuth of 64°.

9.12 Moonrise and moonset: true time and azimuth

Observation would have shown us that the Moon actually rose on November 25, 1980 at $20^h\ 39^m$ at an azimuth of 67°, 34^m later than, and 3° south of, the above simple prediction. The obvious reasons are the Moon's orbital inclination, which alters its declination from the simple ecliptic orbit, and the eccentricity (plus the perturbations), which affects the Moon's longitude and right ascension. The actual declination at the time of the example was 19°34′ (2°36′ south of the ecliptic), and the true right ascension was $7^h\ 45^m$, 28^m or 7° east of the earlier estimate, which act together to delay moonrise from the above estimate by the requisite amount.

At the summer solstice, the Sun has a declination of +23°27′. But because of the Moon's orbital inclination, it can pass the solstice 5°09′ above or below this declination, depending on the longitude of the node. (The maximum value is actually 5°18′ because of perturbations, but in this section we will adopt the average value of the inclination.) Figure 9.25 shows the Moon's orbital path relative to the ecliptic for Ω (the longitude of the ascending node) = 0° (a) and Ω = 90° (b). For the purpose of illustration, the drawings are for an observer at 45° N latitude; the orbit of the Moon relative to the equator and ecliptic will be independent of terrestrial location. When Ω = 0°, the Moon has its largest range of declination, 23°27′ + 5°09′ = 28°36′ north and south of the equator. Once a month we can then see the Moon 5° closer to the zenith than we can ever see the Sun for any particular northern latitude, and two weeks later 5° farther from the zenith than we can ever see the Sun. When these events happen, of course, depends on the time of year and the phase, that is, on the position of the Moon relative to the Sun. As an example, assume that Ω = 0° (Figure 9.25a), it is March 20, and the Moon is at its first quarter phase. The Sun is at the vernal equinox, so the Moon

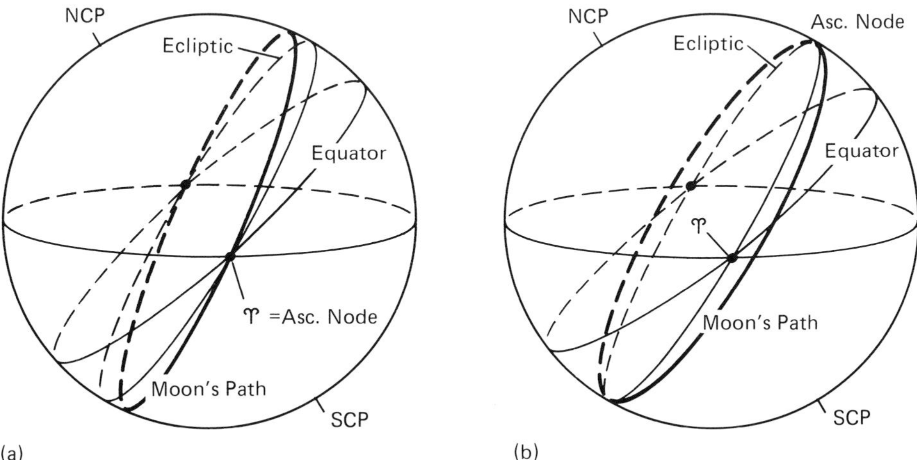

9.25. The Moon's orbital path on the celestial sphere. The Moon's orbit, the heavy lines in the above figures, intersects the ecliptic at the angle of inclination, 5°09'. The longitude of the ascending node, Ω, is 0° in (a) and 90° in (b).

should be passing above the summer solstice, at declination 28°32′. This first quarter Moon will also be circumpolar 5° south of the arctic circle. If it is December 22, and the Sun is at the winter solstice, the maximum declination will occur for *full* Moon. The reader might make up a number of other possibilities.

Next, in Figure 9.25a, mentally move the ascending node to 180°, which means that we simply flip the Moon's orbit 5° to the other side of the ecliptic, to between the ecliptic and the celestial equator. The Moon now has its minimum range of declination, 23°27′ – 5°09′ = 18°18′. On March 21 the first quarter Moon will pass 5° farther from the zenith than the summer solstice Sun, and it cannot be circumpolar until the observer is 5° *north* of the Arctic Circle. In Figure 9.25b, where the node is 90° from the vernal equinox, the Moon will move through the winter and summer solstices, but will pass 5° south of the vernal equinox and 5° north of the autumnal. At Ω = 270°, we see the reverse.

The effect of the inclination on moonrise and moonset is apparent in Figure 9.25, where we need only follow the Moon's daily path for different positions as the Earth rotates. Look first at Figure 9.25b. A first quarter Moon on March 21 will be at the summer solstice, and the Moon will rise and set at the same point on the horizon as will the Sun on June 21. But when the Moon passes the autumnal equinox near full, it rises to the north of 90° azimuth; if it passes the vernal at new on that date, it will rise *south* of that point. And we see the reverse for Ω = 270°. In Figure 9.25a, where Ω = 0°, when the first quarter Moon on March 21 passes 5° north of the solstice it must set north of the point marked by the summer solstice Sun (and the third quarter will set south of that marked by the winter solstice Sun). When Ω = 180°, and the Moon has its minimum declination range, its range of rising and setting azimuths is restricted by a similar amount rather than expanded.

Table 9.3. *Local mean time of moonrise (upper limb), for the meridian of Greenwich*

date/latitude	30°	35°	40°
	h m	h m	h m
1980 Aug. 22	15 41	15 52	16 04
23	16 35	16 45	16 56
24	17 26	17 34	17 44

The difference in the rising and setting points between the Moon and Sun increases with the observer's latitude. Figure A1.1 shows as special cases the rising azimuth for the solstices and that for the Moon when it is at its maximum declination of 28°36′. At the equator, $\phi = 0°$, the maximum difference between the azimuth of moonset and sunset at the solstice can only be 5°09′. At $\phi = 40°$, the difference is 6°.5 (allowing for the shift in sunrise and sunset caused by refraction and semidiameter), and at $\phi = 55°$ it is 11°.5. Hour angles (see Figure A1.2) are similarly affected. The midwinter full Moon in temperate latitudes, when $\Omega = 0°$, with moonrise far to the north (or south), and a transit point high in the sky is a lovely sight.

These phenomena are cyclic according to the regression of the nodes (see Section 9.7) around the ecliptic with its 18.6 year period. Starting with $\Omega = 0°$, we first see the maximum range in lunar declination and in the azimuth of moonrise and set. Four and a half years later, $\Omega = 270°$, and the Moon behaves more like the Sun. After 9 years, $\Omega = 180°$, we see a minimum range of declination and horizon azimuth, returning again to maximum values after 18.6 years have elapsed. The wandering of the nodes has great significance for lunar and solar eclipses, that subject to be examined in the next chapter.

Because of the vagaries of the lunar motion, calculation of the exact time and azimuth of moonrise for any given day is quite complicated, and it is simplest to rely on the tables of the *Astronomical Almanac*. This volume gives the right ascension, declination, longitude, and latitude of the Moon throughout the year. It also gives tables for the time of moonrise and moonset for every day of the year for a variety of latitudes. To find moonrise time for a particular location on Earth, it is necessary to interpolate within the table. As an example, assume we wanted to know when the Moon rose on August 23, 1980 at Los Angeles, latitude 34° N, longitude 118° W. A portion of the table from the *Almanac* is reproduced as Table 9.3, which gives moonrise for an observer at the Greenwich meridian. Los Angeles, however, is almost 8 hours west of Greenwich, and by the time the Moon rose there it would have moved some 4° *east*, and consequently it would have risen later, between the Greenwich times listed for August 23 and August 24. The first step is to find the Greenwich moonrise for the latitude of Los Angeles. On August 23, there is a 10 minute difference between 30° and 35°, so that at 34° moonrise occurred 0.8×10^{m} later than the 30° entry, or at 16^{h} 43^{m}. On August 24, moonrise at 34° was at 17^{h} 32^{m}. Los Angeles is $118°/360° = 0.33$

of the way around the globe from Greenwich. Consequently, moonrise must have occurred 0.33 of the way between $16^h\ 43^m$ (August 23) and $17^h\ 32^m$ (August 24), or at $16^h\ 43^m$ plus 0.33 times the difference between the two, or $16^h\ 43^m + 0.33(49^m) = 16^h\ 59^m$ local time. The standard meridian for Los Angeles is at 120°, so we must subtract 8^m to find the Pacific Standard Time of moonrise, $16^h\ 51^m$.

Once we know the local or standard time of moonrise, we can add the longitude or the zone description to find the Universal Time, so that we can look up the Moon's declination at moonrise. Right ascension and declination are presented in the *Almanac* in the form of "daily polynomial coefficients." For any day, we apply the equation

$$\text{quantity } (\alpha \text{ or } \delta) = a_0 + a_1p + a_2p^2 + a_3p^3 + a_4p^4 + a_5p^5,$$

where p is the fraction of the day in UT, and the a_0 through a_5 are coefficients that are given for every day of the year. Once δ is known, we can then find the rising azimuth from Figure A1.1.

We are not yet done. The Moon's actions and influences on the Earth are profound, both in our daily lives and in their impact on scientific research. We still have three major topics to cover: the tides, eclipses, and calendars, which deserve their own chapter.

Chapter 10

Tides, eclipses, and calendars

We now turn homeward to examine the effects the Moon has on our planet and on its people. Our satellite moves the Earth's waters to create the tides and serves as a natural timekeeper. Through its shadow phenomena – eclipses – it can also excite public interest and bring us outdoors to look back into the heavens.

10.1 Tides

To someone who has always lived in the interior of a country, and to whom the tides are hearsay or textbook descriptions, their reality upon a journey to the coast can be little short of astounding. You spread a blanket at some distance from the waterline only to find it awash a few hours later, or in an extreme example, you tie a rowboat to a pier and return to find it hanging vertically out of touch with the water (Figure 10.1).

Observation shows us that the water at the coast first rises and then falls with a period (the interval between successive highs) of 12 hours 24 minutes. This figure is the average interval between successive passages of the Moon across the meridian as it moves 12°.2 per day eastward from the Sun. There can be no doubt that the tides result principally from the gravitational action of our celestial companion.

A tide is caused by a difference in gravitational force across a body of finite size. The force of the lunar gravity on the Earth will be strongest at that point closest to the Moon, the location where the Moon is directly overhead. It will be weakest directly opposite, where the Moon is at the observer's nadir. The Earth is therefore subject to a stretching force whose effect is to elongate the planet into a slightly oval figure directed at the Moon.

Because of the great strength of the Earth, the difference in its dimensions directed toward the Moon and perpendicular to it (ignoring the oblateness of the Earth) is only a matter of centimeters, and is not detectable except with delicate instrumentation. The major effect is on the Earth's waters (Figure 10.2). Think first of our planet as covered by a very deep ocean that represents a large fraction of the terrestrial radius (Figure

(a)

(b)

10.1. The harbor at Cutler, Maine, with one view at high tide (a) and the other at low tide (b). (Maine Department of Sea and Shore Fisheries.)

10.2a). Because water has no significant strength, it will flow in response to the differential gravity (Figure 10.2b), and will stretch out in a line towards the Moon. We will thus get deep water on *both* the side of the Earth facing the Moon and that facing away, and shallow water in the perpendicular direction, from where the water has flowed.

As the Earth rotates on its axis, an observer will pass first under high water and then under low. If the Moon held its position, the person would experience two high tides and two low tides a day alternately 6 hours apart. But while the Earth rotates, the Moon is also moving in the same direction in its orbit. In 12 hours it moves 6°.09 east of the Sun along the ecliptic, and successive high tides will occur 12 hours 24.4 minutes apart.

The Earth is not covered by a deep ocean, but by a shallow one. As deep as it may seem to us, the maximum depth is only 0.0018 of the terrestrial radius. The sea is but a thin film over the solid planet. The sea is rotating with the Earth and it takes time for the water to flow. As a result, the tidal bulge leads the Moon by a degree that depends upon latitude and the details of the ocean floor and coastline. Instead of the high tides occurring when the Moon transits the upper and lower meridians, they will occur well after meridian passage (Figure 10.2c). The low tides will take place in between.

The pull of the Moon on the tidal bulge, which drags against the Earth's solid body, in turn produces the slowing of the Earth described in Section 6.10 and is the major reason for needing Coordinated Universal Time.

10.2 Tides and the Sun

The Sun also raises tides in the Earth's waters, its huge mass compensating for its greater distance. The tidal effect of the Sun is 5/11 that of the Moon, and it has the

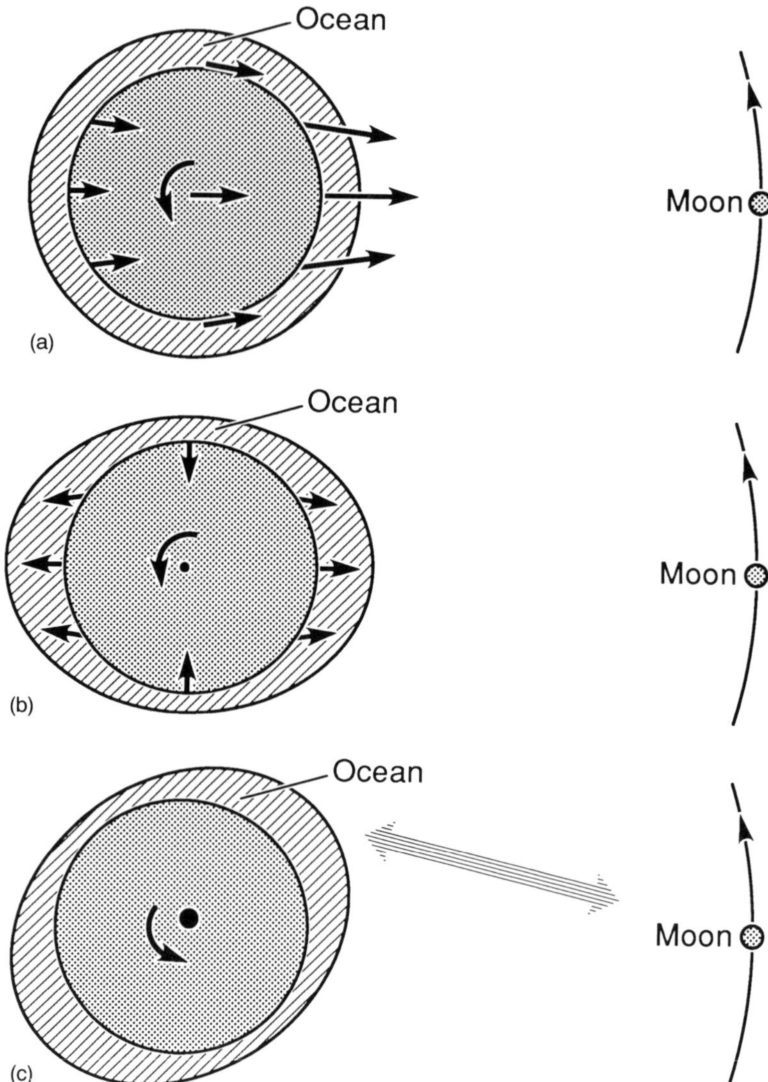

10.2. The creation of the tides. The depth of the ocean is greatly exaggerated. The arrows in (a) show the strength of the lunar gravitational pull at different distances from the Moon. In (b), the force at the Earth's center is subtracted from all the arrows. All that remains is the differential force, which causes water to flow toward the line connecting the Earth and the Moon. As a result of the Earth's rotation and the time it takes the water to flow (c), the tidal bulge leads the Moon.

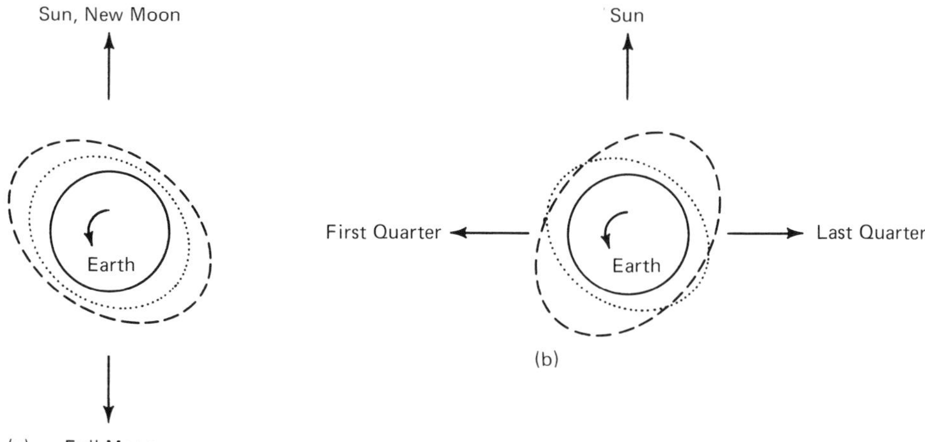

10.3. Spring and neap tides. (a) At new and full Moon the lunar (dashed line) and solar (dotted line) tides act in concert to produce a high observed tide called a *spring* tide, having nothing to do with the seasons. (b) The solar tide partially compensates for and fills in the lunar, to produce a *neap* tide in which the highs are not as high and the lows not as low. Remember that the tidal bulge is dragged ahead as a result of the Earth's rotation.

effect of modulating the lunar tides. When the two are lined up at new and full Moon the tides are over twice as high as they are when the Moon is at its quarters, when the forces partially compensate for each other.

Figure 10.3a shows the solar and lunar tides at new or full Moons, when they add together. The solar and lunar tides act independently. The solar tide adds to the lunar at full moon exactly as it does at new Moon. These highest and lowest tides are called the *spring tides*. In Figure 10.3b the Moon and Sun are 90° apart. The solar tide then works perpendicular to, or against the lunar, reducing high tide, and filling in the low. On the beach we would still experience two high and low tides about every 24 hours and 50 minutes, but their ranges are reduced by roughly half to produce the *neap tides*.

10.3 Tides and orbits

The size of a tide is proportional to the inverse of the cube of the distance to the body causing it. If we could place the Moon twice as far away, the amplitude of the tides would be $(^1/_2)^3 = {}^1/_8$ as great. Thus the range of the tide is quite sensitive to the exact lunar distance and as well as to its orbital eccentricity. From the variation in lunar distance given in Table 9.2, the lunar tide at average perigee will be $(^{406}/_{363})^3 = 1.4$ times greater than at average apogee, a surprisingly high difference. The highest and lowest general tides will therefore take place when full or new Moon coincide with perigee. Because of the advance of the line of apsides at its rate of 41° per year (see Section 9.7), this event (or some approximation to it) will occur roughly twice in a $13^1/_2$ month period.

The eccentric orbit of the Earth has a similar, but smaller, effect. The solar tide is about 10% higher in January when the Earth is at perihelion than it is in July at aphelion. The two orbital effects add together, producing a modulation in the amplitude of the tides with an average cycle time half the period of the advance of lunar perigee. If the January (perihelion) new Moon occurs near perigee, it will be near apogee about 4 years later, when the *full* Moon will be near perigee. These maximum-range tides can be quite astonishing. Sunken ships and other debris that have not been seen in years become exposed. They can also be dangerous because they tempt the curious to venture beyond safety, to the point where they can actually be caught by the inrushing waters.

10.4 Prediction of the tides

The general theory of tides is extremely complicated; the actual prediction of the time of high or low tide at a place on the coast is quite impossible. We state that it will occur rather well past the meridian transit of the Moon, but it is strongly affected by the depth of the sea, the topography of the sea floor, and the shape of the coastline. For any location, we first have to determine the times of the tides empirically, from observation, to know when they occur relative to lunar meridian passage. This interval, called the *establishment of port*, will on the average be constant, allowing subsequent prediction. The variation caused by the lay of the coast can be remarkable. Two towns separated by only a few kilometers can have high tides over an hour apart. The only real constant is the interval of $12^h\ 24^m$ between times of high water at any one location. And even that is only an average, since the exact time of high tide is influenced by the speed and direction of the wind and by barometric pressure.

The height of the tide is similarly theoretically unpredictable and depends strongly upon local topography. The tide in the open ocean is only somewhat under 1 meter and cannot be detected by a ship at sea. In many places at the shoreline, it is not much greater. But at some harbors, where the lay of the land tends to funnel the rising water into the shore, the tidal range can be astounding. At the head of the Bay of Fundy in the US State of Maine, the high spring tide is 15 meters above the low. The tidal range is also strongly affected by winds and air pressure, as well as by the varying distances of the Moon and Sun. The high tide with hurricane winds blowing to shore can be several meters above normal and periodically produces disaster.

In a few locations, even the normal tide can be dangerous. At low tide it is tempting to walk way out to look, perhaps, for shells. Off the northern coast of France, where the sea floor has a very small slope, the tide returns faster than a human can run. In some places, a river bed creates something of a natural funnel to the sea; it may slope up very slowly, and small changes in water depth cannot be adjusted quickly in the river. The high tide moves into it so rapidly that the water piles up. The result can be a literal wall of water, a *tidal bore*, roaring up the channel. The phenomenon is well known on the Amazon and on the River Severn in England as well as on many others.

The daily tidal range must also clearly be affected by the Moon's altitude above the horizon. As is evident from Figure 10.2, maximum tides will take place at locations that are on a line between the Moon and the center of the Earth, the points toward which the water flows. Maximum tides therefore occur where the Moon passes through the zenith, as it will do in the tropics. At lower altitudes the high tide will not be as great.

Since the altitude of meridian transit depends upon lunar declination, so too will the depth of the tide. When the Moon is on the celestial equator, the two altitudes of culmination are symmetrical; that is, the angle of the Moon from the zenith at upper meridian transit is the same as that from the nadir at lower. But at high declination it is not. At 40° N latitude, and with a lunar declination of 23°, the zenith angle of the Moon at upper transit is 17°, while at lower, because the daily path is parallel to the equator, the Moon is 63° from the nadir (see Chapter 2). Therefore the depths of the two daily tides will be different. This situation can reach an extreme at high latitudes, where it is possible to see only one high tide per day rather than two.

Still other variations depend on lunar distance. In some places there may be tides at perigee but not at apogee. Such an "apogean tide" (or lack thereof) produced a major disaster for United States amphibian forces in World War II. They had expected to ride a high tide across an anti-invasion barrier surrounding a South Pacific island. But the tides had not been studied well, the Moon was at apogee, the high water never came, and the invaders were trapped.

It is easy to see now why tidal prediction at a given locale is only possible through experience and observation over a fairly long time. Empirical prediction tables for major ports and towns along the coast are available from the *United States Coast and Geodetic Survey*, from the *Institute of Oceanographic Services* in Great Britain, and from similar agencies in other countries.

10.5 Tidal effects on the Moon

Just as the Moon raises tides in the solid body of the Earth, so the Earth does in the solid Moon. Since our planet is 80 times more massive than our satellite, the terrestrially produced tide is 80 times greater. The result is that, although the Moon causes only a minor slowing of the Earth's spin as a result of the movement of the oceans, the Earth has stopped the Moon's rotation dead (relative to itself) as previewed in Section 9.4. And although the terrestrial shape is only barely (and almost unmeasurably) changed by the Moon, the lunar is rather distorted by Earth along the line that connects the two.

There is another, long-term, but profound effect. The tides raised by the Moon on the Earth cause our satellite to move farther away from us, the result of a phenomenon called the *conservation of angular momentum*. Tie a rock to a string and whirl it about you. The angular momentum (L) is the rock's mass (m) times the length of the string (r) times the revolution velocity (v), or $L = mvr$. The total angular momentum of a closed, isolated physical system must be constant, that is, it must be *conserved*. L cannot be changed without the application of external forces. If we pull in the string, and make r smaller, v *must increase* so as to keep L constant (try it: carefully). The principle is used by a spinning skater, who speeds up when his or her arms are brought inward.

10.4. An eclipse of the Moon. This multiple exposure shows the Moon passing (from right to left) through the shadow of the Earth. (Akira Fujii.)

The Earth–Moon system fits the conservation requirement to a reasonable approximation. Its total angular momentum is the sum of the rotating Earth, the orbiting Moon, what is left of the (sidereal) lunar spin, and the small orbit of the Earth about the barycenter. The disturbing effect of the Sun can, at least in a first approximation, be ignored. The Earth is slowing because of tides and is losing angular momentum, which must go someplace. It goes into the orbit of the Moon. The lunar distance is increasing such that its gain in L equals the amount that our planet is losing. The outward movement is very small, only a few centimeters a year, but inexorable. We have direct evidence that at one time the Moon was only half its present distance from us. In earlier ages, when it was closer, its period was also considerably less, and we can see the effect on the shell growth rates of fossils of animals that live in tidal zones. The effect will cease only when the Earth finally locks synchronously onto the Moon and the two keep their faces pointing to one another.

Physically, the effect is accomplished through the Earth's oceanic tidal bulge. The bulge leads the Moon, and then its weak gravity acts upon the satellite to pull it forward and give it additional angular momentum, thereby increasing the orbit's semi-major axis.

10.6 Eclipses

Like a tree, or yourself, the Earth and Moon cast shadows in the Sun. In both cases, the shadows are long enough to affect the other body. When the Moon passes through the Earth's shadow, cutting the Moon off from its source of light, we see an *eclipse of the Moon* (Figure 10.4). The phenomenon obviously can occur only when the Moon is at its full phase. When the Moon is interposed between the Earth and Sun at its new phase, blocking the latter, we see an *eclipse of the Sun.*

The shadows are shown in Figure 10.5. Because of the different sizes of the Earth and Moon, the terrestrial shadow is some 4 times longer than the lunar. The relative distances and sizes in the figure have to be distorted to draw all three bodies on the same page (the Sun is almost 400 times more distant than the Moon).

The physical lengths of the two shadows can be calculated from the dimensions of the Earth, Moon, and Sun: that of the Earth's is 1.383 million kilometers, with a

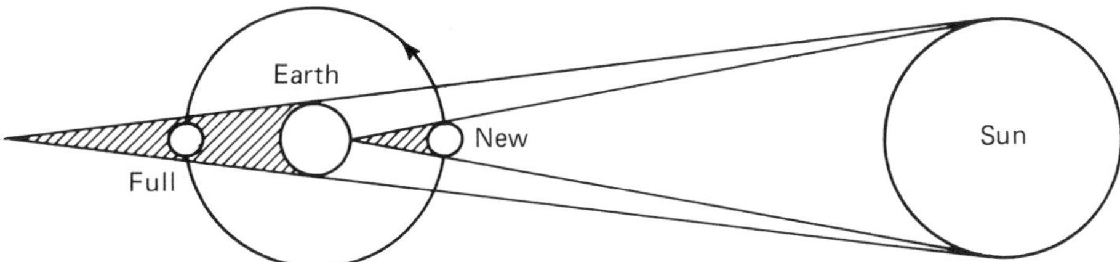

10.5. Shadows of the Earth and Moon. That of the Earth is much longer than the distance from it to the Moon. That of the Moon can touch the Earth, but more often misses it because of the eccentricity of the orbit. The relative distances are grossly distorted.

variation of 23 000 kilometers each way because of the eccentricity of the terrestrial orbit, and that of the Moon's averages 374 000 kilometers, with an analogous variation of 6400 kilometers. Since the Moon's distance averages 384 000 kilometers, it may easily be entirely immersed in the Earth's shadow, which at that point is still 2.8 times the lunar diameter. But that distance, even measuring from the Earth's surface rather than from the center, is greater than the *Moon's* average shadow length, so that the shadow cone can just miss the Earth. The lunar distance, however, can be cut to 363 000 kilometers at mean perigee (and to somewhat less when perturbations are taken into account), and when our satellite is only just a bit closer than average, the shadow can strike the ground. Under the most favorable conditions, the Moon at perigee and the Earth at aphelion, the shadow at the sublunar point (as in Figure 10.5) can have a maximum diameter of 270 kilometers. (When the shadow strikes near the limb, where the Sun would be seen low on the horizon, either at sunrise or sunset, or at polar latitudes, its outline will be an oval with a width of about 215 kilometers but greater length.)

Expressed in a different way, the Moon's angular diameter is on the average slightly less than the Sun's, but under favorable conditions can be slightly greater. In the first case, if the Moon is centered on the Sun, we would see a ring of sunlight around the apparently smaller Moon, a condition called an *annular eclipse* (Figure 10.6). In the second, the Moon can entirely block the light of the bright solar surface, resulting in a *total eclipse* (Figure 10.7). The number of annular and total events over a long period are roughly equal. Because of the curvature of the Earth's surface, the distance from any point to the Moon can vary by the terrestrial radius of 6400 kilometers and occasionally a given eclipse will have both annular and total segments.

Eclipses of the Sun and Moon can be either total or *partial*, depending upon what part of the shadow is involved. The zone of total shadow (the *umbra*) and partial shadow (the *penumbra*) are shown in Figure 10.8. In an umbral shadow all the light is cut off whereas in a penumbral shadow, an observer would see a portion of the light source. If you are in the Moon's umbra you see a total eclipse of the Sun, and if in the penumbra you witness a partial (or even annular) eclipse. In a partial solar eclipse, the Moon seems to the observer to take a bite out of the solar limb (and a partial phase must precede and succeed totality). When the Moon is in the Earth's umbra, and is

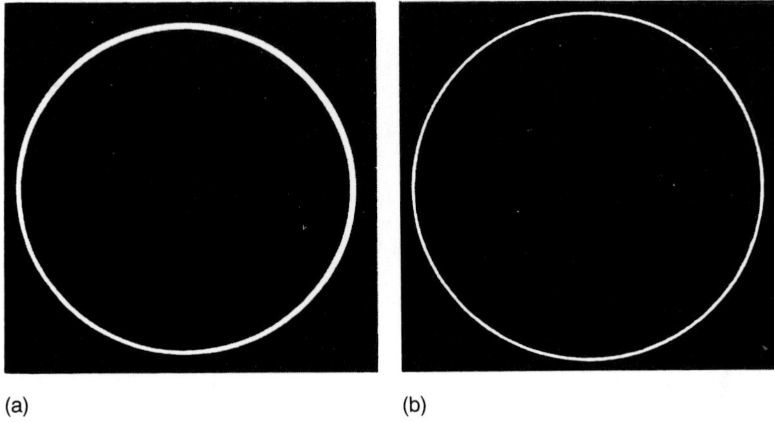

10.6. Annular eclipses of the Sun: (a) the eclipse of July 31, 1962; (b) that of January 25, 1963. In neither case is the Moon close enough to allow the shadow cone to touch Earth, and the lunar angular diameter is less than the solar. In (a) the Moon was closer to apogee and the ring of sunlight larger. The 1962 eclipse lasted for 3^m 25^s, the 1963 one for 0^m 37^s. If the Moon had been a bit closer in 1963 we would have seen a total eclipse. ((a) Lockheed; (b) US Air Force.)

totally eclipsed, a *lunar* visitor would see a total eclipse of the Sun, and when the Moon is in the Earth's penumbra, he or she would see a partial solar eclipse. In the latter case, we on Earth see a *penumbral eclipse* of the Moon, which results only in an overall dimming of the lunar brightness, a phenomenon that is only barely noticeable. We reserve the term *partial eclipse of the Moon* for the times when the Moon is partly in the umbra and partly in the penumbra. Eclipses of the Moon can obviously never be annular.

Even under the most favorable conditions, the size of the Moon's shadow-spot on Earth is so small that few people will see any particular solar eclipse as total. As the Moon orbits the Earth, the shadow sweeps from west to east across the planet, tracing a path at most only a few hundred kilometers wide. And more often than not, the shadow path moves across oceans and remote areas of the globe. The great majority of observers would see the eclipse as partial: the lunar penumbra can be over 6000 kilometers across at the Earth, large enough to embrace North or South America. However, from Figure 10.5, we see that when a lunar eclipse occurs, everyone on the nighttime side of the Earth will see it. Thus even though total (or annular) solar eclipses actually occur more frequently than lunar, an individual will see many more of the latter.

10.7 Eclipse conditions and seasons

If the Moon orbited the Earth exactly in the ecliptic plane, we would see one solar eclipse (somewhere on Earth) and one lunar eclipse (from one terrestrial hemisphere) every synodic month of 29½ days. From experience, we know that is not the case:

10.7. A total eclipse of the Sun. The Moon was closer to perigee than in Figure 10.6 and produced this eclipse of August 31, 1932. The shadow cone touches Earth, the Moon is angularly larger than the Sun, and the corona, the hot extended solar atmosphere, becomes visible. (Naval Research Laboratory.)

ordinarily we witness a total lunar eclipse every couple of years, and a partial solar eclipse even more rarely. The reason is the 5° inclination of the lunar orbit to the ecliptic plane. At new and full phase the Moon usually passes either above or below the Sun or the Earth's shadow. Only when new or full Moon occurs while the Moon is near the ecliptic, or rather near a node, will we have an eclipse.

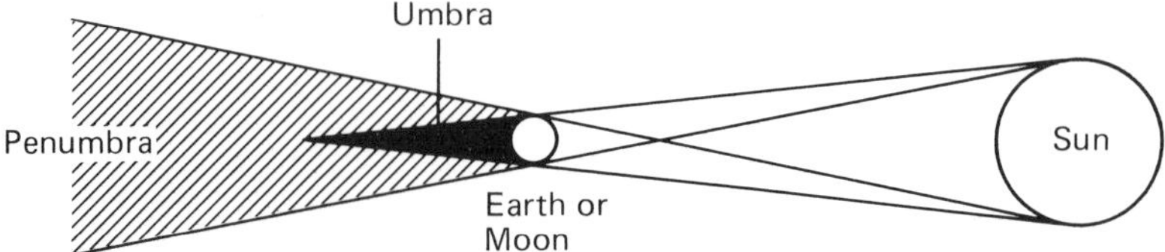

10.8. The parts of a shadow. Any body in sunlight casts a region of total shadow, the *umbra*, and one of partial shadow, the *penumbra*.

Because of the width of the terrestrial shadow at the Moon's distance (1°.4 as viewed from Earth) the Moon does not need to be exactly at a node for an eclipse to occur; it must only be within a specified angle from the node, called the *ecliptic limit*. These are shown for both lunar and solar eclipses in Figure 10.9. In Figure 10.9a, the Moon just grazes the Earth's umbral shadow for a partial lunar eclipse as it approaches the descending node. The Moon must be within the arc L_M, the lunar ecliptic limit, for the eclipse to occur. The shadow's radius is about 3/4° and the Moon's is 1/4°, so the center of the Moon must be about 1° above the ecliptic. The angle of intersection is 5°, and simple trigonometry shows that L_M is about 11°. It varies somewhat because of the eccentricities of the terrestrial and lunar orbits, which change the size of the shadow at the Moon's distance of passage. An exact calculation at perigee and aphelion shows L_M to be 12°.3, which is called the *major limit*. The *minor limit* of 9°.5 occurs at apogee and perihelion.

The limits for a solar eclipse involve an added complication. The Moon is seen on its orbital path only on the line that runs between the lunar and terrestrial centers. Because of parallax (see Section 9.8), the Moon can be seen as much as a degree above

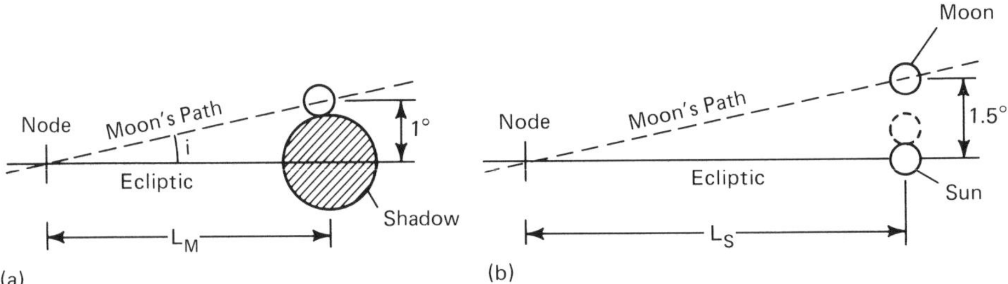

10.9. Ecliptic limits. (a) The ecliptic limit for the Moon. A partial eclipse is just possible when the Moon, grazing the Earth's shadow, is L_M away from the node. The shadow is 3/4° in radius, the Moon 1/4°, so the Moon must be about 1° from the ecliptic. The Moon's orbit intersects the ecliptic at a 5° angle, so L_M is easily calculated to be about 11°. (b) The ecliptic limit for the Sun. A partial eclipse of the Sun occurs when the Moon grazes the upper solar limb. Each is 1/2° in radius. But the Moon can be depressed above or below its true position (dashed circle) by just under 1° because of parallax, and therefore can be 1°.5 from the ecliptic, half again as great as the criterion for a lunar eclipse. L_S is therefore 50% greater than L_M.

or below its orbital path (Figure 10.9b). The right side of the triangle (the angular distance from the ecliptic to the orbital path) can now be as much as $1^1/_2°$, 50% greater than the right side of the triangle for the lunar eclipse. The solar ecliptic limit, L_S, is consequently 50% greater as well. Exact calculations give major and minor values of L_S of 18°.5 and 15°.4. (Parallax is unimportant in the lunar eclipse because the Moon and shadow are affected the same way.)

If the full moon is to be within L_M of one node for either a partial or total lunar eclipse to occur, the Sun must be within L_M of the other. The Sun must also then be within L_S, which is larger than L_M. Consequently, we see both kinds of eclipses only during *eclipse seasons*, when the celestial longitude of the Sun is within L_M (at full moon) or L_S (at new moon) of the longitude of the ascending node, Ω, or that of the descending node, $\Omega \pm 180°$. Because of the regression of the nodes (Section 8.7), the Sun encounters a particular lunar node every 346.62 days, an interval called the *eclipse year*. There are obviously only two eclipse seasons during an eclipse year, and they move back by 19 days (365 - 346) every calendar year.

The eclipse seasons can easily be found from a knowledge of Ω. For example from Table 9.2, at the beginning of 1994, $\Omega = 241°$. The ascending node was therefore at $\lambda = 241°$ and the descending at $241° - 180° = 61°$. The Sun encountered the descending node first. At 0.986° per day, the Sun reached 61° longitude in 62 days from the time of the vernal equinox, or from Table A1.1 (with equinox passage on March 20) on May 21. This date is 140 days past January 1, and at a regression rate of 19.3° per year, the node had moved back by $(^{140}/_{365}) \times 19.3° = 7°$, so the Sun encountered it 7 days earlier, on May 14. (The motion of the node during this 7-day interval is under 1°, so for this rough calculation can be ignored.) A lunar eclipse was then *possible* for 12 days either side of this date, the time it takes the Sun to move through the major ecliptic limit of 12°, or from May 2 to May 26 (possible because the full Moon might not touch the shadow if near apogee). It was *inevitable* if the full Moon occurred while the Sun was within the minor ecliptic limit (9°) of the node, or between May 3 and May 23. (Even *at* apogee the Moon *must* now strike the shadow.) The Moon reached full phase on May 25, was just within the major ecliptic limit, and since it had passed perigee the day before, suffered a partial eclipse (see Table 10.1).

Calculation of the second eclipse season adds a minor complication. The Sun encountered 241° longitude 244 days from the vernal equinox, which from Table A1.1 was on November 19. This date is 322 days past January 1, and at a regression rate of 19°.3 per year, the node had moved back by $(^{322}/_{365}) \times 19°.3 = 17°$. On that basis, the Sun would have encountered it about 17 days earlier. However, in a 17-day interval, the node regresses about 1°, so the Sun actually crossed it *18* days earlier, or on November 1. A lunar eclipse was therefore possible between October 20 and November 13 and inevitable between October 23 and November 10. The full Moon took place on October 19, just outside the major limit, so no eclipse took place. (The Moon did, however, pass through the Earth's penumbra to produce a barely visible penumbral eclipse.)

It takes the Sun 31 days to pass through an arc of double its minor ecliptic limit, which is longer than the Moon's synodic period. Therefore, a solar eclipse *must* take

Table 10.1. *Eclipses of the Moon, 1990 - 2025*[a]

Greenwich	Date and hour		Type	Moon overhead at mid-eclipse		
1990	Feb. 9	19	T	76 E	14 N	India
1990	Aug. 6	14	P	149 E	17 S	Coral Sea
1991	Dec. 21	11	P	159 W	23 N	Hawaii
1992	June 15	5	P	74 W	23 S	Chile
1992	Dec. 9	24	T	3 W	23 N	Algeria
1993	June 4	13	T	165 E	22 S	Loyalty Islands
1993	Nov. 29	6	T	99 W	21 N	Mexico
1994	May 25	3	P	53 W	21 S	Brazil
1995	Apr. 15	12	P	176 E	10 S	Micronesia
1996	Apr. 4	0	T	1 W	6 S	Gulf of Guinea
1996	Sept. 27	3	T	46 W	1 N	Brazil
1997	Mar. 24	5	P	69 W	1 S	Columbia
1997	Sept. 16	19	T	77 E	3 S	Indian Ocean
1999	July 28	12	P	172 W	19 S	Samoa
2000	Jan. 21	5	T	68 W	20 N	Haiti
2000	July 16	14	T	153 E	21 S	Coral Sea
2001	Jan. 9	20	T	57 E	22 N	Oman
2001	July 5	15	P	136 E	23 S	Australia
2003	May 16	4	T	56 W	19 S	Brazil
2003	Nov. 9	1	T	23 W	16 N	Cape Verde Islands
2004	May 4	21	T	52 E	16 S	Madagascar
2004	Oct. 28	3	T	50 W	13 N	West Indies
2005	Oct. 17	12	P	176 E	10 N	Central Pacific
2006	Sept. 7	19	P	76 E	6 S	Indian Ocean
2007	Mar. 3	23	T	13 E	7 N	E Africa Coast
2007	Aug. 28	11	T	158 W	10 S	Central Pacific
2008	Feb. 21	3	T	48 W	11 N	W Central Atlantic
2008	Aug. 16	21	P	44 E	14 S	Madagascar
2009	Dec. 31	19	P	69 E	23 N	W India
2010	Jun. 26	12	P	174 W	23 S	Samoa
2010	Dec. 21	8	T	125 W	23 N	W Mexican Coast
2011	Jun. 15	20	T	57 E	23 S	Madagascar
2011	Dec. 10	15	T	140 E	23 N	Marianas
2012	Jun. 4	11	P	166 W	22 S	Cook Islands
2013	Apr. 25	20	P	57 E	13 S	Madagascar
2014	Apr. 15	8	T	117 W	9 S	Central Pacific

Table 10.1. (*cont.*)

Greenwich	Date and hour		Type	Moon overhead at mid-eclipse		
2014	Oct. 8	11	T	166 W	6 N	Central Pacific
2015	Apr. 4	12	T	180	6 S	Central Pacific
2015	Sept. 28	3	T	44 W	2 N	N Brazil Coast
2017	Aug. 7	18	P	87 E	16 S	Indian Ocean
2018	Jan. 31	14	T	161 E	17 N	Central Pacific
2018	Jul. 27	20	T	56 E	19 S	Mauritania
2019	Jan. 21	5	T	75 W	20 N	Cuba
2019	Jul. 16	22	P	39 E	21 S	Madagascar
2021	May 26	11	T	171 W	21 S	Tonga
2021	Nov. 19	9	P	139 W	19 N	E of Hawaii
2022	May 16	4	T	64 W	19 S	Bolivia
2022	Nov. 8	11	T	169 W	16 N	Central Pacific
2023	Oct. 28	20	P	53 E	13 N	NE African Coast
2024	Sept. 18	3	P	43 W	2 S	N Brazil Coast
2025	Mar. 14	7	T	102 W	3 N	E Cent. Pacific
2025	Sept. 7	18	T	87 E	6 S	Indian Ocean

[a]32 total (T) and 20 partial (P) lunar eclipses are scheduled to occur in the 31-year interval 1990 through 2025. Roughly, each is visible over the terrestrial hemisphere whose pole is located at the indicated latitude and longitude. Saros pairs (Section 10.10) can be seen starting with (1990 Feb. 9 – 2008 Feb. 21), continuing to the end of the table with (2007 Aug. 28 – 2025 Sept. 7).

place at the new Moon closest to the time of solar nodal passage. And indeed there was an annular eclipse on May 10 and a total eclipse on November 3 (see Table 10.2).

10.8 Frequency and prediction

The major lunar eclipse period (twice the major ecliptic limit) is at most 25° long, which the Sun will traverse in 25 days. Since the synodic period of the Moon is 29.3 days, it is possible for the Moon to avoid an eclipse during an eclipse season, as it did in November of 1994. On the average a lunar eclipse will take place three out of four times during a season. There can, however, be only one lunar eclipse during each season. In some years, such as 1997, the Moon will miss the shadow during both eclipse seasons, whereas in years like 1996, it will hit both times. As the node slides backwards, the eclipse season regresses into early January, as it did in 1982, when the Sun passed the descending node on January 5. It traversed the ascending node on July 3, and by the end of the year, since the nodes continued to retreat, the Sun met the descending node again on December 21. Thus it is possible to have as many as 3 lunar eclipses

Table 10.2. *Eclipses of the Sun, 1990–2025*[a]

Greenwich date and time		Type	Path of annularity or totality
1990 Jan. 26	19	A	S Pacific
1990 July 22	3	T	Finland, NW Siberia, E Pacific
1991 Jan. 15	24	A	Indian Ocean, Australia, S Pacific
1991 July 11	19	T	Central Pacific, Central America, Brazil
1992 Jan. 4	23	A	Pacific Ocean
1992 June 30	12	T	S Atlantic
1992 Dec. 24	1	P	. . .
1993 May 21	14	P	. . .
1993 Nov. 13	22	P	. . .
1994 May 10	17	A	N Pacific, US, N Atlantic
1994 Nov. 3	14	T	S Pacific, Central South America, S Atlantic
1995 Apr. 29	18	A	S Pacific, N South America
1995 Oct. 24	5	T	Iran, India, Thailand, W Pacific
1996 Apr. 17	23	P	. . .
1996 Oct. 12	14	P	. . .
1997 Mar. 9	1	T	Siberia, Arctic Ocean
1997 Sept. 1	24	P	. . .
1998 Feb. 26	17	T	Central Pacific, N South America, N Atlantic
1998 Aug. 22	2	A	Indonesia, S Pacific
1999 Feb. 16	7	A	S Africa, Indian Ocean, Australia
1999 Aug. 11	11	T	North Atlantic, Europe, India, Thailand
2000 Feb. 5	13	P	. . .
2000 July 1	19	P	. . .
2000 July 31	2	P	. . .
2000 Dec. 25	17	P	. . .
2001 June 21	12	T	S Atlantic, S Africa, Madagascar
2001 Dec. 14	21	A	Pacific Ocean, Central America
2002 June 10	24	A	Northern Pacific Ocean
2002 Dec. 4	8	T	S Africa, Indian Ocean, Australia
2003 May 31	4	A	Greenland, North Atlantic
2003 Nov. 23	23	T	S Indian Ocean, Antarctica
2004 Apr. 19	13	P	. . .
2004 Oct. 14	3	P	. . .
2005 Apr. 8	21	A–T	S Pacific, N South America
2005 Oct. 3	10	A	N Atlantic, N Africa, Indian Ocean
2006 Mar. 29	10	T	Atlantic, N Africa, Central Asia
2006 Sept.22	12	A	Indian Ocean, S Atlantic, N South America
2007 Mar. 19	3	P	. . .

Table 10.2. (*cont.*)

Greenwich date and time		Type	Path of annularity or totality
2007 Sept.11	13	P	. . .
2008 Feb. 7	4	A	Antarctica, S Pacific
2008 Aug. 1	10	T	N Canada, Arctic, Russia
2009 Jan. 26	8	A	Atlantic, Indian Ocean, Indonesia
2009 Jul. 22	3	T	India, China, Central Pacific
2010 Jan. 15	7	A	Africa, Indian Ocean, China
2010 Jul. 11	20	T	Central Pacific
2011 Jun. 4	9	P	. . .
2011 Jun. 1	21	P	. . .
2011 Jul. 1	9	P	. . .
2011 Nov. 25	6	P	. . .
2012 May 20	24	A	China, N Pacific, United States
2012 Nov. 13	22	T	N Australia, S Pacific
2013 May 10	0	A	N Australia, New Guinea, Central Pacific
2013 Nov. 3	13	A	Atlantic, Central Africa
2014 Apr. 29	6	A	Antarctica
2014 Oct. 23	22	P	. . .
2015 Mar. 20	10	T	N Atlantic, Arctic
2015 Sept. 13	7	P	. . .
2016 Mar. 9	2	T	Indian Ocean, Borneo, Central Pacific
2016 Sept. 1	9	A	Atlantic, Africa, Indian Ocean
2017 Feb. 26	15	A	S South America, Atlantic, Central Africa
2017 Aug. 21	19	T	Pacific, United States, Atlantic
2018 Feb. 15	21	P	. . .
2018 Jul. 13	3	P	. . .
2018 Aug. 11	10	P	. . .
2019 Jan. 6	1	P	. . .
2019 Jul. 2	19	T	Pacific, S South America
2019 Dec. 26	5	A	Arabia, Indian Ocean, Borneo
2020 Jun. 21	7	A	Africa, China, Central Pacific
2020 Dec. 14	16	T	Pacific, S America, Atlantic
2021 Jun. 10	11	A	Canada, Arctic, Siberia
2021 Dec. 4	8	T	S Atlantic, Antarctica
2022 Apr. 30	20	P	. . .
2022 Oct. 25	11	P	. . .
2023 Apr. 20	4	T	Indian Ocean, New Guinea, Central Pacific
2023 Oct. 14	18	A	Mexico, N South America
2024 Apr. 8	18	T	Central Pacific, Mexico, US, Atlantic

Table 10.2. (*cont.*)

Greenwich date and time		Type	Path of annularity or totality
2024 Oct. 2	19	A	Cent. Pacific, Central South America, Atlantic
2025 Mar. 29	11	P	. . .
2025 Sept. 21	20	P	. . .

[a]26 annular (A), 24 total (T), 1 annular-total (A-T) and 28 partial only (P) solar eclipses are scheduled for the 31-year interval from 1990 through 2025. The path of each eclipse begins in the first-listed geographic area, sweeps sequentially across other regions indicated, and ends in the last-listed area. The partials are seen only at higher latitudes, with maxima generally near the poles. Saros pairs (Section 10.10) can be seen starting with (1990 Jan. 26 – 2008 Feb. 7), continuing to the end of the table with (2007 Sept. 11 – 2025 Sept. 21).

(including partial and total) in a year, as in fact happened in 1982, for only the third time in the twentieth century.

The situation for the Sun and solar eclipses is rather different. Because the Sun is eclipsed during each eclipse season, there must be a minimum of 2 solar eclipses per year. And since the arc of the solar ecliptic limit is so long, it is also possible for an eclipse to occur at the beginning of an eclipse season and for another to take place at the end. Moreover, this phenomenon can take place twice per eclipse year, during each eclipse season. Furthermore, if the first eclipse season is in early January, 2 solar eclipses can occur then, 2 more in July, and a 5th in late December, by which time the node will have regressed enough to allow another. Thus we see how common solar eclipses really are. They are considered rare only because of the tiny size of the shadow. There may indeed be as many as 5 per year, but at any given place on Earth we would see totality on the average but once in 300 years. (This simple statistic does not mean much however; some locations could go thousands of years without one, yet there is one location in the US state of Tennessee where there will be 2 within 7 years: 2017 and 2024.)

If two solar eclipses occur during one eclipse season, one at the start and one at the end, the Sun must be very near the node at full moon, and a total lunar eclipse in between the two solar eclipses is inevitable. Thus there may be as few as 2 eclipses per year (both solar) and as many as 7 (5 solar and 2 lunar). This rare occurrence happened in 1935 and will not take place again until 2160. A somewhat more common combination of 7, 4 solar and 3 lunar, happened in 1982, and will again in 2094. In the latter case we observe the three lunar as described earlier, a solar eclipse at the end of the first eclipse season in January, two during the summer season, and then the last at the beginning of the season that has crept backwards into December.

The actual prediction of an eclipse, the times at which it occurs, and the region of visibility is fairly straightforward, at least in principle, and depends upon our knowledge of the orbits of the Moon and Earth. Armed with the data of Table 9.2, with the longitude of the ascending node, the argument of perihelion, and the periods of the phases, nodal regression, and advance of perihelion, the reader can do a fairly accurate job of calculating the date of an eclipse. The exact time of day at which the Moon contacts the Earth's shadow, or crosses the solar disk, is beyond our capability here, however, since we must know the orbits in more detail than are given in the table.

The solar eclipses are the more difficult of the two to handle. In the case of the lunar, we need consider only the Moon moving into a large shadow, which everyone in the nighttime hemisphere will see. But for the solar, we have to consider a tiny shadow spot projected onto a spherical, rotating globe. We not only have to predict when the eclipse will occur, but *where*, as the umbral shadow sweeps in a narrow path across the planet.

It can, and has, been done with surprising accuracy. Not only can we predict eclipses thousands of years into the future, but we can "postdict" them into the past as well. The latter calculations have been quite important in historical dating, wherein it is possible to establish the date and time of an event from a reference to an historical eclipse. The calculation of eclipses is highly accurate. But when we go too far into the past or future and try to be very precise, we run into that great unknown: the rotation speed of the Earth. The discussion in Chapter 6, and the difference between Universal Time and ephemeris (or atomic) time (see Section 6.11), now takes on new and very real relevance. We can precisely predict an eclipse a thousand years ahead in ephemeris time, or atomic time, our constant time standards. But since the Earth is slowing down in an unknown way, Universal Time, our civil standard, will be out of synchrony with AT or ET, and the correction to TAI, for example, will not be known. It does little good to predict an eclipse to the second in TAI, and not know UT minus TAI to an hour. All we know is that in the distant future the people observing an eclipse will see it happen earlier in UT than our prediction in TAI.

But hold this concept for a moment, turn it around, and see how useful it is when we project eclipses into the past. We can calculate when in TAI an ancient eclipse should have occurred. In several cases we know from written records the time at which it actually did take place. The difference between the two tells us how the rotation speed of the Earth has changed over the interval. It is from these data that we in fact first discovered that the day is slowly getting longer by almost two-thousandths of a second per century.

Tables 10.1 and 10.2 give dates and times for coming lunar and solar eclipses between 1990 and 2025. They are taken from a fundamental reference called the *Canon der Finsternisse* (*Canon of Eclipses*) by T. Oppolzer. The book provides data on eclipses between 1207 BC and AD 2162.

10.9 The path and duration of a solar eclipse

Figure 10.10 shows a solar eclipse from a different perspective, wherein we look at the Moon and Earth with the eye in the ecliptic plane; north is towards the top. If an eclipse of the Sun occurs with the Moon well north of the ecliptic plane (position *A*), the eclipse will take place in the northern hemisphere. (Remember, though, that the

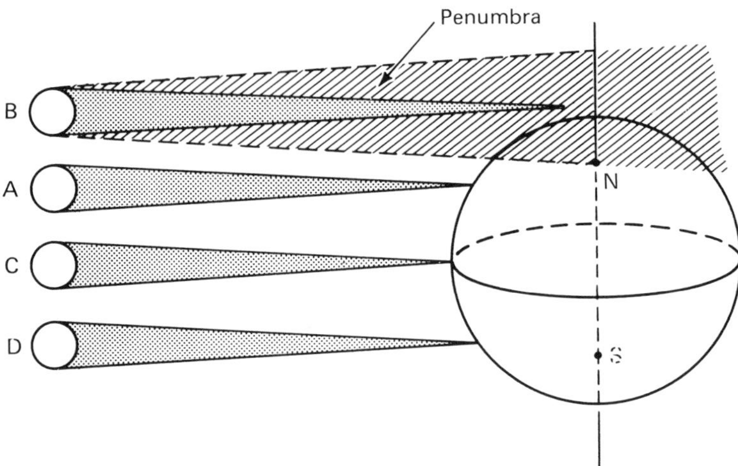

10.10. Locations of solar eclipses. The viewer's eye is in the plane of the ecliptic. The Earth is at an equinox, so the Sun is overhead at the equator and the rotation axis appears tilted out of the page. If the Moon is north of the plane, position *A*, the shadow will generally fall in the northern hemisphere. At *B* the umbra misses the Earth and a partial eclipse would be seen in the arctic. At *C* the shadow falls near the equator, and at *D* it falls in the southern hemisphere. The scales and shadow lengths are greatly distorted.

Earth's equator will not necessarily be in the ecliptic plane, so that if the Moon is north of, but close to, the ecliptic, the shadow could fall in the southern hemisphere.) At *B*, near, but just inside of, the extreme limit diagrammed in Figure 10.9b, the Earth just misses the umbra but does catch the penumbra. Only those people living in north polar latitudes can then see the Moon take a bite out of the Sun, for a partial solar eclipse (which will nowhere be total). If the Moon is exactly in the ecliptic plane (position *C*), the eclipse will take place in the tropics, and if it is south of the plane (*D*), it will be seen only in the southern hemisphere. The first two configurations, with the Moon north of the plane, will be visible if the eclipse occurs either as the Moon approaches the descending node or recedes from the ascending (and vice versa for *D*).

Since the Moon orbits the Earth counterclockwise, the shadow path must sweep from west to east across the globe. The direction of the path on Earth will roughly parallel the path of the ecliptic at the time of the eclipse. In spring, the ecliptic at the Sun runs from southwest to northeast, and so will the track of the eclipse. In summer or winter, the path will be roughly along a parallel of latitude, and in the autumn it will move northwest to southeast. These are broad generalizations. The actual path is modified by rotation and is a curve on the Earth, and the details depend upon where on our planet the shadow strikes as well as the altitude of the Sun at the time. A typical eclipse path, which includes times and locations, is shown in Figure 10.11. These charts are given annually in the *Astronomical Almanac*.

The duration of an eclipse at a given location depends upon the orbital velocity of

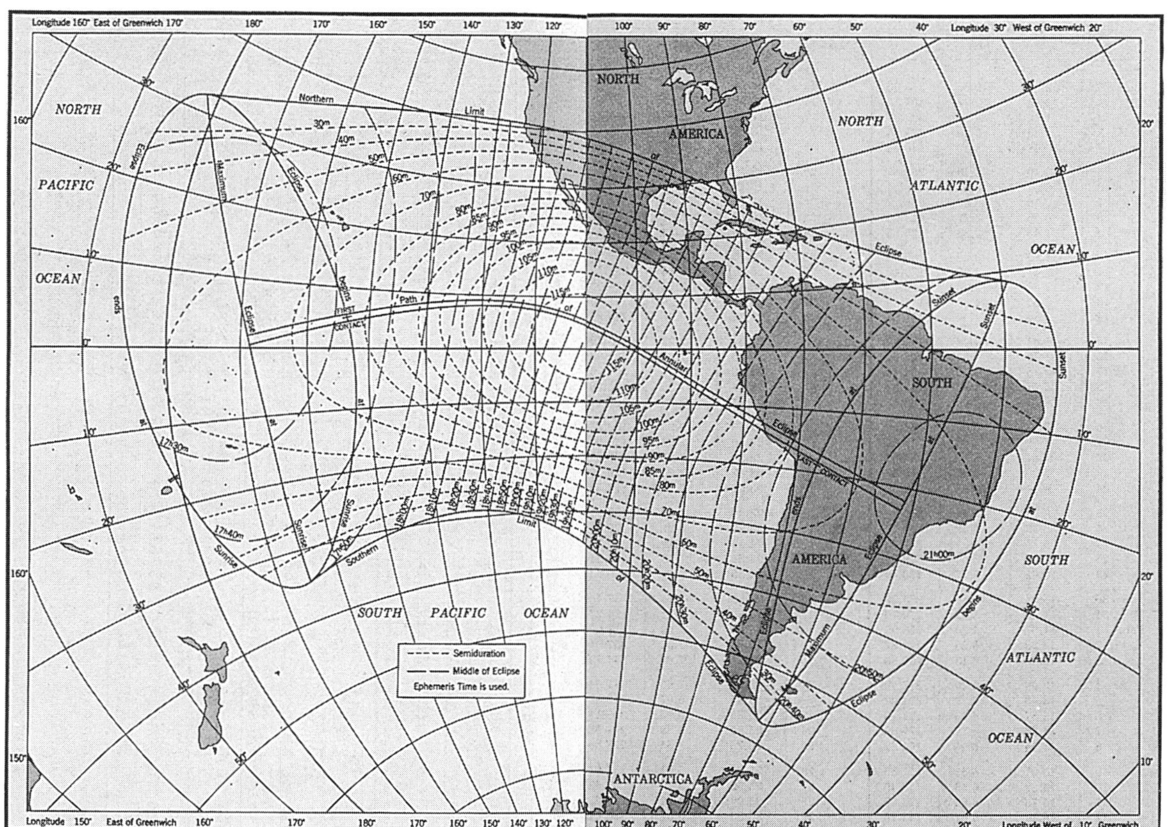

10.11. The path of the annular eclipse of August 10, 1980, taken from the *Astronomical Almanac*. The duration of the partial phases is shown on the dashed lines.

the Moon minus the rotational velocity of the Earth at that point, and the size of the shadow-spot, which in turn depends on the distances of the Moon from the Earth and the Earth from the Sun, and on the degree of elongation of the shadow caused by the Earth's curvature. At the equator, with the Moon at the zenith, the shadow speeds along at about 1700 km/hour. It will move much faster – around 8000 km/hour – if the shadow falls near the Earth's terminator, that is at sunrise and sunset, but the higher speed is offset by the shadow's elongation. Since the observer's rotational speed decreases toward the poles, the shadow speed increases, and eclipses become shorter. Because the cross-section of the shadow is small at best, the duration of a total solar eclipse must necessarily always be very short. Under the most favorable combination of shadow speed and size, the event at best lasts but 7^{m} 40^{s}, the interval decreasing to zero if the eclipse is annular. From our earlier discussions it is obvious that lunar eclipses last much longer.

10.10 The saros

By coincidence, 223 synodic months, a period called the *saros*, is only 0.46 days short of 19 eclipse years. The conditions that produce a given eclipse will thus be almost exactly repeated at intervals of 223 synodic months = 6585.32 days = 18 years 11 days. The word "saros" is of Chaldean origin and the cycle has been known since ancient times. If an eclipse were to occur at a given terrestrial location today, there will be another 18 years 11 days later displaced by only a few degrees of latitude, but because of the additional third of a rotation of the Earth, 120° of longitude to the west. After three saros periods of 54 years, an eclipse will again occur at about the same longitude.

If the original eclipse takes place near the ascending node, shortly before the Moon leaves the allowed zone bounded by the solar ecliptic limit, it will be partial and will be seen near the north pole (see Section 10.9 and Figure 10.10). Since the saros period is about ½ day less than 19 eclipse years, at the next eclipse in the series the Sun will be about ½° west of its original position within the ecliptic limit, the Moon will be a bit closer to the node, and the eclipse will occur farther south. As the series progresses, the umbra will sometimes strike the Earth, providing us with total eclipses, or will just miss, giving us the annular variety. After about 30–35 saros periods, the Moon will have moved through the entire ecliptic limit, and after about 600 years the last eclipses in the sequence will again be partial ones, now seen near the south pole. If the first eclipse in a sequence occurs near the *descending* node, it will first be seen at the *south* pole, and subsequent eclipses will creep to the north, each one again about 120° longitude to the west. The displacement in latitude between successive events averages about 3°. Each eclipse that takes place within the 18 year period is part of a different saros cycle, so that there are a good part of a hundred going on at any one time.

The saros is usually directly associated with solar eclipses, but the concept is applicable to the lunar as well. The major difference is that the eclipses do not move north or south, since those of the Moon are always visible over an entire hemisphere at a time. Those in a series do, however, start and end as partial.

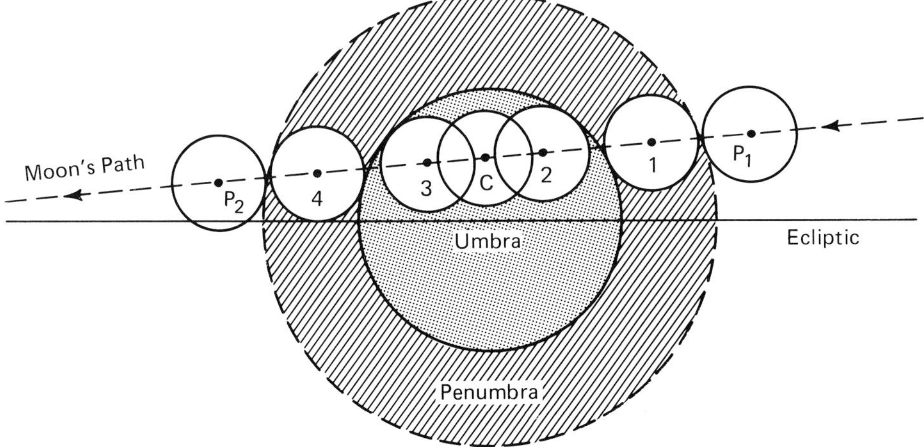

10.12. Lunar eclipse contacts. P_1: The Moon enters the penumbra; 1: *first contact*, the Moon enters the umbra; 2: *second contact*, the Moon becomes immersed in the umbra; *C*: central eclipse, where the Moon will be darkest; 3: *third contact*, the Moon begins to leave the umbra; 4: *fourth contact*, the Moon leaves the umbra; P_2: the Moon leaves the penumbra.

10.11 The phenomena of the lunar eclipse

Most people have probably seen a lunar eclipse, but they have never really *watched* one. They are of minimal research interest to the professional astronomer, but they are attractive and intriguing for the public and highly instructive. There are several specific instances associated with an eclipse as the Moon enters and leaves various parts of the Earth's shadow. These times are predicted well in advance, and are available from the *Astronomical Almanac* as well as from other publications. They are defined and described in Figure 10.12, which shows the Moon's path through the Earth's shadow at a typical eclipse. The penumbral phases, between P_1 and 1, and between 4 and P_2, are almost imperceptible. The Moon goes through a period of very gradual darkening as it approaches point 1, and only just before the Moon enters the umbra will the observer note that the left-hand lunar limb appears noticeably darker than the right. At the Moon's distance the penumbra is $2^\circ.4$ wide, about 1.7 times the width of the umbra. The angle from the edge of the umbra to the edge of the penumbra is about $^1/_2{}^\circ$, so that the Moon will just about fit into it without touching the umbra.

The important predicted times are the four *contacts* and the moment of central eclipse. The eclipse first really becomes noticeable at point 1 (first contact), when the Earth's umbra takes its initial bite. We then see the shadow progress across the Moon during the partial phase until the entire body is immersed at second contact and totality begins. At point 3, the partial part of the eclipse is repeated in reverse. Note that the shadow's outline is circular: the historical proof that the Earth is a sphere.

From Figure 10.12 it is clear that the duration of a lunar eclipse depends upon the

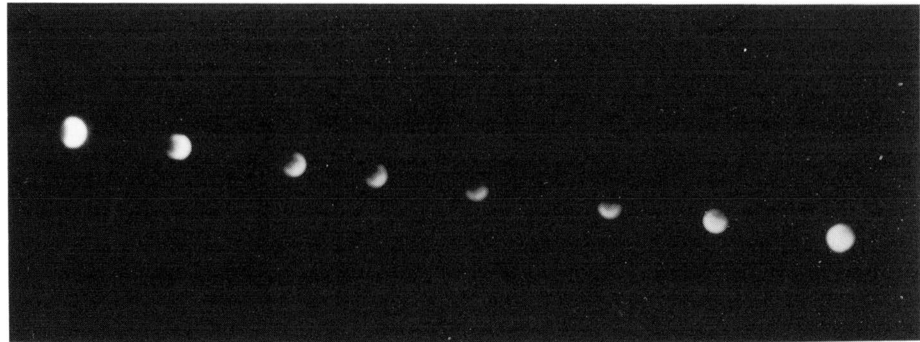

10.13. The lunar eclipse. Multiple exposures show the totally eclipsed Moon passing through the Earth's shadow during the total eclipse of March 13, 1960. Unlike the drawing in Figure 10.12 the sequence goes from left to right because of diurnal motion. Note the easy visibility of the Moon in totality (center) and the variation in the brightness of the shadow. (G. H. Newsom.)

Moon's proximity to a node. For a *central eclipse,* which occurs when the Moon is *at* a node, totality will last just over an hour and a half as the Moon moves through the 1°.4 wide shadow at 33 minutes of arc per hour. The eclipse pictured in Figure 10.12 will be about two-thirds as long. Each portion of the penumbral phase typically lasts about an hour. If the angle between the shadow and the node is sufficiently large, there will be no totality, and the eclipse will be only partial as the Moon grazes the umbra without full immersion.

During totality the Moon does not ordinarily disappear from view, as the umbra of the Earth is not truly black. Our atmosphere refracts and scatters a considerable amount of sunshine into the shadow. The sunlight is highly reddened in its passage through the air, the same phenomenon that causes a red sunset (see Chapter 13), and the totally eclipsed Moon will glow a dull brick-red. An observer standing on the Moon would see a "new" Earth, with the Sun behind it, surrounded by a bright reddish halo. The visibility of the Moon in its total phase is critically dependent on the state of the atmosphere. Heavy cloud cover on the Earth can prevent light from getting through, darkening the eclipse. Volcanic activity is more important. Dust and aerosols spewed out in a violent eruption blanket the Earth, cause colorful sunsets, and again deepen eclipses. Between the two effects the eclipsed Moon will occasionally disappear.

On the common occasions when it is visible, the disk of the eclipsed Moon is not uniformly illuminated. If the Moon is not at the shadow's center, it must be brighter toward the limb nearer the shadow's edge where more scattered and refracted light falls. The progression of the lunar disk through the terrestrial shadow is seen in Figure 10.13, which allows us to see the variation in brightness within the darkened cone. Even for a central eclipse, the illumination can vary from one hemisphere to the other as a result of non-uniform cloud cover on Earth.

Visibility during totality is perhaps the most interesting feature of an eclipse because of what it can tell us about the terrestrial atmosphere. Otherwise, there is little of importance. The temperature of the lunar surface can be measured to determine the rate at which it cools during eclipse, which gives clues to its nature; occultations of faint stars (Sections 9.10) are much more observable; and the darkened surface has been searched unsuccessfully for bright explosions caused by meteor impacts.

10.12 The phenomena of the solar eclipse

The eclipsed Moon may be scientifically unrewarding, but the eclipsed Sun is a profoundly different matter. The phenomena on view during a solar eclipse are of such great significance that astronomers have spent months in preparation and have traveled to the wildest parts of the globe to examine an event lasting but a few minutes. The sights to be witnessed by even the casual bystander will probably be remembered and reviewed for a lifetime.

First a warning: **DO NOT EVER ATTEMPT TO VIEW THE SUN DIRECTLY**, unless it is actually in complete total eclipse. Even a swift glance at the Sun can burn the human eye, producing irreparable damage and a permanent blind-spot. It does not matter that the Sun may be in partial eclipse; the total amount of light is cut down, but the brightness of the visible solar surface does *not* change. The eclipse itself is not hazardous, as so many people believe; the danger involves the publicity surrounding an eclipse that tempts people, particularly children, to have a look. Ophthalmologists and optometrists have seen the images of the partially eclipsed crescent of the Sun burned into young retinas. Unless you truly know what you are doing, do not attempt to make and use a filter to view the Sun. Even if the solar disk is darkened and made comfortably visible with a homemade device, radiation at other, invisible, wavelength bands can pass through it and burn the eye without the viewer being aware, because the retina has no nerves that transmit pain. Sunglasses and smoked glass do not work, welders' glasses are usually of the wrong density, and darkened (overexposed and overdeveloped) film is not dark enough. Moreover overexposed modern film negatives may look black to the eye but can pass ultraviolet radiation. The way to view the partial phases of the eclipse is with specifically designed commercial solar filters or by pinhole projection. Simply punch a tiny hole in a piece of cardboard, or the end of a box, and let the image of the Sun be projected against a wall, the ground, or the other end of the box. It works beautifully.

A solar eclipse is highly significant because it allows us to see the faint and delicate outer atmospheric layers of the Sun. These radiate so weakly that they are blotted out by atmospheric scattering of sunlight, and except through the application of specialized telescopic techniques (or from space), they can only be viewed during totality. The bright visible yellow disk of the Sun is called the *photosphere*, from the Greek *phos* for "light," and will be explored briefly in Chapter 13. Surrounding the photosphere is a thin red layer called the *chromosphere* (Figure 10.14), now from the Greek *chromos* for "color." And finally around that is a pearly white halo, several times the diameter of the photosphere, known as the *corona* (Figures 10.7 and 10.15), Latin for *crown*. The corona has no edge but just gradually fades away until it blends with the darkened sky. Just as totality begins, the narrow red chromosphere pops into view, and then the

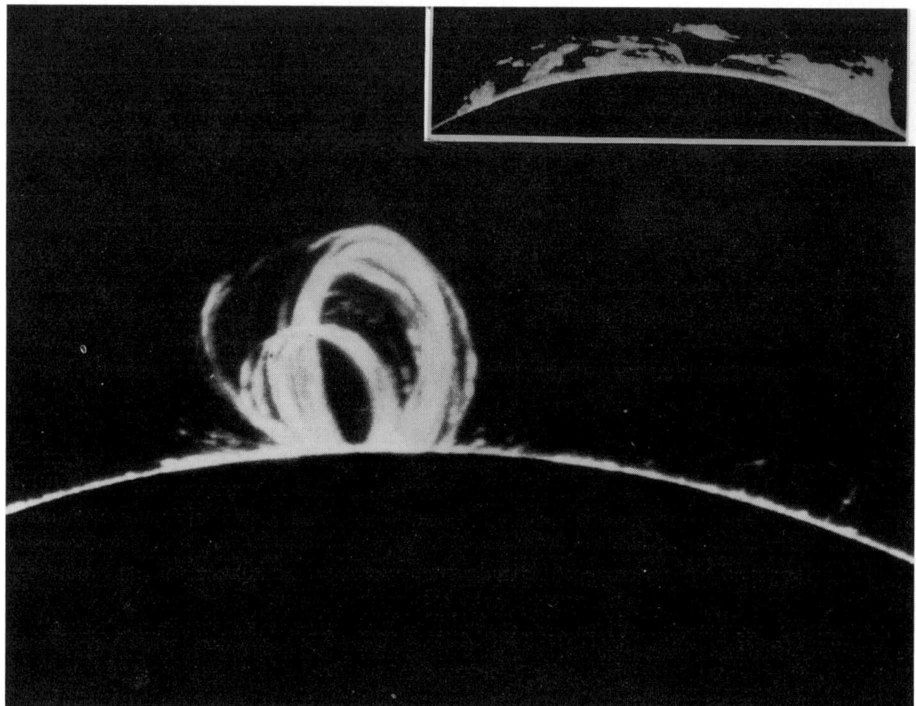

10.14. The solar chromosphere, viewed here not during an eclipse but with a special telescope. The main photo and the inset show two different kinds of prominences that extend upward from chromosphere and intrude on the hot corona above, seen in Figure 10.15. (National Optical Astronomy Observatories; inset: University of Colorado High Altitude Observatory.)

awesomely beautiful corona. It was, and still is, a sight to inspire fear, wonder, mystery, and curiosity about the workings of the heavens.

Today, we know that the corona is an energetic manifestation of the Sun's magnetic field and is linked to a great, hot wind of gas that blows steadily from the Sun, and that the chromosphere is the layer of transition between it and the stable photosphere. The chromosphere is so thin that its visibility at any eclipse depends to some extent on the angular diameter of the Moon. The appearance of the corona can change considerably with time. Its structure is highly dependent upon the level of solar magnetic activity, which in turn is broadly dependent upon an 11-year solar magnetic, or sunspot, cycle, a subject to be addressed in Chapter 13. The corona will appear almost round at the time of maximum solar magnetic activity (Figure 10.15) and will develop long streamers roughly parallel to the Sun's rotational equator (and the ecliptic) during low magnetic activity (Figure 10.7). In addition, during maximum activity the observer can often see the *prominences* displayed in Figure 10.14, huge red sheets of gas hanging above the photosphere.

Many minor, but no less interesting, sights also await the eclipse-watcher. The event

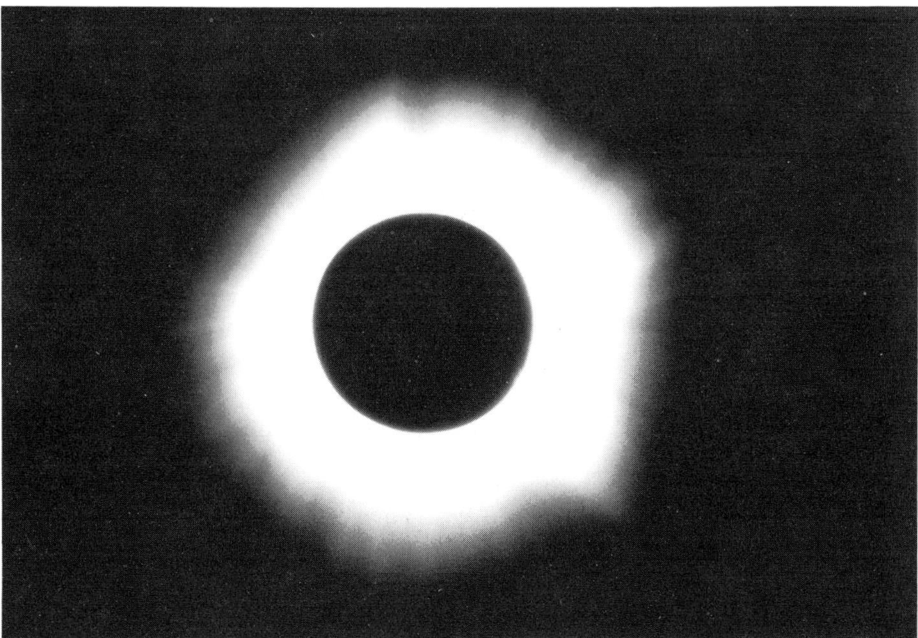

10.15. The solar corona near maximum magnetic activity seen during the eclipse of March 17, 1970. Compare the shape with that at minimum, pictured in Figure 10.8. Fine streamers lying along magnetic lines of force can be seen at top and bottom, at the poles. (National Optical Astronomy Observatories.)

begins with *first contact* as the Moon takes the first bite out of the Sun and the lunar penumbra moves across the observer. For a while, little is noticed, but as the Moon moves farther across the solar disk, the sky begins to darken, taking on a strange cool blue hue. As totality approaches, a person standing under a tree sees the ground covered with tiny crescents: the gaps between the leaves make hundreds of pin-hole cameras that dapple the Earth with images of the Sun. At the last moments before totality, the slim solar crescent produces waves of moving shadows, called *shadow bands*, a still poorly understood atmospheric phenomenon. At this time, animals become confused and birds begin to roost as if night is coming on. The approaching umbra can be seen in the distance as a darkening of the atmosphere, a kind of twilight. Totality is preceded by *Baily's beads* (Figure 10.16): the last remnants of sunlight shining between the mountains on the limb of the new Moon.

Finally, the moment of a lifetime, *second contact*, the Sun is blotted out, and the chromosphere and then the corona appear but only briefly as the tiny shadow spot flies along at 1700 km/hour or more. The sky turns quite dark. The planets and the brighter stars come out, as the amount of light is but that of a couple of full Moons. Then it is over. At *third contact* the viewer suddenly sees a brilliant flash of photospheric light at the Moon's western limb, which with the rapidly disappearing corona creates the *diamond ring* (Figure 10.17), and the eyes must be quickly averted. The phenomena that

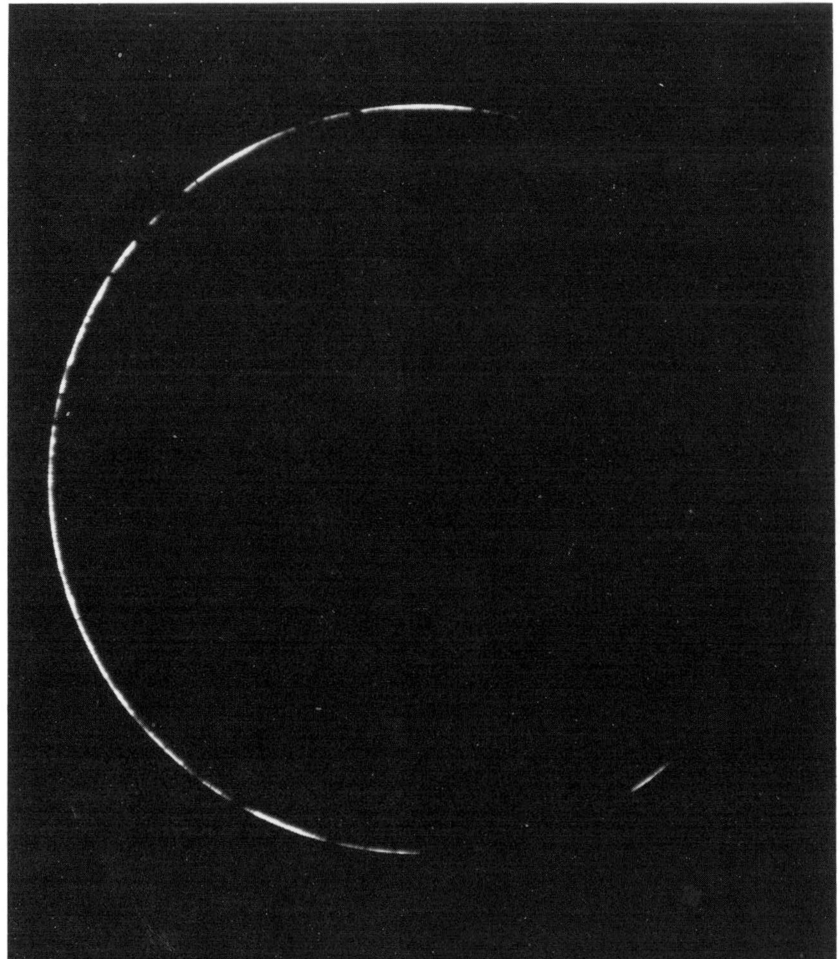

10.16. Baily's beads, sunlight peeking between the irregularities of the lunar limb. (Yerkes Observatory.)

precede totality now reappear in reverse order, until at *fourth contact*, some two hours after the first, the eclipse is over, perhaps for the world traveler to be repeated some years hence.

10.13 The Moon and the calendar

Although no longer a part of our modern Roman-Latin western calendar, the Moon's easy visibility and regular phases made it a foundation-stone of the calendars of most early cultures, and it still makes its appearance in many religious calendars. In the most basic kind of calendar people can simply count lunations and refer to a time past as

10.17. The diamond ring. The first bright flash of emerging sunlight is seen together with the inner corona. (National Optical Astronomy Observatories.)

"so many Moons ago." The significance of these passing lunations is such that full Moons are given names that reflect their seasons, the "harvest Moon" of September and the "hunter's Moon" of October being but two examples. Several others, many drawn from Native American folklore, are given in Table 10.3. The fact that there can be 13 full moons in a year is of little relevance in such an informal system.

In contrast, the broad subject of calendars and date-keeping can be remarkably complicated. We may divide most calendars into three broad groups: purely solar, purely lunar, and combined luni-solar. The modern solar calendar, the Gregorian, was discussed in Section 6.7. That problem is a relatively simple one to solve, since we need deal with only one period, the Earth's tropical year of 365.2422. . . days. It is so close to 365¼ days that we can make do very well by adding a day to the ordinary 365 day year once every 4 years, with a minor adjustment in 3 out of 4 century years (dropping the extra day), so that the average Gregorian calendar year comes out to 365.2425 days. The difference of only 0.0003 days (25.92 seconds) per year produces an error of one day in 3000 years and is ignored.

A calendar based entirely upon the Moon will be more complicated, since the lunar synodic period is not easily reconcilable to whole numbers. Since there is not an even number of annual lunar periods (12.37 lunations per tropical year), a strictly lunar calendar will rapidly get out of step with the seasons. Nevertheless, such calendars are still in active use, the best known being the Islamic. The 12 months of the Moslem

Table 10.3. *Names of full Moons*

January:	Old Moon, Moon After Yule, Wolf Moon, Winter Moon, Deep Snow Moon
February:	Snow Moon, Hunger Moon, Wolf Moon, Trapper's Moon, Crust of Snow Moon
March:	Sap Moon, Crow Moon, Lenten Moon, Fish Moon, Worm Moon, Snowshoe Breaking Moon
April:	Grass Moon, Egg Moon, Sprouting Grass Moon, Pink Moon, Planter's Moon, Maple Sugar Moon
May:	Planting Moon, Milk Moon, Mother's Moon, Flower Moon, Budding Plant Moon
June:	Rose Moon, Flower Moon, Strawberry Moon, Stockman's Moon
July:	Thunder Moon, Hay Moon, Buck Moon, Summer Moon, Midsummer Moon
August:	Green Corn Moon, Grain Moon, Sturgeon Moon, Harvest Moon
September:	Harvest Moon, Fruit Moon, Fall Moon, Wild Rice Moon
October:	Hunter's Moon, Dying Grass Moon, Harvest Moon, Falling Leaves Moon
November:	Frosty Moon, Beaver Moon, Hunter's Moon, Freezing Moon
December:	Long Night Moon, Moon Before Yule, Cold Moon, Christmas Moon, Descending Cold Moon

Compiled by Randolph E. Schmid of the Associated Press.

calendar, each of which begins at the first sighting of the new Moon, alternate between 30 days for the odd months and 29 for the even, which provides an average close to the true $29^{d}.531$ synodic period.

Twelve lunations take 354.367 days. A *lunar year* of 354 days, the closest whole number, then falls short by 0.367 days. An approximate solution would be to *intercalate* (or add) an extra day every 3 years, but that leaves us with a calendar that is still too short by 0.10 day in the 3 year interval. A better reconciliation is made by adding 11 days over a 30 year period, which results in a shortfall of only 0.012 days, which is one day in 85 cycles (2500 years), more than satisfactory. The Moslem calendar adds an extra day to the last month in years 2, 5, 7, 10, 13, 16, 18, 21, 24, 26, and 29 of the cycle. The beginning is 622 AD, the time of Mohammed's escape from Mecca, the Moslem year being called *Anno Hegirae* (AH), "in the year of the flight."

The obvious problem with a strictly lunar calendar is that it is not synchronous with the seasons. With a 354-day year, spring will fall on the average 11 days later in the calendar each year. Or, alternatively, the Moslem new year will fall 11 days (or 10 or 12 depending upon whether it is a Moslem or Gregorian leap year) earlier in the Gregorian calendar each year, with a period of return of $^{365}/_{11}$ or about 33 years.

A reconciliation of a lunar calendar with the seasons raises the level of complexity.

Not only must one have leap years to synchronize the average calendar month with the synodic month, one must also intercalate extra days into the 354-day lunar calendar to match the 365¼-day solar calendar. In 3 years, the lunar calendar falls short of the solar by 365.242 minus 354.367 times 3, or 32.63 days, so that very roughly in that period we should intercalate an extra lunar month of 29.53 days. Such a year, with 13 lunations, has 384 days and is called an *embolismic year*. This intercalation would still be too short by 3 days every 3 years, however, and is not satisfactory. A better solution, used in early Babylonia, is an 8-year period (called an *octaeteris* by the Greeks) in which there will be 98.96 lunations, so that we need to have 3 years (99 lunations minus 8 × 12) with extra months. This calendar will be too long by 1.6 days in 8 years. Later, the Babylonians used an 11-year cycle with 136.05 lunations, which needs 4 embolismic years (136 − 11 × 12). This one is too short by 1.5 days over 11 years, only marginally better.

The best solution is the 19-year period of the Metonic cycle (see Section 9.2 and Table 9.1). In 19 solar years there will be 234.997 lunations. If we round the number to 235, we must intercalate 235 − 19 × 12 = 7 months. The luni-solar calendar will now be long by only 0.09 days in the 19-year period, or but one day in 220 years, a good solution.

The most widely known example of this type of system is the Hebrew or Jewish calendar. In its basic form it has the common alternating pattern of months with 30 and 29 days, again with the months starting when the Moon is first visible past new, and with the new year beginning about the time of the autumnal equinox. As is necessary for a lunar calendar, extra days must be added to make the average length of the months correspond to a lunation. However, because of religious constraints that prohibit some of the holidays from falling on certain days of the week, this addition is done in a complex manner: the second and sixth months, which normally have 29 days, may be increased to 30, and the third, usually with 30 days, may be decreased to 29. The individual common years may thus have 353, 354, or 355 days, although the net effect is synchronization with the lunar synodic period. The embolismic month, (called *Adar II*), always of 29 days, is added 7 times in the 19-year Metonic period, in the middle of the year following the sixth month *Adar*. Because synchronization with the seasons and the Sun is done by adding increments of a month, the new year cannot be made to coincide with the equinox, but moves around relative to the Gregorian calendar, falling anywhere from September 6 to October 5. The other Jewish holidays shift around in a like manner.

This study of the moon opens the door to the Solar System at large. We now turn to the planets, to examine their motions and their significance to humanity.

Chapter 11

The planets

To the naked eye, the stars are fixed in space for our lifetimes. The only changes we see in them are those produced by the occasional variable or exploding star and by the motions of the Earth on which we live. Superimposed on this distant backdrop, however, are the nearby bodies of our Solar System, which move across the sky through the constellations (Figure 11.1) and change their appearances as they orbit the Sun, or in the case of the Moon, our Earth. The Solar System is complex and consists of a number of different kinds of bodies. We will survey the major ones – the nine *planets* – in this chapter, and examine the smaller ones in the next.

11.1 Organization and orbits

The ancients knew of five bright planets, from the Greek *planetai*, meaning "wanderers." These, with the Sun and Moon, make up the classical seven moving bodies of the sky, perhaps originating the old notion of that number's magic nature. Three much fainter planets were discovered in modern times, and with our own Earth (for the last 400 years counted among them), they total nine. In order outward from the Sun, they are Mercury, Venus, our Earth, Mars, Jupiter, Saturn, Uranus, Neptune, and Pluto (Table 11.1).

The five bright planets are commonly referred to as the *ancient planets*. The first two are closer to the Sun than Earth and are thereby also known as the *inferior* planets, rendering the outer six the *superior*. The first four, small, rocky bodies with minimal atmospheres, are called the *terrestrial* (Earth-like) *planets* because of their physical similarity. The next quartet, called the *Jovians* for their likeness to Jupiter, are much larger, more massive, and possess thick atmospheres. Tiny Pluto is unique and defines its own class.

The solar family, the remnant of the Sun's formation 4½ billion years ago, is impressively regular and symmetrical, providing a clue to the system's origin, a subject to be explored in the next chapter. Two of the most significant symmetries are the planets' nearly circular elliptical orbits and their flat distribution (Table 11.2 and Fig-

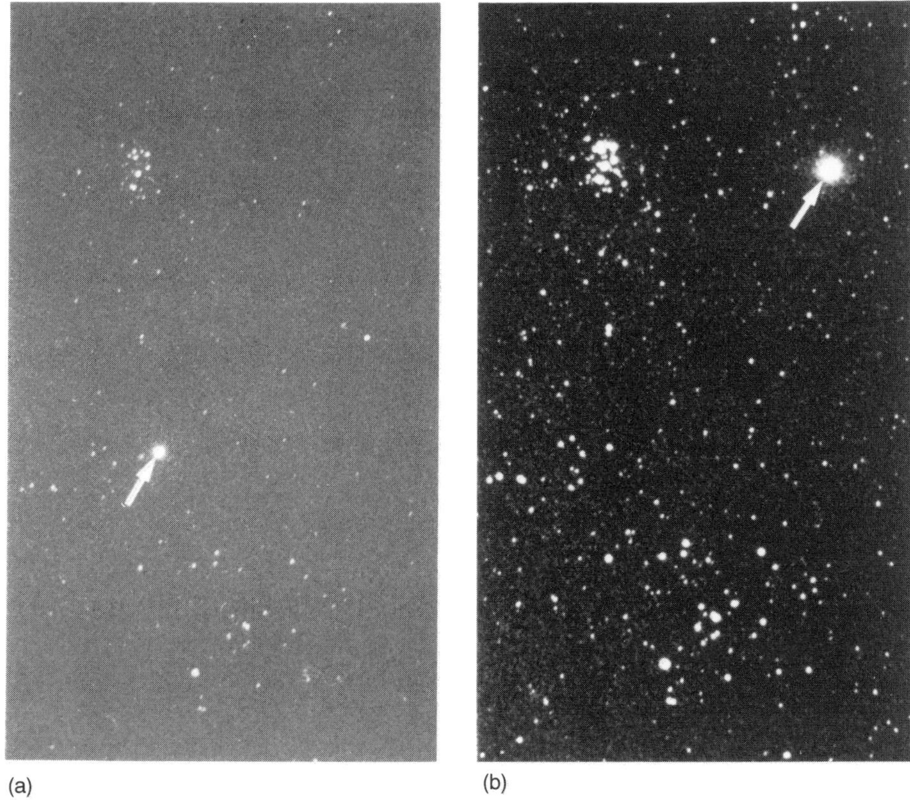

11.1. The motion of Jupiter. The two photos show the planet's motion in the constellation Taurus between October 1988 (a) and January 1989 (b). The Pleiades are at upper left and the Hyades are below. The bright star near the bottom is Aldebaran. (Author's photographs.)

Table 11.1. *Planetary organization*

Planet	Discovery	Orbit	Physical
Mercury	ancient	inferior	terrestrial
Venus	ancient	inferior	terrestrial
Earth	—	—	terrestrial
Mars	ancient	superior	terrestrial
Jupiter	ancient	superior	Jovian
Saturn	ancient	superior	Jovian
Uranus	modern	superior	Jovian
Neptune	modern	superior	Jovian
Pluto	modern	superior	—

Table 11.2. *Orbital elements and other planetary data*

Planet	m	a (AU)	e	i°	Ω°	ω°	P	n°/day	v (km/s)	P_{syn} (days)	R
(1)	(2)	(3)	(4)	(5)	(6)	(7)	(8)	(9)	(10)	(11)	(12)
Mercury	−1.8	0.387	0.206	7.00	48.1	29.1	88^d	4.09	48	116	23
Venus	−4.7	0.723	0.007	3.39	76.5	54.8	225^d	1.60	35	584	42
Earth	—	1.000	0.017	0.00	—	—	1.000^y	0.99	30	—	—
Mars	−2.8[a]	1.524	0.093	1.85	49.4	286.3	1.88^y	0.52	24	780	73
Ceres[b]	+7.9	2.77	0.077	10.6	80.1	73.9	4.61^y	0.21	18	466	85
Jupiter	−2.9	5.20	0.048	1.31	100.2	274.2	11.9^y	0.083	13	399	121
Saturn	+0.0	9.54	0.056	2.49	113.5	341.9	29.5^y	0.033	10	378	138
Uranus	+5.5	19.2	0.047	0.77	74.0	98.0	84.0^y	0.012	7	370	152
Neptune	+7.7	30.1	0.009	1.77	131.5	289.4	165^y	0.006	5	367	158
Pluto	+14	39.4	0.249	17.1	109.9	113.0	249^y	0.004	5	367	162

The first two entries are the planets' names and the maximum brightnesses (minimum magnitudes). The next six give the orbital elements (Section 8.6) for the planets and the largest asteroid, Ceres. The other entries are n: mean motion along the orbit in degrees per day; v: mean orbital velocity; P_{syn}: synodic period expressed in days; R: number of days the planet spends in retrograde motion each year as viewed from Earth.

[a]at favorable opposition

[b]largest asteroid (Chapter 11)

ures 11.2 and 11.3). Only the orbits of Mercury and Pluto – and to a degree Mars – are sufficiently eccentric to make the Sun readily appear off-center. All orbit the Sun in planes inclined only a few degrees to that of the Earth's. The largest angle of 17° is made by distant Pluto's orbit. The next one down is Mercury's, at only 7°. As a result, the planets' apparent paths as viewed by us from Earth are close to that of the Sun, or the ecliptic, so the planets are generally found within the constellations of the zodiac. (Pluto, however, can stray rather far from the zodiac, for example passing deep into Cetus where it will be toward the end of the twenty-first century). The planets therefore lend even more importance to this band of constellations, their positions and motions giving rise to astrology, the occult system of magic in which the stars and planets are said have the power to determine and predict human destiny.

The planets, from the definition of the word, must move. All orbit the Sun in the same direction. If we look down on the Solar System from the north ecliptic pole, the usual convention, they revolve counterclockwise about the Sun, the same direction in which the Earth rotates and revolves. As viewed from Earth, the planets will have more complicated movements, since we watch from a platform that is also moving, but the counterclockwise revolution gives them a principal, or average, motion toward the east against the stellar background.

Distances in the Solar System are commonly expressed relative to the semimajor axis (a) of the terrestrial orbit, called the *astronomical unit*, AU. The AU is also the

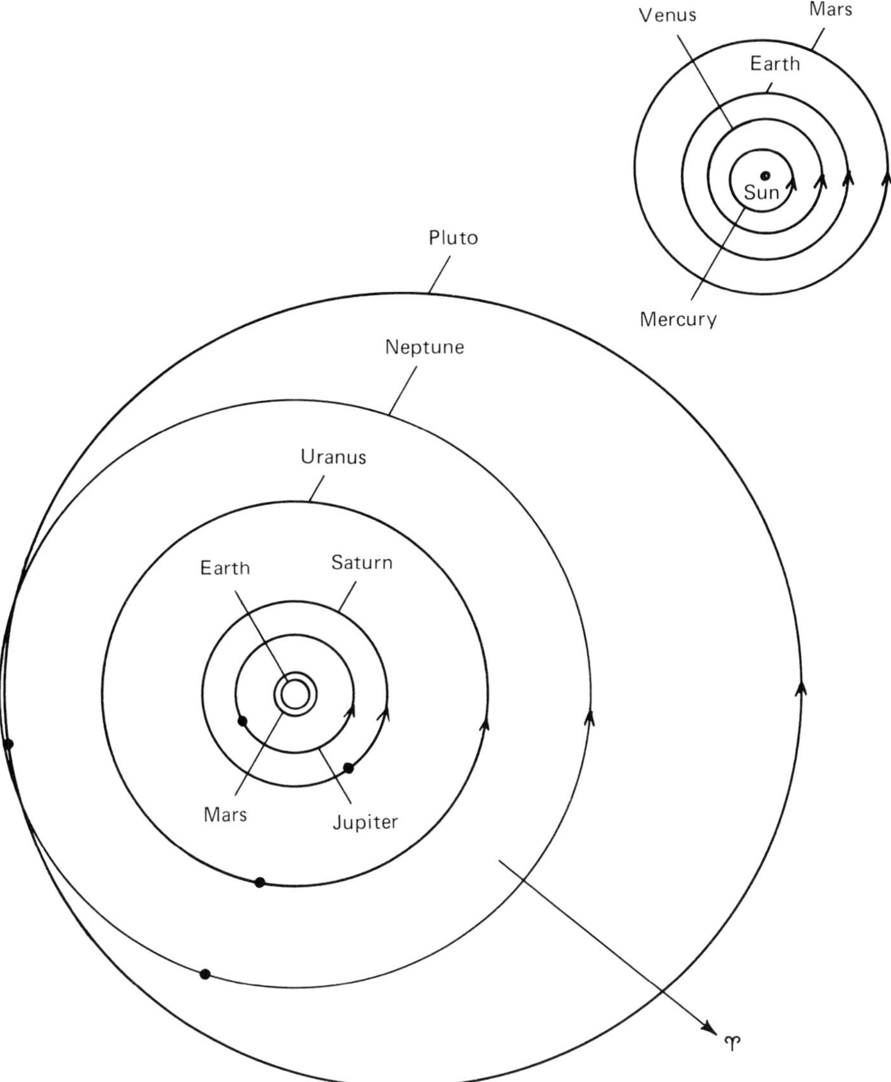

11.2. An overhead scale view of the Solar System. The planets all orbit counterclockwise looking down from the north. The terrestrial planets are so tightly bunched at the center that their orbits are difficult to see, so they are shown expanded at the upper right. The eccentricities and orientations of the orbits of Mercury, Mars, and Pluto are properly shown relative to the direction to the vernal equinox, which is indicated by the arrow. The positions of the outer planets are shown for 1995.

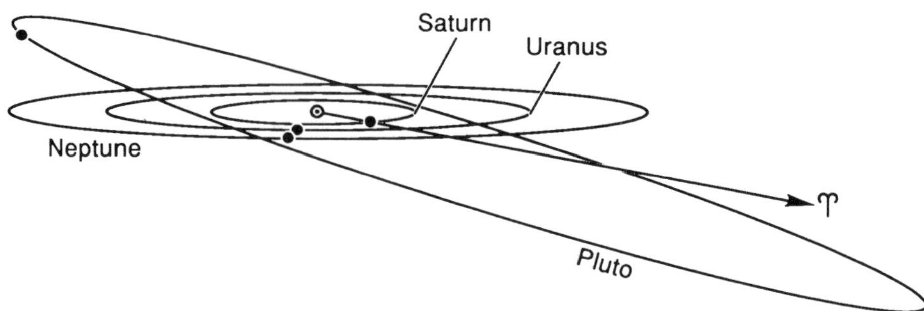

11.3. A side view of the Solar System. Only the outer planets are shown. The orbits are not all in the same plane. Pluto in particular has an inclination of 17° and on the left passes well above the orbit of Neptune. The positions of the outer planets are shown for 1995.

average distance of the Earth from the Sun, and is equal to 149.5979 million kilometers. Planetary distances range from only 0.4 AU for Mercury to 39 for Pluto (column 3 in Table 11.2). They are symmetrically arranged, with the four terrestrials all close to the Sun, within 1.5 AU, and the Jovians at great distances starting with Jupiter at 5.2 AU and proceeding to Neptune at 30.1 AU. The semimajor axes of the first three Jovians closely double one another, from 5 to 10 to 19 AU. The difference in distance from the Sun is the reason for the physical differences between the terrestrial and Jovian planets, as will become apparent in the next chapter.

The distances follow a curious numerical progression discovered in 1772 by an obscure mathematics professor named Johann Titius. It was popularized by his contemporary, the German astronomer Johann Bode (see Section 4.4 and Figure 4.5), and has been known as *Bode's law* ever since. A modern version begins with a geometric progression of numbers that starts at 0, goes to 3, then doubles. Below each write a string of 4s:

0	3	6	12	24	48	96	192	384	768
4	4	4	4	4	4	4	4	4	4

Add the columns, divide by 10, and compare to the actual distances of the planets from the Sun in AU:

prediction:	0.4	0.7	1.0	1.6	2.8	5.2	10.0	19.6	38.8	77.2
distances:	0.4	0.7	1.0	1.5		5.2	9.5	19.2	30.1	39.5
	M	V	Earth	M		J	S	U	N	P.

Compare with the true values in Table 11.2, and note that only Neptune and Pluto really fail. The gap at 2.8 AU was filled in the year 1800 with the asteroid Ceres, to be discussed in the next chapter. There is no physical explanation for the rule and it may be simply coincidental.

Pluto does not fit the general Solar System pattern at all. Its eccentricity of 0.25 and its proximity to Neptune allow it to come closer to the Sun than Neptune for a portion of its path, where it travels between the years 1979 and 1999. Their orbits, however, are differently inclined so that the do not collide (see Figure 11.3). Pluto's renegade behavior suggests that it may not really belong to the family of planets and ought to be classified somewhat differently, the subject to be amplified also in the next chapter.

These three orbital elements (see Section 9.6) – the planetary semimajor axes (a), eccentricities (e), and inclinations to the ecliptic (i) – are summarized in Table 11.2. The elements of orientation, the longitude of the ascending node, Ω, and the argument of perihelion, ω, are listed next. The angles are randomly distributed and have no particular relation to one another. The perihelia of the planets are not readily detectable, with the exception of Mars, which becomes noticeably brighter when nearest the Sun. If you know the location of the ecliptic, the nodes of some of the planets are crudely detectable by eye. Of the superior planets, Saturn's node is especially easy to find as the planet can be found as much as $2^1/_2°$ from the solar path. Since the ascending node, at $\Omega = 114°$, is in Gemini, Saturn will be farthest north of the ecliptic when it is in Virgo, and farthest south when in Pisces. Although Mercury has a high inclination, its relation to the ecliptic is more difficult to see since it is always viewed in bright twilight.

The rate of motion of a planet depends upon its *sidereal period*, or the time it takes the body to make a full 360° orbit of the Sun. As for the Moon, it is the interval between successive passages, or conjunctions (see Section 9.3) of a planet with a given star, after accounting for changes in direction caused by the varying position of the Earth. The sidereal period depends entirely upon the planet's distance from the Sun. Not only do the more distant planets move more slowly (in kilometers per second) along their paths because of the weakened effect of solar gravity, but they have farther to go to circumnavigate their orbits.

The sidereal periods are given in column 8 of Table 11.2, expressed in Earth years or days, and the average angular and true orbital speeds in columns 9 and 10. The range in periods is quite impressive. Mercury, swift and close to the Sun, takes but 88 days to orbit, while distant plodding Pluto requires 249 years to make the journey; neither Neptune nor Pluto has made a complete orbit since its discovery, although Neptune is now coming close (achieving a closed circuit in 2011). The bunching of the terrestrials is evident in their short sidereal periods; note the great gap between those of Mars and Jupiter. The average angular rates of motion (column 9) can be found simply by dividing 360° by P (in days). The actual daily rates can deviate wildly from the averages, however, as we will see. The average speed in orbit (column 10) is found by dividing the orbital circumference in kilometers by the period in seconds. The speeds at any moment depend slightly upon a planet's eccentricity and its actual distance from the Sun.

11.2 Names and identification

Everyone has seen a planet at some time or other, even if not aware of it, since four can outshine all but the Sun and Moon. Their apparitions are steeped in the lore of millennia. All are individuals and you learn to recognize them as you would the faces of friends, allowing observation of their passages over the course of your lifetime. The last three are at or below the limit of naked-eye vision, but the five bright ancient planets are easy to know. They are always found in or near the constellations of the zodiac and move against the stellar backdrop, so that over an interval of several nights or weeks their character becomes quite apparent (see Figure 11.1). On any given night, during which any motion is indiscernible, they can be discriminated because they do not twinkle in the manner of stars. Twinkling (see Chapter 13) is caused by the passage of starlight through the Earth's turbulent refracting atmosphere, and it is diminished for planets because their angular diameters are large compared with those of stars.

Each of the ancient planets' characteristics is expressed by its name. The names we use are of Roman gods, which were derived from their Greek counterparts. At one time the planets truly represented these deities for much of humanity, and we on Earth commonly believed we could see gods' dispositions and our fates by watching the planets' changing aspects in the sky, the beginning of astrology, which is practiced even today. Mercury, Greek *Hermes*, was the gods' courier, portrayed flying on winged feet to carry their words. And the planet represents him well, so swift and elusive that few, without real effort, ever see it. Since Mercury is so close physically to the Sun, it never gets angularly very far away either, and always sets or rises in twilight: it can never be seen in a completely dark sky. Because of its short period, daily motion is very rapid; thus it can be observed for only a few days at a time near favorable angular separation from the Sun, shortly after sunset or before sunrise. An encounter with a gleaming jewel near a clear flat horizon in the growing dusk is probably Mercury.

It may, however, be Venus, which for a short time will be similarly close to the Sun in angle. But several days of observation will tell the difference, since Venus will remain longer than Mercury in the morning sky, or will move on successive evenings into the darker night. Named after Greek *Aphrodite*, representing the goddess of love and beauty, Venus is outstanding as the loveliest and most graceful of the planets. Since it is between the Earth and Sun, like Mercury it follows or precedes the Sun in the sky, and is seen only in the evening or morning. But because it is farther from the Sun than Mercury, the angular separation is generally larger, and it can be seen in full darkness, up to three hours after or before sunrise from typical temperate latitudes. Because of a perpetual reflecting cloud cover and its proximity to Earth, it outshines all the other planets, glowing a brilliant cream-white. Only the Moon and Sun outshine it, and at its brightest, it is the only celestial object, other than a rare bright meteor or comet, that is visible in full daylight. At night in a very dark location its gleam can cast shadows. *Everyone*, at some time or other has seen Venus. More than any other body, the planet is identified as the *evening or morning star*, though on occasion when Venus is not visible, Jupiter, Mars, or even Sirius may usurp that distinguished position. As the morning star in western literature it has been called *Lucifer*, the archangel in Christianity that fell from heaven.

From love to war, we move to Mars, *Ares* in Greek, the name derived from the planet's orange-red color. Mars varies in brightness more than any other planet; because of its orbital proximity to Earth it will be brilliantly visible for long periods, with only Sirius as a stellar rival, and then for months will be quite inconspicuous. Its color gives it away: there are few stars with such a hue, only Aldebaran and Antares in the zodiac.

The latter especially is sometimes confused with the planet by the beginning observer: its name "Ant-ares," means "like Ares, or Mars."

Next outward we find stately Jupiter, god of the heavens, king of the gods. Its name comes from its Greek counterpart *Zeus*, *Zeus-pater*, Zeus the father. Even though quite distant – five times farther from the Sun than Earth – its great physical size makes it still very bright, exceeded only by Venus and Mars at its most luminous. Jupiter's sidereal period of 12 years makes it move through one zodiacal constellation per year, perhaps accounting for its choice as ruler of the skies. The combination of brilliance and motion caused the ancients to name the physically largest of the planets, with a diameter 11 times Earth's, as lord.

Saturn, the Roman god of agriculture, is generally identified with the Greek *Cronus*, the father of Zeus. Saturn is the dimmest and slowest of the ancient planets, moving with half the angular speed of Jupiter, perhaps thereby seeming in a supreme or fatherly position. It takes about 2½ years for Saturn to drift across the average constellation of the zodiac. Although never dim, it has many stellar rivals, and would appear about halfway down a list of first magnitude stars.

The last three planets were discovered in modern times, even though Uranus can actually be seen with the unaided eye. It was occasionally recorded as a star before William Herschel gave it permanent recognition in 1781. Uranus, a Greek name, personifying heaven itself, was husband to Gaea (Earth), and father to Cronus, thus continuing the Jupiter–Saturn progression. Neptune, found in 1846, is the Greek *Poseidon*, god of the sea. Finally, Pluto was discovered in 1930. He was god of Hades, or the underworld, and is identified with the Greek *Orcas*. The planet is certainly aptly named, for it is dim and hard to find. We will look closely at the discovery of these "modern" planets in Section 11.13.

All the planets are represented by symbols, some of which are clearly tied to the nature of the associated god. These, with the symbols used for the zodiacal constellations and planetary aspects (see the next section), are presented in Table 11.3. The ones employed in biology for male and female are the same as used for Mars and Venus, who personified masculinity and femininity to our ancestors.

11.3 Planetary aspects

As the planets and Earth orbit the Sun, all with various speeds, each will continuously change its position in our sky relative to the Sun and to the other bodies of the Solar System. There are certain points in a planet's orbit, diagramed in Figure 11.4, that are particularly important. These have astrological as well as astronomical significance and will be discussed again in Section 11.18.

They are best defined through the concept of *elongation*, which is the angle at any given time between a planet and the Sun as viewed from Earth. When a planet is seen within 180° east of the Sun it is said to be in *eastern elongation*, and within 180° west, *western elongation*. Note these directions in Figure 11.4 relative to the rotation of the Earth; a planet on the right hand side of the diagram is west of the Sun and vice versa. The drawing shows the orbit of the Earth as a dashed circle, and the orbits of both

Table 11.3. *Astronomical symbols*

Planets		Aspects		Zodiacal constellations	
Sun	☉	conjunction (180°)	☌	Aries	♈
Mercury	☿	opposition (0°)	☍	Taurus	♉
Venus	♀	quadrature or square (90°)	□	Gemini	♊
Earth	⊕	sextile (60°)	⚹	Cancer	♋
Moon	☽	semi-square (45°)[a]	∠	Leo	♌
Mars	♂	semi-sextile (30°)[a]	⚺	Virgo	♍
Jupiter	♃	quintile (72°)[a]		Libra	♎
Saturn	♄	biquintile (144°)[a]		Scorpio	♏
Uranus	⛢ ♅	trine (120°)[a]	△	Sagittarius	♐
Neptune	♆	sesquiquadrate (135°)[a]	⚼	Capricornus	♑
Pluto	♇	quincunx (150°)[a]	⚻	Aquarius	♒
		ascending node	☊	Pisces	♓
		descending node	☋		
		vernal equinox	♈		
		retrograde	℞		

[a]used in astrology; see Section 10.17.

superior and inferior bodies as solid circles. When a planet is seen directly toward, or in line with the Sun, with an elongation of 0°, it is said to be in *conjunction* with the Sun. An inferior planet can be in conjunction at two places in its orbit, once when physically between the Earth and Sun, and again when the Sun is between it and Earth. We call these two instances *inferior* and *superior conjunction* respectively. A planet opposite the Sun in the sky, that is with Earth between the Sun and the planet, with an elongation of 180°, is in solar *opposition* to the Sun. Superior planets come into both opposition and conjunction as they and the Earth move in their orbits; inferior planets obviously can never be at opposition. We encountered both of these concepts in Section 9.2 with the phases of the Moon: full Moon occurs at opposition, new Moon at conjunction. Opposition and conjunction, for which the Sun, Earth, and planet (or Moon) are all on the same line, are jointly referred to as *syzygy*.

These definitions are not quite so simple as stated because the planetary orbits are not all in the same plane. Thus at conjunction a planet usually passes either north or south of the Sun. The term conjunction technically refers to the point of closest angular separation, when the Sun and planet have the same celestial longitude (see Section 3.9): the celestial latitudes at this moment can differ by several degrees. Opposition takes place when the longitudes differ by 180°. We can also have conjunction and opposition in right ascension, when α(Sun) and α(planet) are respectively equal or 12^{h} apart. The *Astronomical Almanac* always refers to planetary configurations based on longitudes; those defined by right ascension should be explicitly so stated.

When an inferior planet is at a tangent point in its orbit as viewed from Earth, it

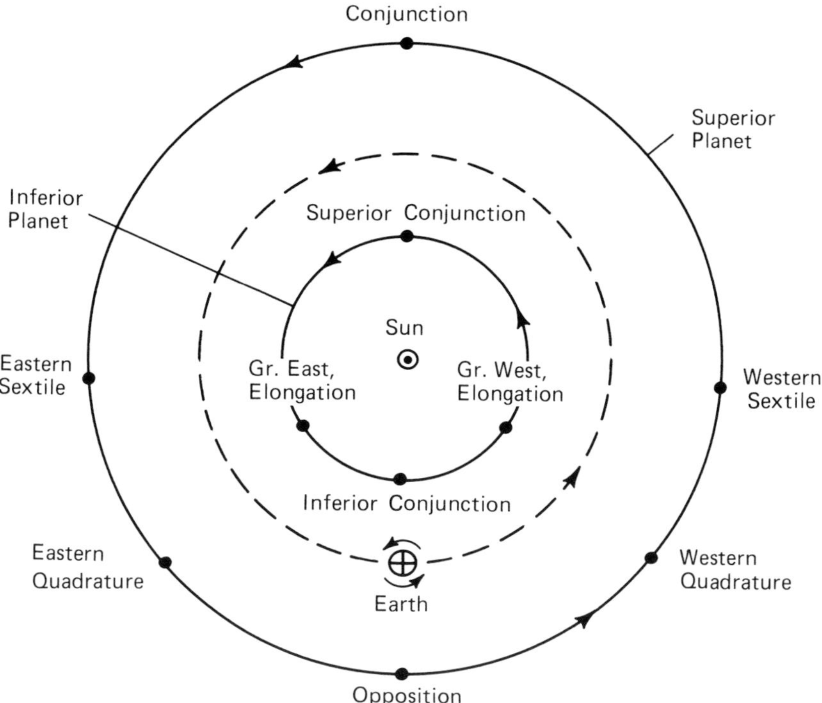

11.4. Aspects, or configurations, of the planets. The Earth's orbit is shown by a dashed circle, and inferior and superior planets (inside and outside Earth's orbit) by solid ones. Planets seen from Earth in the direction of the Sun are in conjunction with it, in the opposite direction, in opposition; 90° and 60° separations are called quadrature and sextile. The inferior planets pass through two conjunctions and are angularly most distant from the Sun at greatest elongation.

achieves maximum angular separation from the Sun. These instances are called *greatest elongations*, either eastern or western, when the planet will be visible in the evening or morning sky, respectively. The average value of the maximum elongation depends upon the semi-major axes of the orbits of the Earth and planet. For Venus it is 46°.3, and Mercury 22°.8, and thus they are both exclusively morning or evening objects as discussed above. Since all orbits are actually eccentric, and the planets vary somewhat in their solar distances, any given maximum elongation can differ from the mean. The variation is small for Venus since its orbit is the most circular of all, and Earth's is only third after Neptune in the order of increasing eccentricities (see Table 11.2): maximum elongations vary between 44°.9, with Venus at perihelion and Earth at aphelion, and 46°.3 when the configuration is reversed. Such a difference is hardly detectable to the casual observer.

The variation for Mercury is considerably greater owing to its much larger eccentricity, second highest after Pluto. Between perihelion and aphelion, greatest elongation will vary between 17°.6 and 28°.3, which adds considerably to the planet's elusive nature. Compounding Mercury's visibility problem even further is the variation in the slant angle of the ecliptic to the horizon at twilight at different times of the year. For example, at sunset in September in the northern hemisphere the ecliptic arches low to Sagittarius on the meridian (see Figure 3.6a), and at any value of elongation, Mercury will be close to the horizon. Unfortunately, the orientation of Mercury's orbit is such that the larger maximum elongations occur close to the poorer ecliptic viewing angles. The ecliptic effect is generally the more important of the two, however, and the best time to view the planet is still in spring evenings or autumn mornings (see Figure 3.6c), when the ecliptic is pitched high to the horizon.

Two other frequently used configurations pertain to the superior planets only. *Quadrature* refers to an elongation of 90°, and *sextile* to one of 60°. Quadrature is of more use astronomically since it means that a planet is near the meridian at sunset (eastern quadrature) or sunrise (western), or that it rises or sets roughly near midnight. The first and third quarter Moons respectively occur at eastern and western quadratures. Sextile, along with several other angles, is used primarily for making astrological predictions (see Section 11.18).

When the terms opposition, conjunction, etc. are used by themselves, they generally tacitly refer to aspects between a planet and the Sun. But any of the configurations can also refer to angles between any two planets, a planet and the Moon, or between any orbiting body and a star. Conjunctions between two planets are fascinating to watch, simply because of the juxtaposition of two bright, colored bodies in the sky. The best known of these is the *grand conjunction* of Jupiter and Saturn, which since Jupiter must chase the slower-moving Saturn, occurs only once every 20 years. These planetary conjunctions and their periods of repetition are also important with regard to calculations of gravitational perturbations, and thus to the prediction of position.

Along with the planets, planetary aspects are associated with traditional symbols, as are the constellations of the zodiac (Table 11.3). The origins of some – a box for quadrature, rippling water for Aquarius, horned heads for Aries and Taurus – are obvious, but others are lost. We have continually made use of the Aries symbol for the vernal equinox, also given in Table 11.2. The symbols can be used as a shorthand to express planetary aspects. An opposition of Mars and the Sun can then be written ☍ ♂☉, and a grand conjunction in Virgo, which happened in 1983, as ☌♃♄. Such notation is no longer in common use.

11.4 Synodic periods

As in the case of the Moon (Section 9.2), planets have synodic periods (see column 11 of Table 11.2) as well as sidereal, and they mean the same thing: the interval between successive conjunctions or oppositions with the Sun. To the Earth-bound viewer, the synodic periods also indicate for each of the planets the span between the times of optimum observation. A synodic period is the "lapping time," the interval between the moments when the Earth first passes a more slowly moving superior planet, orbits, and catches up with it again (or the reverse in the case of an inferior planet).

The meaning of the synodic period is diagramed for Jupiter in Figure 11.5, where

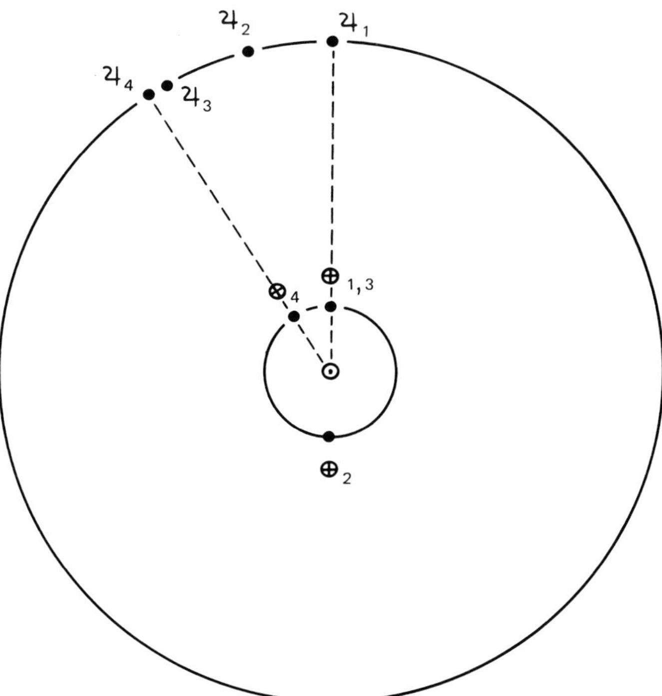

11.5. The synodic period of a planet. The numbers give successive times. In one year, 1 to 3, Jupiter (♃) moves through $^1/_{12}$ of its orbit, and it takes the Earth another 34 days to catch up with it again.

the numbers represent instants of time. At 1 Jupiter is in opposition to the Sun, as viewed from Earth. In 6 months, at 2, the Earth has moved through half of its orbit, but Jupiter has moved as well and is not yet quite in conjunction with the Sun. After a full year has elapsed, at 3, Jupiter has marched $^1/_{12}$ of the way around its path, and the Earth must travel for another 34 days to bring it into opposition again, at 4, to make the synodic period 399 days.

The longer the sidereal period of a superior planet, the shorter will be the synodic. If a planet were to be so far away as to appear stationary against the stars, the synodic period would be that of the stars themselves, 365.24 days. Neptune and Pluto come very close: with annual motions of only about 2° per year, the synodic periods are each 367 days. A planet that has a sidereal period close to that of Earth, however, will have a long synodic one. Since Mars is moving rapidly in its orbit, it takes the Earth 780 days, some 26 months, to catch up with it again after it once passes opposition. Consequently, we see Mars in the sky for a long interval as we slowly catch up, pass, and pull away from it; then for about a year it stays near the Sun and is difficult to find.

The situation is reversed for the inferior planets, where the faster one, Mercury, has the shorter synodic period, only 116 days. Note that it must pass inferior conjunction two or three times per year and either eastern or western greatest elongation five or six

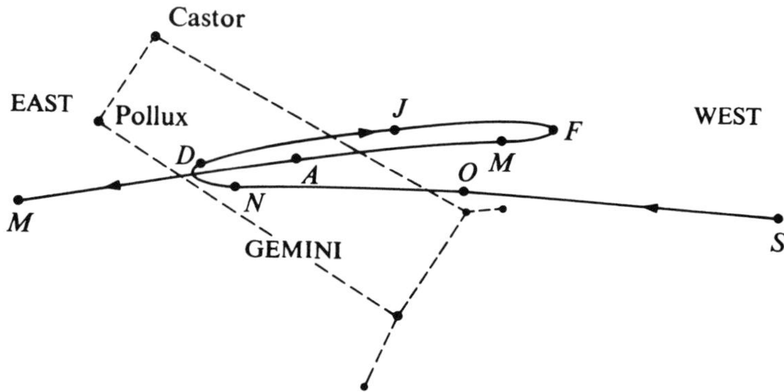

11.6. Observed retrograde motion, exemplified by Mars in Gemini during 1960 and 1961. The position of the planet on September 1, 1960 is labeled *S*. The other letters represent the first of succeeding months, ending on May 1, 1961. Retrograde motion lasted between November 1960 and February 1961, when the planet was *stationary*. Opposition took place on December 30, 1960. (*Principles of Astronomy: A Short Version*, S. P. Wyatt and J. B. Kaler, 2nd ed., Allyn and Bacon, Boston, 1981.)

times. Even though elusive, we are given ample opportunity on a regular basis to see it. Venus, however, is more like Mars, with a sidereal period close to that of Earth, and a long synodic period.

11.5 Retrograde motion

The movement of the planets against the stellar backdrop is easily noticed: Mars averages half a degree per day, and the inferior planets considerably more (see column 9 of Table 11.2). The dominant motion is toward the east, or counterclockwise, through the stars, divorced of course from the westward daily motion caused by terrestrial rotation. But once every synodic period, and fascinating to watch, the planet appears to stop, reverse its direction, and begin to move westward against the stellar backdrop. This backwards, or *retrograde* motion (as opposed to *direct*, or *prograde*) is always centered on opposition (Figure 11.6).

The explanation of this phenomenon is difficult if – as was commonly believed in early times – the Earth is at the center of the Solar System, but it is readily transparent if we understand that the planets all revolve about the Sun. The retrograde interval is caused by a planet being overtaken by the Earth, or vice versa, as the two bodies orbit at different physical speeds.

For a superior planet, it is like your passing a car with both autos going north: the other seems to be southbound if you consider yourself to be stationary. In Figure 11.7, the planet is still moving eastward at point *a*. Retrograde motion begins at point *b*, when the component of the Earth's angular motion across the line of sight to the planet becomes greater than the planet's rate of movement. At this position, Mars or Jupiter will stop its easterly motion, and will appear momentarily *stationary*, whence they com-

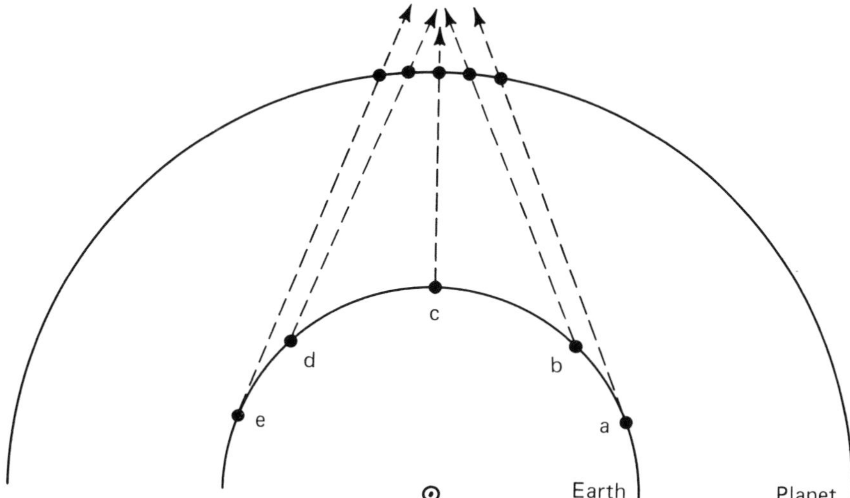

11.7. The explanation of retrograde motion. At points *a* and *e* the Earth appears to be at greatest elongation as seen from the planet. Retrograde motion occurs between *b* and *d*, when the direction of motion of the planet appears reversed. The Earth is inferior to the planet, and to a visitor on that body, the Earth will also be in retrograde motion between *b* and *d*. Point *c* represents opposition of the planet for an Earthly resident and terrestrial inferior conjunction for a planetary citizen.

mence a westward reversal. The backward motion relative to the zodiac continues until stationary point *d*, when the direct track is again taken. The changes in celestial latitude during retrograde motion that cause the distinctive loop seen in Figure 10.5 result from the Earth and planet orbiting in somewhat different planes. The average amount of time that each planet spends in retrograde motion is given in the last column of Table 11.2.

The two inferior planets must be viewed somewhat differently. Even in direct motion through the zodiac, the Sun is always gaining on the six superiors, but Mercury and Venus will first move ahead of the Sun, then fall back, in an oscillatory pattern. Between the times of greatest western and greatest eastern elongations (see Figure 11.4) the two will advance eastward relative to both the Sun and stars. At greatest eastern elongation, when they are seen as evening stars at maximum separation from the Sun, they begin to lose ground to the Sun, although maintaining a direct motion through the zodiac. Then as the inferior planets approach inferior conjunction and swing between Earth and Sun, they enter a brief retrograde interval that ends shortly after they become visible in the early morning hours. The Sun then continues to pull ahead of them until greatest western elongation, and the process reverses.

Table 11.4 gives the time periods between various positions of the inferior planets. The numbers for Mercury can differ considerably from one apparition to the next as a result of its high orbital eccentricity. Note Venus's very short retrograde interval of only six weeks. It disappears very quickly from the evening sky following reversal of

Table 11.4. *Venus-Mercury timetable*
From inferior to superior conjunction read upward and substitute "end retrograde" for "start retrograde," and "western" for "eastern".

	Vernus	Mercury
superior conjunction		
	7 months	5 weeks[a]
greatest eastern elongation		
	5 weeks	
greatest brilliancy		10 days
	2 weeks	
stationary (start retrograde)		
	3 weeks	11 days
inferior conjunction		

[a]Can vary by over a week because of Mercury's eccentricity

direction. Mercury's is even shorter: once it starts to move backward we lose it from the evening sky within a veritable blink of the eye.

11.6 Phases and brightness

All the planets vary in brightness as they move through their synodic periods, some considerably so. The cause for the superiors is almost solely variation in distance. The brightness of a light source is dependent upon the inverse of the square of its distance: if you move it twice as far away, it will appear 1/4 as bright. Mars' distance from the Earth changes from about 0.5 AU at a typical opposition, when it is obviously brightest, to 2.5 AU at conjunction. The planet's distance from us varies by a factor of 5 and its brightness by a factor of 25. On the average, it is some 3.5 magnitudes fainter when it is opposite the Sun than when in retrograde. This great drop in brightness causes Mars to be somewhat elusive between solar conjunction and the sextiles. Since the other superior planets are much farther away, their range in luminosity is much less. Jupiter changes by just under one magnitude and Saturn by only half a magnitude.

Mars's orbital eccentricity lends it some additional distinction. The distance from the Earth at opposition depends upon whether the planet is near aphelion or perihelion at the time (the Earth's smaller eccentricity somewhat exaggerates the effect since the terrestrial aphelion point roughly coincides in time with that of Martian perihelion). At perihelion, Mars is only 56 million kilometers away from Earth; if at aphelion, it is nearly twice as distant, some 100 million kilometers (Figure 11.8). The close approach is called a *favorable opposition*, and happens only once about every 17 years, when opposition occurs closest to August 29. The variation in maximum brightness at opposition is about $1^1/_4$ magnitudes. The eccentricity and resulting variation in orbital speed also means that the intervals between successive oppositions can be 20 to 30 days longer

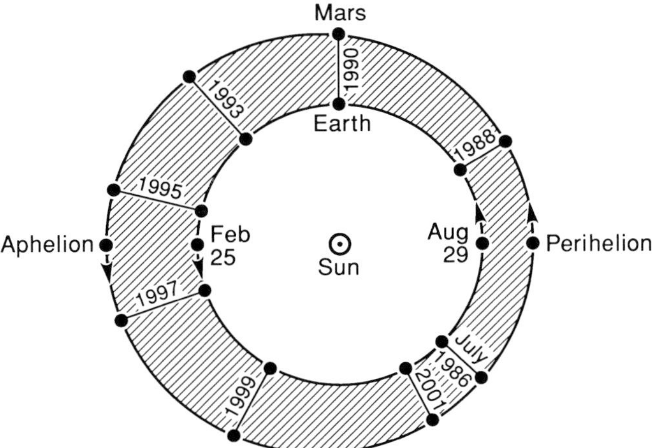

11.8. Oppositions of Mars from 1988 to 2001. At a favorable opposition that occurs in late August, the red planet is as bright as it can be because it is then at perihelion and closest to Earth.

or shorter than the mean synodic period of Table 11.2. The phenomenon is obviously of great significance to telescopic observers trying to see fine detail on the planetary surface since the angular size of the apparent disk is about double at favorable opposition than what it is at least favorable. Southern viewers are generally in the best position to see a favorable opposition, since Martian perihelion occurs at a celestial longitude of 336° ($\Omega + \omega$ in Table 11.2), in Aquarius, at a declination of −9°. Note that this longitude is also that of the Sun on August 29 plus 180°.

The brightnesses of the inferior planets are controlled not only by their variations in distance by their *phases* as well (Figure 11.9), which require telescopic power to see. Look again at Figure 11.4. At superior conjunction Venus and Mercury must appear "full"; at inferior, we can see only the nighttime side, and they will be "new." A half-phase appears at the elongations, and between them and inferior conjunction we view a crescent. Phase correlates with angular diameter, the planet being smallest at superior conjunction and largest at inferior; the variation in apparent size for Venus is huge, a factor of six. Maximum brilliance for the two planets must occur at the time of largest illuminated *angular area* in square seconds of arc.

Venus varies in distance by a somewhat greater factor than Mars, but as it grows in angular diameter, it shrinks in phase. As a result, the phase change moderates that caused by the changing distance, and Venus does not vary much in brightness over the course of its orbit (generally from around apparent magnitude −3.9 to about −4.6). It reaches maximum brilliance in the crescent phase, about midway between the elongations and inferior conjunction (respectively about five weeks after and before greatest eastern and western elongations). After maximum, it passes very quickly from the evening sky, and a few weeks past inferior conjunction, reigns over the morning as the sequence of events of Table 11.4 is reversed.

Mercury behaves quite differently. Because it is so close to the Sun, its variation in

11.9. The phases of Venus from near superior conjunction at upper left to near inferior conjunction at lower left. Maximum brightness occurs near the moment when the planet looks about like the image at lower right. (Lowell Observatory.)

distance is much less than that of Venus, about a factor of 2.3. As a result, its brightness is controlled largely by its changing phase. It is brightest at superior conjunction (when invisible), with a magnitude of about −2. It then dims to roughly magnitude 0 as it approaches greatest elongation and then fades quickly to third or fourth magnitude near inferior conjunction.

The superior planets go through phases as well. From Figure 11.4, we see that near quadrature we are able to view some of a planet's nighttime side. Since it is so close, Mars can appear markedly gibbous, which contributes some to its magnitude variation. The Jovians, however, are all too far away for there to be any significant observable effect.

11.7 Transits and cycles

Both Mercury and Venus have orbits that are inclined to that of the Earth, and at inferior conjunction, they usually pass above or below the Sun, appearing as the slimmest of crescents (which are dangerous to look at because of their proximity to the Sun). Like the Moon, the orbital paths take the planets across the ecliptic twice each sidereal year at the nodes. If inferior conjunction takes place near nodal passage, we will see the body cross the face of the Sun in a *transit*. The phenomenon is analogous

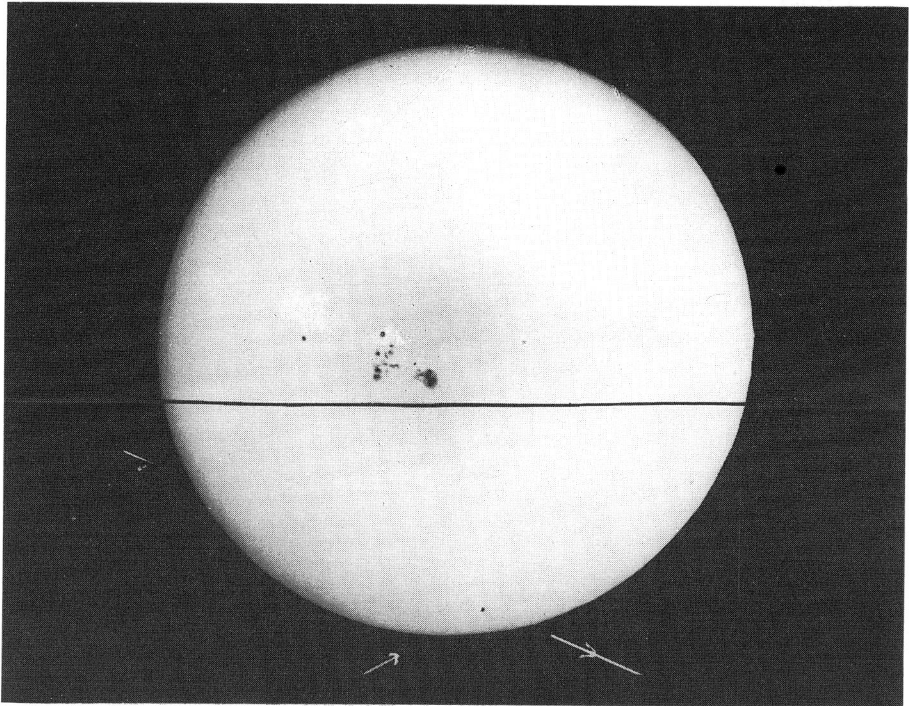

11.10. The November 14, 1907 transit of Mercury. The path of the planet is indicated by the white line, and the arrow points to its position. (Yerkes Observatory.)

to a lunar eclipse of the Sun, and similar, though smaller, ecliptic limits can be defined. Mercury is so distant and tiny that its transit is a telescopic affair (Figure 11.10), but that of Venus is viewable with binoculars (as it then has maximum angular size), with of course proper solar filtration.

Mercury's nodes are projected against the solar center on November 9 and May 7, so transits can take place only near those dates. Because Mercury's perihelion is aligned with the November nodal passage, when the planet is closer to the Sun and the chance for a transit consequently greater, the autumn events are over twice as common: they are separated by intervals of about 7 or 14 years depending upon where within the ecliptic limit the planet attains conjunction. Two May events can be separated by 14 years, and must occur every 46. The dates of the last three transits of the twentieth century are (were) on November 12, 1986, November 5, 1993, and November 15, 1999.

Because of Venus' greater synodic period and distance from the Sun, transits are much less likely. They can happen only within a day or so of June 7 and December 8, and for each date currently come in pairs 8 years apart separated by 243 years, with those of the other date appearing in between. The last ones took place in December of 1874 and 1882, and the next are set for June 8, 2004 and June 6, 2012.

With the advent of the space program and modern detection techniques, transits have lost much of their scientific significance. In the past, they were important for the measurement of planetary diameters, and precise timing was used to improve the knowledge of the orbit. At one time, they were also important in the establishment of longitude. A transit of Venus or Mercury can be predicted accurately in Greenwich Time, and therefore the planets may be used as clocks. Observations of the events at remote locations establish GMT (or UT) and simultaneous determinations of local times yield longitudes. On his first voyage of exploration, Captain James Cook observed the Venusian transit of June 3, 1769 and the November 9, 1769 transit of Mercury, to which the names "Point Venus" in Tahiti and "Mercury Bay" and "Mercury Island" in New Zealand still attest.

The eight-year interval separating the Venusian transit pair is a result of a curious cycle between the orbit of that planet and that of Earth. Eight terrestrial years equals five synodic periods of Venus to within one day. Thus after an eight-year interval, Venus will repeat its aspects almost exactly on the same dates. The phenomenon provides a time-marker over a long interval and even found its way into a South American native calendar.

11.8 Old theories of planetary motion

The idea that the planets revolve about the Sun (in a *heliocentric* system) rather than the Earth (a *geocentric* system) is usually credited to Aristarchus of Samos (310–250 BC). A remarkable individual, a mathematician, he also determined the time of the summer solstice and measured the ratio of distances between the Earth and the Sun and the Earth and the Moon. But in the absence of solid evidence, geocentric theories that set the Earth into an exalted position were in competition, and for over two millennia held sway.

Starting with roughly Homeric times some 500 years before Aristarchus, many schemes were invented to explain celestial motions. In this era, the Earth was conceived of as a flat disk encircled by the River Oceanus. The stars were set onto a crystal sphere, the Sun set to the west, and somehow moved around below north to rise again. These ideas culminated in a highly structured plan developed by Eudoxus (the "celebrated one," 409–365 BC). He was an Athenian from Asia Minor, and represented a transition from a time of speculation into a time of science. In his universe, constructed of a system of 27 nested homocentric spheres, he made a genuine attempt to satisfy the observed motions of the heavens.

In the center was the sphere of the Earth. The Moon was mounted on the equator of the innermost of three nested spheres, each of which rotated about us on different axes at different rates so as to account for lunar motion. The Sun was similarly set onto three more spheres external to the Moon, and each of the five planets employed four spheres apiece. The final one held the stars. He quite likely did not believe his scheme to be physically real, but viewed it as a mathematical device that could be used successfully to predict planetary motion.

During the next few hundred years Aristotle added another 22 homocentric spheres to refine the predicted motions, and Apollonius of Perga (circa 230 BC) invented the

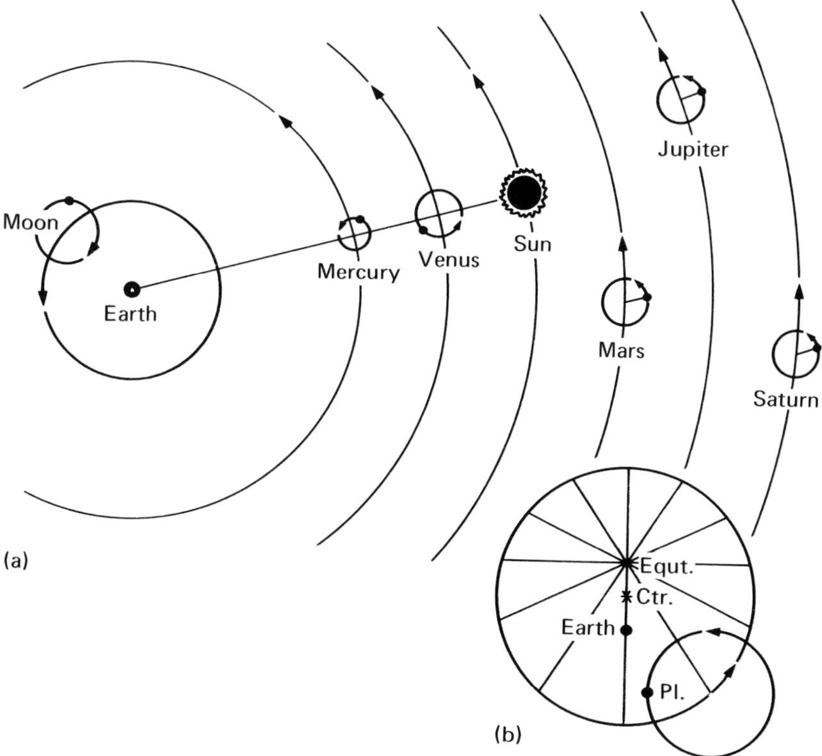

11.11. The Ptolemaic system. The planets revolve on epicycles (a) at their synodic periods, which in turn encircle the Earth on deferents at the sidereal period. Only the Sun has no epicycle. The angular motion of the epicycle is constant as seen from the *equant*, labelled "equt." (b). The observer is opposite the center from the equant. ((a) From *Principles of Astronomy: A Short Version*, S. P. Wyatt and J. B. Kaler, 2nd ed., Allyn and Bacon, Boston, 1981; (b) from *A History of Astronomy*, A. Pannekoek, George Allen and Unwin, 1961.)

eccentric, which displaced the Earth from the center of motion, thus explaining the inequality of the seasons (see Section 6.2). The culminating geocentric theory was synthesized about 140 AD by Claudius Ptolemaeus – Ptolemy – an Alexandrian Greek, in his great book, the *Mathematical System of Astronomy*, known also as the *Syntaxis* (meaning "compositions"). The eighth-century Arabians who translated and preserved it called the book the *Almagest*, from *al Magisti*, "the greatest."

Building upon earlier concepts, such as that of the eccentric, he disposed of the rotating spheres and adopted orbits known as *deferents* that encircled the Earth (Figure 11.11a). In this scheme, a planet revolves with its synodic period on a sub-orbit called an *epicycle*, which is centered upon the deferent. The epicycle travels about the Earth in the sidereal period. As the planet moves along the deferent against the *primum mobile* – the outer starry sphere – it appears to pass through retrograde when it travels the inner portion of its epicycle once each synodic period.

The Ptolemaic system is an ingenious device that mathematically explains the observations reasonably well. There are a number of details to the scheme. The epicyclic centers of Mercury and Venus are always between Earth and Sun in order that they not stray far from the latter. For Mars, Jupiter, and Saturn, the line from the center of the epicycle to the planet stays parallel to the terrestrial–solar axis to correlate retrograde with opposition. The lunar epicyclic period is equal to its orbital (sidereal) period, with the phases controlled as always by the relative positions of the Moon and Sun, and the epicyclic motion is backward relative to that of a planet. Although the Moon does not move retrograde, it has a highly variable angular motion, which requires this explanatory mechanism. The starry sphere rotates slowly against the equinoxes with a (then-thought) 36 000-year period to reproduce some of the effects of precession. As a fine-tuning device, the Earth is placed off-center, and the epicycles move along their deferents uniformly as seen from the *equant* (Figure 11.11b), equally, but oppositely, off-center. Finally, the whole system rotates about the fixed Earth once per day to simulate daily motion. Never mind that the scheme had little to do with reality: it more or less worked. Astronomers could predict planetary positions with some kind of precision, which for hundreds of subsequent years is all that seems to have mattered.

11.9 The revolution

The modern view was ushered in by Nicolaus Copernicus (1473–1543), who – possibly influenced by Aristarchus – finally and forever placed the Sun at the center of the Solar System and demonstrated how simply planetary motions could be explained by having the Earth move. The results (Figure 11.12) appeared in his great treatise, *De Revolutionibus Orbis Celestium* ("On the Revolutions of the Celestial Spheres"), said to have been handed to him on the day of his death in 1543. Much of the book involves the development of the mathematics needed to explain the subject. Prior to Copernicus's time, the planets were placed in distance from the Earth according to their sidereal periods. Copernicus, however, determined accurate relative distances for the first time by calculation. Figure 11.13 shows the planet Venus at greatest western elongation. Since Earth–Venus–Sun form a right triangle (one with a right angle of 90°), Copernicus could easily calculate the relative lengths of the sides, and the distance Sun–Venus in terms of the distance Sun–Earth (or in modern terms, the distance of Venus in AU). The calculation of the relative distances of the superior planets is similar, but more complicated, and involves observing the planet at the time that the *Earth* would appear to be at maximum elongation as seen from the other body. The results are essentially the modern values, all determined from naked-eye observations.

Though there was at the time no way of knowing which theory – Copernican or Ptolemaic – was correct, the Copernican had the advantage of simplicity even though Copernicus still clung to the idea of circular orbits and consequently had to keep the concept of epicycles to explain the irregularities of planetary movements. However, because the heliocentric theory displaced the Earth from the center of the Universe,

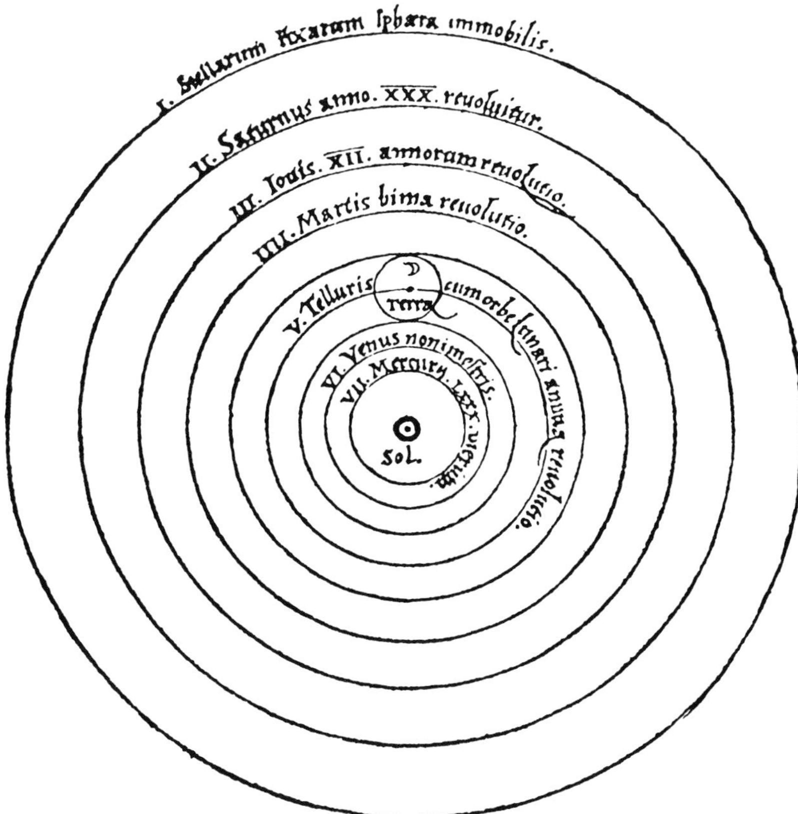

11.12. The Copernican view of the Solar System, taken from *De Revolutionibus*. It was drawn deliberately out of scale to enable placement on a page. Names of the bodies are given in Latin. *Telluris* refers to the Earth, about which the Moon is in orbit. Surrounding the Solar System is the celestial sphere of the "fixed stars."

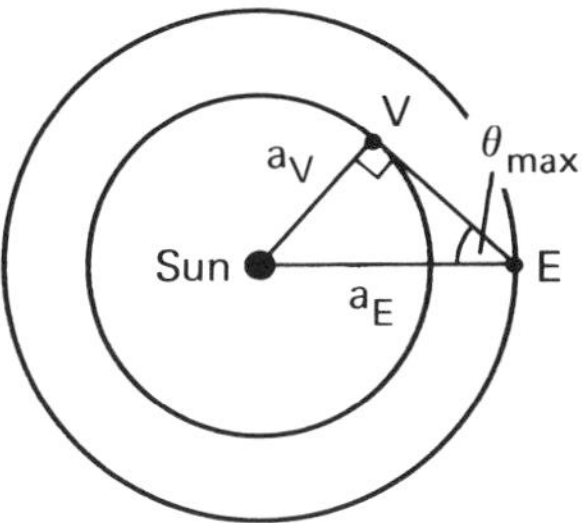

11.13. Copernicus's method for obtaining the distance to Venus and Mercury. He determined the angle at which the planet is at maximum elongation, E_{max}. Since Earth–Venus–Sun is at that time a right triangle and $a_{Earth} = 1$ AU, the distance of Venus is easily found to be 0.72 AU.

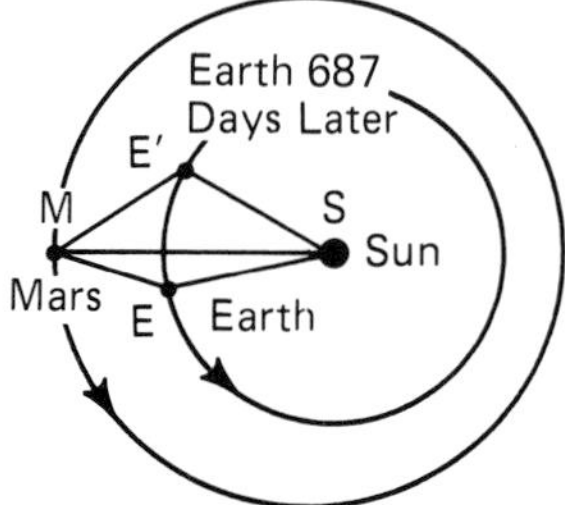

11.14. Kepler's measurements of planetary orbits. Mars is observed from Earth at intervals of the Martian sidereal period, 687 days. The angles are known and distances can be computed.

the Catholic Church banned the volume, and it remained on the *Index of Prohibited Books* until 1837, long after the heliocentric theory was firmly shown to be correct.

Dramatic support for Copernicus was provided by Galileo when, in 1609, he turned the first telescope to the heavens. Two observations in particular confirmed the Copernican view: the discovery of Jupiter's satellites, which orbit that body in the same way that the planets were now said to circle the Sun; and the sighting of the full set of phases of Venus, which is not possible in the Ptolemaic geocentric universe, but which find ready explanation if both Venus and Earth are in solar orbit. For his views, Galileo was convicted of heresy in 1633 and sentenced to house arrest after publicly recanting them.

The rest of the story involves four principal individuals. The first was the last of the great naked-eye observers, Tycho Brahe, who worked in the latter half of the sixteenth century, shortly before Galileo's development of the astronomical telescope. From an island observatory, Uraniborg ("Castle of the Heavens"), granted him by the King of Denmark, Tycho pursued positional measurements of the stars and the planets with instruments that included a 19-foot wall-mounted quadrant and a 6-foot sextant.

Toward the close of the century, Tycho's quarrelsome personality finally disenchanted the new king and he was forced to leave Denmark for his final destination, Prague. In the year 1600, he hired a new assistant, Johannes Kepler, who had left his native Germany as a result of religious persecution. Tycho died the next year, whereupon Kepler succeeded to his academic position and acquired the precious positional data.

Copernican theory still could not accurately predict planetary positions. Kepler set out to find how the planets truly move about the Sun by using Tycho's observations, particularly those of Mars, to modify the theory. He first tried circular orbits and adopted Ptolemy's equant, but with the Sun replacing the Earth. By setting up pairs of observations at intervals of a Martian sidereal year, in which the Earth would be at two different places (Figure 11.14), he could use the principles of geometry to determine the offsets of the Sun and the equant from the center. Then with similar pairs at intervals of an Earth year, in which Mars was now at two different places, he could find the analogous offsets to the Martian orbit. Computation of position still showed differences with observation larger than he believed Tycho could have made. Kepler

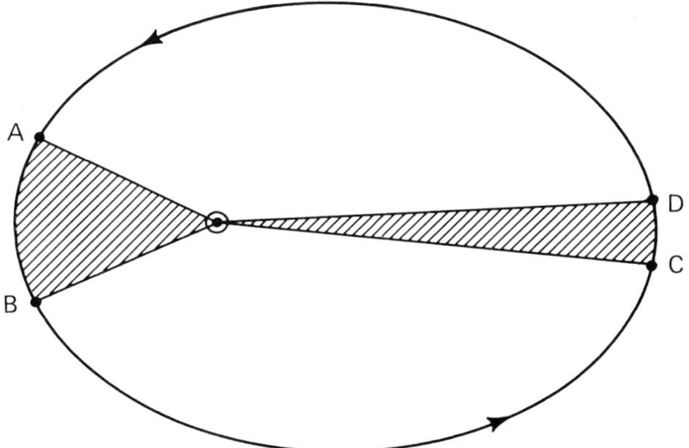

11.15. Kepler's second law. A planet moves between points A and B in a certain period of time. Over that same interval the distance covered from C to D is such that the shaded regions have the same areas. The planet must therefore move faster when it is closer to the Sun.

then used pairs of observations like those of Figure 11.14 to measure the distances of Mars from the Sun at various positions, and found, as he wrote in his *New Astronomy*, "it is as if I awoke from sleep and saw a new light": the orbit is an *ellipse*. His measurements of the eccentricities of Earth's and Mars's orbits were remarkably accurate, practically the same as the modern results. Tycho's data on the other planets were not good enough to determine orbital shapes, but it seemed quite reasonable to suppose that they travel on elliptical orbits too.

The work took three years of calculation and was completed in 1604. In 1609 he was finally able to publish the first two of what would eventually be his three *laws of planetary motion* (introduced in Section 6.2 and repeated here).

1. *The law of ellipses*: The orbit of a planet is an ellipse with the Sun located at one focus.
2. *The law of areas*: The radius vector (the line that joins a planet to the Sun) sweeps out equal areas in equal amounts of time.

The first law accounts for the variation in the distance of a planet from the Sun and the second for the variation in velocity along its orbit. As a planet moves closer to the Sun along its path, it must speed up so that the area covered by the radius vector stays the same (Figure 11.15). The second law quantifies one of the reasons for the equation of time discussed in Section 6.3.

Kepler then set out to examine the relations among the different planetary orbits. Ten years after his publication of the first two laws, he established his third law of planetary motion.

3. *The harmonic law*: The square of the orbital period of a planet is proportional to the cube of its mean distance from the Sun.

Expressed more simply, Kepler's third law states that

$$P^2 = a^3,$$

where the period P is in years and the semimajor axis, a, is expressed in astronomical units.

The law is delightfully simple. We watch Jupiter move and determine from its annual rate of motion that it will take 12 years to orbit the Sun. Twelve squared is 144, which is then also a^3. The distance of the planet from the Sun must be the cube root of 144; a short trial and error calculation shows that $a = 5.2$. A short look at any newly discovered body in the Solar System will reveal the length of its orbital semimajor axis.

Kepler's laws allow us to predict planetary positions with precision. He knew that the Sun is somehow responsible for planetary movement, but did not know why. That was left for Isaac Newton and Albert Einstein.

11.10 Newton's laws and gravity

Isaac Newton derived Kepler's three laws of planetary motion from three far more fundamental rules of nature, *Newton's laws of motion*, coupled with his law of gravity. They were published a half-century after Kepler's work, in 1687, in Newton's great work, *Philosophiae Naturalis Principia Mathematica* ("Mathematical Principles of Natural Philosophy"), commonly called the *Principia*.

Newton's laws of motion collectively represent a system of mechanics that describes how things move. To understand the laws, we need to give definitions of some terms that relate to motion, definitions that are somewhat different from those used in common speech. *Speed* is the rate of change of the position of a body with time, typically expressed in meters per second, m/s, or kilometers per hour, km/hr. *Velocity*, which is not a synonym for speed, incorporates both speed and *direction*. Two bodies moving at the same speed in opposite directions have different velocities. An *acceleration* is any change in *either* speed or direction. If you step on either the accelerator or the brake of a car, you accelerate the vehicle because you change its speed. If you drive around a curve at a constant speed, you accelerate as well, because you change direction. Acceleration is expressed as a rate of change of velocity with time, or in meters per second per second, or m/s/s. A *force* is anything that produces an acceleration. If you push a bowling ball you apply a force and accelerate it; if you jump from an airplane, the force of gravity causes you to accelerate toward the ground. *Mass*, measured in kilograms, kg, is commonly thought of as the amount of matter in a body. It is actually defined as the degree to which a force is resisted, or by the amount of force needed to give it a specific acceleration.

Newton's laws of motion now state that:

1. Left undisturbed, a body will continue in a state of rest or uniform straight-line motion.
2. The size of an acceleration is directly proportional to the force applied and

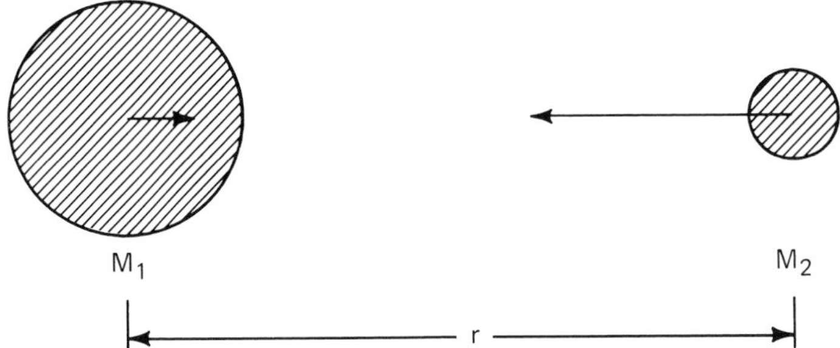

11.16. Mutual acceleration of gravitating bodies. Mass M_1 is larger than M_2, so M_2 accelerates proportionately more than M_1. They are attracted by equal and opposite forces.

inversely proportional to the mass of the body being accelerated, and the direction of the acceleration is the same as the direction of the force.

3. For every force applied to a body, there is an equal force exerted in the opposite direction (or action equals reaction).

The second law is usually expressed as acceleration (A) equals force (F) divided by mass (M),

$$A = F/M,$$

or as

$$F = MA.$$

Planets in orbit about the Sun must be subject to a continuous force to keep them from flying off into space, the force of gravity. Newton quantified this mysterious force of nature (see Section 9.5), and discovered that the acceleration (g) that it imparts to one body falling toward another is equal to a constant (G, the *constant of gravity*) times the mass of the attracting body divided by the square of the distance between the centers of the two. In Figure 11.16, two bodies, one of large mass M_1 and another of small mass M_2, are separated by distance R. Gravity will cause M_2 to be accelerated toward M_1 at a rate GM_1/R^2, and M_1 to be accelerated toward M_2 at GM_2/R^2 (bodies act gravitationally as if all the matter were concentrated at the center). The smaller mass will move with the larger acceleration. Now couple these to Newton's second law, $F = MA$ (or $F = Mg$). Mass M_2 in Figure 11.16 is subject to a force $F = M_2(GM_1/R^2)$ and mass M_1 to $F = M_1(GM_2/R^2)$, which are identical. The forces between the two are equal and opposite.

Apply these concepts to the Earth. The acceleration of gravity is $GM_{\text{Earth}}/R_{\text{Earth}}^2$, where R_{Earth} is the terrestrial radius. We can measure the downward acceleration of a falling body, and with the known radius of the Earth (and G determined by laboratory experiment) we can measure the Earth's mass. If you jump off a table, you accelerate downward. The Earth must jump upward as well, but with an acceleration reduced by the ratio of masses (and which is not detectable). When you stand on Earth, the downward force on the ground equals the upward force of the Earth on you. Both forces are expressed by $F = GM_{\text{you}}M_{\text{Earth}}/R_{\text{Earth}}^2$. F is your *weight*. By social convention, mass and weight at the Earth's surface are numerically set equal to each other, but they are not the same. Clearly your weight depends on your distance from the center of the Earth. If you stand on a chair you weigh less, and if you could double your distance from the center of the Earth by standing on an imaginary platform 6400 km high, you would weigh only $1/4$ your weight at the surface. Your mass, however, stays the same. Your weight on another planet would depend on the planet's mass and radius. On Mars, $1/10$ the mass but $1/2$ the radius of Earth, you would weigh $(1/10)/(1/2)^2 = 4/10$ as much as you do here. On the Moon with $27/100$ the terrestrial radius and $1/80$ the mass, your weight would be $1/6$ of normal and you could jump 6 times as high (imagine a lunar basketball hoop set 60 feet – 19 meters – above the floor); a golf ball could be driven nearly two kilometers.

11.11 Generalization of Kepler's laws

From his laws of motion and the law of gravity, Newton could derive the sizes and shapes of the planetary orbits. That is, he could re-derive Kepler's laws, but in a more elegant generalized form, in which the original rules are special cases. *Kepler's generalized laws* deal with the mutual orbits of *any* two bodies, not just the Sun and a planet. The first states that:

1. The orbit of each body is a conic section with the center of mass at the focus of each orbit.

Or if we place our center of reference at the Sun (as we did for the Moon in Figure 9.19), we can say that:

1. The orbit of a body in the Solar System is a conic section with the Sun at a focus.

The *conic sections* (Figure 11.17) are a family of curves, one member of which is Kepler's original special case, the ellipse. They are generated by slicing a cone with a plane set at different angles (Figure 11.18). The reader can easily produce them by cutting a conical paper cup. The circle and parabola are actually limiting cases in which the cone is cut parallel to one of its parts (base and side respectively); the ellipse (see also Figure 6.1) and hyperbola, however, are not so restricted and come in an infinite variety of shapes. The circle, of course, is simply an extreme type of ellipse. The parabola and hyperbola (Figure 10.15) differ from the ellipse in that the curves are not closed, but open-ended; the parabola is the special case that separates the ellipse family from the hyperbolae.

A body in a hyperbolic (or parabolic) solar orbit will pass around the Sun but once, and will never come back. It is said to have a speed that exceeds (or in the case of the parabola, equals) the body's *escape velocity*. This critical value is that speed beyond which a planet has an insufficient gravitational pull ever to bring the departing body to rest. At the surface of the Earth it is 11.2 km/s.

The second generalized law is simply a statement of the conservation of angular momentum (see Section 10.5), that is:

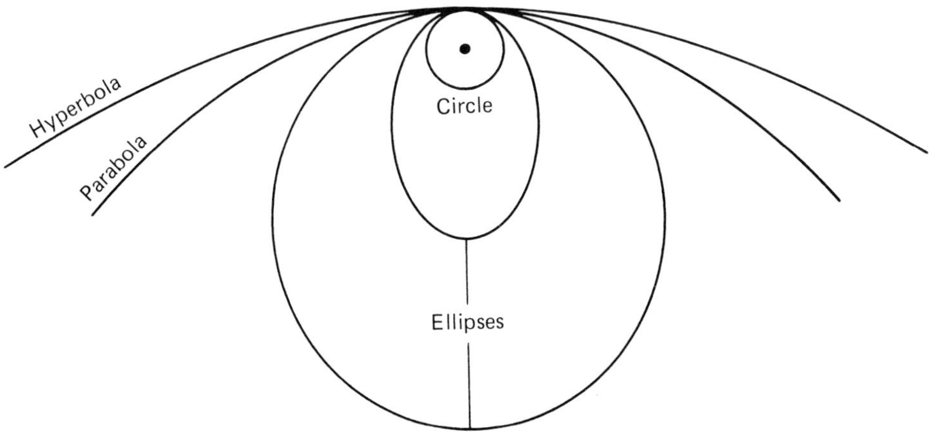

11.17. Conic sections. The sections consist of the circle, the ellipse, the parabola, and the hyperbola. The parabola and hyperbola are open-ended.

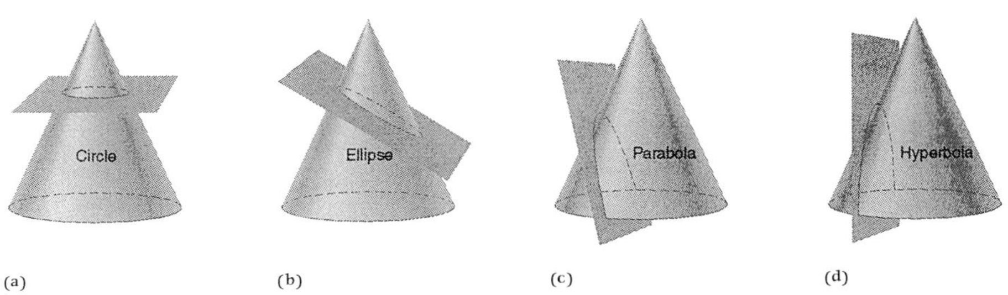

11.18. Construction of the conic sections. The curves are produced by the intersection of a cone and plane set at various angles to the base: (a) circle (plane parallel to base); (b) ellipse; (c) parabola (plane parallel to side); (d) hyperbola. (From *Astronomy!*, J. B. Kaler, HarperCollins, 1994.)

2. *angular momentum* is always constant in an undisturbed system.

The tidal slowing of the Earth results in the Moon moving farther away from us. In the case of planetary orbits, as the radius vector decreases, speed increases so as to give us Kepler's simple equal area law.

The third generalized law incorporates mass. Two bodies always have mutual orbits about a center of mass (see Section 9.5) and the relative sizes of the orbits depend upon the relative masses. Therefore, one simply cannot exclude the masses of the system's components in an orbital calculation. Kepler's third law in its original form makes no mention of mass, yet it is obvious that if one could increase the mass of the Earth, and thereby magnify the gravitational attraction, that our planet's orbit *must* change. The generalized third law states that:

3. $$P^2 = \left(\frac{4\pi^2}{G}\right) \frac{a^3}{(M_1 + M_2)} .$$

G is the constant of gravity, so $4\pi^2/G$ is but a constant as well. P is now the period in seconds, a is the semimajor axis of the orbit of one body about the other in meters, and M_1 and M_2 represent the masses of the two bodies in kilograms. That is, we now use physical units rather than units expressed relative to the Earth (like years and AU).

Kepler's $P^2 = a^3$ relation holds for the Solar System because in $M_1 + M_2 = M_{\text{planet}} + M_{\text{Sun}}$, M_{Sun} so dominates that the sum is nearly the same for all planets, and is very close to being constant. The Sun is 330 000 times more massive than the Earth and 1000 times more so than all the planets combined. If we construct the third law first for a planet and then for the Earth, and divide the first by the second, we simply derive $P^2 = a^3$ in years and AU respectively.

The beauty of this law is that it can be used to *derive* masses: if we know P and a in seconds and meters, the sum of masses in kilograms is easily found. Since M_{Earth} is inconsequential compared to M_{Sun}, the solar mass is derived immediately from the parameters of the terrestrial orbit. We similarly find the mass of Jupiter from the orbits of its satellites. If the masses are more comparable, as they are in the Earth–Moon system, we find their sum from Kepler's third law, and their ratio by locating the center of mass, which together yield the individual values. The method is even used to find the masses of stars that orbit each other.

11.12 True orbits

Unfortunately, Kepler's laws, even in their elegant generalized form, do not actually work in practice in the Solar System, since each planet is pulled, however weakly, by every other, thus rendering perfectly elliptical orbits impossible. The problem is the same as that for the Moon (see Section 9.7) in that we must overlay the elliptical orbits that the planets have at any specific moment – those we would see if external influences should disappear – with a set of perturbations. Each of the orbital elements in Table 11.2 therefore has a variation, or set of variations, both periodic and secular, associated with it. The planets *are* more isolated than the Moon, however, and the perturbations are very much smaller. Those of the inner planets are essentially indiscernible with the human eye; those of the outer can reach one degree, though the periods are enormously long.

Even though small, the perturbations of the terrestrial orbit are profoundly significant. The *Milankovich cycles* involve a 100 000-year periodicity in the eccentricity of the Earth's orbit and a 41 000-year 3° variation in the obliquity of the ecliptic (see Section 5.10). These conjoined with the 26 000-year precession cycle of the equinoxes may trigger long-term climate changes on Earth and may be responsible for the ice ages that periodically glaciate much of the terrestrial surface.

The problem of tracking the planets with the highest precision is an immensely complex one. For two bodies, we have a beautifully simple set of rules. For just three bodies mutually pulling on one another it is mathematically proven that no such rules exist. The difficulty in calculation increases geometrically as additional perturbers are added. We can only compute the paths numerically and consider first the relative positions of the bodies at any moment, compute the gravitational forces, determine the direction in which all the planets will go, reconsider the new relative locations after a given time-step, and so on.

Obviously, this job is eminently suited to high-speed computers, and they have been enormously successful. But so were the diligent nineteenth- and early twentieth-century astronomers who calculated orbits with approximate, but hugely complex "perturbation formulae." The triumph of this mathematical science, and indeed of Newtonian mechanics itself, was in the discovery of Neptune.

11.13 Uranus, Neptune, and Pluto

Uranus was discovered by accident in 1781 by Sir William Herschel while surveying the skies with a telescope of his own making. Herschel is commonly considered to be the founder of modern observational astronomy and was responsible for the discovery of an immense number of clusters and nebulae (including those that would eventually be known as galaxies). The object he chanced upon was clearly non-stellar since it displayed a tiny disk. He called it a comet, but within a year it became clear that the orbit was nearly circular, and from the period he saw that it lay well beyond Saturn and was, with certainty, a new planet. With this unprecedented discovery arose the problem of a name. Herschel originally called it "Georgius Sidus," George's Star, after King George III. But nationalistic names rarely survive in an international subject. Eventually the name of Saturn's father, Uranus – the heavens personified – was adopted.

The planet was a natural laboratory for the developing science of celestial mechanics. Attempts at predicting its observed position repeatedly failed, even taking into account the perturbations induced by the other known planets. By 1845, astronomers began to suspect that the problem lay with a heretofore undiscovered planet outside the Uranian orbit. Urbain Leverrier in France and John Couch Adams in England therefore began to calculate where the unseen body must be to explain the deviations. They found the same result at almost the same time. Adams was first, but Johann Galle in Berlin, to whom Leverrier sent his predicted position a short time later, actually found the eighth-magnitude planet in 1846. He needed only to point his telescope to the calculated position.

Even after Neptune's orbit was known, Uranus still seemed to refuse to behave and the same technique was tried to find yet another planet. The calculations were finished in 1915, but this time, the discovery did not follow so quickly. Finally, in 1929, Percival Lowell, who had founded the Lowell Observatory in Arizona to study the so-called canals of Mars, hired Clyde Tombaugh to look for the mysterious planet. The search technique called for taking pairs of plates at the point in the sky opposite the Sun. The

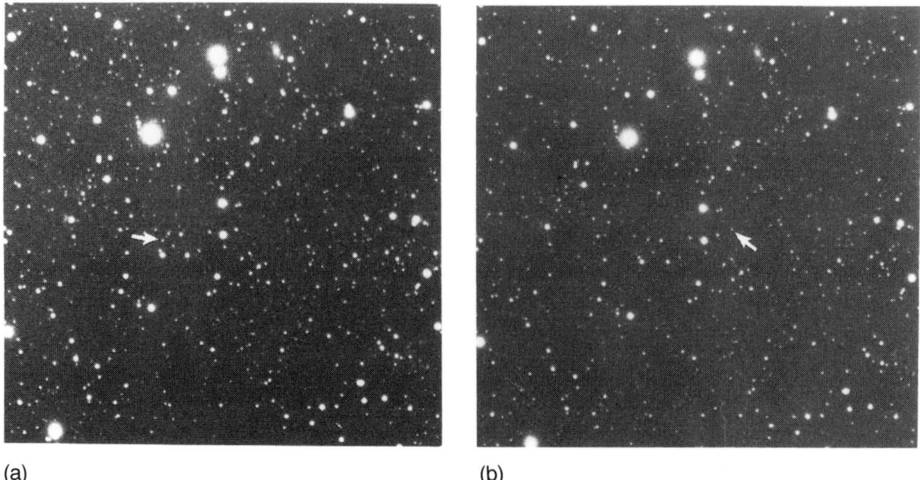

11.19. The motion of Pluto, shown here in Tombaugh's discovery plates of January 23, 1930 (a) and January 29, 1930 (b). (Lowell Observatory.)

plates were then examined by lining them up in a binocular microscope so that one eye looked at each. By "blinking" the plates – looking first at one then the other – Tombaugh could see a moving object appear to jump back and forth (Figure 11.19). At opposition all Solar System bodies are moving retrograde, and the distance could be calculated by how much it was displaced as a result of the Earth's orbital motion. The result was the discovery of dim Pluto. But then, after all the labor, later calculations clearly showed that the mass of this dim fifteenth-magnitude body (10 000 times fainter than the human eye can see) is too low to have caused the observed Uranian perturbations: its discovery had been accidental.

Little Pluto, however, does have a close relation with Neptune. Long-term orbital calculations show that its average period (that found by smoothing out the various perturbations) is $^3/_2$ that of its large cousin. Pluto seems to be trapped in a gravitational *resonance* with Neptune. Resonances occur when two bodies have orbital periods that are simple multiples of each other, the forces allowing either entrapment or large perturbations depending on the situation. As a result, Pluto is always near aphelion when in conjunction with Neptune (see Figures 11.2 and 11.3), ensuring that even if Pluto's inclination changes as a result of other perturbations, the two planets still cannot collide.

Improved knowledge of the planetary masses has solved Uranus's problem. As we will see in the next chapter, there may be hosts of small bodies beyond the orbit of Pluto, but there is no evidence of any planetary-sized body. We seem to have reached the end of the planetary system.

11.14 The final step

One of the perturbations of any planetary orbit is the rotation of the line of apsides, the major axis of the elliptical orbit that connects aphelion and perihelion. The perihelion of

Mercury advances steadily, that is, ω and ω' increase over the years. Since the Mercurian orbit is so elliptical, the rate of motion is quite large and accurately observable, about 570 seconds of arc per century, or nearly 6 seconds of arc per year. However, about 1870, Urbain Leverrier, the co-discoverer of Neptune, calculated with Newtonian gravitational mechanics that the advance should be roughly 40 seconds of arc per century less. With the greatest care in computing the planetary perturbative forces, the shortfall was eventually placed at 42″.8 per century. Several speculations were put forward to explain the discrepancy. Perhaps a cloud of meteoric particles surrounded the Sun and affect Mercury's orbit. Maybe the theory of gravity is incorrect: instead of a simple inverse square law, perhaps the exponent is actually 2.000 000 15, which satisfied the motion but made no physical sense.

The correct explanation was found by Albert Einstein, developer of the theory of mechanics called *relativity*. The subject was briefly introduced in Section 6.11, where it is involved with the definitions of precise time, since time is distorted because the Earth is in motion.

The need for relativity theory arises because of the finite velocity of light and its independence of the velocity of the observer. Newtonian mechanics is truly valid only at zero speed; the closer the motion of a moving body approaches the speed of light, the more the Newtonian predictions are in error. Relativity provides an accurate view of the world at all speeds, although at the velocities encountered in daily life, so far below that of light, Newton's laws are an excellent approximation, essentially indistinguishable from Einstein's.

There are two aspects to relativity. *Special relativity*, developed in a scientific paper that appeared in 1905, involves uniform relative motion and distortions of space and time that develop as a body approaches the speed of light. For example, a clock held by a friend moving past you would appear to tick more slowly than it would to the person holding it. *General relativity*, born in 1914, includes accelerations. It is really a theory of gravity that predicts that space (or rather, *spacetime*, where the three dimensions of space are conjoined with the one dimension of time into a four-dimensional structure) is distorted by mass, the distortion felt as gravity.

Mercury is close to the Sun and is the fastest-moving planet. Although its orbital speed is far below that of light, its elliptical path continually moves it through different curvatures of spacetime. The result is the missing 43 seconds of arc per century. Mercury thus provides a primary proof for the validity of the theory.

11.15 Chaos

Hold a sharp pencil and balance it as best you can on its point. Let it go and note the direction in which it falls. Every time you try the experiment, the pencil will fall a different way that is impossible to predict. The direction of fall is exquisitely sensitive to initial conditions, the exact angle at which the pencil is held. Many other physical systems behave in the same manner. A small disturbance will quickly be greatly magnified, resulting in a completely different result.

In the eighteenth and nineteenth centuries, scientists tended toward a deterministic philosophy of the Universe: given the positions of all the atoms and their velocities, it should be possible to predict the state of the Universe for all time in all places. The inclusion of relativity does not change that idea. However, for many physical systems it is *not* possible to know the requisite initial conditions; it is then not possible to make predictions even into the short-term future. These concepts are part of the growing mathematical science of *chaos*. A chaotic system is one that is inherently unpredictable. The weather affords a prime natural example. Even if we have perfect models for the behavior of the atmosphere, we cannot ever specify the input data – temperatures, wind velocities, etc. – with sufficient precision. Even if we had detectors every cubic meter of atmosphere, something in between them would happen that would grow to destroy the prediction.

Pluto may provide an astronomical version of chaotic behavior. Though apparently locked into a gravitational resonance with Neptune, its motion is very sensitive to the exact gravitational pulls of the other planets, and therefore to their exact masses and distances. Sample calculations show that if you could shift Mars by one millimeter, in 80 million years Pluto would shift from its previously predicted position by over 10 000 km. The error in Mars's positioning is vastly greater than 1 mm and the Jovian Planets have a far greater influence. The orbits of the other planets are almost certainly chaotic too, but on unknown time scales. There are, therefore, apparently limitations on what we can do, limitations that nevertheless add even more intrigue to the Solar System, to the sky, and to nature itself.

11.16 Spaceflight

On October 4, 1957, the world changed forever with the launching of the first artificial satellite. Artificial moons are now so common that we barely give them thought any-more. They are used to scan weather patterns, to locate minerals, spy on those we mistrust, and bring us telephone calls and television pictures. We have broken them free of the Earth, sending spacecraft to explore the Moon and other planets, and have even launched four of them entirely out of the Solar System into interstellar space. You can often can see a high-flying satellite drifting along in the early night when it can still reflect some sunlight downward; you watch for a while and then see it wink out as it hits the Earth's shadow.

The principles that govern spacecraft are the same that rule the orbits of the planets, that is, Kepler's laws in their generalized form (see Section 11.11). We can easily scale the orbit of an artificial Earth satellite to that of our Moon. For simplicity, assume circular orbits. Also at first assume that the Moon and the satellite are massless (the Moon has a mass only about 1% that of the Earth). Therefore $P_{sat}{}^2/P_{Moon}{}^2 = a_{sat}{}^3/a_{Moon}{}^3$. The Moon's sidereal period is $27.3 \times 24 \times 60 = 3.93 \times 10^4$ minutes. Therefore, $P_{sat}{}^2 = 1.55 \times 10^9\ (a_{sat}/a_{Moon})^3$. To allow for the actual mass of the Moon, multiply the result by 1.01. The semimajor axes always refer to the Earth's center, so to find the semimajor axis of a satellite with a specific altitude above the Earth's surface, add the terrestrial radius of 6370 kilometers. The minimum period (at zero kilometers altitude) is 85 minutes; the minimum realistic period, high enough to get above most of the atmos-phere (say 150 kilometers) is 88 minutes. At this height, the satellite must travel at a speed of 7.9 km/s, fast indeed. It is not surprising that spaceflight is so difficult and expensive, as it requires an immense force to accelerate a package to the huge velocity required to sustain an orbit. The major United States launch facility, Cape Canaveral, is located on the east coast to take advantage of the 0.3 km/s terrestrial rotation speed: even that little extra kick saves substantial fuel.

As we proceed outward, the period must increase. There is a special position 35 800 kilometers (5.6 Earth radii) above the Earth's equator for which the period is exactly 24 hours. Such a *geosynchronous* (or *co-rotational*) satellite revolves in harmony with the Earth's rotation, and for a terrestrial viewer hangs motionless in the sky. This location is ideal for communications satellites that can then be accessed by fixed ground-based antennas. The satellite dishes that point upward to such visually undetectable orbiters have become quite familiar.

We watch satellite-transmitted television signals with little thought of where the signal may be coming from. But twice a year, the Sun, which produces a great amount of interfering radio radiation lets us know. The geosynchronous craft are above the Earth's equator, but are so close that parallax is significant. From the northern hemisphere, they will appear south of the celestial equator; for New York, latitude 40° N, δ(satellite) = −6°.3. On the dates when the Sun has this declination, March 4 and October 11, it will interrupt communication to the city when the transmitter is in solar conjunction, at an hour angle that depends on the city's and satellites' relative longitudes. These dates will move closer to the equinoxes for viewers closer to the equator. For a southern-hemisphere observer the interruptions will occur symmetrically on the other sides of the equinoxes.

The launch of a spacecraft to rendezvous with another body such as the Moon or a planet is an exercise in orbital mechanics. We do not simply fire a craft directly at the Moon, but must place it in an elliptical (or hyperbolic) orbit about the Earth such that it and the Moon arrive at the same place at the same time, of course allowing for the simultaneous gravitational influence of the Moon.

The problem of a journey to another planet is similar. Given a limited amount of rocket power, we launch our craft into a minimum-energy orbit: an elliptical path with the Sun at one focus, a track that is tangential to both the Earth's and planet's orbits (Figure 11.20) in which we ignore the gravitational effects of the planets themselves. The semimajor axis of the path is half the sum of the semimajor axes of the planet's and the Earth's orbits. The travel time is half the period, which is found from $P^2 = a^3$. It therefore takes 0.72 years to go to Mars, 2.7 years to Jupiter, 6 years to Saturn, and 37 years to Pluto. For a trip to Mars, which has a sidereal period of 1.88 years, the craft must be launched with Mars $({}^{0.72}/_{1.88}) \times 360 = 135°$ behind its rendezvous point. Any attempt to shorten the trip requires greater rocket power and more fuel, with a concomitant reduction in the precious payload.

The successful flybys of the major planets have required great ingenuity and skill in calculation and in the use of high-speed computers. All the while the craft is approaching its target it is subject to constantly changing forces produced by all the other planets. We must continually update the path, which requires several mid-course rocket firings. The problem is complicated by the substantial travel time of light. To effect a

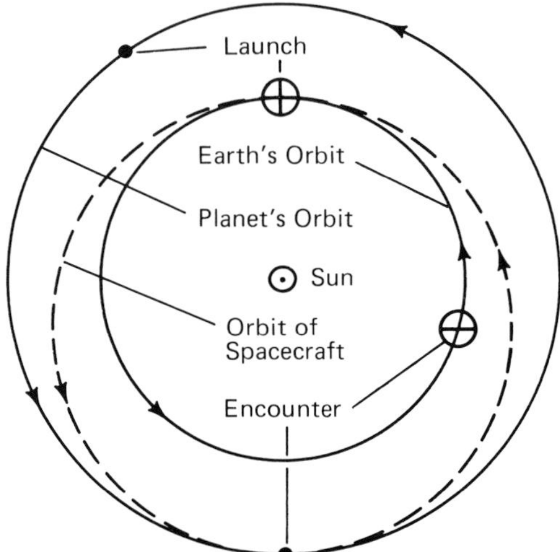

11.20. A minimum-energy orbit on a trip to Mars. Mars' orbit is circularized for simplicity. The spacecraft is flown into orbit around the Sun when the Earth and Mars are at the positions indicated by "Launch." The spacecraft leaves Earth on an elliptical path whose aphelion touches Mars' orbit. The craft and Mars then arrive at the same place at the same time.

maneuver at a specific moment while the craft is closing in on Saturn, we must send the command signal 1½ hours in advance; and we will not know of a successful result until we get a return signal 1½ hours after that!

Yet with all the complications, we have shot our spaceships literally through keyholes in space, maneuvering close to Jupiter's satellites, and taking close-up photographs of Saturn's rings. For example, Mercury and Venus were visited by US *Mariner 10*, Venus by USSR *Veneras 8* through *16* (*9* and *10* landed), US *Pioneers*, and *Magellan*; Mars was attended by US *Mariners 6* and *9* (the latter orbited), the *Viking 1* and *2* landers, and USSR *Mars 2*. The outer planets have been examined by US *Pioneers 10* and *11* and *Voyagers 1* and *2*. And this list is nowhere near complete. We will look at several results of these explorations in the next section.

The spacecraft carry a variety of scientific detectors in addition to the familiar imagers (which are similar to video cameras), for example radiation and particle detectors and magnetometers. *Pioneer Venus* and *Magellan* carried radars to map the Venusian surface, and the *Viking* landers had automated laboratories for soil analysis as well as the necessary shovels with which to dig. The champion of all of them is *Voyager 2* (Figure 11.21), which took a magnificent four-planet tour that began in 1977. The craft visited Jupiter in 1979; the gravity of that planet was used to speed it up and direct it toward Saturn (as it did for *Voyager 1*), which it passed in 1981. Another gravity assist shot it to Uranus for a 1986 flyby, and finally Uranus' gravity hurled it to Neptune in 1989. Along with *Voyager 1* and *Pioneers 10* and *11*, it has been hurled out of the planetary system, never to return. The *Voyagers* will continue to send back inform-

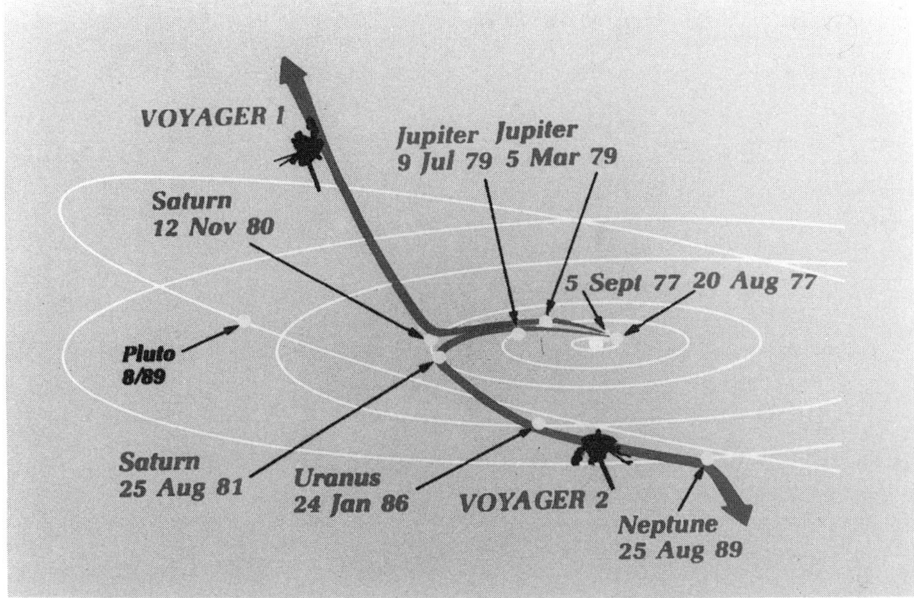

(a)

(b)

11.21. The *Voyager* spacecraft. (a) Orbits and encounter times; *Voyager 1* was sent north of the ecliptic plane by Saturn and never visited Uranus or Neptune. (b) The craft itself. The large dish is the communications antenna. Above it are the imaging devices, and below it is a small power reactor. The long rods are particle and field detectors. (NASA/JPL.)

Table 11.5. *Physical aspects of the planets*

Planet	Angular diameter (seconds)	Diameter[a] (km)	M_{Earth}	Mass M_{Earth}	Rotation period[b]	Magnetic field Mag_{Earth}	Gravity g_{Earth}	Satellites[c]
Mercury	11–5	4 900	0.38	0.055	59^d	0.01	0.38	0
Venus	60–10	12 100	0.95	0.81	243^dr	none	0.90	0
Earth	—	12 700	1.00	1	$23^h\ 56^m$	1	1	1
Moon	32.9′–29.4′[d]	3 500	0.27	0.012	$27^d.3$	fossil	0.17	0
Mars	18[e]–4	6 800	0.53	0.11	$24^h\ 36^m$	none	0.38	2
Jupiter	47–32	138 400	11.2	318	$9^h\ 55^m.5$	14	2.4	16
Saturn	19–16[f]	114 800	9.45	95.3	$10^h\ 40^m.5$	0.71	0.93	18
Uranus	4	50 500	4.01	14.5	$17^h\ 14^m$ r	0.74	0.79	15
Neptune	2	49 100	3.88	17.1	$16^h\ 07^m$	0.43	1.12	8
Pluto	0.067–0.110	2 400	0.19	0.0022	$6^d.39$ r	?	0.04	1

[a]Mean of equatorial and polar diameters.
[b]r denotes retrograde (backward) rotation.
[c]Known satellites; the Jovian planets probably have many more small ones exclusive of ring particles.
[d]Angular diameter of the Moon is in minutes of arc.
[e]Maximum is 25″ at favorable opposition.
[f]44″–36″ for the ring system.

ation on conditions outside the planetary system until well into the twenty-first century.

Spacecraft of the future will surely carry humans aboard, as those of the past did to the Moon. Our Solar System is slowly becoming familiar as we visit its members with ever greater care and depth, making it more and more our home.

11.17 Telescopic views: physical natures of the planets

This book primarily wraps itself around the naked-eye view of the sky and the movements of celestial bodies. But the planets are so fascinating telescopically that we should at least take a brief look at their physical characteristics, some of which are summarized in Table 11.5.

What can we see when we apply some telescopic (and space probe!) power to these worlds? The most distinctive aspect of the inferior planets is their phases, which we examined in Section 10.7. Otherwise, little can be seen from Earth (Figures 11.9 and 11.22 inset). Space probes have shown that Mercury is covered with craters very much like our Moon (Figure 11.22), but none of these can be seen from Earth. All we can detect from here are subtle changes in shading. Because the view is so poor, we did

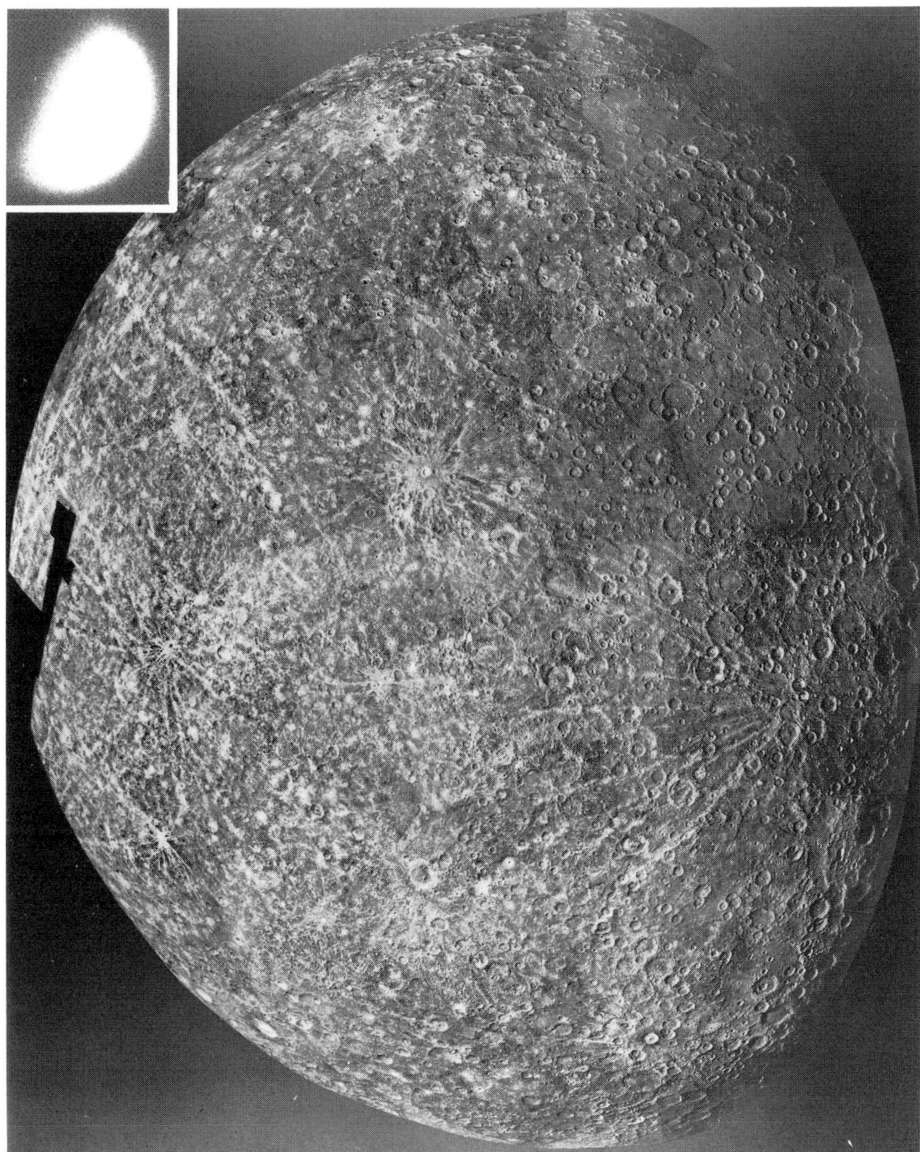

11.22. Mercury. The *Mariner 10* flyby showed a heavily cratered surface reminiscent of the Moon. The view from Earth, seen in the inset at upper left, is exceedingly poor. (NASA.)

11.23. Venus. Clouds hide the surface in this *Mariner 10* flyby image; the dark areas are simply lower clouds. The Y-shaped structure, created by winds, is always present and can just barely be seen from Earth. (NASA.)

not even know the rotation period of the planet until it was measured by radar in 1965. It is tidally locked to the Sun, as the Moon is to Earth, except that the rotation period is $^2/_3$ that of the orbital because of the high eccentricity. The nearly airless surface is hostile, with ground temperatures reaching nearly to 430 °C at noon and plunging to almost −200 °C at night. There is evidence for ice caps at the poles, where craters can permanently shield the ground from sunlight.

Venus is no more benign. Its lovely aspect from Earth belies its true nature. In contrast to Mercury, the planet has a remarkably thick atmosphere, 100 times denser than Earth's, composed almost entirely of carbon dioxide (Figure 11.23). The planet is covered perpetually by clouds so thick that the surface rocks cannot be seen from the outside, clouds that give the planet its creamy color and its great brilliance. The clouds are noxious, however, composed in part of sulfuric acid. The atmospheric carbon dioxide produces a greenhouse effect that lets in sunlight and traps re-radiated infrared radiation, resulting in an enormously high surface temperature of 470 °C.

Radar transmitted from Earth and orbiting spacecraft (particularly the enormously successful *Magellan* mission) has shown us the surface (as has a pair of Soviet landers) and given us the planet's rotational characteristics. It spins very slowly in the backward (clockwise) direction, taking 247 Earth days to go around. Venus is dominated by low rolling plains onto which are set a few relatively high volcanic rises (Figure 11.24a). It does not have deep ocean basins and continents like the Earth and only relatively small chains of mountains. Set into the plains are hundreds of volcanic features, including several great shield volcanoes (Figure 11.24b) that are reminiscent of the Hawaiian Islands and that are produced by hot rising columns of internal rock, and low volcanic rises unlike anything on Earth. Impact craters abound (Figure 11.24c). From the number per unit area and the known rate of current cratering on the Moon, we estimate that the Venusian surface is only about half a billion years old, suggesting that the planetary crust periodically turns over volcanically, thereby renewing itself.

Mars (Figure 11.25a) has always been the most telescopically fascinating of the planets, since its surface can be directly and easily examined through the thin atmosphere (1% the density of ours, and only 1/10 000 that of Venus). It is intriguingly Earth-like, and for decades was suspected of harboring life. From here we can see permanent large dusky markings that allow measurement of a rotation period of just over 24 hours. We also see polar caps that wax and wane with seasons that are rather like ours, since the obliquity of the planet's ecliptic is 24°. Equatorial noontime temperatures can almost be comfortable in human terms, though the air is far too thin to support animal life.

The most historically interesting features are the so-called "canals," a network of fine lines first noted in 1877 (Figure 11.25b). Some astronomers and much of the general public believed them to be artificial, true canals designed by an intelligent race to bring water from the poles into the arid temperate and tropical deserts. High-resolution observation from Earth, direct views by passing spacecraft, and various orbiters ended any speculation. Mars appears truly lifeless: the canals were an illusion caused by the eye stringing subtle features into lines; the air is made principally of carbon dioxide, and so largely are the polar caps (dry ice underlain with water ice); the seasonal changes are caused not by vegetation but by wind patterns shifting the sandy soil (winds that can envelope the planet in fearsome dust storms).

The spacecraft revealed the real wonders of the planet. The southern hemisphere is covered with ancient impact craters. The northern was geologically active. There we see great volcanic rises topped by massive volcanoes extending as high as 35 km above the plains. The largest, Olympus Mons (Figure 11.26a), set at the edge of the largest rise, the 3000-km-wide Tharsis bulge, is 400 km across. Radiating to the east of Tharsis is the 3000-km-long Valles Marineris (Figure 11.26b), a fault or crack in the Martian crust caused by the uplifting of Tharsis. Perhaps the most surprising discovery is that

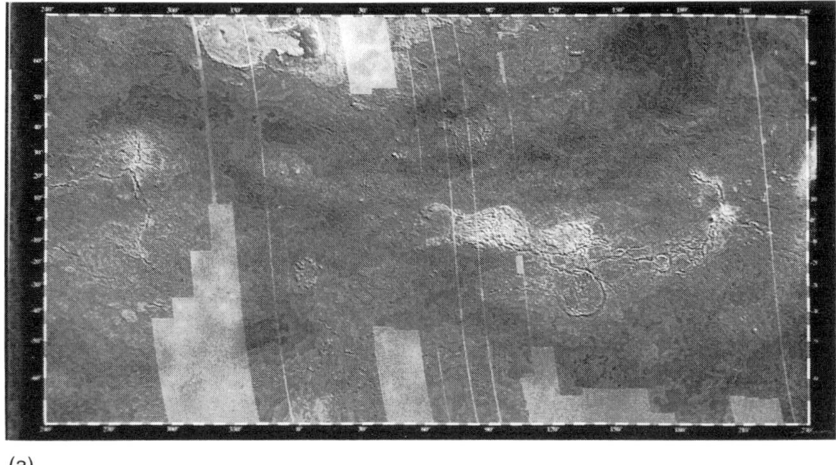
(a)

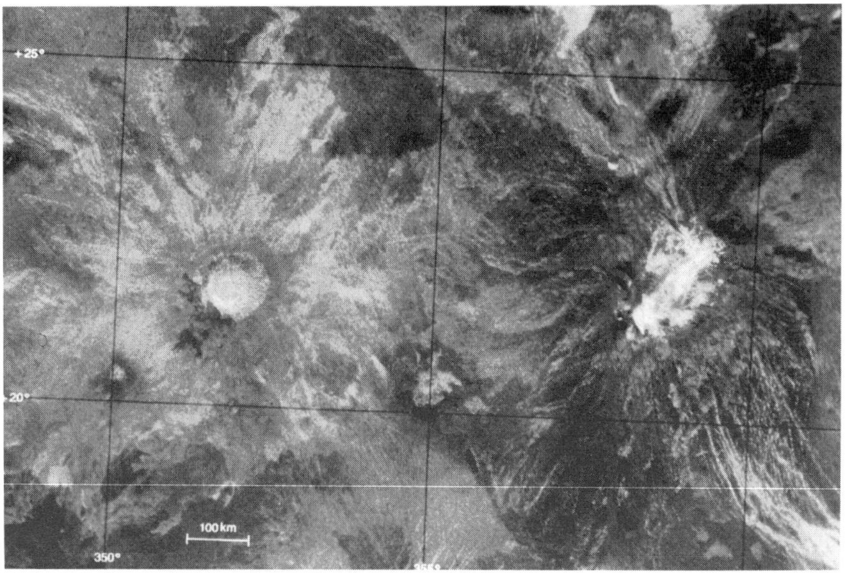

(b)

of stream and river beds, now dry, in which water once flowed (Figure 11.26c), in some cases in vast amounts, so much that Mars is thought once to have had large oceans, which in turn suggest a once thick atmosphere. Yet *Viking* found a bleak landscape (Figure 11.26d) and no life.

From Mars, we take the great leap to the Jovian planets. Jupiter and Saturn are giants, with diameters respectively 11.2 and 9.5 times the terrestrial diameter and masses 318 and 95 times greater than Earth's. They are made primarily of helium

(c)

11.24. The surface of Venus. (a) A radar image made by *Magellan* shows a northerly rise at the top, Ishtar Terra, a larger equatorial rise to the right of center, Aphrodite Terra, and several others set on low rolling volcanic plains. (b) Magellan radar gives us an overhead view of the volcanic plains to show two great volcanoes, Sif and Gula Mons, about 1000 km apart. (c) We see three impact craters set into the plains that seem nowhere more than half a billion years old. (NASA/JPL.)

and hydrogen that surround high-temperature rocky cores and in that sense are more comparable to the Sun than to Earth. The telescope immediately shows us Jupiter's parallel cloud bands (Figure 11.27a), made of ammonia droplets and a variety of other chemicals, and the noticeable flattening caused by its rapid (10 hour) rotation. The *Great Red Spot* – a spinning storm alive for over 300 years – is often quite visible, and is an extraordinary sight up close (Figure 11.27b). The largest four of its 16 known satellites (the Galileans, so named in honor of their discoverer: Io, Europa, Ganymede, and Callisto, to be discussed in Chapter 12) accompany the planet in an ever-changing array. These bodies, in a very real sense Jupiter's family of planets, are easily visible in steadily held binoculars.

Saturn (see Figure 5.1) has its cloud belts too, and several viewable satellites, one of which (Titan) is easily viewed in only a small telescope. But its real claim to attention is its magnificent ring system (Figure 11.28). Visible even with very low power, the rings stretch out well over two planetary diameters. From Earth, we see three distinct rings, with the outer two separated by a pronounced dark gap (the Cassini division); the *Voyager* spacecraft showed thousands of individual ringlets that blend into one when viewed at a great distance. Twice during Saturn's journey around the Sun it will

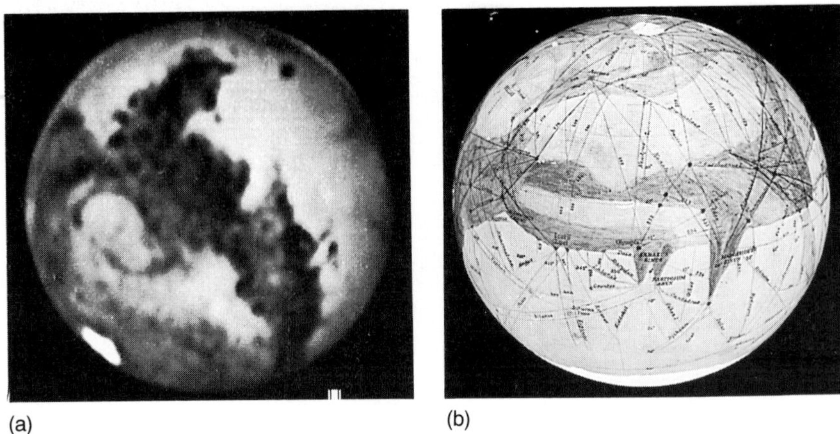

(a) (b)

11.25. Views of Mars from Earth. (a) A superb image taken during the 1988 favorable oppositions shows numerous dark markings and the south polar cap. (b) Lowell's drawn globe of Mars shows a network of "canals" that are not really there. The views are nearly the same: try to match the features. ((a) Pic du Midi Observatory; (b) Lowell Observatory.)

(a)

11.26. Mars up close. (a) Olympus Mons, 20 kilometers high, the greatest volcano of the Solar System; (b) Valles Marineris, a geologic crustal crack; (c) dry stream beds, where water once flowed, in a field of meteor craters; (d) a view from the surface taken by *Viking 1*. (NASA/JPL.)

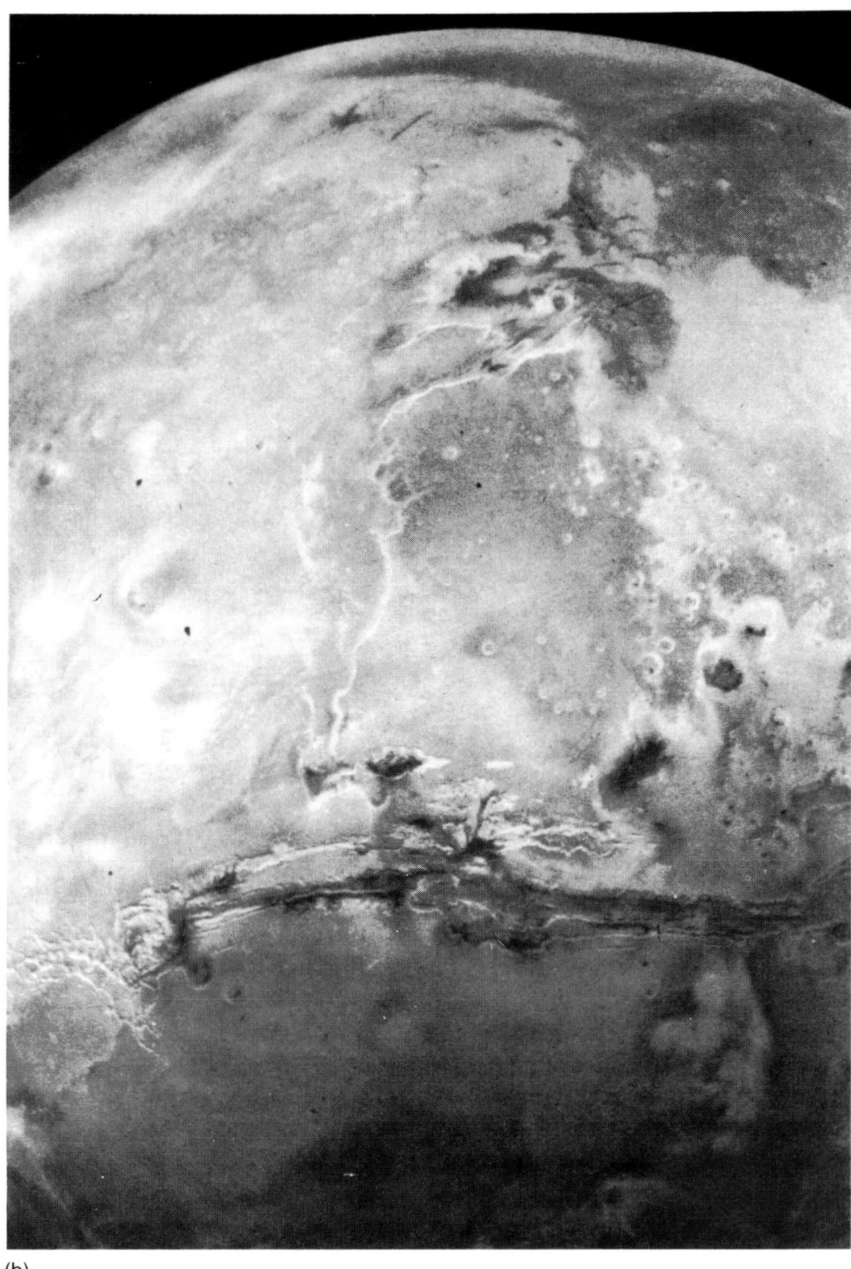

(b)

11.26 cont.

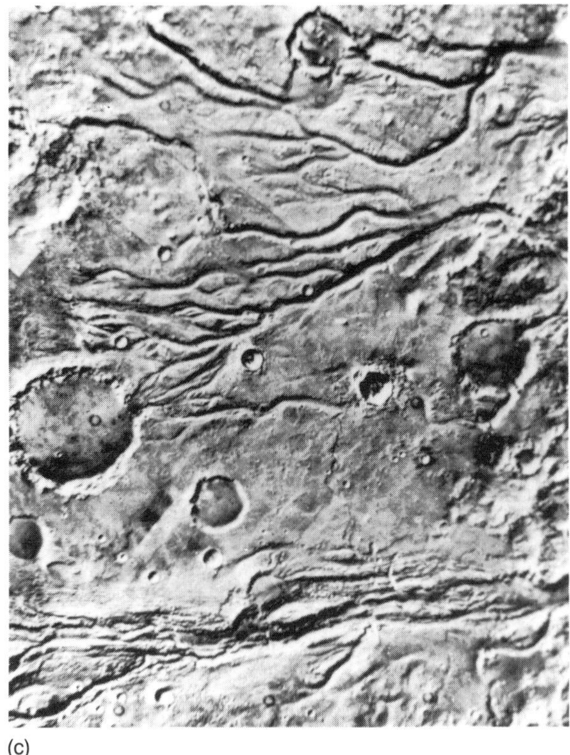

(c)

(d)

11.26 cont.

(a)

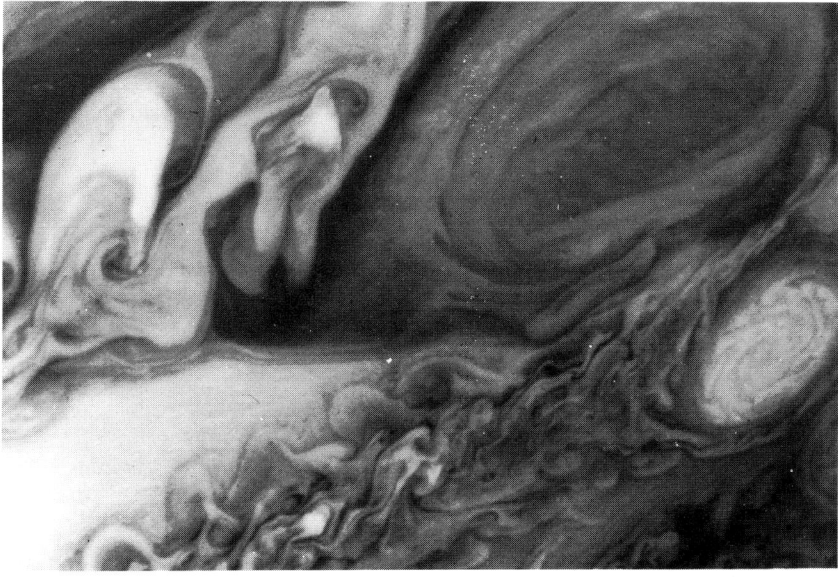

(b)

11.27. Jupiter. (a) A view from Earth shows the cloud belts; (b) a close-up view from space reveals the intricate details in the clouds surrounding the Great Red Spot. ((a) National Optical Astronomy Observatories; (b) NASA/JPL.)

(a)

(b)

11.28. Saturn. (a) The Cassini division in the rings is easily noticeable, as are the cloud belts in this global *Voyager* image. The view from Earth is similar except that the detail is somewhat blurred by our atmosphere. The cloud belts are fainter than Jupiter's, and several satellites hover below. (b) Details of the rings by *Voyager 2* showing hundreds of ringlets and spoke-like features apparently caused by the action of Saturn's magnetic field on dust particles. (NASA/JPL.)

present its ring plane (which is also the plane of its equator) to the Earth, and the rings disappear from view (see Figure 5.1)! They are no more than a few tens of meters thick and are composed of countless fragments of ice coated rocks, each in its own orbit.

The Cassini division is caused by a resonance. Any particle within the narrow band has a period about the planet half that of Saturn's satellite Mimas. The particle and Mimas therefore continually come into the same configuration. Instead of being trapped, however (as Pluto is by Neptune), gravitational perturbations build to the point where particles are ejected from the zone, rendering it relatively clean.

Farther outward, Uranus and Neptune are different from their larger Jovian cousins. Their diameters and masses are respectively about 4 and 15 times the values for Earth. They have a much lower percentage of hydrogen and helium and relatively larger rocky cores. They really form their own class of planet and are crudely intermediate in construction between the two great Jovians, Jupiter and Saturn, and the terrestrial planets.

Our view of these planets from Earth is poor. Uranus (Figure 11.29a) is only 2 seconds of arc in diameter, so that few features would be visible on its gaseous, cloudy surface. The planet rotates in about 16 hours, but the axis is tipped nearly on its side, to an angle of 98° relative to its ecliptic: technically, like Venus, it rotates backward. Even its system of satellites (five intermediate sized and several smaller ones), which orbit equatorially, is tipped over, suggesting a violent collision during or near the time of formation. *Voyager 2* found a nearly featureless cloud cover (Figure 11.29b). Uranus also displays a set of rings (Figure 11.29c) that were discovered from Earth during the occultation (see Section 9.10) of a star by the planet: the star was seen to blink momentarily as each invisible ring passed over it. They are the reverse of Saturn's: instead of broad rings with narrow gaps, we see several narrow rings separated by broad gaps. These too are tilted with the Uranian equator.

Neptune (Figure 11.30), also cloud covered, again shows us little from Earth, except for two satellites and a smudgy cloud feature. The *Voyager 2* view, however (Figure 11.30 insets), revealed the spectacular *Great Dark Spot* (which, from *Hubble Space Telescope* observations, has since disappeared), a storm feature similar to Jupiter's Great Red Spot, and irregular rings that had already been detected by occultation experiments. Ring systems, once thought to be the province of Saturn alone, are now found to be common. Not only do Uranus and Neptune have rings, but even Jupiter has one, albeit thin and nearly invisible.

We have actually learned a surprising amount about tiny, distant Pluto (Figures 11.19 and 11.31a and b) because of its satellite Charon, which was discovered in 1978. Charon and Pluto are synchronously locked with a period of about 6 days, and the orbit and planetary rotation are retrograde (tilted at 122°), the third such planet in the Solar System. Between 1985 and 1991, the two mutually eclipsed each other (Figure 11.31c), allowing the determination of an accurate orbit, radii, and masses. Pluto's mass is only 1/400 that of Earth, 20% that of the Moon. The little planet is not at all like the Jovians but more like the terrestrial planets except that it appears to be about 40% water ice. There is a thin carbon monoxide–methane–nitrogen atmosphere (a hundred-thousandth the pressure of Earth's) that at aphelion freezes and falls to the surface as

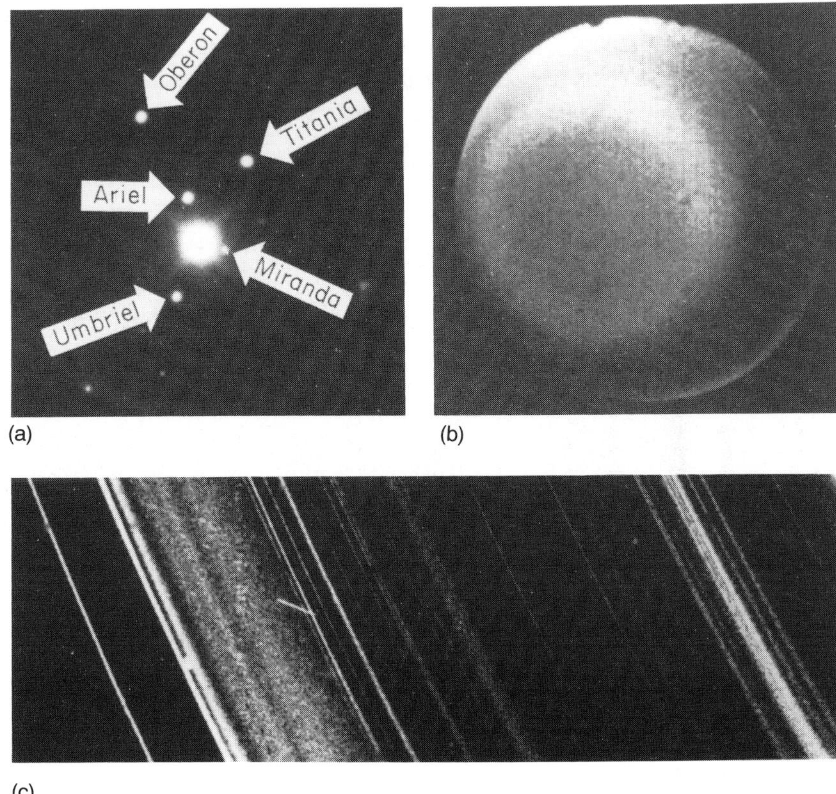

11.29. Uranus. (a) The view from Earth shows us nothing but the satellites. The faint spikes are an artifact of the telescope. (b) The view of the planet by *Voyager 2* exhibits little more than a featureless cloud deck; great contrast enhancement shows some banding concentric with the pole (which nearly faces the observer) and a bright cloud to the upper right. (c) The rings are narrow, with wide gaps between. ((a) National Optical Astronomy Observatories; (b) and (c) NASA/JPL.)

a light snow. Charon has a mass some 10% that of Pluto. The system is more of a double planet than even the Earth–Moon pair. Pluto falls into its own class. As we will see from the next chapter, it has more the characteristics of the debris of the Solar System than it does an actual planet; our own Moon is better qualified.

11.18 Astrology

The planets, their positions and their motions, have long been used in systems of fortunetelling collectively called astrology. The subject is ancient: since the gods created and placed the stars, these celestial lights therefore must indicate the gods' dispositions and plans. Astrology grew in parallel with astronomy, and is to this science what alchemy is to chemistry and phrenology (the "reading" of bumps on the skull) is to

11.30. Neptune. The main photo shows the view from Earth with its satellites Triton immediately next to it and Nereid to upper right. *Voyager 2* images are seen in the insets, which show clouds, the Great Dark Spot, and thin rings. (Main photo: Yerkes Observatory; insets: NASA/JPL.)

psychiatry. Of all forms of magic and mysticism it is by far the most widely believed, as evidenced by columns in so many otherwise reputable newspapers.

Various societies have their own versions of this art of soothsaying. Here we look only at the Western form, that consistent with our conventional constellations and planetary names. In the simplistic "Sun-sign" astrology, our personalities, appearances, and destinies are controlled by the position of the Sun at the time of our birth: those of us born under the sign of Capricorn are stubborn, those under Leo powerful and generous. This simplistic astrology, however, is only a distillation of a vastly more complex "art" of forecasting. In "real" astrology, our lives and fortunes are influenced by the locations of the Sun, Moon, and planets within the signs of the zodiac, by their positions relative to the horizon and celestial meridian, and by their angles relative to one another.

The astrological *signs* of the zodiac took their names and qualities from the constellations in which the Sun was found some 2500 years ago. The majority of astrologers, the *tropical astrologers*, ignore precession (see Section 5.5). The sign of Aries still begins at the vernal equinox and overlays the astronomical figure of Pisces. When you were

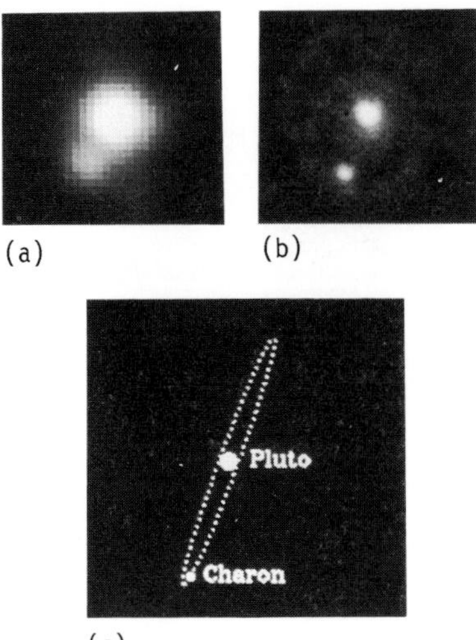

11.31. Pluto. (a) From the ground, Charon is barely resolvable from Pluto. (b) The *Hubble Space Telescope*, however, separates the two nicely. (c) During the latter part of the twentieth century, the satellite's orbit was presented edge-on, resulting in mutual eclipses and transits between. ((a) United States Naval Observatory; (b) Space Telescope Science Institute.)

born, the Sun was actually roughly one constellation to the west of that corresponding to your sign. Although the stars are supposed to guide our lives, they apparently change their meanings every two or so millennia as the equinox shifts to a new constellation. The *sidereal astrologers* take the shift into account, but seem to be largely scorned by the others.

From the locations of celestial bodies the astrologer establishes a *horoscope*, which can come in a variety of forms: *natal* for one's birth (the individual being called the "native"), *horary* for predicting the fate of the moment, *electional* for deciding when a move is auspicious, and others. The first is the most common. For that, the exact time (to the minute) and longitude of birth must be known. This information fixes the celestial sphere, the positions of the signs (not the constellations) relative to the observer, and the positions of the planets (including the Moon and Sun) in the signs. Each planet has specific characteristics associated with it: Jupiter is generous and jovial, Mercury quick, intelligent and changeable, and Mars war-like, from its bloody color. Venus and Jupiter are generally of good nature and are called the *benefics*; Mars and Saturn are contrary and represent the *malefics*; Uranus and Neptune incline to the latter.

The planetary influences are modified according to the natures of the containing signs, and each planet is the *ruler* of a given sign in which it has its most powerful effect. The signs are divided into several groupings with common, specifically defined characteristics. (The details are too many to describe completely; what is important is the sense of complexity.) These sets all start with Aries, and are the *polarities*

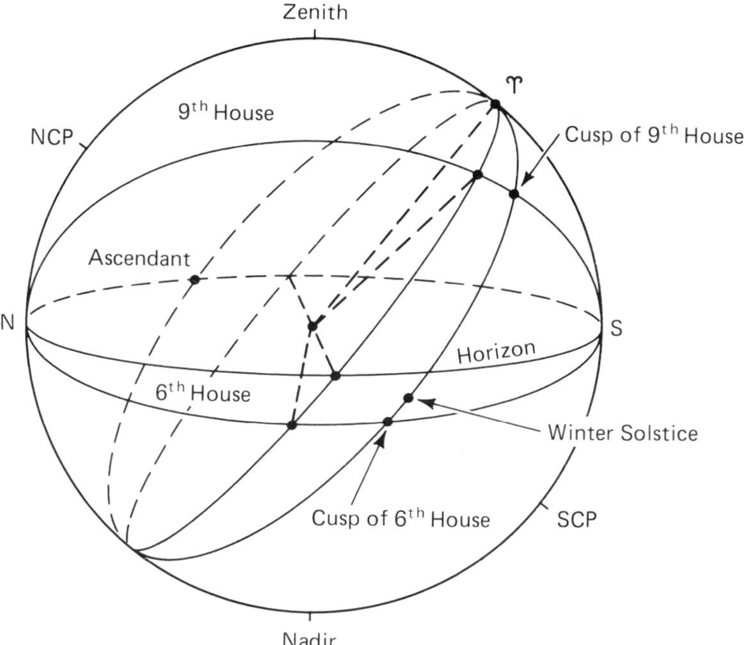

11.32. The houses of astrology. Those presented are calculated in the system of Regiomantanus (a fifteenth-century astronomer and astrologer), which is one of several. Twelve equidistant points are marked counterclockwise around the celestial equator beginning at the east point of the horizon. Twelve great circles are then drawn from north to south through each of them. The intersections of these great circles with the ecliptic (here shown with the vernal equinox at upper culmination) define the unequally spaced cusps. The first cusp is at the ascendant and the first house below the horizon. Only the sixth and ninth houses are illustrated here.

(alternating strong, weak, strong, etc.), the *qualities* (cardinal, fixed, mutable, cardinal, etc.), or *quadruplicities* (because there are four signs assigned to each), and the *elements* (fire, air, water, Earth, fire, etc.) or *triplicities* (because each has three assigned signs). Gemini, for example, is a strong, mutable water sign. According to the rules, Venus in Aries suggests attractiveness; in Taurus musicality; in Gemini kindliness.

Next, the influences of the planets are further modified by their *aspects*, or the angles between them (see Section 11.3 and Table 11.3). They are ranked by their power: opposition and conjunction have stronger effects than semi-square. Trine and sextile are supposed to have good effects, opposition and square, bad. Conjunction can go either way depending upon the situation; conjunction with the Sun, called *cumbust*, can destroy a planetary influence. These aspects cannot, of course, be exact. An astronomical conjunction occurs only at a specific instant; but astrologers consider an aspect to be in force as long as the planets are within certain defined limits of the actual value, called the *orb*, typically 3°. As examples of the effects, Venus in good aspect with Mars suggests that the individual's feelings are easily aroused; in bad aspect that there will

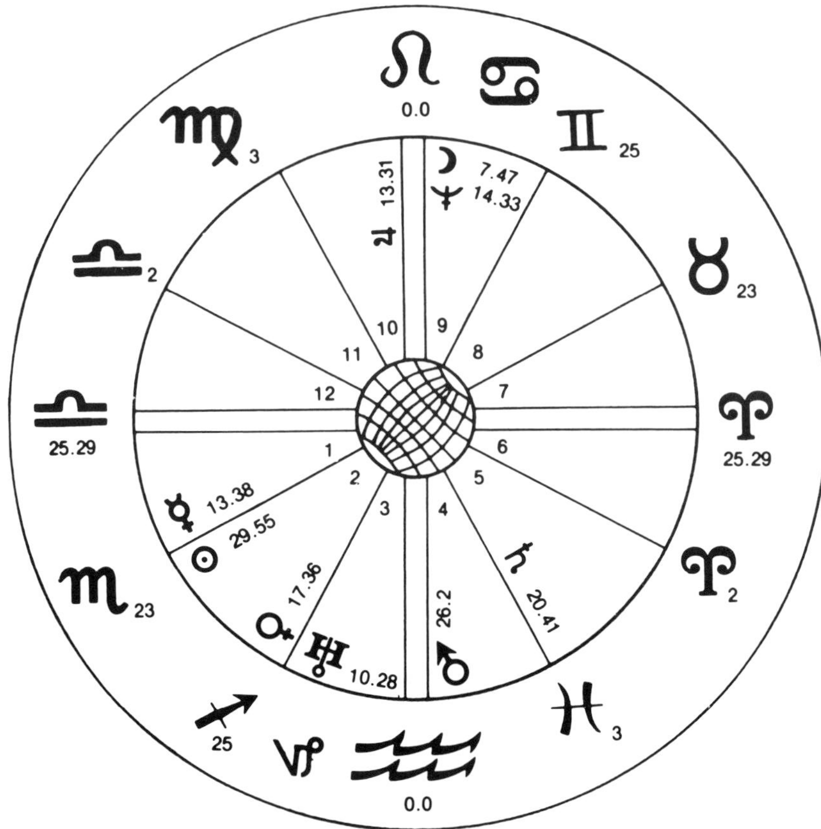

11.33. A natal horoscope. This figure is constructed for November 23, 1907, 09^h 00^m GMT for longitude 75°58′ W, latitude 40°43′ N. (From *Objections to Astrology*, B. Bok and L. Jerome, Prometheus Books, Buffalo, NY, 1975.)

be lack of peace in one's life. The number of combinations is very large and often contradictory, hence the "skill" required of the astrologer in deriving a general interpretation of all the influences.

Finally, planetary influences are channelled to the individual through the *houses* (Figure 11.32). The twelve houses are divisions of the ecliptic that are fixed not to the stars but to the observer: the constellations, planets, and signs rotate past and through them on their daily paths. They are numbered counterclockwise, the first having its beginning or *cusp* at the *ascendant*, the point at which the ecliptic rises from the eastern horizon. The fourth and tenth start at the lower and upper meridians, and their cusps are known as the *Imum Coeli* (I.C.) and *Medium Coeli* (M.C., or *mid-heaven*) respectively. There are a variety of ways of constructing the houses, which gives them different relative proportions on the ecliptic. Each house represents a certain aspect of life: the first the physical self, the second possessions, the twelfth sorrows. Each of the planets plays upon these qualities according to their natures, the signs they are in, and their aspects relative to other bodies. The first house is by far the most important, and great significance is attached to the sign of its cusp at the time of birth: we might hear of "a Gemini nature (a person whose Sun sign is Gemini) with Scorpio rising."

The entire picture is cast or erected into the horoscopic figure illustrated in Figure 11.33. The ascendant is to the left and the spokes mark the cusps. The zodiacal sign at each cusp is indicated around the circumference of the circle and the positions of the planets within both the signs and the houses are shown by their symbols. The unequal division of houses is quite obvious.

The interpretation of such constructions by a serious astrologer is enormously complex and time consuming, and it requires long study to acquire the necessary expertise. Unfortunately for these practitioners (and for the pocketbooks of the recipients), none of it works. When asked what forces could cause such influences, the astrologers fumble and refer to gravity, tides, and radiation. However, the gravitational forces produced on a newborn by the birth attendants (not to mention the mother) are far greater than those of any of the planets, and the planetary radiations (except for that of the Sun) are feeble indeed. The astrologers then resort to vague "cosmic vibrations" that remain undetected by scientists, and we again see the subject as mystical and pseudo-religious nonsense.

However, as strange and as illogical as it may seem, the system of astrology must still be dealt with in a manner consistent with other hypotheses. After all, as the astrologers might say, we are still a long way from truly understanding nature; ideas must not be dismissed so summarily. To counter this argument, a number of proper scientific experiments have been conducted to ascertain whether or not there are relations between personality and time of birth, and whether or not astrologers have the capability to predict what a person may be like on the basis of his or her horoscope. All the experiments are null, that is, the results are what would be expected if personality, skills, etc., and birth moment were entirely unrelated. It is a false system from both theoretical and experimental points of view, a vapor that vanishes in the light of investigation.

Chapter 12

The small bodies of the Solar System

We tend to think of the Solar System as consisting of the Sun, the planets, and little else. To be sure, the Sun dominates, containing 99.8% of the system's mass. The planets take about half the remainder. Of this 0.1%, Jupiter has an impressive 71%, the Earth a mere 0.22%, not much given its supreme importance to us. The other 0.1% consists of a vast number of other, smaller, bodies. They are of great significance to the Earth and to our understanding of how the Solar System was created.

12.1 Satellites

Imagine you are Galileo and have just completed your first primitive telescope. At night you turn it to the sky's brightest bodies and discover that the Moon, our satellite, now known to be revolving about us as a result of mutual gravity, has craters and mountains. You look at Jupiter and find it accompanied by four bright star-like objects that go around it. If Jupiter can make bodies orbit, so can the Sun: Copernicus was right!

The satellites of the Solar System (Table 12.1) are as diverse as the planets themselves. The first distinction to be made is between the Jovian and terrestrial planets. In general, the Jovians have regular systems with numerous satellites that were formed as part of planetary formation; the terrestrials do not. Mercury and Venus have no satellites that we have ever been able to see. The Earth has one, our Moon, described in detail in Chapter 9. It has a high mass relative to its parent planet and orbits not in its planet's equatorial plane as most satellites do, but in the *ecliptic* plane. It is now believed to have formed as a result of a collision between the primitive Earth and another large freely orbiting body. The Earth–Moon system thus constitutes something of a double planet; it is in fact probably fair to call the Moon the fifth terrestrial planet.

Mars has two satellites, but these are odd too. As large as our Moon is relative to its planet, the Martian satellites are small, mere rocks circulating above the arid planetary surface (Figure 12.1). Irregular blocks of rock measuring averaging 14 and 24 kilometers

Table 12.1. *The principal satellites of the planets.*
None of these, except for our own, can be seen without optical aid. The table below lists the names, semimajor axes (a) relative to the parent planets' diameters, their periods (P) in days, their radii (r) in kilometers, and where appropriate, some brief commentaries.

Name	Semimajor axis a (planetary radii)	P (days)	Diameter (km)
Earth			
Moon	30	27.3	3476
Mars			
Phobos	2.8	0.32	24
Deimos	6.9	1.26	14
Jupiter			
Amalthea	2.5	0.50	220
Io	5.9	1.77	3630
Europa	9.4	3.55	3138
Ganymede	15.0	7.16	5262
Callisto	26.3	16.69	4800
plus 11 other small bodies 10 to 200 kilometers in diameter that extend out as far as 165 Jovian diameters.			
Saturn			
Mimas	3.1	0.94	390
Enceladus	4.0	1.37	500
Tethys	4.9	1.89	1050
Dione	6.3	2.74	1120
Rhea	8.8	4.52	1530
Titan	20.3	15.95	5150
Hyperion	24.6	21.3	285
Iapetus	59.1	79.3	1440
Phoebe	215	550[a]	220
plus at least 9 other small bodies 15 to 200 kilometers in diameter.			
Uranus			
Miranda	5.1	1.41	470
Ariel	7.5	2.52	1160
Umbriel	10.4	4.14	1190
Titania	17.1	8.71	1580
Oberon	22.8	13.5	1520
plus 10 other small bodies 40 to 60 kilometers in diameter.			

Table 12.1. (*cont.*)

Neptune			
Triton	14.3	5.88[a]	2700
Nereid	223	360	340
plus 3 other bodies a few tens of kilometers in diameter.			
Pluto			
Charon	16.3	6.4	1180

[a]orbits retrograde

across, *Phobos* and *Deimos* ("fear" and "panic," appropriate attendants to the god of war) are exceedingly difficult to see from Earth, and were not found until 1877 (by Asaph Hall at the US Naval Observatory). They have so little mass that their weak gravity is insufficient even to force them into spherical shapes. Their surfaces are badly beaten and cratered, showing that they suffered through the heavy bombardment of the early Solar System (see Section 9.4).

The Martian moons have a history akin to the strange early ideas about the planet itself. Kepler predicted them in 1610 on numerological grounds in that Mercury and Venus were seen to have none, the Earth one, and Jupiter at the time seemed to have four. Jonathan Swift, in *Gulliver's Travels*, a 1726 political satire, later wrote that the inhabitants of his mythical island "Laputa" had discovered a pair of satellites revolving about the planet. Hall's discovery confirmed these guesses.

The Jovian satellite systems are very different. With the exception of the companions of Neptune, they are miniature Solar Systems with the planet acting as the moons' suns. The major satellites orbit in the planets' equatorial planes (just as the planets themselves orbit in the solar equatorial plane) and in the directions of the planetary spins.

The best system belongs to Jupiter (Figure 12.2), accompanied in its orbit around the Sun by the *Galilean satellites* Io, Europa, Ganymede, and Callisto. All but Europa are larger than our own Moon; Callisto is as big as Mercury, and Ganymede, the largest satellite in the Solar System, is 10% larger. All could hold their own as terrestrial planets. They are remarkably bright, typically fifth magnitude. Were it not for Jupiter's proximity and glare, they could be seen with the naked eye. As it is, they are readily visible in steadily held binoculars. In a small telescope the Galileans are endlessly entertaining as they orbit (with periods between about 2 days for Io and 2 weeks for Callisto), continuously changing their aspects. Sometimes they will be evenly split east and west (as in Figure 10.2), sometimes all four will lie in a line to one side. With care you can watch them disappear behind the planet, be eclipsed by Jupiter's shadow (which projects somewhat to the side of the planet when not at opposition), or transit across the surface.

The satellites are highly individualistic. Figure 12.2 also shows views taken by the *Voyagers*. The inner two, Io and Europa, have average densities (derived from

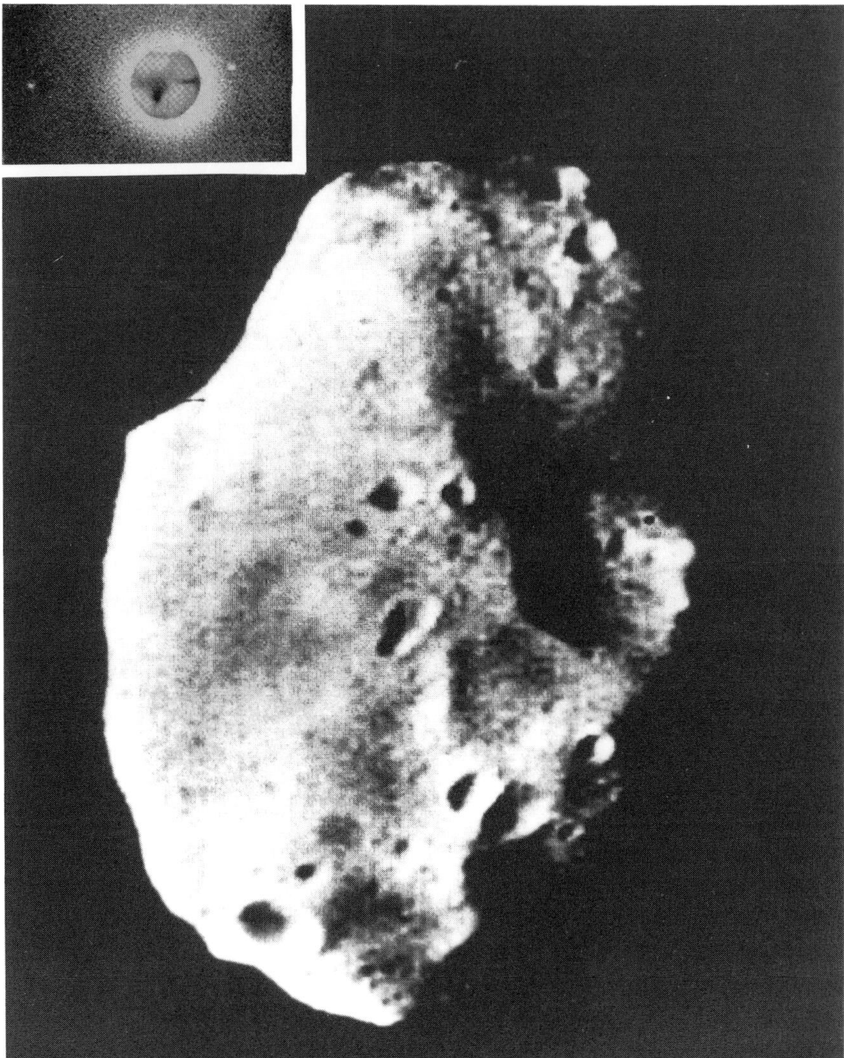

12.1. The Moons of Mars. The inset shows the view from Earth of Phobos and Deimos. A short exposure of Mars is set into the middle. A close-up of Phobos (imaged by *Viking* spacecraft) is shown in the large image. Note the cratering. The two satellites, averaging only 23 and 13 kilometers across, appear to be captured asteroids. (Inset: Lowell Observatory; main photo: NASA/JPL.)

their measured masses and radii) similar to the Moon and are made mostly of silicate rock. Ganymede and Callisto, however, have densities only about two-thirds the lunar value, and are likely to be half water ice surrounding silicate cores. Callisto and Ganymede are heavily cratered, though Ganymede shows signs of geologic activity. Europa has a bright icy surface and is the smoothest body known

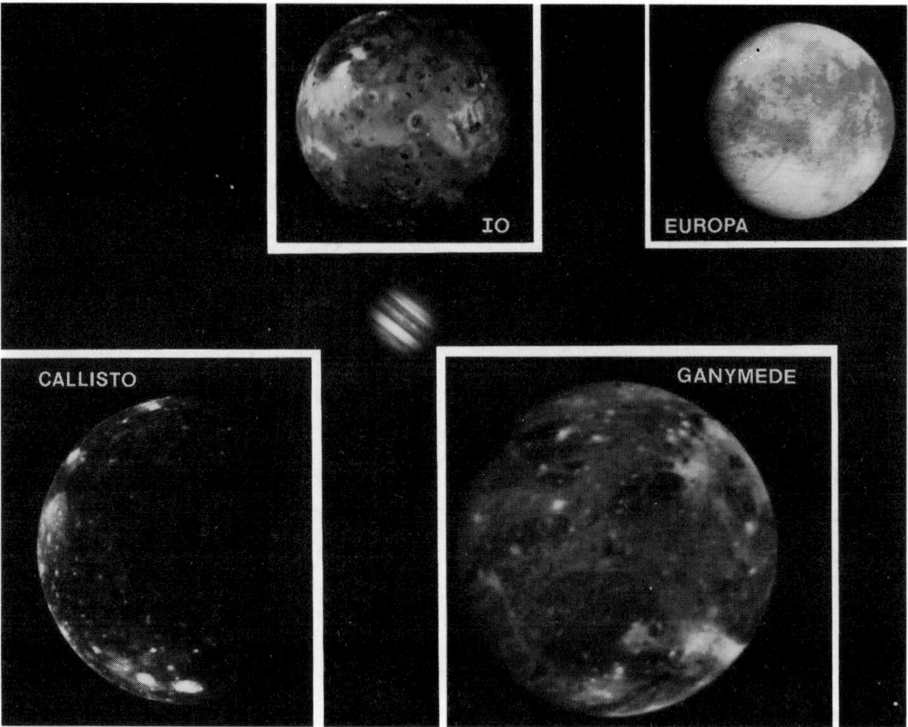

12.2. The Galilean satellites of Jupiter. The large picture shows the view from Earth with two satellites up and to the left and two others down and to the right. The insets show close-ups imaged by one of the *Voyagers*. Io is covered with sulfur-spewing volcanos, Europa with ice; the other two are heavily cratered. All except Europa are larger than our Moon. (Insets: NASA; center: Lick Observatory.)

in the Solar System, with features no more than 50 meters high, and therefore appears to have been heated in the relatively recent past. Io, covered with active sulfurous volcanoes, is one of the truly bizarre characters of the Solar System. Its interior is heated as a result of changing tidal deformation as it is pulled back and forth in its orbit by the other Galileans.

The other dozen known satellites are very small and break down into three quartets. The first is inside the orbit of Io. Even the largest of them, little Amalthea, is only 200 kilometers in diameter and very difficult to see. The others were discovered by *Voyager*. Four more small bodies, only a few tens of kilometers across, orbit six times farther away than Io, and four more twice as far as that, the latter taking some two years to make a full circuit orbiting backwards.

While Jupiter displays four large Moons and many smaller ones, Saturn gives us one large, some small, and a host of intermediate-sized bodies a few hundred to 1500 kilometers wide (Figure 12.3). The planet also holds the record for number, a total of

18 discovered satellites, not counting the moonlets that make up the rings. The largest, Titan, is easily visible in a small telescope; in a larger instrument it is set into the swarm made of its lesser kin.

Again we see great individuality. Titan (Figure 12.3a), the second largest of all the satellites and another rival of Mercury, has a thick atmosphere and is blanketed by methane haze. Its density indicates that it is half water ice. The intermediately sized satellites may be as much as three-quarters water. Mimas (Figure 12.3b), Tethys, and Dione are heavily cratered, but the latter two show evidence of geologic activity and resurfacing with water ice. Rhea also seems to have been resurfaced, but not to the degree of Enceladus (Figure 12.3c), which is icy-bright, indeed the brightest body in the Solar System. Hyperion, outside Titan's orbit, is the largest irregular body in the Solar System, and rotates chaotically as a result of gravitational pulls from the other moons; it actually has no rotation period. Iapetus (Figure 12.3d) displays one bright and one dark side made of material that it seems to have swept up in orbit. Little Phoebe is similar to Jupiter's most distant satellites, takes 1.5 years to circle its host, and again orbits backward.

Uranus (see Figure 11.29) presents us with yet more contrast. Here we find no big moons, but 5 of only intermediate size (ranging from about 500 to 1600 km in diameter) plus 10 known tiny ones. These (see Section 11.17) all orbit in the tipped Uranian equatorial plane, making it possible (twice a Uranian sidereal year) to see the whole system face-on. The small moons have orbits that are all interior to those of the larger ones. The densities of the larger five indicate they are about half water ice, containing more rock than the larger satellites of Saturn. Oberon, the most distant, and Umbriel have old cratered surfaces, whereas Titania is partially covered with extruded water ice. The two inner larger satellites, Ariel (Figure 12.4a) and Miranda (Figure 12.4b), have been subject to considerable activity, probably as a result of tidal heating. Ariel is covered with rift valleys and water ice, and weird Miranda's surface has an oddly shaped grooved terrain that indicates a once-considerable level of internal heat and activity.

Neptune again has a large satellite, Triton (see Figures 11.30 and 12.5), three-quarters the size of our Moon, two of intermediate size, Nereid (seen in Figure 11.30) and Proteus (discovered by *Voyager 2*), and five smaller ones. Nereid, like Phoebe of Saturn and Jupiter's outer four, is far away from the planet and requires a year to orbit. Triton orbits retrograde, has a light atmosphere that precipitates to the polar cap as nitrogen snow, and a surface that shows clear evidence for activity. Triton's retrograde orbit suggests that it is a captured body. It is similar in size and density to Pluto. These bodies may represent only two of a great many more that are currently beyond our telescopic view. Finally, Pluto's satellite Charon (see Figure 11.32) is about half the diameter of Pluto itself, making it the largest moon relative to its planet in the Solar System.

The satellites may be numerous, but even rolled together they do not amount to much. Their combined masses are little more than 10% the mass of Earth. They are vastly outnumbered by the asteroids.

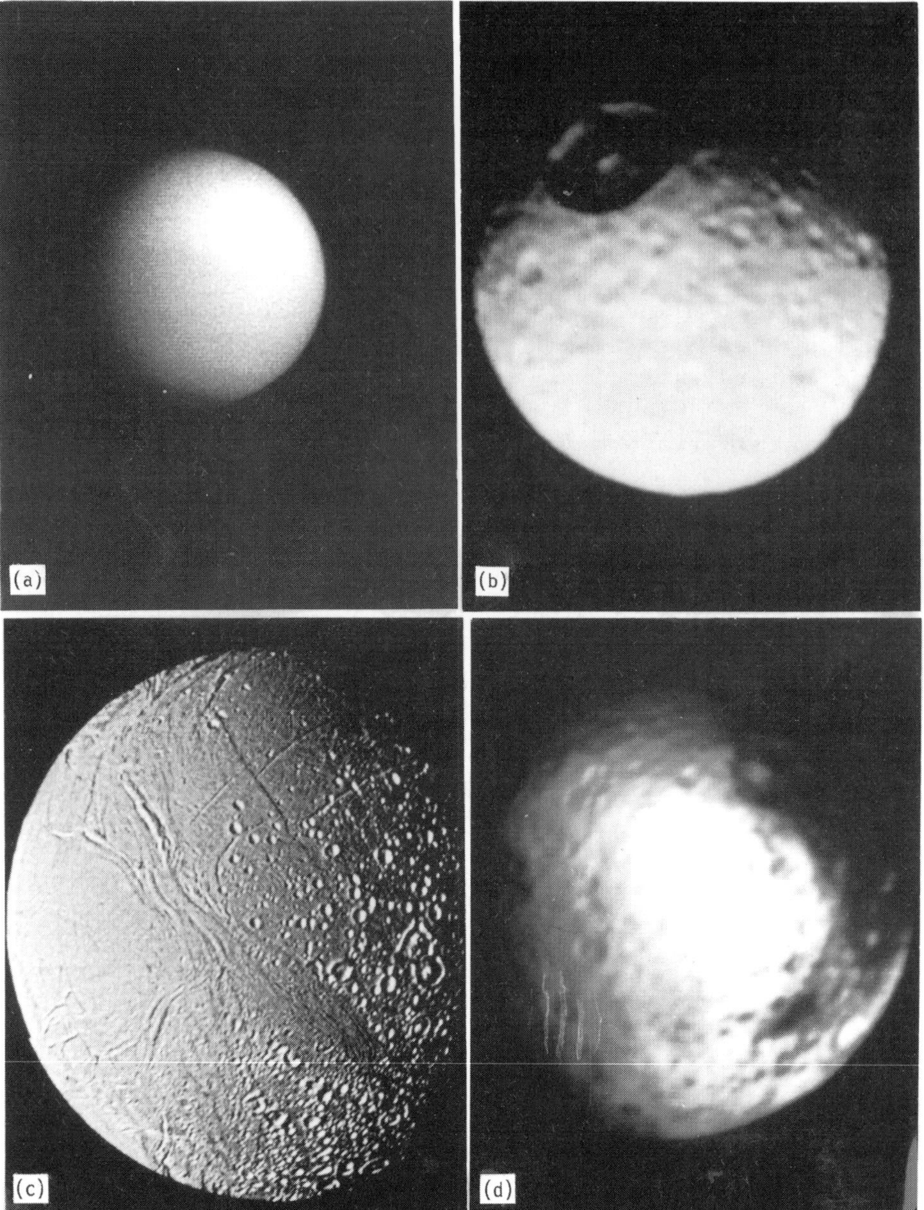

12.3. Some satellites of Saturn as imaged by the *Voyagers*. (a) Titan has a hazy atmosphere that hides its surface; (b) Mimas is heavily cratered; (c) Enceladus is grooved; (d) Iapetus is two-faced. (NASA/JPL.)

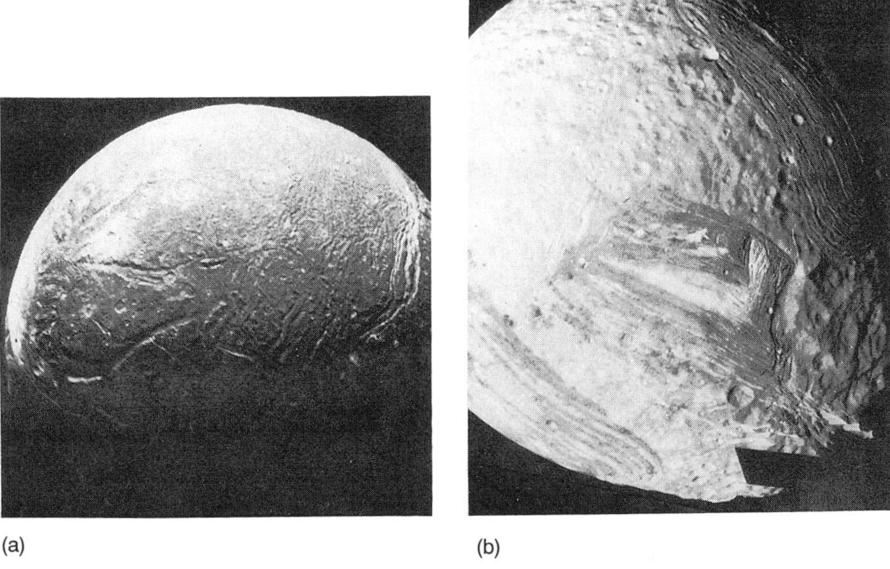

(a) (b)

12.4. Two satellites of Uranus as imaged by *Voyager 2*. (a) Ariel is grooved and bright with ice. (b) Miranda exhibits oddly shaped features. Here we see a close-up of the "Chevron," caused by internal forces or even by collision. (NASA/JPL.)

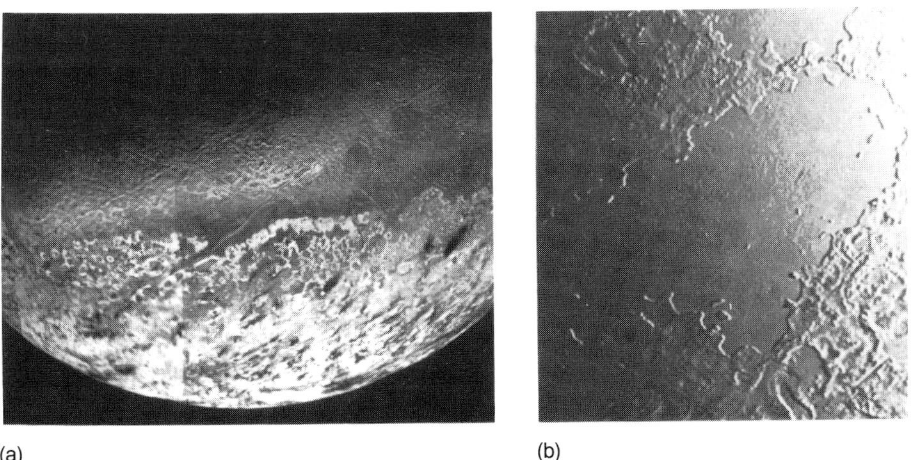

(a) (b)

12.5. Neptune's large satellite Triton. (a) A wide-angle *Voyager 2* image shows a nitrogen-snow polar cap. Grooves and other features provide evidence of geologic activity. (b) We also see basins that have been flooded with a mush of water, methane, and ammonia. (NASA/JPL.)

12.2 Asteroids

Ceres, the first known *asteroid* (meaning "star-like"), or *minor planet* (also, though rarely, *planetoid*), was discovered on January 1, 1801 by the Italian astronomer Giuseppe Piazzi, and fits neatly into the space predicted by Bode's law at 2.7 AU (see Section 11.1). Ceres is also the largest asteroid, with a diameter of 940 kilometers, roughly a quarter the size of Earth's Moon. Its discovery was quickly followed by those of Pallas, Juno, and Vesta at about the same distance from the Sun. Their small sizes render them faint. Ceres is seventh magnitude and only Vesta is visible to the naked eye, reaching sixth magnitude at opposition.

Over 20 000 asteroids have been discovered. They are easy to find on exposures taken within the zodiac (Figure 12.6), as their motions cause them to leave short streaks on the image; they are so common as to have been referred to as the "vermin of the skies." Over 6000 now carry names and have known orbits (Figure 12.7a), the vast majority between Mars and Jupiter and most in a *main belt* between 2.1 and 3.2 AU. Their proper names are preceded by their discovery number, so Ceres and Vesta become 1 Ceres and 4 Vesta. A mere 25 or so have diameters larger than 200 kilometers, and there are about 100 between 100 and 200 kilometers. The numbers increase rapidly with diminishing size, with a good million greater than 2 kilometers across and billions no larger than small rocks.

The semimajor axes of the orbits are bunched in distinct bands Figure 12.7b) with separations (the *Kirkwood gaps*) reminiscent of those in Saturn's rings (see Section 11.17). The gaps are caused by resonances with Jupiter and occur where the orbital periods would be simple fractions of the giant planet's period. The gaps are not apparent in plots of asteroid positions (as in Figure 12.7a) because the orbital eccentricities keep the small bodies moving through the gaps and smear them out.

The asteroids are also arranged in a variety of families. Some families represent the remains of larger fragmented bodies that share physical properties and are still moving more or less together. Other families involve only orbits. The *Trojan asteroids* have been captured into Jupiter's *Lagrangian points*, stable orbital positions that lead and follow the planet by 60°. Of more practical interest are the *Amors*, which approach the orbit of the Earth, the *Apollos*, which cross it, and the *Atens*, which have semimajor axes *less* than that of the Earth. Collectively, they are known as *AAA objects*, or AAAOs. All are on elliptical orbits that take them out to the main belt and all have the potential, given the inevitable gravitational perturbations, of colliding with the Earth. Some asteroids have apparently been captured by other planets to become the tiny satellites of Mars and the backward-moving outer moons of Jupiter.

Close-ups of asteroids (Figure 12.6 inset and Figure 12.1) show them to be battered and broken. They are the fragments of once larger bodies that have ground themselves down by constant collision over the 4.5-billion-year lifetime of the Solar System. The original bodies were still small, however, as the combined masses of the asteroids amount to only about a thousandth that of Earth and could never have made anything close to a respectable planet. Most of the mass of the Solar System not tied up in the Sun and planets is not in the satellites or asteroids but in the comets.

12.6. Asteroids. During the exposure of the main image, taken from Earth, two asteroids moved in orbit, allowing their dim glows to trail across the photo. Most asteroids are found this way. The inset shows an image of 951 Gaspra taken by the *Galileo* spacecraft during its trip to Jupiter. (Main image: Yerkes Observatory; inset: NASA; JPL.)

12.3 Comets

With the exception of our Moon and Vesta, the planetary satellites and asteroids are telescopic objects, commonly read about but perhaps never seen. Not so the *comets*. A nearby bright one can offer an awesome sight (Figure 12.8). In the public mind they are commonly confused with *meteors*, bits of rock in the atmosphere that streak brightly across the heavens and are gone within seconds. The comets, however, are distant and orbit the Sun, so that like the planets they move only slowly against the starry background. Occasionally, one will be visible to the naked eye for weeks or even months. Unlike the planets, however, their orbits are highly eccentric, which take them hundreds, thousands, or even tens of thousands of astronomical units from the Sun. From

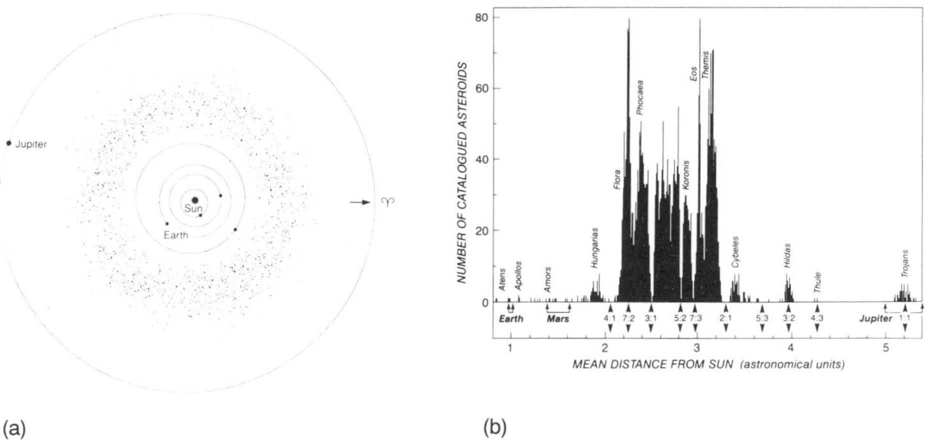

12.7. Asteroid orbits. (a) The map shows the positions of 5,011 named asteroids on May 1, 1992. Note the Trojans 60° ahead of and behind Jupiter and the AAAOs near the Earth's orbit. (b) The distribution of orbital semimajor axes shows the Kirkwood gaps and several families. The ratios of the orbital period of Jupiter to those of asteroids in the gaps are shown across the bottom. ((a) © copyright 1995, L.-A. Mcfadden and J. Bytof, University of California, San Diego; (b) R. Binzel, MIT.)

Kepler's third law, we therefore see that the orbital periods can be very long, some measured in millions of years.

A typical comet (Figure 12.9) consists of a small, star-like *nucleus* surrounded by a fuzzy ball of gas, the *coma*, the two together referred to as the *head*. Streaming outward from the head is the *tail*, sometimes almost non-existent, sometimes extremely bright and long, often curved and wide, and so delicate that stars are easily seen through it. Closer examination reveals two separate tails, one quite straight, the other gracefully curved.

A comet nucleus is pictured as a "dirty snowball," small rocks and fragile clumps of dust grains embedded in a matrix of ices, mostly water typically only a few kilometers across. When it is far from the Sun, beyond Jupiter, it flies through space undisturbed and nearly invisible. But, if it approaches within a few astronomical units of us, the solar heat turns the ice into a gas that blows out through cracks in a crusty surface and surrounds the nucleus as the coma. Energetic sunlight then ionizes the gas, stripping the electrons from its atoms, making it glow. The *solar wind*, the steady stream of electrically charged atomic particles (protons and electrons) from the Sun (described in Chapter 13), also carries with it the solar magnetic field. When the wind encounters the coma, it blows the gas back away from the Sun, wrapping the magnetic field around it like a giant wind sock (see Figure 12.9). This *gas tail*, which consists of hydrogen and molecules like carbon monoxide and cyanogen always points away from the Sun no matter what the comet's direction of motion; it can precede the comet just as well as follow it.

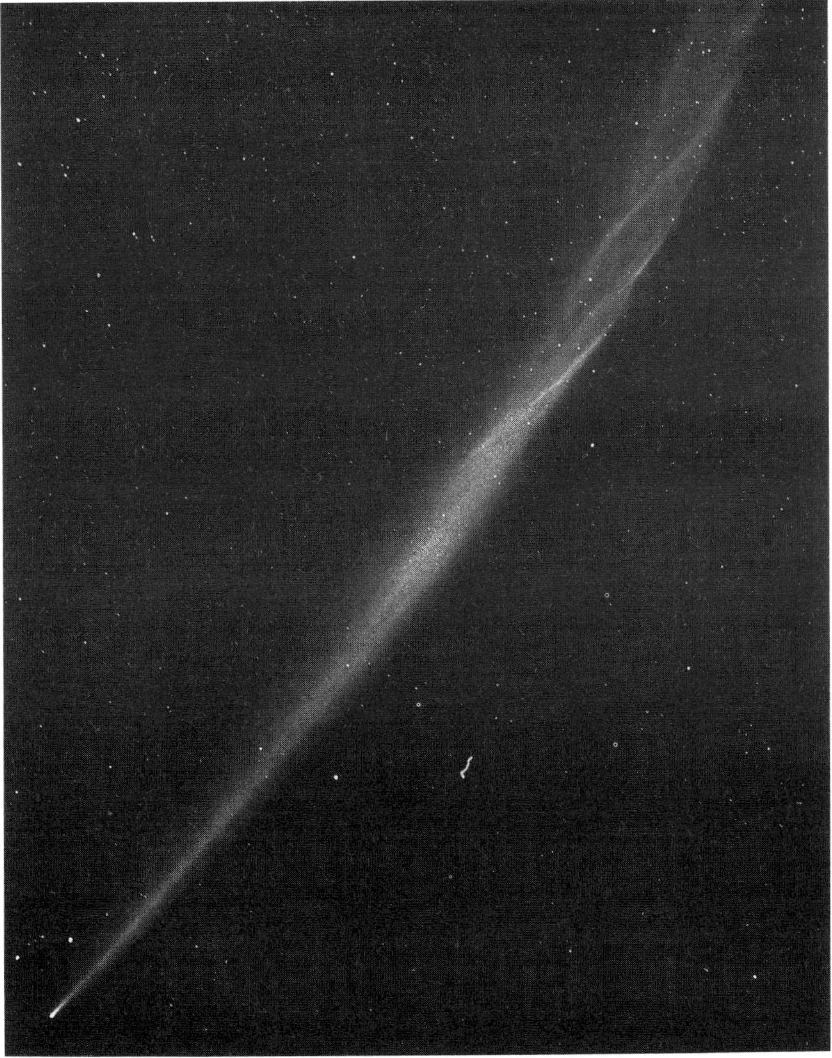

12.8. Comet Ikeya–Seki, 1965 VIII, a sungrazer with a magnificent gas tail. (Lick Observatory photograph.)

With the melting of the ice, solid grains are also released. These are pushed by sunlight, but since they have more mass and inertia than the gas, they are influenced by the orbital motion and do not move straight back. As a result, comet tails usually have two visible components, the straight gas tail and a curved *dust tail* (Figures 12.9 and 12.10) that reflects sunlight. The prominence of one kind of tail relative to the other depends upon the comet's dustiness.

As creatures of the Sun, of its wind and magnetic field, the comets make excellent

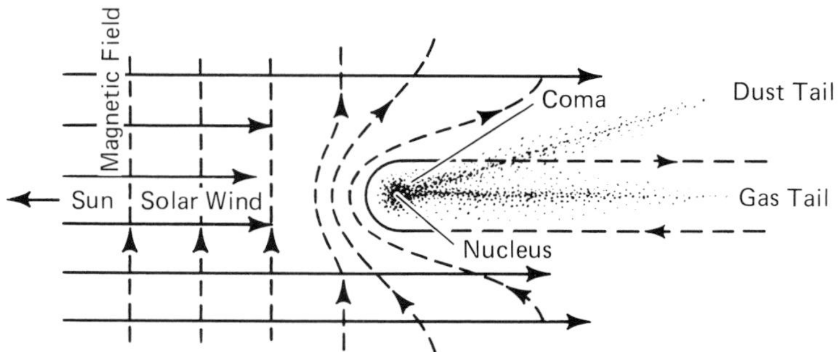

12.9. The parts of a comet. The coma is an excited molecular gas driven off the tiny nucleus by sunlight. The solar wind sweeps the gas backwards, wrapping the solar magnetic field around it to form the gas tail. As the comet's nucleus evaporates, small grains are released to form the dust tail.

probes of the interplanetary environment. Fluctuations in the wind are reflected in the appearance of the tail, and when the comet passes from one region of the solar magnetic field into another, where the field's direction is reversed, the tail is snapped off, a new one growing immediately.

The melting process is erratic and unpredictable. The gas and dust are spewed out of the nucleus in fountain-like plumes, which act as rockets, slightly altering the comet's orbit. Sometimes gas streams away furiously, at other times it can nearly stop as different layers of the nucleus are exposed. These variations are easily seen as puffs of gas in the tail that move rapidly outward away from the Sun. All this activity makes comets great fun to watch as they change their appearances almost from night to night.

There are two basic varieties of comets, *short period* and *long period*, divided at about 200 years. Short-period comets have been observed at more than one perihelion passage, the long-period variety but once. Those of short period tend to orbit the Sun in the usual counterclockwise direction and to stick (within broad limits) to the ecliptic plane. In contrast, the long-period comets have random orbits, half going retrograde, and they have no respect for the ecliptic. Several new long-period comets are discovered every year, most by amateur comet hunters. The majority are telescopic, but once every decade or two we will see a large, spectacular one. In older times such apparitions were believed to foretell doom, war, and destruction; the abrupt appearance of such a spectacle was greatly feared.

Comets are named both after their discoverers and in order of their discovery within a certain year, for example Comet Morehouse 1908c, or Arend–Roland 1956h. After the orbits are known, they are renamed in order of perihelion passage, wherein Arend–Roland became 1957 III. The names of short-period comets are preceded by P/, for example, P/Encke.

In either case, whether short period or long, any comet with a perihelion close to the Sun is doomed to destruction, since at each solar passage, it will lose one or two

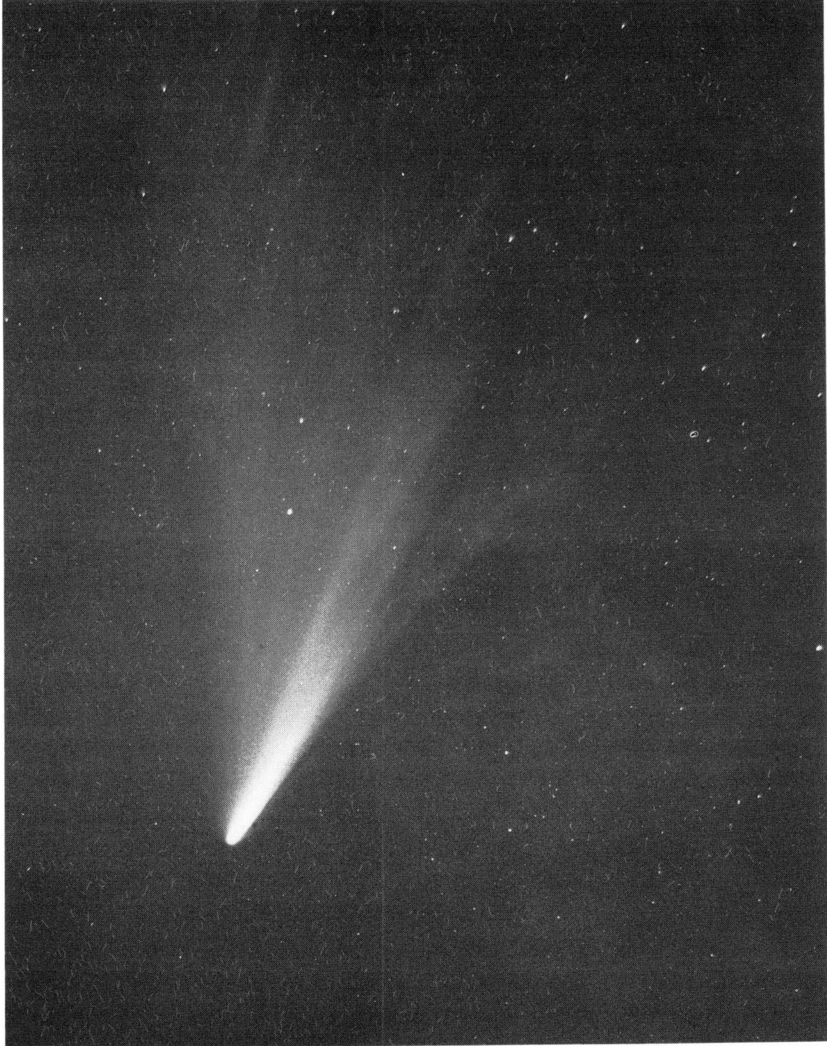

12.10. Comet West, 1976 VI, with its great fan-shaped dust tail. (Lick Observatory.)

percent of its mass. Comets have even been seen to disintegrate as they orbit. Comet West (Figure 12.10) broke into four pieces in 1976 and Biela's comet, which had a 6-year period and was last seen in 1846, simply vanished. Short-period comets must lose their mass quickly, and as a result, they are generally smaller and fainter than their long-period counterparts. After a time, nothing may be left but a dark nucleus. Some half of the AAAO asteroids may be defunct short-period comets.

Since comets are ephemeral objects, but are still seen in abundance 4.6 billion years after the Solar System was born, there must be vast reservoirs that contain enormous numbers of them. Trillions of long-period comets are believed to dwell in the vast *Oort comet cloud* (after astronomer Jan Oort) that extends in all directions some 100 000 AU from the Sun. Most have orbits that never take them near us, and remain forever invisible. But once in a while a nearby passing star or interstellar gas cloud will gravitationally disturb the comet cloud and send a few individuals inward toward the Sun on nearly parabolic paths. Over the aeons an enormous number will develop the narrow elongated orbits that take them every few thousands or millions of years into the inner Solar System. The short-period comets, however, seem to come from a thick disk called the *Kuiper belt* (after Gerard Kuiper) that lies in the plane of the ecliptic to several hundred AU from the Sun.

The best known of all comets is Halley's comet (P/Halley), seen in Figure 12.11 at its 1910 apparition. This brilliant comet has an aphelion near Pluto's orbit (Figure 12.12) and an average period of 76 years. Edmund Halley did not discover it, but using Kepler's laws as generalized by Newton determined that the comets of 1531, 1607, and 1682 were one and the same. We now know that nearly 30 historic comets seen over the previous two millennia had simply been Halley reappearances.

This great astronomical body has worked its way into popular lore and almost epitomizes astronomy. People live and die with it, and it is considered a great feat to have been able to see it twice. Its 1910 apparition (see Figure 12.11) was spectacular. The Earth was well placed close to the comet's perihelion at the time of arrival so that it was viewed near its brightest, and the tail stretched across much of the sky. The passage was so close that the Earth actually passed through the tail. Since cyanogen had earlier been found in the gas, there was widespread fear that humanity would be poisoned. But the tail is so vacuous that no effect could be noticed: it is bright all out of proportion to its mass, which is but a fraction of a percent that of the already small nucleus.

The 1985–86 passage (Figure 12.13) was much poorer. Halley came into perihelion near inferior conjunction, so it was brightest when least visible. Even when it was at its best it was placed at such low declinations that it was a special disappointment to northern hemisphere viewers who had waited in some instances decades to see it. In the southern hemisphere, however, the view was still quite wonderful. More importantly, this passage offered the possibility of a close-up view with spacecraft. An armada of probes from Europe, Japan, and the Soviet Union headed toward rendezvous in March of 1986 when the comet crossed the ecliptic plane (it is very expensive to launch a spacecraft into an inclined orbit). The triumph of the missions was close-up imaging of the comet's nucleus, which showed an 8×16 kilometer mass spewing fountains of gas and dust (Figure 12.14). It too is melting away, someday to vanish, or to appear as only a dark orbiting rock.

Halley's is hardly the only famous comet. Its 1910 reappearance was rivalled by the Great January Comet also of 1910. Two of the century's best were Ikeya–Seki, 1965 VIII (Figure 12.8), with its stunning gas tail, and lovely West (Figure 12.10) with its broad dust tail. Ikeya–Seki was a "sungrazer," passing within 500 000 kilometers of the solar surface, one of a string of them that goes back a century, and which may once have been all one huge comet. Also of note were Arend–Roland (1957 III), which

12.11. Comet Halley as seen in 1910. The break in the tail occurred when the comet passed into a reversed section of the Sun's magnetic field. The bright image is Venus and the long streaks are city lights. The stars are short streaks because the comet moved while the exposure was taken. (Lowell Observatory.)

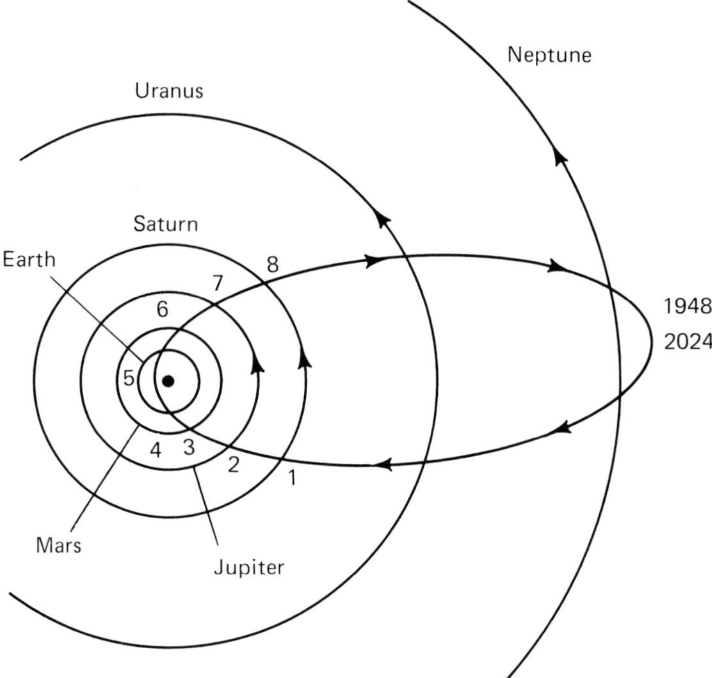

12.12. The orbit of Halley's comet. The orbit is inclined to the ecliptic by 162°, so that it moves retrograde, but here is projected onto the Solar System's plane. The planetary orbits are not to scale. At perihelion the distance of the comet from the Sun is less than that of the Earth; at aphelion the comet recedes to beyond Neptune. The positions are: (1) mid-1983; (2) late winter 1985; (3) summer 1985; (4) winter 1985; (5) February 1986 (perihelion, when the comet and Earth were on opposite sides of the Sun); (6) spring 1986; (7) autumn 1986; (8) summer 1988. The last aphelion was in 1948, the next will be in 2024, followed again by perihelion in 2061. The tail begins to develop outside the orbit of Mars.

although not that bright, broke a long dry spell and had a sunward-pointing spike, a dust-sheet seen edge-on, and Bennett (1970 II). Hardly "great," but eminently notable was Kohoutek (1973 XII), famous for what it did *not* do. Remarkably bright for its distance at discovery, and wildly anticipated by the news media, its crusted surface never allowed it to come close to expectations. The nineteenth century was better, with great comets in 1811, 1843 (which had a tail over two AU long and a nucleus visible in daylight!), 1858 (Donati's), 1861, 1874 (Coggia's), and 1881 (another daylight comet, Figure 12.15a, and the first to be photographed). Notable in the eighteenth century was De Chéseaux's comet of 1744 (Figure 12.15b) with its six well-defined tails, and in the seventeenth, the brilliant daylight comet of 1680. Truly bright ones are seen only once every decade or two. But they are unpredictable, so be aware: a new one may be found tonight.

12.4 Meteors

Tiny asteroids and the rocky debris of disintegrating comets, called *meteoroids*, pervade the Solar System and are constantly colliding with Earth. As a meteoroid enters the air some 50 kilometers above the ground at a speed of tens of kilometers per second, it heats up by generating an atmospheric shock wave, and a *meteor* (Figure 12.16) is seen to streak across the sky. We do not actually witness the particle itself, but the glow generated by the ionization of the atmosphere produced by the fierce heat of entry. As direct evidence of this phenomenon, the tube of hot air takes time to return to normal, and after a brighter meteor we can usually see an afterglow or *train* that may last for some seconds or even minutes.

The vast majority of meteors arise from the cometary debris and are extremely common. On any moonless night a dedicated observer might see a dozen or more. As bright as they may be, however, the particles are mostly very small, mere fragile clumps of dust a few millimeters across. Their energy comes not from their masses but from their ferocious speeds. More rarely, a larger one hits us. One a few centimeters across may even reach the ground, becoming a *meteorite*. These are all asteroidal in nature; the cometary particles are too small and delicate.

The best time to view meteors is in the early morning hours of a dark, moonless night. The observer is then on the side of the Earth facing in the direction of orbital motion so that meteoroids are swept up, yielding a basic count of about three per hour. Because they collide head-on with the speeding Earth, they are swift, white, and hot. In the evening we see far fewer, since meteoroids must catch up with us, and in contrast move more slowly, are cooler, and consequently redder.

The intimate relation between comets and meteors is seen in the *meteor showers* that occur frequently during the year. The disintegrating comets leave trails of dusty, gritty debris in their wakes. If the Earth passes through, or near, a cometary orbit, we will see a marked increase in the number of meteors. Since the paths of the particles are parallel to one another, we on Earth see them appear to radiate from a particular point in the sky, much as parallel railroad tracks seem to converge to a point on the horizon. The location of this *radiant* among the stars depends on the relative motion of the Earth through the meteoroid stream, which will be nearly the same at every passage. Individual showers are named after the constellations that contain the radiants, hence the Lyrids or the Geminids. During a shower the individual meteors can be seen all over the sky; the radiant is identified by tracing their paths backward to see where they meet. The best place to look is always near the zenith, where atmospheric absorption of light is minimized.

Dozens of showers are known. They range from those that require real diligence to see, with yields of only a few per hour or night, to showpieces such as the January Quadrantids (named after the defunct constellation Quadrans), August Perseids (Figure 12.17), and December Geminids, which have typical early morning dark-sky counts of more than one meteor per minute. These and a few other well-known showers are listed in Table 12.2 along with the dates, radiants, rough typical hourly rates, and the parent comets. Some showers occur just during the day and are known through radar

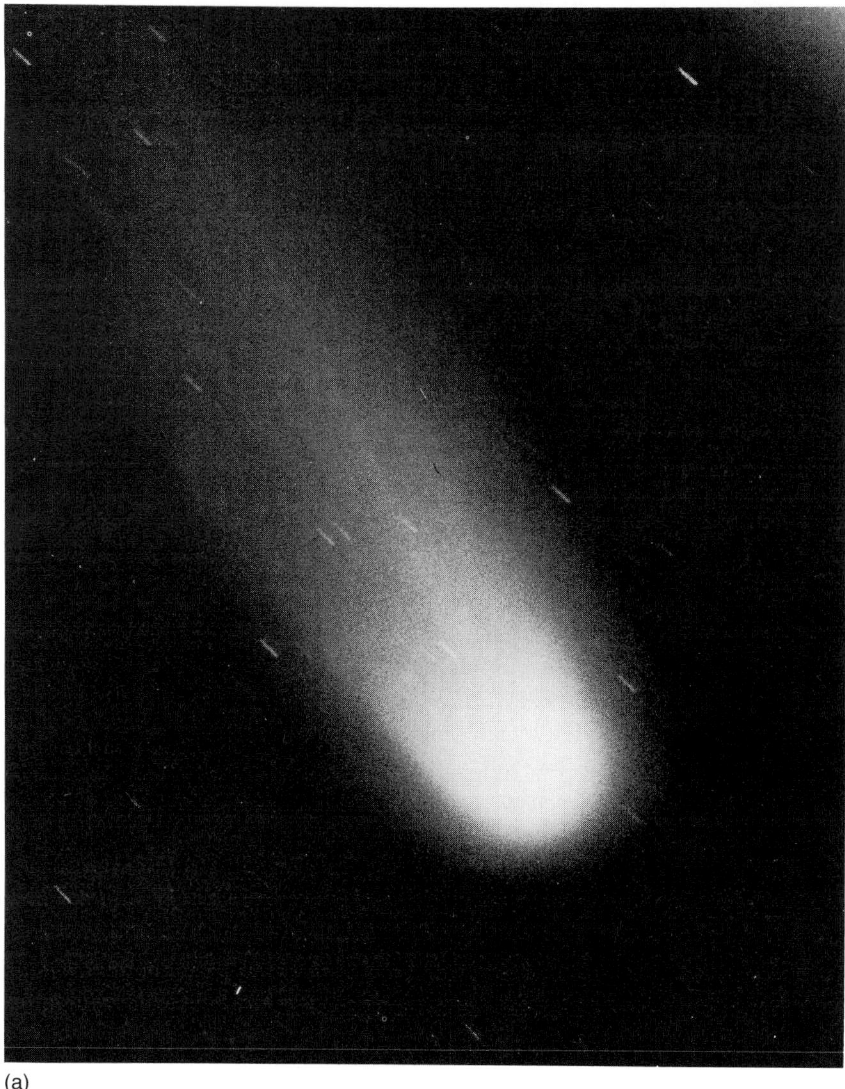

(a)

detection. A few are sharply peaked over intervals of only a few hours, others are spread out for days or even weeks, depending upon the width of the meteoroid stream. As the ages pass, a distinct shower ceases to be recognizable as such, and we see random or sporadic meteors, all of which develop from the spreading swarms.

The debris in a comet's orbit is irregularly distributed and can be highly localized into a concentrated volume that orbits the Sun. Consequently, there can be considerable variation from one year to the next. The most extreme case would be a comet that has recently disintegrated, yielding only a loosely organized volume of tiny rocks. If the

(b)

12.13. Halley's return in 1986. (a) A close-up of the head and inner tail made on March 6, 1986. Note the streamers. (b) The full view made on March 17, 1986 shows both gas and dust tails. ((a) Mt. Laguna Observatory, San Diego State University and the University of Illinois; (b) University of Chile.)

Earth encounters this concentration, which can happen only at infrequent intervals related to the comet's orbital period, we will see a truly spectacular event – a *meteor storm* – in which thousands of meteors can be seen per hour, much like a celestial snowfall. Since the concentrated volumes are very small, these great showers will be very brief, occurring over only an hour or two.

The most famous meteor storms are the Draconids, Leonids, and Andromedids (or Bielids). The first, related to the short period comet Giacobini-Zinner (Figure 12.18)

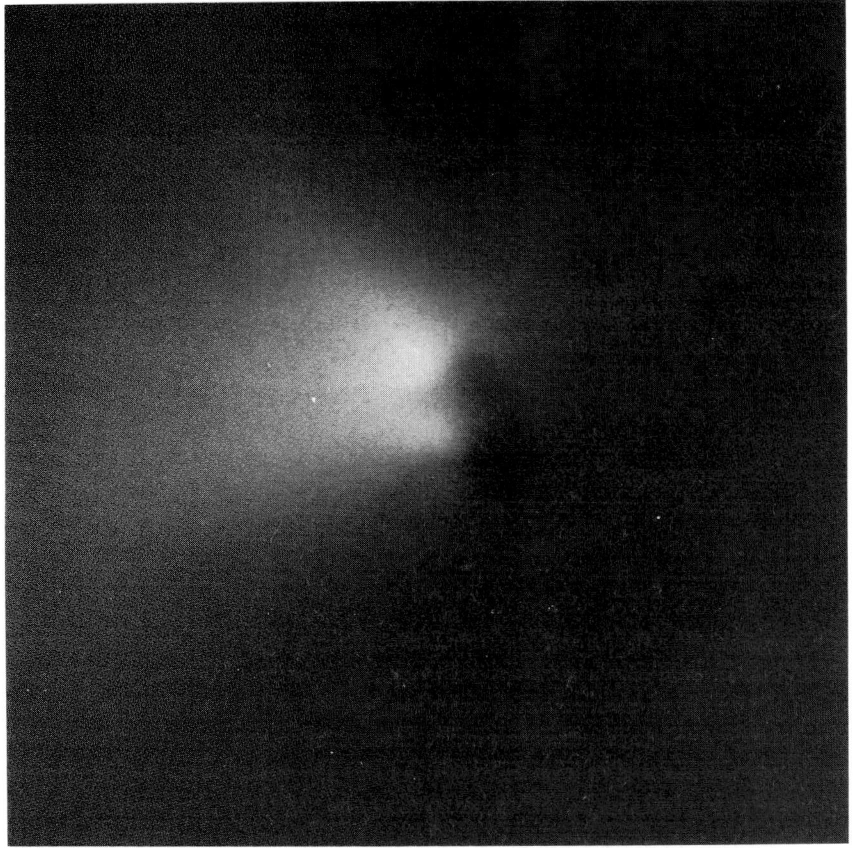

12.14. A close-up of the nucleus of Halley's Comet taken with the European *Giotto* spacecraft at a distance of 26 000 kilometers. It is only 8 × 16 kilometers across and is spewing forth bright jets of gas under the heating action of sunlight. (European Space Agency/ Max-Planck Institut für Aeronomie, courtesy H. U. Keller.)

and sometimes called the Giacobinids, produced great displays in 1933 and 1946 in which rates of over 10 000 per hour were witnessed. Jupiter has perturbed the orbit of the pack of meteoroids, and it has not been seen since. The Leonids (Figure 12.19), related to Comet 1866 I, are even greater. They nominally recur every 33 years and were observed in 1799, 1833 and 1866. Again, Jupiter changed the swarm's path, and we missed it in 1899 and 1933. But then the planet undid its damage, and in 1966 we were again on a collision course: a lucky few saw a fall estimated at 150 000 per hour that extended over only a brief four-hour interval. The Andromedids or Bielids are no longer seen but are historically important. In 1845 Biela's comet split into two parts, each of which returned in 1852; then both simply disappeared. But in 1872, at the predicted time of the comet's appearance there occurred a spectacular meteor storm. And thus it was shown that meteors are the spawn of comets.

At the opposite end of the meteor family, but no less spectacular, are the lone large

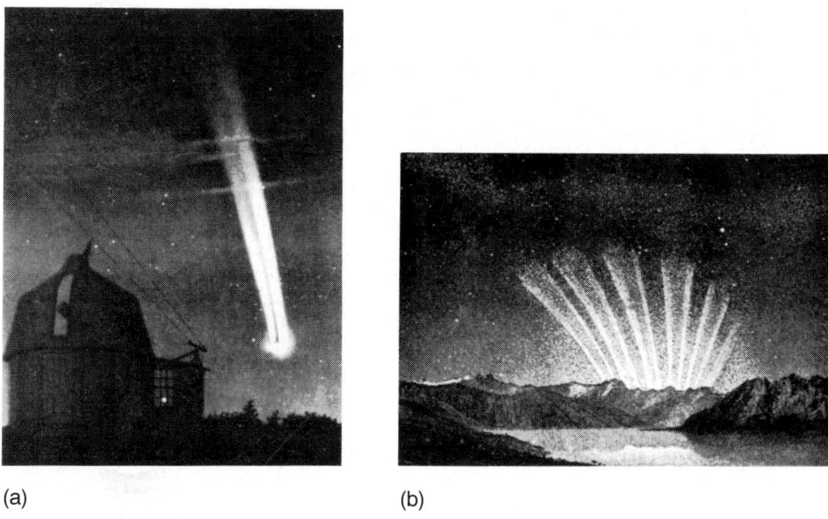

(a) (b)

12.15. Two historic comets: (a) The Great Comet of 1881, whose nucleus could be seen in daylight; (b) De Chéseaux's comet, rising with its six tails. Note Altair, in Aquila, and Delphinus.

12.16. A common meteor. We see the ionization trail of a small piece of rock that is perhaps 50 kilometers up. Note the sudden burst of light, the result of the meteoroid beginning to break up. (Yerkes Observatory.)

12.17. The Perseid meteor shower. The meteors are seen to radiate from the constellation Perseus. (Miroshi Hayashi.)

fireballs, or *bolides*, that blaze across the sky in a second or two, sometimes from horizon to horizon, and can light up the ground as bright as the full Moon. They may be seen over an area many hundreds or even thousands of kilometers across as they bore through the Earth's atmosphere. Very rarely, one may even be glimpsed in daylight (Figure 12.20). If we are lucky enough to be near the end of the path, we may see it break up, or explode in a shower of debris. If the bolide is large enough, we may even hear the sonic boom some minutes later caused by its supersonic speed. Occasionally, the meteor may leave an ionization train behind that can endure for 10 or 20 minutes. Although a typical fireball may be appear to be close – observers commonly insist that it landed in a neighbor's yard – it is really quite distant, perhaps 50 kilometers high and one or two hundred kilometers away when seen near the horizon. Some are associated with comets, but the majority are likely to be small asteroids that end their lives in a sudden flash. We also sometimes see "artificial meteors" caused by the spectacular re-entry into the atmosphere of orbiting space vehicles. They go much slower than the natural variety, and may take tens of seconds to cross the sky.

The meteors provide a rich source of useful enjoyment for the amateur astronomer. Most of our knowledge of showers comes from dedicated observers who spend countless hours charting paths, determining radiants, and measuring hourly count rates. And what we know of bolides usually comes from the random lay person who just happens to be looking up at the right time and who is able to describe the event to the astronomer who collects the data.

12.5 Meteorites

On occasion, a large meteoroid a centimeter or more across survives its atmospheric passage, slamming into the ground to become a meteorite (Figure 12.21). Meteorites

Table 12.2. *Major nighttime meteor showers*

Shower	Dates	Approximate Radiant α	δ	Hourly Rate[a]	Associated comet	Comet period
Quadrantids[b]	Jan. 3 ± 1	$15^h\ 20^m$	+49°	30	?	—
Lyrids	Apr. 21 ± 1	18 00	+33	8	1861 I	415
Eta Aquarids	May 4 ± 2	22 20	0	10	Halley	76
Delta Aquarids	July 30 ± 12	22 40	−10	15	?	—
Perseids	Aug. 12 ± 11	3 00	+58	40	1862 III	105
Draconids	Oct. 8 ± 2	17 40	+54	20[c]	Giacobini–Zinner	7
Orionids	Oct. 20 ± 5	6 20	+15	15	Halley	76
Taurids	Nov. 5 ± 18	3 30	+16	8	Encke	3
Andromedids	Nov. 14 ± 7	1 40	+44	10	Biela[d]	—
Leonids	Nov. 17 ± 2	10 10	+22	6[c]	1866 I	33
Geminids	Dec. 13 ± 3	7 30	+32	50	3200 Phaeton[e]	—
Ursids	Dec. 22 ± 2	14 00	+78	12	Tuttle	—

[a]The listed rate is approximate only. It is an indication of the maximum number of shower meteors seen per hour by an observer with the radiant overhead on a clear moonless night.

[b]Named after the defunct constellation Quadrans (section 5.5 and Table 5.3).

[c]These entries are rates for typical years; great displays of Draconids and Leonids have been witnessed in certain years: see the text.

[d]Comet broken up and lost.

[e]Originally classified as an asteroid; now known to be a defunct comet.

litter the Earth, which accumulates some 800 over 10 grams per day. Nevertheless, few of us will ever be aware of one landing nearby. Although houses and cars have been struck by them, there is no confirmed record of anyone being killed. It is quite safe to stand outside. The champion geographic area is Antarctica. The meteorites that have accumulated over several million years ride the glaciers and are concentrated into vast meteorite fields in which they are there for the plucking. Several expeditions have brought back tons of extraterrestrial material.

Meteorites are small asteroids, AAAOs, that have collided with the Earth. (A handful of exceptions are meteorites that came from violent impacts on the Moon or Mars that hurled planetary debris into solar orbit.) We can therefore use them to study asteroidal natures and compositions firsthand. There are two basic kinds of meteorites (and therefore asteroids), *stones* (Figure 12.21a) and *irons* (Figure 12.21b), actually an alloy of iron and nickel. (A small fraction, the *stony-irons*, are mixed). The stones are by far the most common, but are hard to recognize unless actually seen to fall, so a disproportionate number in museums are of the metallic variety. The stony meteorites include the *chond-*

12.18. Comet Giacobini–Zinner. This short-period comet, (6.6 years) with a lance-like gas tail, produced the great Draconid or Giacobinid meteor storms. It was also the first comet explored by spacecraft, visited by the US *International Cometary Explorer* in 1985. (US Naval Observatory.)

rites, so named for their *chondrules*, small spherical inclusions of what appear to be solidified droplets (see Figure 12.21a). The most intriguing is a group called the *carbonaceous chondrites*, which have high proportions of carbon.

All meteors are old, the chondrites the oldest. The ages are determined by examining radioactive isotopes of various chemical elements. Such an isotope will decay to another at a known rate, so that by comparing the amount of the daughter product with that of the parent, we can derive the age since the rock solidified. The meteorites are all

12.19. The Leonid storm of 1833 as captured in an old woodcut.

about 4½ billion years old, the very oldest 4.6×10^9 years, which is the greatest age that can be found anywhere and which is therefore taken as the age of the Solar System. These ancient asteroids, along with the comets, appear to be among the original condensates of the raw material out of which the Solar System formed.

Very rarely, a really large meteorite – perhaps as big as a house – will strike. These, which come but once in tens of thousands of years, carry so much energy that when they are suddenly stopped by the ground they excavate craters just like those on the

12.20. A bolide or fireball meteor. These are large enough to reach the Earth as meteorites. This particular one was photographed *in daylight* above the western mountains of the United States. It fortunately skipped off the atmosphere and went back into space.

Moon. About 200 of these blemishes, sometimes called *astroblemes*, are recognized, and there must be many more. However, unlike those on the Moon, terrestrial craters are quickly hidden by vegetation, eroded by wind and water, and destroyed by the tectonic processes that raise mountains. All that may be left is cracked and broken rock beneath the surface soil, or a basin that is nearly unrecognizable except from space.

We see evidence of only the most recent impacts or those that have somehow been preserved (Table 12.3). On Arizona's northern plain, near the town of Winslow, we find Meteor Crater, a hole 1.2 kilometers in diameter and 200 meters deep (Figure 12.22). Fragments of nickel-iron lie all around. About 50 000 years ago, an asteroid some tens of meters across (a million tons of iron) created it, and it is still quite fresh looking. In Canada we find lake-filled Chubb crater, 3 kilometers wide, and 100-km-wide Lake Manicouagan. The violence of such events, which would dwarf even the largest thermonuclear bombs, is unimaginable. The great Rieskessel (giant's kettle) in southern Germany was created by a body perhaps a kilometer across and weighing a thousand-million tons; it may be 5 kilometers deep, although little is left now but cracked rock. The debris in such an explosion was blown out of the atmosphere, whence it re-entered to form odd melted-looking frozen droplets called *tektites* (Figure 12.23), which are found over much of Europe. There are other tektite fields in Indonesia, Australia, and the United States. Among the largest craters is the Chicxlub structure in Mexico's Yucatán peninsula, some 200 kilometers from edge to edge and 65 million years old.

This crater has been identified with the extinction of the dinosaurs and many other species that also took place 65 million years ago. The awful violence of the event raised

(a)

(b)

12.21. Meteorites. (a) A stony chondrite. Note the spherical chondrules. (b) An iron meteorite from Canyon Diablo in the United States. (Author's photographs from: (a) Bruce Dod collection; (b) Smithsonian Institution.)

so much dust that the climate changed dramatically. The principal evidence is a layer of rock rich in iridium that corresponds to the time of the great mass extinction. Iridium is rare on Earth but relatively common in meteorites.

Could a large strike happen again? It did, not long ago. This last great collision occurred in Siberia in 1908, when a forested area 60 kilometers wide was laid flat. There is no large central crater, and the event is believed to have been caused by a stony asteroid that exploded in the atmosphere. Like volcanoes and earthquakes, the meteors produce periodic natural disasters over which we have no control, at least not as yet.

12.6 The zodiacal light

The plane of the ecliptic is pervaded by a fine dust made of cometary effluvia and the debris of asteroidal collisions. We can see the effect of this material best just after

Table 12.3. *Some terrestrial meteorite craters*[a]

Name and location	Latitude and longitude		Diameter
Odessa, Texas, USA	31°45′ N,	102°29′ W	170 m
Wolf Creek, Australia	19°18′ S,	127°46′ E	880 m
Meteor (Barringer) Crater, USA	35° 0′ N,	111° 0′ W	1.2 km
Steinheim, Germany	48°02′ N,	10°04′ E	3.4 km
Chubb ("New Quebec"), Canada	61°17′ N,	73°40′ W	3.4 km
Deep Bay, Saskatchewan, Canada	56°24′ N,	102°59′ W	10 km
Charlevoix Structure, Quebec, Canada	47°32′ N,	70°18′ W	54 km
Rieskessel, Germany	49° N	10°37′ E	24 km
Manicouagan, Canada	51°23′N	68°42′E	100 km
Sudbury, Ontario, Canada	46°36′ N,	81°10′ W	200 km

[a]Principal source: *Meteorite Craters and Impact Structures of Earth*, P. Hodge, Cambridge, 1994.

12.22. Meteor Crater, near Winslow, Arizona, USA. The pit is 1.2 kilometers across. (Yerkes Observatory.)

twilight ends in the evening, or just before it starts in the morning. If it is very dark, with no Moon, you may see a conical band of light stretching from the horizon halfway or more up the sky: sunlight reflected from countless tiny grains. This glow, the *zodiacal light*, is sometimes referred to as the "false dawn," since it makes the sky appear to brighten an hour or so before twilight is due to start. A superb view of it is shown in the all-sky photo in Figure 4.19. Most people are not aware of its existence, but once seen, it can appear so bright that we are astonished that we could have missed

12.23. A glassy tektite, the splashed debris of a meteorite strike, found in the US state of Georgia. (Author's photo from the Bruce Dod collection.)

it. But the least artificial lighting or moonlight washes it away again. The best time to view the phenomenon is when the ecliptic is most nearly perpendicular to the horizon: spring evenings and autumn mornings in either hemisphere since both the seasons and the tilt of the ecliptic are reversed.

Under the best conditions, the zodiacal light can be seen to stretch all the way around the ecliptic, becoming steadily fainter as the elongation from the Sun increases. However, as we approach the antisolar point it brightens again as a result of the increased ability of the dust particles to scatter light in the backwards direction. This *counterglow*, or *gegenschein* can be seen as a faint large patch several degrees across on the ecliptic opposite the Sun. It is best viewed at midnight when it culminates the meridian, and from the northern hemisphere in late autumn or late winter when it is still high in the sky but does not compete with the Milky Way, as it does near the time of solar winter solstice passage. From southern latitudes the prime seasons are the same. The sky must be extremely dark to see this ghostly light, with no Moon or artificial lighting.

12.7 The origin of the Solar System

With the components of our planetary home introduced, we can look briefly at how we believe it came to be. Evidence from our own Solar System and from other stars powerfully indicates that stars condense out of cold dense clouds of interstellar matter, hydrogen and helium already rich in molecules and solid grains. (Look at the Coalsack

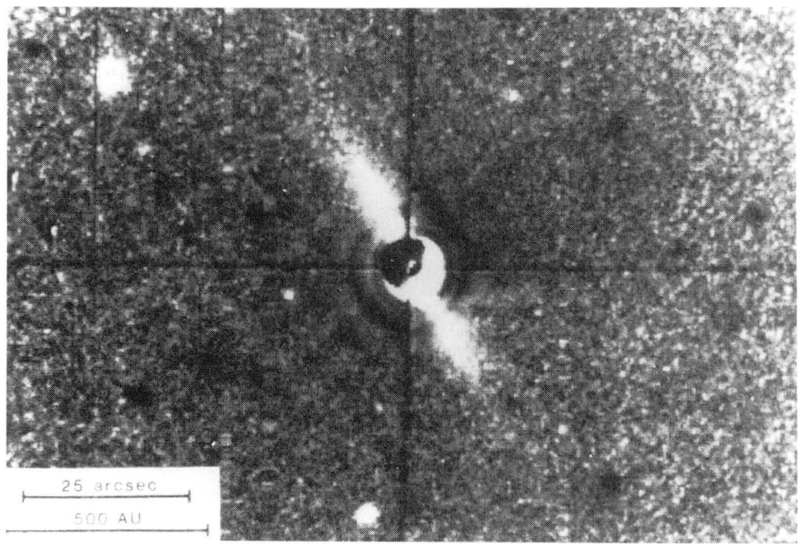

12.24. Beta Pictoris and its surrounding cold disk. The star itself is partially hidden behind an occulting plate in the telescope. The disk is some 500 AU across. Would our system look like this from a distance? Are there planets buried within? (B. A. Smith and R. J. Terrile, Mt. Wilson and Las Campanas Observatories, Carnegie Institution of Washington.)

in Crux in Figure 4.7 and at the great dark structures in the Milky Way in Figure 4.19.) Various compressional processes produce dense knots within these clouds that contract under their own gravity. Because of the conservation of angular momentum, some of the infalling matter spins out into a disk with the greatest density at the center. Eventually the central mass becomes hot enough to sustain thermonuclear fusion and a star is born, surrounded by a rotating dark cold disk of gas and dust.

During the formation of the Sun the dust grains in the cold disk began to agglomerate into larger ones as they collided and as various gases condensed onto them. They ultimately grew into larger solid bodies, planetesimals, the remains of which are the asteroids and comets of today and perhaps the Plutos and Tritons. The planetesimals further agglomerated into planets. The planets in the inner region of the Solar System were too close to the Sun and too hot to retain water. However, in the outer reaches – Jupiter and beyond – it was cold enough for water to freeze, and the resulting solid bodies (like the outer satellites) are rich in ice. The masses of the planets depended on the distribution of matter in the circulating disk. The heat of sunlight also drove the lighter materials away from the terrestrial planets, leaving behind the rocky cores. The Jovians were so far away from the Sun that they were able to accumulate lighter gases. The planets, hot from formation, melted and differentiated, the heavier metals – mostly iron and nickel – settling to their centers. The giant planets gravitationally developed their own disks of cold gas and grains, and out of these formed the satellite systems.

In the early days of the Solar System the new solar wind cleared much of the remaining dusty gas away. Much of the larger solid material was swept up by the just-formed and cooling planets, leaving behind cratered, heavily scarred surfaces. On the Earth the early craters have long since been eroded away, but they are still there to see on less active worlds.

A great deal of debris is still left. Many cometary bodies still exist within the disk-shaped Kuiper belt well outside the boundaries of the Jovian planetary system. Several are now known with orbits comparable to that of Pluto. Jupiter and Saturn hurled great numbers of comets out of the Solar System altogether, but the weaker gravities of Uranus and Neptune ejected trillions to form the Oort cloud. Jupiter threw bodies into the asteroid belt, the resulting collisions breaking apart a few larger differentiated asteroids, from which came the rock and iron asteroids of today and the stony and iron meteorites that litter the ground.

We do not appear to be unique. Other stars have cold disks around them that may relate to our Kuiper belt (Figure 12.24). Four and a half billion years after the creation of our planetary system we are here to see and reconstruct it: as remarkable a happening as was the formation itself.

Chapter 13

Light and the atmosphere

In earlier chapters we thought of the sky as a surface, the inside face of a gigantic sphere centered upon us. Various celestial events and movements unfold themselves across this great vault, which seems infinitely far away. Yet the sky is actually three dimensional: it has depth that is not generally appreciated, and the many phenomena that we have encountered take place at vastly different distances. The sky truly begins at the observer's eye and extends outward in all directions. In this chapter we now move inward from the stars and planets and examine the play of light upon our atmosphere. In doing so, we will also explore some properties of light itself as well as the principal instruments that astronomers use to explore the heavens.

13.1 Light

If left alone, light travels in a straight line from place to place at an astounding 299 792 km per second (the speed called *c*). What is this strange substance? We cannot ever see the light itself in its flight, we can only note its effect as it strikes our eye. It is perhaps most simply described as a traveling electromagnetic disturbance, as fluctuating and alternating electrical and magnetic fields. We actually think of light as existing in two forms that depend on the phenomenon to be described. In some ways it acts as an up-and-down wave, much as one finds on the surface of a pond. At other times its behavior is more akin to a stream of tiny particles. These two concepts are embodied into a particle called a *photon*, in a loose sense a piece of light, a very short segment of a wave.

The photon is the fundamental transmitter of energy in the Universe. The amount contained by one depends on its *frequency* or *wavelength*. The frequency ν (sometimes f) is the number of individual wave crests within the photon that an observer would count passing by per second (Figure 13.1). It is expressed in cycles per second, or *hertz*, Hz (after the German physicist Heinrich Hertz). The wavelength (λ) is the physical distance between individual wave crests expressed in meters or some convenient fraction thereof. Frequency and wavelength are related through the speed of the wave,

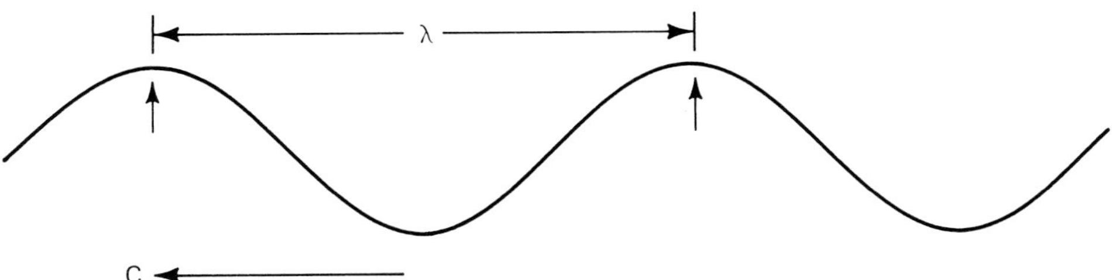

13.1. Light. A photon moves to the left at the speed of light, c. The separation between crests is the wavelength, λ. An observer at the vertical arrows would count waves at a frequency of ν crests per second.

which must be the number of crests times the distance between them, or $c = \lambda\,\nu$. The energy, E, transported by a photon is directly proportional to frequency, or $E = h\nu$, where h is a constant of proportionality called *Planck's constant*. Since $c = \lambda\,\nu$, energy is inversely proportional to wavelength, or $E = hc/\lambda$.

13.2 The spectrum

There are no real limits in nature on the wavelength or frequency of light, nor on the energy that can be carried by a photon. These quantities range over huge factors, their array being called the *electromagnetic spectrum*. Different segments of the spectrum go by different names (Figure 13.2). Except for their energies, they are conceptually all the same, however, and *all* move at c. We are most familiar with the optical radiation that excites vision in our eyes, but which occupies only a short section, a mere octave, of wavelength out of a range that spans a factor of nearly 10^{20}. In the short-wave part of the spectrum, we generally employ small units of wavelength, either the ångstrom, Å (10^{-8} cm), *the nanometer*, nm (10^{-9} meter, 10^{-7} cm, or 10 Å), or the micrometer, or *micron*, μm (10^{-6} meter, 10^{-4} cm, or 10 000 Å). The human eye is sensitive to wavelengths from just under 4000 Å to somewhat over 7000 Å. Within this span we loosely cite photon energies by *color*: violet at 4000 Å, through blue (4500 Å), green (5000 Å), yellow (5500 Å), orange (6000 Å), and red (6500 Å and beyond). These divisions are for broad convenience only: they all blend gradually into one another and the marvel that is our eye can separate thousands of shades. The pure spectral colors are distinct from physiological color, wherein the eye can be fooled. Green, for example, can be produced by mixing blue and yellow, but that does not mean that green photons (the photons are not green, the color refers only to wavelength) are actually created, only their effects. The discussion here involves only true energy, frequency, and wavelength.

The numbers involved can be astonishing. The frequencies of optical waves are very high: 5.5×10^{14} Hz at 5500 Å. Planck's constant, h, is measured in the laboratory to have a value of 6.6×10^{-34} joule seconds in the common meter–kilogram–second system of units. Consequently, the energy carried by this yellow photon is a minuscule 4×10^{-19} joule. The joule is a unit of energy; one watt, a unit of power, corresponds to an

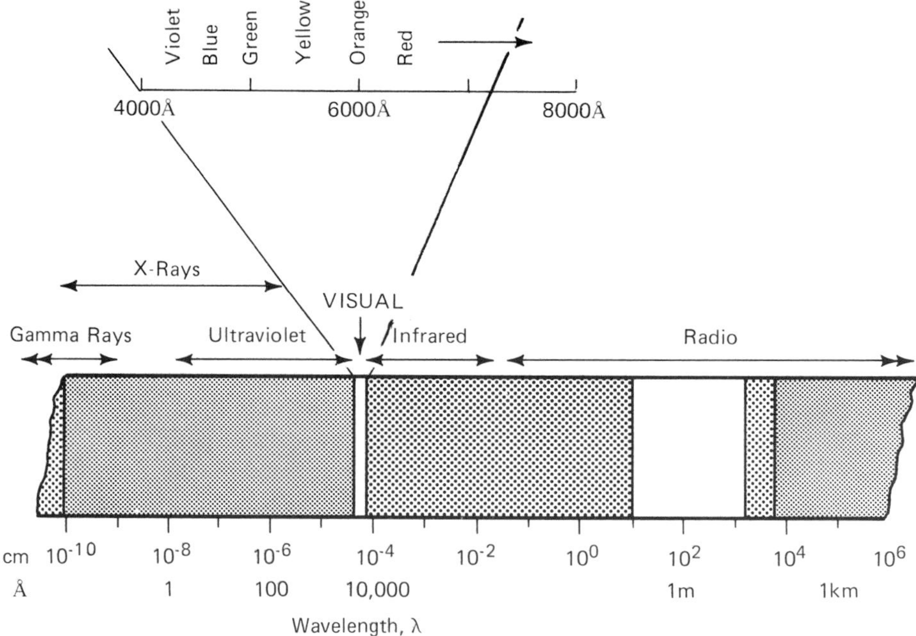

13.2. The electromagnetic spectrum. Arranged according to wavelength are gamma (γ) and X-rays, ultraviolet, optical, and infrared light, and finally radio waves. Wavelengths are expressed in centimeters, and for shorter waves also in aångstroms, where 1 Å = 10^{-8} cm. One meter and kilometer are also noted. The optical region is expanded above to show the wavelengths of the principal colors. The transmission of radiation through the Earth's atmosphere is approximately shown by the degree of shading. The atmosphere is quite transparent at optical wavelengths, and fortunately very opaque at higher energies (lower wavelengths). The infrared and radio show several bands of clarity through which we can look to view the Universe. (Adapted from *Principles of Astronomy, A Short Version*, S. P. Wyatt and J. B. Kaler, Allyn and Bacon, Boston, 1981.)

energy flow of one joule per second. A hundred watt light bulb, allowing for a distribution of photons among the wavelengths, therefore produces somewhat over 10^{20} photons every second; the Sun, with a radiant power of 10^{26} watts, generates over 10^{44} of them in that short interval of time.

As we pass to both higher and lower energies, the photons become invisible. Shorter than violet lie the *ultraviolet* (UV) rays. These begin to become dangerous to living organisms. Some of the longer-wave UV radiated by the Sun (the *near ultraviolet*) sneaks through our atmosphere and causes tans and sunburns on the skin. The shorter *far ultraviolet* is quite hazardous and can quickly cause severe burns; fortunately none of it can penetrate our atmospheric blanket. Farther along we reach the highly energetic *X-ray* photons between roughly 1 and 100 Å that can penetrate human tissue, even metals. And at the highest energies we find the *gamma* (γ) *rays*, exceedingly dangerous radiation produced in copious amounts in the solar interior and in nuclear bomb explosions.

In the other direction, beyond the red, we encounter the *infrared* (IR). We are sensitive to the shorter waves, the *near IR*, which we feel as heat. Most of the mid and *far* (long-wave) *IR* is again blocked, and in fact trapped, by the Earth's atmosphere, producing the *greenhouse effect*, which keeps our planet warm enough for habitation. Gradually, the IR merges into the radio part of the spectrum, which is vast and includes everything longer than about one millimeter. It is broken into several subdivisions according to use. The shortest and most energetic waves are called *microwaves*, used in radar and in ovens where they are absorbed by water; they can be hazardous to human tissue. Television and FM radio are broadcast in the few-centimeter to a meter region, followed by short-wave radio. The standard AM band and then long-wave radio complete the picture. Much of the radio spectrum penetrates the atmosphere, allowing the science of radio astronomy. However, in the so-called short-wave radio bands the atmosphere is a good reflector (actually refractor), which allows us to bounce transmissions off it, making long-distance radio communication possible.

We therefore see a vast variety of photons with a multiplicity of traits and uses. Physically, they are all the same, all carriers of energy, but each wavelength has its own unique characteristics.

13.3 Reflection of light

Electromagnetic radiation travels in a straight line except when it encounters some kind of barrier. Then its direction may be changed and bent by a variety of interactions that must be examined before we can look at the wealth of atmospheric phenomena the sky has to offer. The simplest is *reflection*. If a beam of light approaches a reflecting surface – a mirror or a pool of water – it bounces off at the same angle with which it arrived (Figure 13.3a). Consequently, an image reflected from a smooth surface will be perfect except for reversal. Wavelength is immaterial, and therefore there will be no color distortion. If the surface is uneven, the rule still applies, but the numerous changes in pitch angle will roughen the image: ripples in a pool will produce ripples in a face, and a concrete roadway will generate only a diffuse light from the Sun.

13.4 Refraction

We encountered this phenomenon in Section 7.2, where we had to examine the effect it had on the position of the Sun in the sky, and only some elaboration is needed here. The speed of light, or of any kind of radiation, depends on the material through which is traveling. Light achieves a velocity of c only in a vacuum. In water the speed drops by roughly 20% to 240 000 km/s. The speed is mostly dependent on density. In glass, it is down to 187 000 km/s, and in diamond it fairly creeps at 130 000 km/s. Even in our tenuous air, the effect is measurable, with the speed of light diminished by 0.002%. This reduction in velocity causes a ray of light to alter its direction upon entering one substance from another. The change in direction will be toward the perpendicular to

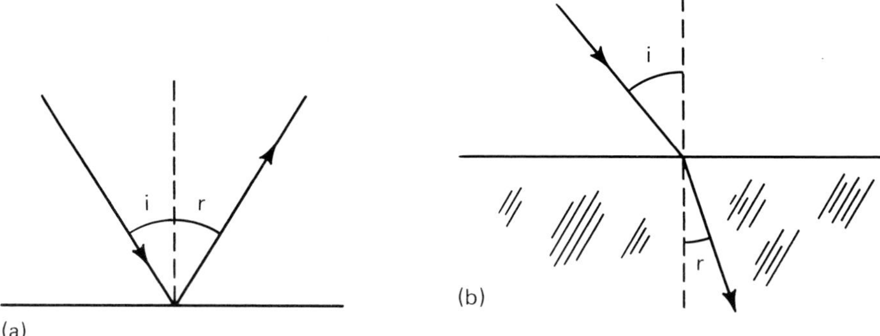

13.3. Reflection and refraction of light. The dashed line (the *normal*) is perpendicular to the surfaces at the points of incidence of rays of light. (a) In reflection, the angle of reflection, r, equals that of incidence, i. (b) In refraction, if the light enters a dense substance from a tenuous one (as in the drawing), it bends toward the normal and the angle of refraction, r, is less than i. If the light path and thus r and i are reversed, that is, if the ray flows from a more to a less dense substance, r is greater than i.

the surface if the light goes from a tenuous to a dense material (Figures 13.3b, 7.2, and 7.3) and away from it if the light beam moves from a dense material into a tenuous one. That is, in Figure 13.3b, the path stays the same no matter which way the light is going.

The change is mathematically predictable. Every non-opaque substance has an *index of refraction*, n, the ratio of c to the speed of light in that material. The amount of bending, $(i - r)$ in Figure 13.3b (or $r - i$ if the direction and the lettering of the angles are reversed), thus depends strictly upon the angle of incidence, i, and n, all three increasing or decreasing in concert (trigonometrically, $\sin r/\sin i = n$; see Appendix A3.1). The result is the distortion of any object seen through another.

13.5 Dispersion

Unlike reflection, refraction is sensitive to wavelength. Look again at the ray of light of Figure 13.3b, redrawn in Figure 13.4a. This time it is white light composed of many rays of all colors. The speed of light in matter depends not only on the nature of the substance but also upon wavelength. With few exceptions, n increases as λ gets shorter: upon striking water from air, violet light bends more than blue, and green more than red. The effect is to spread, or *disperse*, a ray of white light into its component colors, or its *spectrum*. Refractive dispersion (or the degree of spread at a given wavelength) decreases as we proceed toward longer wavelengths: the red end of the spectrum is compressed and the blue end will be more spread out. Light will be equally dispersed upon exiting a substance, as seen in Figure 13.4b, which demonstrates the effect of a standard dispersing device, the prism.

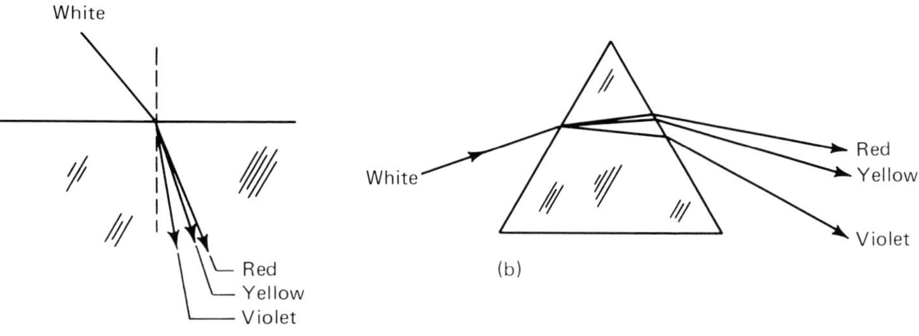

13.4. Dispersion of light. (a) The refraction of light depends on its wavelength, so that in the blue component of the incoming white ray (in which the colors are mixed) is bent more than the red. (b) The white light enters the prism from the left, and is dispersed both on entering and again upon exiting the glass.

13.6 Diffraction

Light is also bent, or *diffracted*, as it passes a barrier. In Figure 13.5a, light of one color (*monochromatic*) impinges upon a plate with two very narrow slits (a and b) cut onto it. We might expect that on a wall some distance away we would see two long lines, images of each aperture. Instead we find something quite different, and at first inexplicable, a bright line centered *between* the slits, and then parallel bands or *fringes* fading away on each side (Figure 13.6).

Each of the slits acts as if it were an original source of light with the two new sets of waves expanding outward together in step, or in *phase*. The center position, c_o in Figure 13.5a, is equidistant from each slit, and the two waves must fall on top of each other with their energies adding together, as in Figure 13.5b. The phenomenon, *constructive interference*, produces a bright patch of light. Now we move away from c_o, say upward in Figure 13.5a, and as we do we move farther from each slit, but less far from b than a. At point $d_{1/2}$ on the wall the difference in distance is exactly one-half of a wavelength. The waves from the slits then fall shifted from one another by one-half a wavelength, as in Figure 13.5c. Now, they cancel, the crest of one filling the trough of the other. The effect, *destructive interference*, results in a zone of darkness. A little bit farther upward (c_1 in Figure 13.5a), the difference in distance will be exactly one wavelength, and once again we get constructive interference and another bright line or fringe. This pattern of light and dark will continue as we pass through the $d_{3/2}$ dark zone positioned between c_1 and c_2, c_2 (light), $d_{5/2}$ (dark), etc., the fringes gradually fading away. The identical effect must be seen on the other side of c_o as well. In between c_o and $d_{1/2}$, and between $d_{1/2}$ and c_1 in Figure 13.5, we get partially constructive interference, and a gray area; that is, the fringes do not have sharp edges, but fade into one another, looking like the pattern in Figure 13.6.

A single slit produces the same kind of fringes. The opening in Figure 13.7 can be

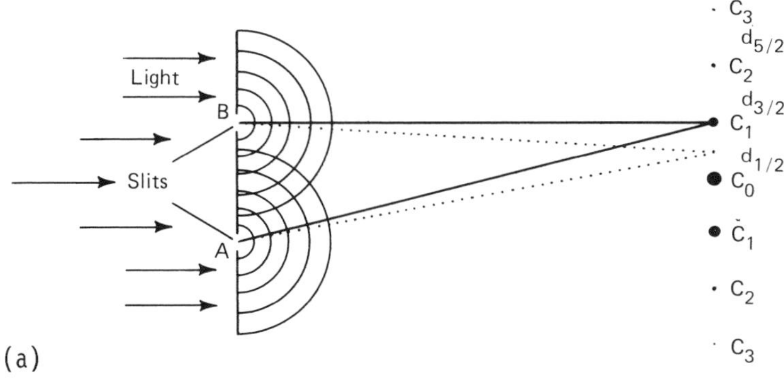

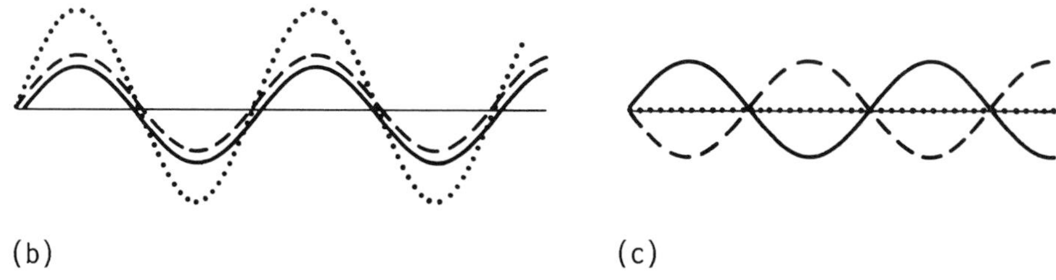

13.5. Diffraction of light. (a) A beam of light encounters a plate with two vertical slits (*A* and *B*) placed perpendicular to the page. Position c_o is equidistant from the slits so that the light waves arrive in phase and constructively interfere as in (b), producing a bright spot. At position c_1, the distance from slit *A* is λ less than to slit *B*, the waves (solid lines) are offset by λ, and we again see constructive interference and a bright spot, as we will also at c_2, c_3, etc. At $d_{1/2}$ (dotted lines) the difference in distance to the two slits is $\frac{1}{2}\lambda$, the waves are shifted by $\frac{1}{2}\lambda$, and destructively interfere and cancel as in (c), resulting in a dark spot. Destructive interference will also occur at $d_{3/2}$, $d_{5/2}$, etc. The resulting image will then be a series of bright and dark bands, or *interference fringes*, parallel to the slits.

broken into imaginary pairs of slits $a - a$, $b - b$, etc., each of which will produce a pattern and which will sum to produce diffuse fringes. The effect can be seen by holding two fingers close together near the eye. Just before they touch, you will see a dark-light fringe pattern between them.

Diffraction will also disperse light. Instead of monochromatic light in Figure 13.5, imagine a purple beam consisting of red and blue. The wavelength of red light is longer than it is for blue. The position on the wall for which the red waves are off-shifted by one wavelength (the c_1 position on the figure) must be a bit farther away from c_o than it is for the blue waves. The first bright fringe, as well as all the others, will then be colored blue on one side, red on the other. Next imagine white light of all colors. Each fringe of Figure 13.6 will now be a tiny spectrum, just as we would see if light were

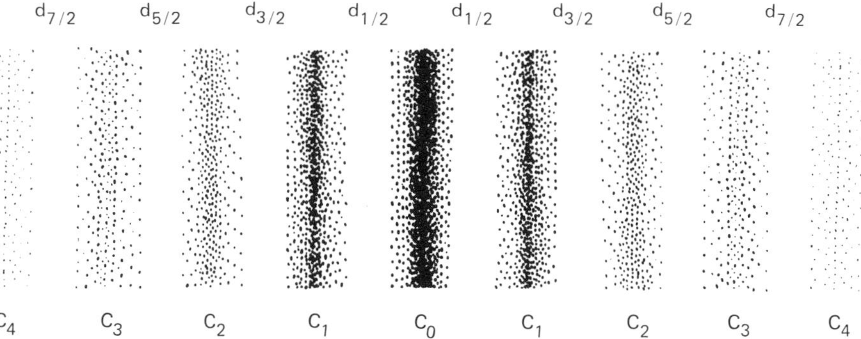

13.6. Interference fringes. That at c_0 is centered between the two slits of Figure 13.5. If the incoming light is a mixture of colors, the colors will be separated to produce colored fringes, that is, the interference pattern for the longer-wave red light will be shifted farther away from the central position than will the pattern for the blue light. The more slits there are, the better the color definition. The result is a large number of individual spectra.

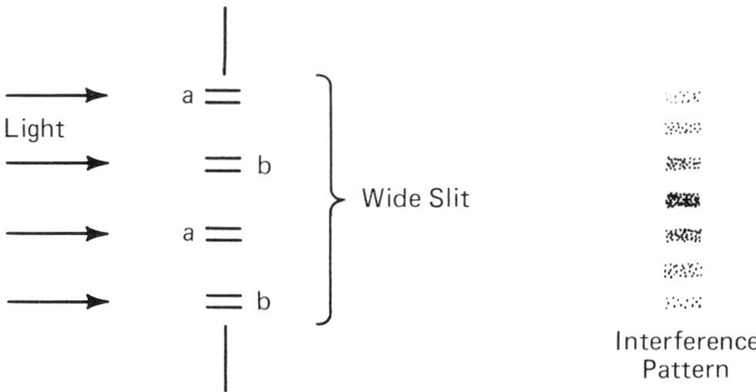

13.7. The single slit. It can be broken into an infinite set of virtual slit pairs, exemplified by *a* and *b*, which add together to produce interference fringes.

dispersed by a refracting prism, except that the colors are reversed, with red bent the most instead of blue.

If we add more slits the color separation becomes clearer since the positional requirement for perfect constructive interference at each wavelength becomes more stringent. Diffractive dispersion with hundreds or thousands of slits will produce magnificent spectra, one at each position marked c_1, c_2, c_3, etc. in Figure 13.6. Each fringe or spectrum is called a spectral *order* (c_1 is the first order, c_2 the second, etc). The higher the order number, the longer or more spread out the spectrum. Diffractive spectra become complicated by the overlapping of orders. A wavelength of 3500 Å in the second order overlaps with 7000 Å in the first. The separation for first and second orders is no problem visually, but the higher orders will blend and be confused with

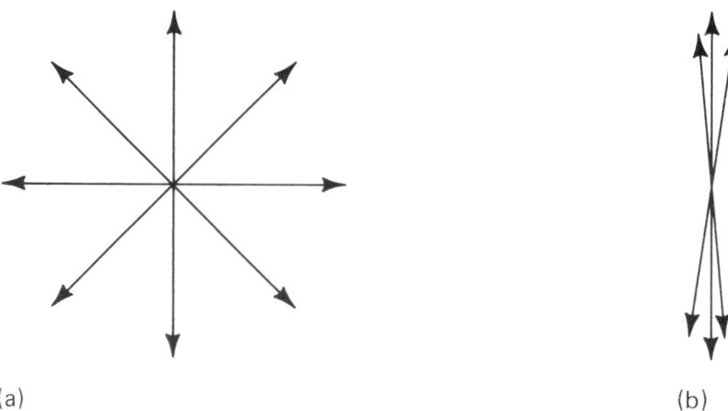

13.8. Polarized light. (a) Non-polarized (or randomly polarized) radiation has waves oscillating in any plane, as seen in the approaching beam. (b) Polarized radiation has waves that vibrate over a restricted band of planes or angles.

one another. The whole effect is easily seen by looking at light reflected from the finely grooved surface of a compact audio disc.

Diffraction spectra are produced for astronomers by diamond-ruled *gratings*, which have almost completely replaced the glass prisms originally used for such purposes. Unlike prismatic spectra, these have uniform color distribution, that is the colors are not jammed together in the red. The orders can be discriminated from one another by the use of filters.

13.7 Polarization

An ordinary beam of light consists of countless individual waves that can oscillate in any plane. If you could stand in the ray and see the photons as they approach you, some would be vibrating up and down, some from side to side, and others at any specified angle in between (Figure 13.8a). However, it is possible through a variety of natural and technical processes to restrict the plane of vibration, so that all the waves oscillate in the same direction (Figure 13.8b). Such radiation is said to be *polarized* (more specifically, *plane-polarized*). Radiation reflected at an angle from a surface is always partially polarized, that is, more light oscillates in one direction than another; the percentage concentrated into one general direction increases along with the angle of reflection.

The transmission of light through various substances will also produce polarization. A regular molecular or crystal structure can filter out the photons that oscillate in all but a narrow range of directions. Materials like mica are very effective. The most common examples are polarizing sunglasses. The effect is immediately seen by taking the sunglasses apart (use a cheap pair of clip-ons) and looking through them both while rotating one with respect to the other. When they are crossed, one blocks most of the light passed by the other and the scene turns dark, making an effective filter (but *not* one dark enough to look at the Sun with – do not try it). Properly designed polarizing

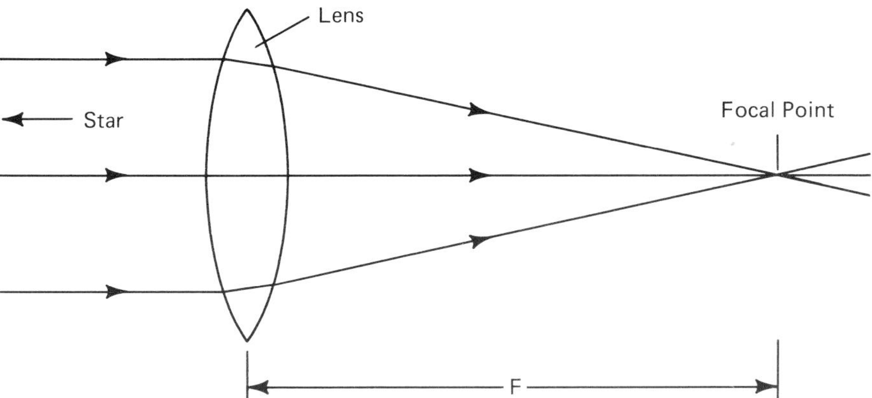

13.9. The refracting telescope. The lens gathers light from a star, which is refracted to a focal point a distance *F* away from the lens, where it creates an image.

filters are used in modern sextants (see Figures 8.5 and 8.6) so that an altitude of the Sun can be obtained. Since reflected light, such as the glare from a roadway, is polarized, we eliminate it by placing the sunglass lenses in such a direction as to filter it out, greatly improving driving visibility. Another example is laser light, which is very highly, really almost perfectly, polarized.

Radiation – none of the above discussion is necessarily confined to optical waves – can be *circularly polarized* as well. If we could again watch a beam of approaching light, we would now see the preferred plane of polarization rotate in either one direction or the other: if counterclockwise (looking in the direction of motion) it is called "right-handed," and vice versa.

13.8 Telescopes

Astronomical telescopes nicely illustrate some of these concepts. They are among the simplest of scientific instruments. The entire purpose of the telescope is to gather a large amount of light, which is funneled to other devices for analysis (for example, a camera or a *spectrograph*, which contains a diffraction grating for the purpose of generating a spectrum). There are two kinds of telescopes: one makes use of the principle of refraction, the other of reflection. The refractor is the older, and provides the initial example. Light waves from a star fall upon a curved convex (curved outward) *lens* in Figure 13.9 and are bent on both entrance and exit toward a common *focal point* a *focal length*, *F*, away. There they cross to produce an *image* of the star.

Telescopes are referred to by the apertures or diameters (D) of their lenses or *objectives*, expressed in centimeters, meters, or inches. The larger the objective, the more light it can collect and the brighter the image will be, and conversely, the fainter the star that we are able to detect. The *light-gathering power*, which directly controls the apparent brightness of a star, depends on the surface area of the lens and thus on the square of the aperture.

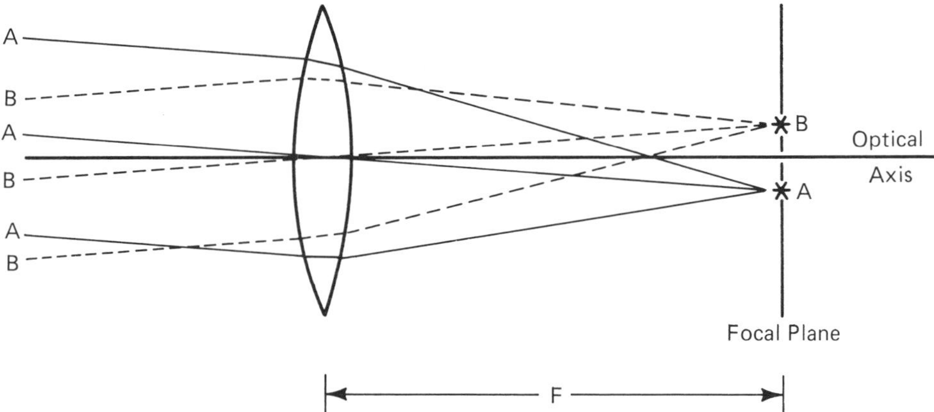

13.10. The focal plane. Stars *A* and *B* lie above and below the optical axis. Parallel light rays from each then enter the lens respectively downward and upward, and are imaged in reverse order on the focal plane.

If we have two stars in the field of view, their images are both formed on the *focal plane*, parallel to the lens. In Figure 13.9 the star is on the *optical axis*, the line perpendicular to the lens at its center. In Figure 13.10, the telescope looks at two stars that lie above and below the optical axis. Their images are both formed at a distance *F* from the lens, but in reverse: astronomical telescopes always yield images that are upside-down and left-for-right. A photographic plate (or modern electronic detector) placed in the plane can record the image. The lens and detector (or other instruments) are rigidly assembled into a tube that can point to different parts of the sky (see Figure 3.21) to complete the instrument (Figure 13.11).

To view the stars directly, we must use a second lens or *eyepiece*, which is ordinarily placed beyond the focus, and whose purpose is to render the light rays parallel for focusing by the eye. With the eyepiece, we examine the image formed on the focal plane, and magnify the scene, that is, we see the stars separated by a wider angle (Figure 13.12). Note that the image is still reversed. The degree of magnification, the *magnifying power* (commonly just called the "power"), of the telescope is the ratio of the apparent angle between the stars as seen through the eyepiece to that seen directly in the sky. It depends on the ratio of the focal lengths of objective and eyepiece, or $M = F_{obj}/F_{eyp}$. The magnifying power can easily be changed by replacing the eyepiece with one of longer or shorter focal length. A concave (curved inward) eyepiece placed in *front* of the focal plane has the same effect. This latter system, used by Galileo in the first telescopes, has the advantage of a right-side-up image and is still used in inexpensive opera glasses. However, the power cannot be raised very high and the field of view is small.

13.9 Reflecting telescopes

We can image the sky by reflection from a curved mirror (Figure 13.13a) to create a *reflecting telescope* (Figure 13.14). The light is reflected from a thin surface of aluminum

13.11. The Yerkes 40-inch (1-meter) refractor, the largest in existence, located in Williams Bay, Wisconsin (USA). The entire observing floor moves up and down so as to place the observer at the focus. (Yerkes Observatory.)

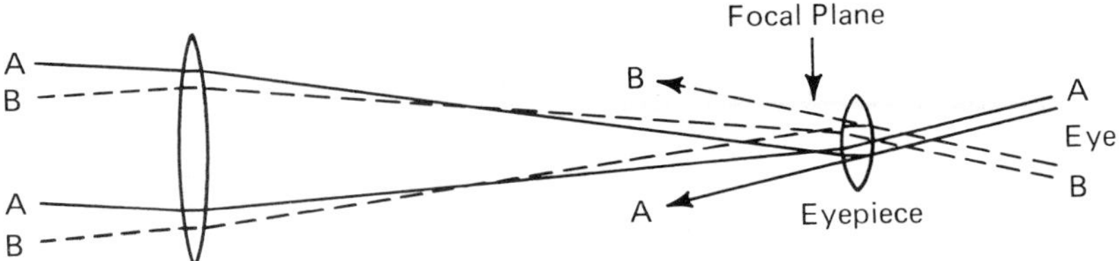

13.12. The eyepiece. The lens at the right renders the light rays again parallel so that they can be focused by the lens of the eye. It also magnifies the view. The images of stars *A* and *B* are seen projected outward separated by a wider angle, but reversed.

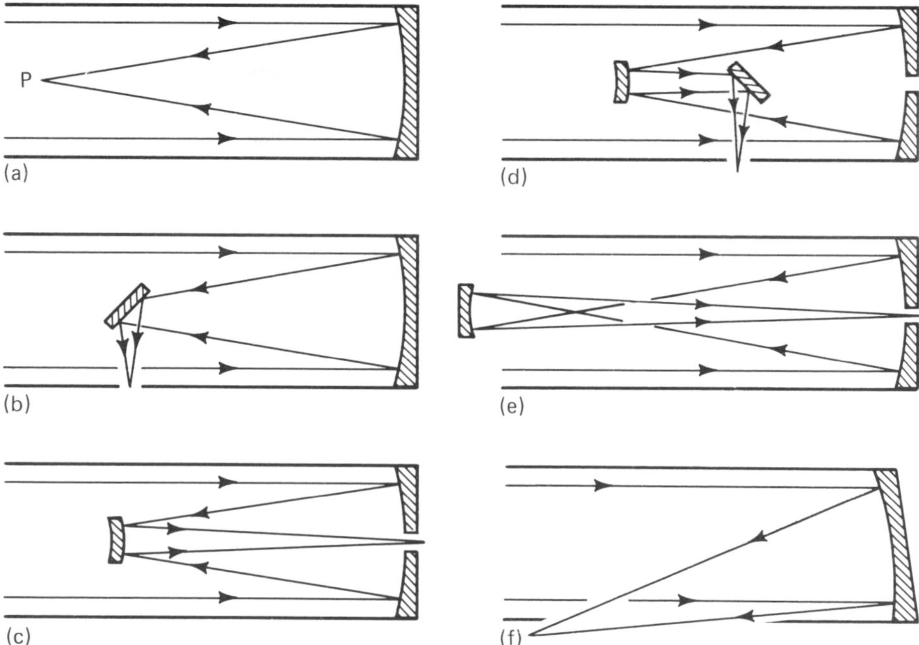

13.13. The reflecting telescope. The paraboloidal mirror focuses the light. (a) The *prime focus* is denoted by *P*; (b) the *Newtonian* uses a flat secondary to reflect the light to the side; (c) the *Cassegrain* employs a convex, hyperboloidal secondary to extend the focus and send the light through a hole in the primary; (d) the *Nasmyth* is a hybrid of Cassegrain and Newtonian that sends light to the side to a fixed position on the telescope's axis; the common *coudé* has a similar arrangement except that the light is again reflected by additional mirrors to a fixed position within the observatory; (e) the *Gregorian* uses a concave ellipsoidal secondary beyond the prime focus; (f) the *Herschelian* (*shiefspiegler*) dispenses with a secondary and uses a tilted primary.

deposited on the front side of a ground and polished glass or ceramic concave surface. For the light to be brought to a proper focus, the mirror must be in the shape of a paraboloid, the figure generated by rotating a parabola (see Figure 11.17) around its axis. The focal plane of a single mirror (again, the *objective*), called the *prime focus*,

13.14. The five-meter Palomar reflector in California seen pointing to the zenith. The telescope tube is an open grid to decrease weight. The primary is at the bottom. The telescope moves in hour angle by rotating the yoke seen to the right. The instrument can be used at the prime, Cassegrain, and coudé foci by moving different mirrors into the light path. (Palomar Observatory, California Institute of Technology.)

appears in front of the mirror, and unless the reflector is very large, the act of observation would block most or all of the incoming light. Consequently, we ordinarily add a small secondary mirror, usually in front of the prime focus, to send the light to a more convenient location.

The variety of reflectors is shown in Figure 13.13. The prime focus (Figure 13.13a) is actually used in the largest instruments, sometimes with the observer installed in a small cage; it is also commonly employed in specialized photographic instruments. The most common form of small reflector, however, is the *Newtonian* (named after its inventor, and the inventor of the reflecting telescope itself, Isaac Newton), which uses a flat secondary set at a 45° angle before the prime focus (Figure 13.13b) to deflect the light to the side of the instrument. The most common professional telescope design is the *Cassegrain* (Figure 13.13c), which uses a convex (hyperboloidal) secondary to extend the focal length and send the light straight back and through a hole drilled in the objective. There are also variants of this system called the *Nasmyth* (Figure 13.13d), which employs a third mirror to send the light to the side along the telescope's rotation axis, and the *coudé*, which by means of additional mirrors extends the Nasmyth focus to a fixed position in a room below the instrument. The *Gregorian* (Figure 13.13e) is a less common system that employs a concave ellipsoidal secondary placed beyond the

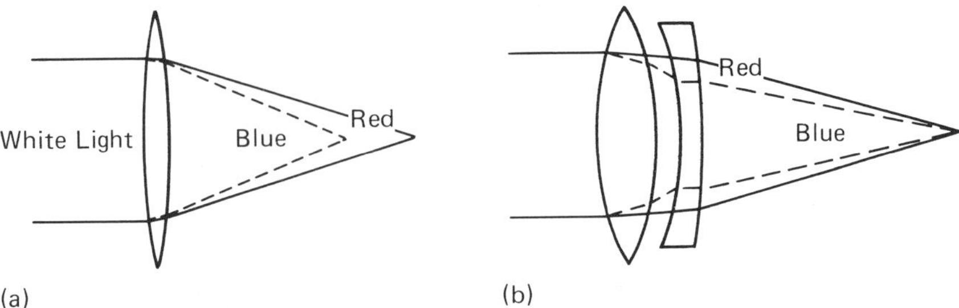

13.15. Chromatic aberration and the achromatic lens. (a) The component colors of the white light, here represented by red (solid) and blue (dashed), are focused onto a line along the optical axis. (b) The *achromatic* lens consists of two single lenses, a convex one made of soft (lower refraction) glass and a concave one of hard glass, which together can bring any two colors, again here red and blue, to a focus.

prime focus. Another unusual design is the *Herschelian*, or *shiefspiegler* (Figure 13.13f), in which the primary mirror is tilted to place the prime focus outside the tube. It has the advantage of avoiding losses of light caused by obstruction and secondary reflection, but produces distortion in the images.

13.10 Telescope aberrations

All optical systems are afflicted with problems or *aberrations*. The worst is caused by dispersion in the lens of the refractor. Since short waves are refracted more than long ones, a simple lens cannot bring a beam of white light to a single focal point, and the focal length depends on color (Figure 13.15a). An eyepiece focused on the red image yields a violet blur and vice versa. This problem of *chromatic aberration* can be controlled (but not eliminated) by placing two lenses of different curvatures and refractive indices together into an *achromatic* (meaning "color free," something of a misnomer) double lens, or *doublet*, which can bring any two selected colors together (Figure 13.15b). Most astronomical refractors are of this sort, and are designed to focus red and yellow, which will still produce a violet blur. A three-lens triplet can focus three colors, a quadruplet four, but these are not common since the additional glass absorbs more light, rendering the image dimmer. This aberration is also diminished by increasing the *focal ratio*, the focal length divided by the aperture, which is why refractors are generally so long. Before the invention of the achromatic lens, this solution led to ponderous lengthy instruments that were extremely difficult to point and use (Figure 13.16). Modern refractive materials are also available that suppress refractive dispersion.

Chromatic aberration is entirely avoided with the reflector, although it is afflicted with aberrations of its own. The principal one is *coma*, which causes images to become progressively worse away from the optical axis. Fortunately, it can nearly be eliminated by figuring the mirrors into more complex shapes and by adding corrective optics. Chromatic aberration and coma plus several other aberrations (*astigmatism*, in which different zones of the mirror or lens produce different foci, and various types of

13.16. A seventeenth-century refractor constructed by Hevelius (Johannes Hevelii) in the early 1600s. The huge focal length was needed to suppress chromatic aberration. (From Hevelius *Selenographia*, Danzig, 1647.)

distortions) are overcome with hybrid systems, such as the *Schmidt reflector* (Figure 13.17), which uses a large non-focusing correcting lens to change the ray paths before they arrive at a spherical primary mirror. This telescope was used to create the Palomar Observatory Sky Survey illustrated in Figure 8.3.

13.11 Resolving power

Second only in importance to light gathering power (see Section 13.8) is the related ability of a telescope to resolve fine detail. The finite aperture of a telescope acts as a large slit that produces a diffuse central image and diffraction fringes. Since the aperture is circular, so will be the fringes. Diffraction smears out the light, so that even though stars are point-like, their images will be disks (analogous to the central fringes in Figures 13.5, 13.6, and 13.7). The phenomenon ultimately limits the ability of any telescope to resolve fine detail.

The larger the instrument, the smaller the diffraction disk, and the better the view. The *angular resolution*, or *resolving power*, of an optical telescope, the minimum angular distance in seconds of arc between two distinctly separated stars (defined as the separ-

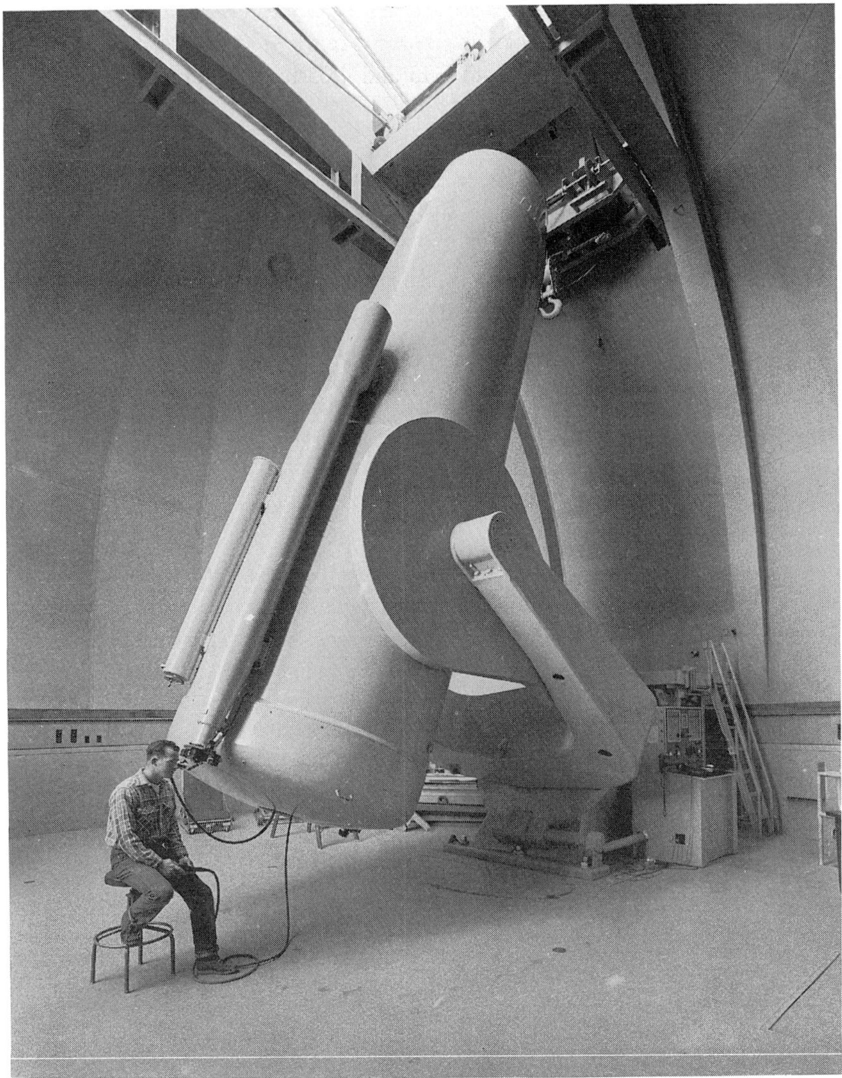

13.17. The Palomar Schmidt telescope. A 48-inch (1.2-meter) correcting lens distorts and sends the waves to a spherical 72-inch (1.8-meter) mirror, resulting in an excellent wide-field image with little aberration. A typical photograph is shown in Figure 8.3. (Palomar Observatory, California Institute of Technology.)

ation whereby the diffraction disk of each star falls into the first dark fringe of the other), can be expressed as $5''/D$ inches = $13''/D$ cm, where D is the telescope aperture. In practice, this formula works only for small instruments; for larger ones the resolution is effectively limited by the turbulence of the Earth's atmosphere (see Section 13.23) to somewhat under one second of arc.

Reflecting telescopes are subject to yet another diffraction effect, one produced by the struts that hold the secondary mirror. These supports are usually in the form of a cross, so that bright stellar images on photos taken by reflectors will commonly display crossed *diffraction spikes* (see Figures 4.13 and 4.25 for examples).

The problem of resolving power is more serious for radio telescopes (Figure 13.18), which like their optical counterparts, are designed to collect radiant energy with a paraboloidal reflector and focus it onto a detector, now a radio antenna. Since diffraction effects depend on wavelength, and since radio waves are a million or so times longer than optical, the diffraction disks are huge. We can only diminish them to achieve reasonable resolution of detail by making the telescope concomitantly large. To achieve just the resolution of the human eye at a wavelength of one centimeter, the telescope would have to be 125 meters across; at a wavelength of one meter it would need to be 100 times bigger, or 13 kilometers! The biggest radio reflector is 300 meters in diameter (which then approaches human resolution at short wavelengths), as opposed to 10 meters for the largest optical telescope (and one meter for the largest refractor: see Table 13.1).

The problem can be overcome by electronically linking two or more separated radio reflectors together to synthesize the effect of a single instrument, the two acting like the double slit in Figure 13.5. Such a device, called an *interferometer*, allows us to observe a diffraction pattern of the part of the sky being observed, from which the actual view can be reconstructed. If we link several telescopes together into an *array*, we get a still-better view with more detail. The prime example of an array is the *Very Large Array* (Figure 13.17) in New Mexico, which produces the effect of a telescope 43 kilometers in diameter and very high resolving power that can be superior to that achieved optically. Specialized techniques, which employ observations synchronized by atomic clocks, even allow observations by telescopes separated by thousands of kilometers, enabling them to resolve to the thousandth of a second of arc. These *very long baseline interferometers* (VLBIs) are also used to establish the rotation characteristics of the Earth and, consequently, time standards (see Sections 6.10 and 6.11). A dedicated array of ten radio telescopes (the *Very Long Baseline Array*) now covers the United States from the Virgin Islands to Hawaii.

13.12 Observatories

Telescopes are ordinarily mounted on twin axes that are aligned to the celestial poles and equator, calibrated so they can be set precisely in hour angle and declination (see Section 3.17 and Figure 3.21), although occasionally one may be designed to operate in altitude and azimuth with conversions made through the astronomical triangle by computer. In either case they are equipped with drive motors that automatically track a star across the sky at the sidereal rate. Optical instruments are nearly always housed in a rotating dome with a movable shutter through which the telescope points (Figure 13.20). This traditional design allows quick access to the sky, at the same time providing protection to the telescope and its attendant devices.

Older observatories are usually found near the institutions with which they are associated, but during this century the emphasis has been to place them on mountain

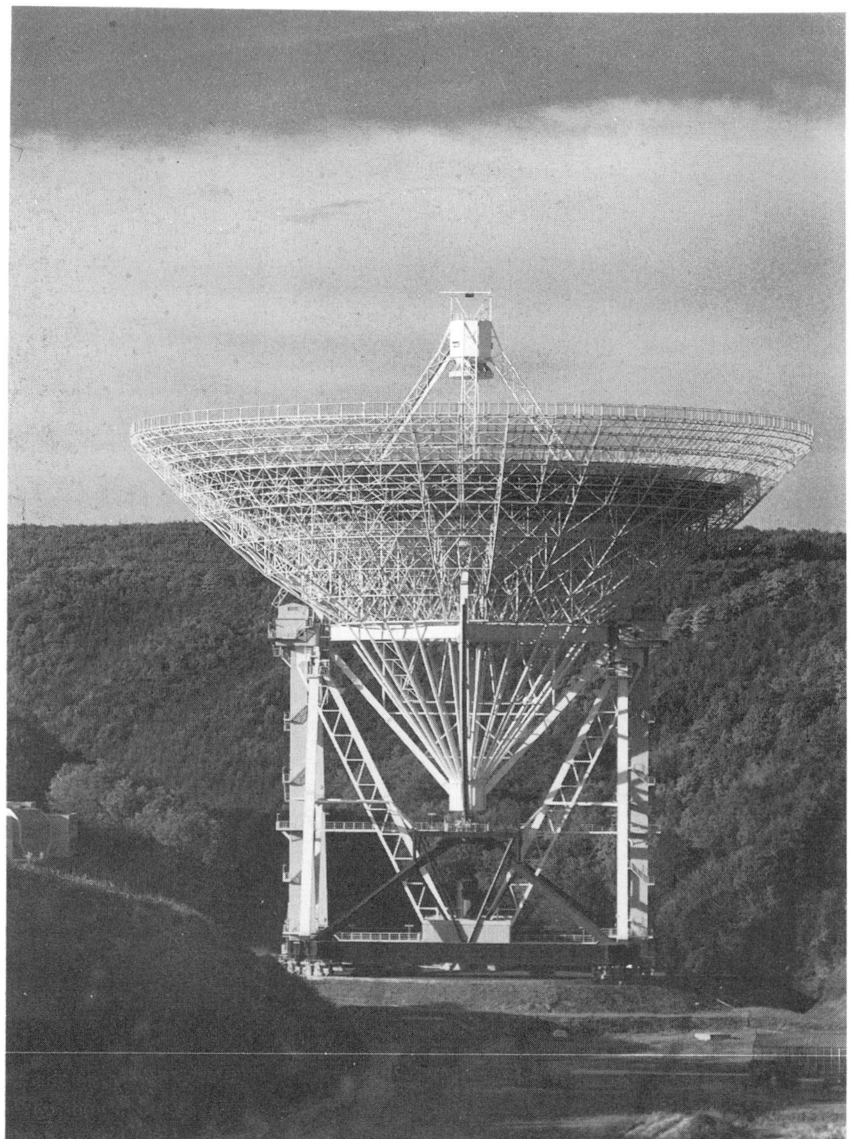

13.18. A single-dish radio telescope. The parabolic reflecting surface, 100 meters across, focuses the radio waves to an antenna mounted atop the four girders that are supported on the dish. This instrument, located near Bonn, Federal Republic of Germany, is the world's largest fully steerable radio telescope. (Max Planck Institute for Radio Astronomy.)

Table 13.1. *The largest telescopes*[a]

Refractors

40-inch (1 meter), Yerkes Observatory, University of Chicago, Williams Bay, Wisconsin, USA

36-inch (0.9 meter), Lick Observatory, University of California, Mt. Hamilton, California, USA

Reflectors

16-meter Very Large Telescope optical array,[b,c] European Southern Observatory, LaSilla, Chile

11.8-meter Large Binocular Telescope, University of Arizona, Arcetri (Italy) Astrophysical Observatory, Mt. Graham, Arizona.[b]

11-meter spectrographic telescope, McDonald Observatory, University of Texas and Pennsylvania State University, Mt. Locke, Texas.[b]

10-meter Keck Telescope, California Institute of Technology and the University of California, Mauna Kea, Hawaii, USA

10-meter Keck II, Mauna Kea, Hawaii[b]

8.3-meter Subaru Telescope, National Astronomy Observatory of Japan, Mauna Kea, Hawaii[b]

8-meter Gemini Telescopes, USA, UK, Canada, Chile, Brazil, and Argentina, two instruments, Mauna Kea, Hawaii, and Cerro Pachon, Chile[b]

8-meters National Optical Astronomy Observatories, Kitt Peak, Arizona, USA, and Cerro Pachon, Chile[b]

8-meters University of Arizona, Mt. Graham, Arizona[b]

6.5-meter Magellan Project, Las Campanas Observatory, Carnegie Institution (US), Las Campanas, Chile[b]

6-meter Caucasus Mountains, Russia

5-meter Hale Observatory, Palomar Mountain, California, USA

4.5-meter equivalent, Multiple Mirror Telescope, University of Arizona and the Smithsonian Center for Astrophysics, Mt. Hopkins, Arizona, USA[d]

4.2-meter Royal Greenwich Observatory, Canary Islands, Spain

4-meter National Optical Astronomy Observatory, Cerro Tololo Interamerican Observatory, Chile

3.9-meter Anglo-Australian Telescope, New South Wales, Australia

3.8-meter National Optical Astronomy Observatories, Kitt Peak National Observatory, Arizona, USA

Radio[e]

43-km Very Large Array (VLA) Interferometer, National Radio Astronomy Observatory, New Mexico, USA

20-km Westerbork Radio Synthesis Observatory Interferometer, Westerbork, The Netherlands

Table 13.1. (*cont.*)

100-meter single dish, Max Planck Institute for Radio Astronomy, Bonn, Fed. Rep. of Germany
100-meter single dish, National Radio Astronomy Observatory, Green Bank, West Virginia, USA[b]
300-meter single fixed dish, Arecibo, Puerto Rico

[a]As of 1994

[b]Planned or under construction.

[c]Array of four 8-meter telescopes with an effective light-gathering power of one 10-meter telescope.

[d]To be replaced by a single 6-meter primary

[e]The largest interferometer had one component in geosynchronous orbit and others on the ground, the components separated by 42 000 km, about three Earth-diameters.

13.19. A radio interferometer. The *Very Large Array*, near Socorro, New Mexico, USA, consists of 27 25-meter radio telescopes that simulate a single telescope that is up to 43 kilometers in diameter. The individual telescopes can be moved to create simulations of different sizes. (National Radio Astronomy Observatory.)

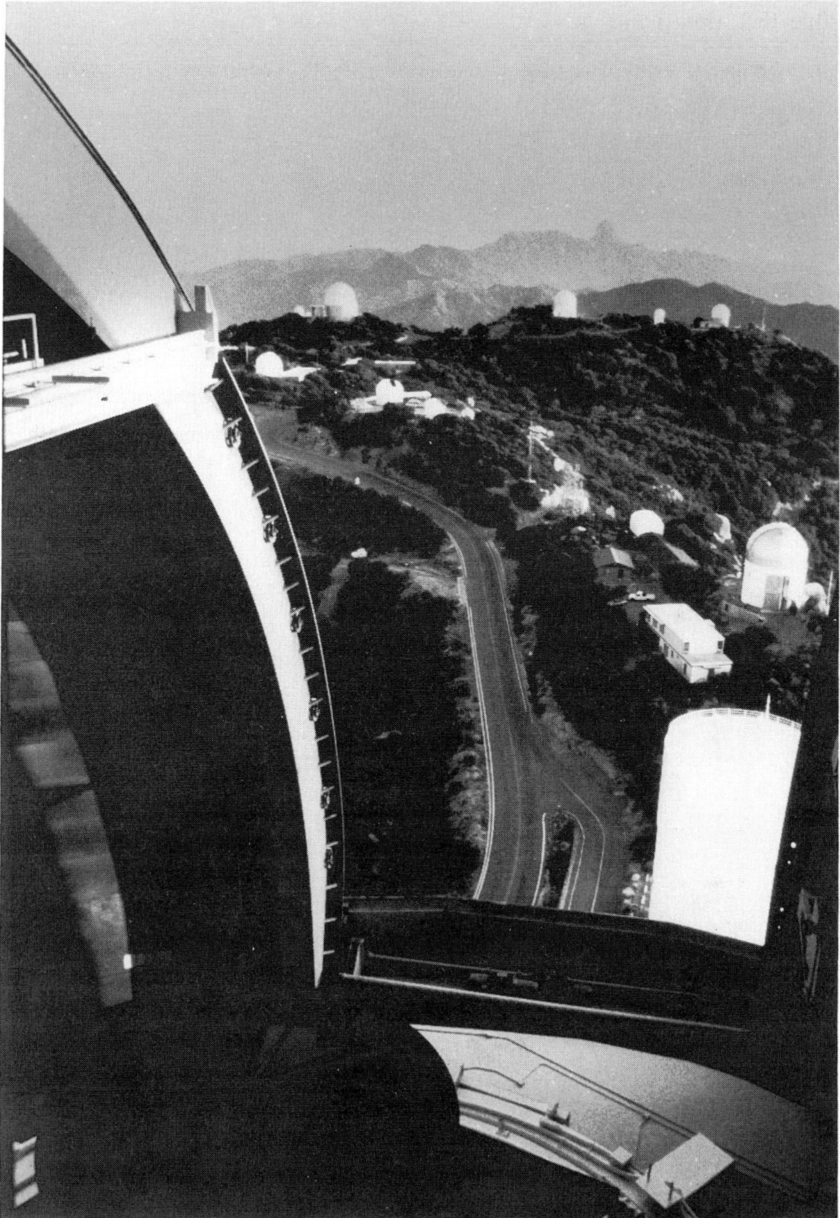

13.20. The Kitt Peak National Observatory. The photograph was taken through the dome-slit of the four meter telescope, and shows the numerous telescopes on the 2000-meter high ridge. (National Optical Astronomy Observatories.)

tops, both to acquire locations as dark as possible and to rise above some of the Earth's absorbing and disturbing blanket of atmosphere (see Section 13.23). Telescopes of different design will frequently be found clustered in prime locations.

The choice between reflector or refractor for the observatory is quite clear, and was made long ago. Astronomical reflectors can be made vastly larger than lenses, resulting in far brighter images, a more distant view, and better detail. And for the same size, the reflector is by far the cheaper. The world's largest refractor, the 1-meter (40-inch) Yerkes Observatory telescope, was built in 1893, and it will almost certainly remain unsurpassed. Refractors must have at least two lenses of high optical quality, and there are four surfaces to grind and polish. The instrument is long and requires a large and expensive housing. The mirror has but one surface, and the interior may have flaws and bubbles, since the light need not pass through it. Moreover, short focal ratios are easy to produce, yielding a relatively smaller device. Most important, lenses and mirrors both sag under the influence of the Earth's gravity. Mirrors can be supported from the back, so large size means less distortion. The locations of the largest instruments as of 1994 are listed in Table 13.1.

Even standard reflectors have their limitations, which are currently somewhere around 8 meters. Modern design is now leaning toward segmented mirrors in which individual mirror components are directed to a single focus to give the effect of a single large mirror. The 10-meter Keck telescope, now the world's largest, is built this way.

No matter what the design, all telescopes are strongly affected by their environments: they can only collect radiation that passes through to the ground. To study the parts of the spectrum that are blocked – particularly the short wavelengths, which carry vital astrophysical information – we have to use orbiting telescopes in space. Among the most successful of these have been the spectrographic 0.4-meter *International Ultraviolet Explorer* (IUE, Figure 13.21a), the *Infrared Astronomical Satellite* (IRAS, Figure 13.21b), and a variety of high-energy X-ray telescopes. The 2.3-meter *Hubble Space Telescope* (Figure 13.21c), launched in 1990 and repaired in 1993, provides a spectacular view of the Universe. A selection of astronomical spacecraft is listed in Table 13.2.

13.13 Amateur telescopes

Discussion of huge and sophisticated instruments makes the telescope and the observatory appear beyond the means of the average person. On the contrary, a great deal of affordable optical equipment, designed to bring in wonderful views of the sky, is available. Start with simple binoculars, paired refractors that allow you to look with both eyes. They are described by a pair of numbers that give magnifying power and aperture. The minimum for nighttime viewing is generally a pair of 7×50 (7 power and 50-mm lenses) that allow you to see the moons of Jupiter, clusters, nebulae, and the star clouds of the Milky Way. The bigger the better, at least up to the point where you have difficulty holding them steadily: the view through 11×75s is quite amazing.

Small telescopes (Figure 13.22) are readily available from a variety of mail-order houses (look for advertisements in popular astronomy magazines). Both refractors and

Table 13.2. *Some astronomical spacecraft*[a]

Name	Launch, status	Function[b]	Region or wavelength	Countries
High Energy Astronomical Observatory (Heao2, Einstein)	1971–72	I, S	X-ray	US
International Ultraviolet Explorer (IUE)	1976 – active	S	UV: 1100–3000 Å	US, UK, ESA[c]
Infrared Astronomical Satellite (IRAS)	1983	I, S	IR: 12, 20, 50, 100 microns	NS
Cosmic Background Explorer (COBE)	1989–91	I, S	microwave radio	US
Röntgen Satellite (ROSAT)	1990 – active	I, S	X-ray	Ger, US
Hubble Space Telescope (HST)	1990 – active	I, S	UV through IR	US
ASTRO	1990	I, S	UV	US
Compton Gamma Ray Observatory (CGRO)	1991	I, S	Gamma ray	US, ESA
Extreme Ultraviolet Explorer (EUVE)	1992	I, S	UV: 70–760 Å	US
Advanced X-ray Facility (AXAF)	planned	I, S	X-ray	
Space Infrared Telescope (SIRTF)	planned	I, S	IR	

[a]As of 1994; planetary probes (e.g. *Voyager*) are not included
[b]I denotes imaging capability; S spectral capability
[c]European Space Agency

reflectors can be obtained, though reflectors are much less expensive. Even a modest 3-inch allows you to see to fainter than tenth magnitude, bringing dozens of celestial sights within view. Again the bigger the better. Simple reflectors of over 16-inch aperture are available, as are Cassegrain instruments with clock drives and even computer controls.

Whatever the level of sophistication, you need only remember that astronomical observing is a learned skill. Use the telescope in the daytime first to align the finder and to get used to focusing. Then at night, examine bright, easily recognizable objects like the Moon and any bright planets that are making their appearances. The views through binoculars are right-side-up, but remember that those through astronomical telescopes are inverted. You move the telescope while looking through it and objects seem to go in the wrong direction. With patience and practice you get used to it and can settle back to a lifetime of enjoyment as the telescopic sky parades before you.

(a)

(b)

(c)

13.21. Space telescopes. (a) The *International Ultraviolet Explorer* is a 0.4-meter spectrographic instrument, which orbits the Earth geosynchronously; (b) The *Infrared Astronomical Satellite*, which orbited Earth in 1983; (c) The *Hubble Space Telescope*, a 2.2-meter reflector launched into near-Earth orbit in 1990. ((a) NASA; (b) NASA/JPL; (c) Space Telescope Science Institute.)

13.14 The blue sky

These examples of optics lead to the real subject of this chapter. The light from our Sun is a mixture of all colors. Solar rays appear yellowish to us simply because yellow photons dominate, but red, green, blue, and violet are present as well, as are those beyond the limits of our vision. The interplay between sunlight and the air provides some of our most common sights, as well as rare spectacles of extraordinary beauty.

Certainly the most familiar aspect of our sky is its remarkable pure blue color, which ranges from the deep, dark hue seen at the zenith from a mountain top to the palest and most delicate of shades near the horizon. The color is caused by another optical phenomenon, *scattering*. Clusters of molecules in the Earth's atmosphere are comparable in size to the wavelengths of visible light. Under these conditions, solar photons can simply bounce off them and change their directions of motion. The efficiency of the process hinges critically upon wavelength: it depends on $(1/\lambda)^4$, so that violet light at 4000 Å is 2^4 or 16 times as likely to be scattered as deep red at 8000 Å. The effect is that the red and yellow rays of sunlight pass through the air relatively unhindered, whereas the blue and violet photons are much more likely to be scattered around and therefore can arrive at our eye from any direction (Figure 13.23). The violet, however, is overwhelmed by the blue since there are more blue photons in sunlight and the eye

13.22. The basic amateur Newtonian reflector, here a 6-inch (15.2 cm), can offer spectacular views of the sky. (Author's photograph.)

is much more sensitive to the somewhat longer rays. So the sky attains its blue shade from a Sun that appears yellowish to the eye; it would be blue no matter what the color of our star. The change in the shade of blue to lighter hues toward the horizon is a result of the increased thickness of air. From a mountain top or airplane, above which the air is thinner, the color of the overhead sky is even richer, almost approaching violet. The Moon produces the same effects: a very clear sky under a full Moon will actually attain a noticeable blue-black color.

Scattering, like reflection, causes partial polarization (see Section 13.7), and so the light from the blue sky is somewhat polarized. Because the origin of the light is the Sun, the direction and magnitude of polarization are symmetrical about it. The plane of the polarized component of skylight is perpendicular to the line to the Sun, and the maximum amount is seen at an angle 90° away. You can spend an interesting few minutes by examining the effect with a lens from a pair of polarizing sunglasses; simply rotate the plastic and watch the sky lighten and darken at different positions relative to the Sun (much the same effect is seen in light scattered from cumulus clouds).

13.15 The red Sun: atmospheric extinction

Since molecular scattering preferentially extracts the shortest wavelengths from a ray of sunlight, the complement to a blue sky must be a reddened Sun. The effect is exacerbated by water aerosols – mists and fogs – and must be greatest at sunrise and

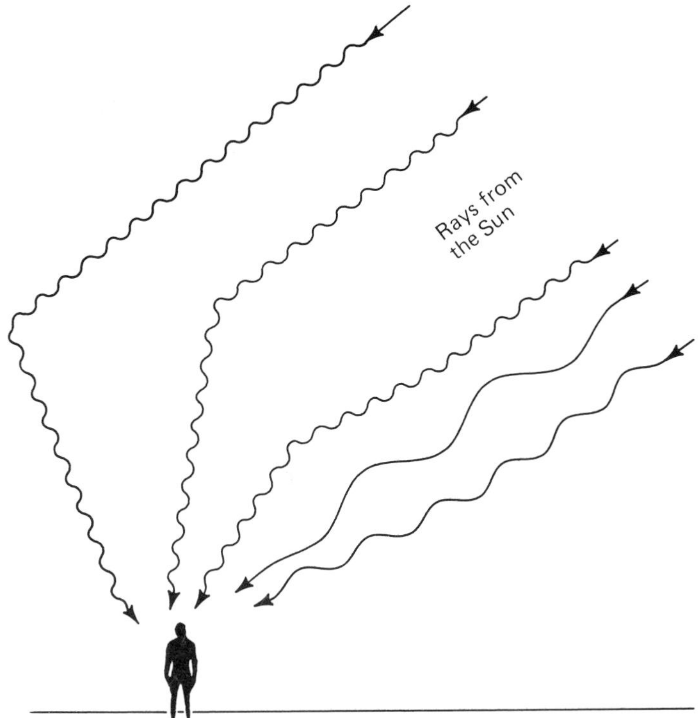

13.23. Atmospheric scattering. Short-wave blue sunlight scatters, or bounces, off molecules, and can arrive from any direction, giving the sky its blue color. The red longer-wave light penetrates directly to the observer with much less probability of being scattered.

sunset when the path length through the atmosphere is longest. All objects in the sky suffer dimming and reddening, including the stars, an effect known as *atmospheric extinction*. Figure 13.24 shows the *extinction curve* for an object at the zenith in a clear sky, where the dimming in magnitudes is expressed as a function of wavelength.

The extinction at the zenith is very small in the infrared, and first increases slowly as the wavelength drops to about 5000 Å. At the peak sensitivity of the eye, it is still roughly only 0.2 magnitudes. But then as the short wavelength limit of vision is approached and scattering really gets its grip, the curve rises dramatically. In the near ultraviolet, absorption by the ozone that resides in a layer some 30 kilometers high becomes important and the air turns completely opaque. Astronomically, the effect is very significant, as we must correct our observations for atmospheric extinction if we are to know the true brightnesses and colors of celestial bodies. Stellar magnitudes are always adjusted for the effect and are given as if seen "outside the atmosphere." The rapid rise in extinction to shorter wavelengths is critical to life on Earth, as the effect blocks the damaging high-energy ultraviolet rays of sunlight. Without it we would be burned to death in only a few seconds if otherwise unprotected.

The degree of extinction at any color depends upon the thickness of air, called the *airmass*, that the light must traverse, one of the chief reasons that astronomers place

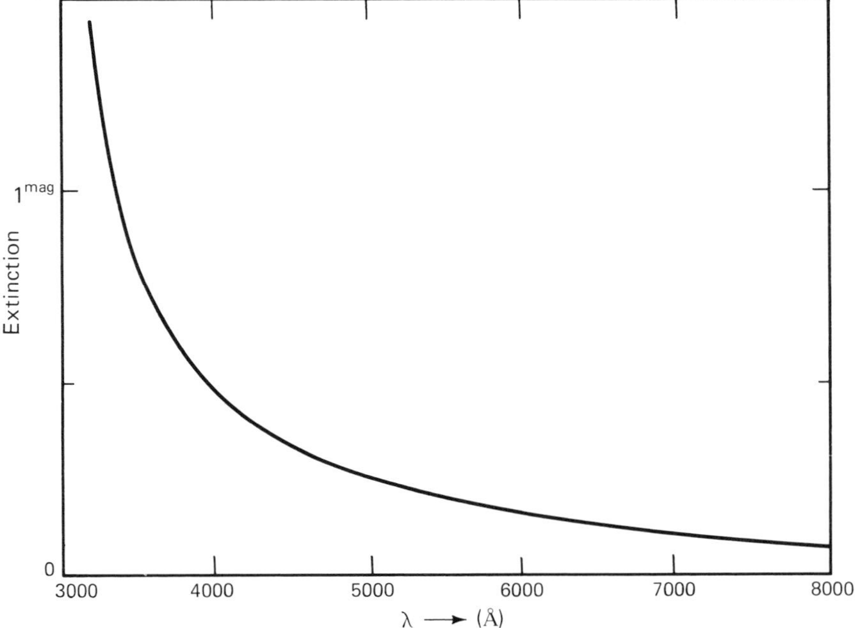

13.24. Atmospheric extinction in magnitudes as a function of wavelength for a clear sky at the zenith.

observatories on the highest mountains. As the Sun or a star sets, the atmospheric thickness increases, and so does extinction, especially in the blue part of the spectrum. At higher astronomical altitudes (i.e. low zenith distances) we can consider the air to be a layer with its top parallel to the ground (Figure 13.25a). Simple trigonometry (Appendix Section A3.1 and Figure A3.4) then shows us that the path length has twice the zenithal (or unit) airmass at 30° elevation (star *2*), and thrice at 19° (star *3*). Below

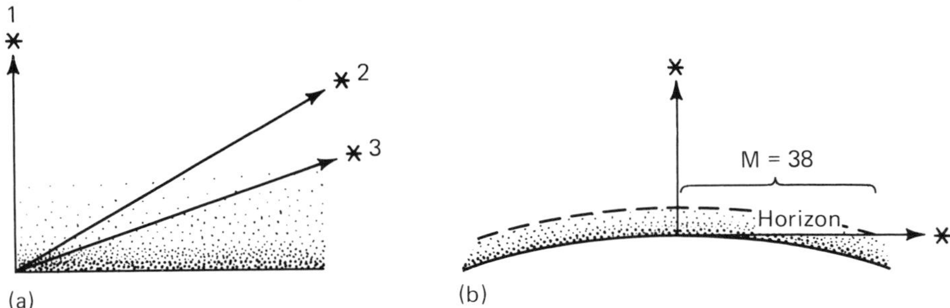

13.25. Pathlengths in the Earth's atmosphere. (a) In a plane-parallel atmosphere, that is, one whose top is parallel to a flat Earth, star 2, with a zenith distance of 60°, is seen through twice the airmass as star 1 overhead; star 3, with z = 19°, is seen through triple the airmass. (b) Because of the curvature of the Earth and its atmosphere (not drawn to scale), the maximum airmass is 38 times that at the zenith.

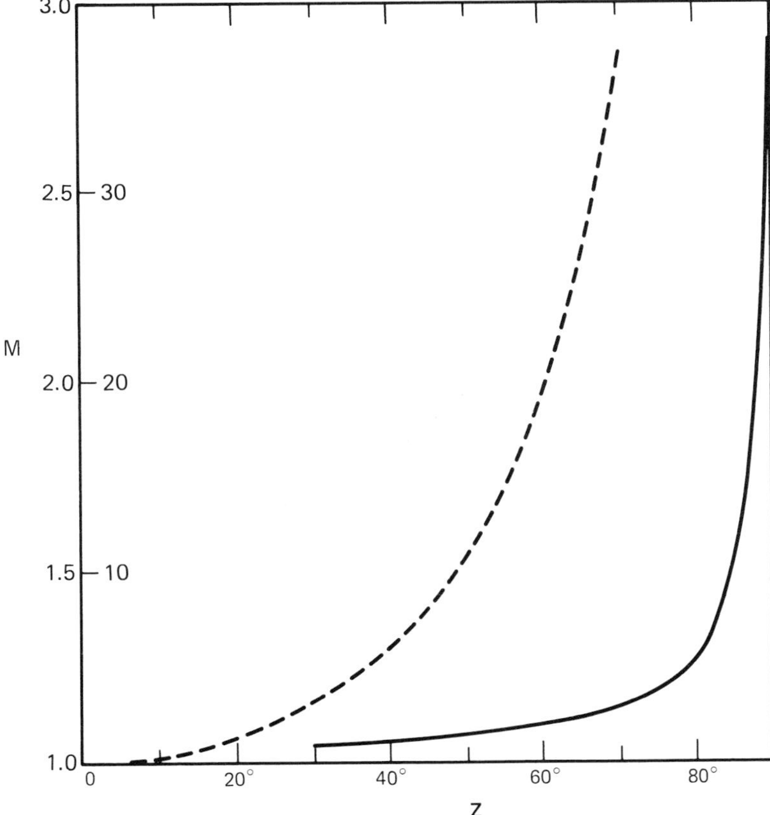

13.26. Airmass (M) as a function of zenith angle, z. Because of the large variation in extinction, the graph is separated into two parts. The airmass values given outside the left axis go with the dashed curve, and those inside it with the solid curve. At 0° (overhead) we look through a unit airmass (see Figure 13.25a), and below about 30° the airmass increases linearly with a trigonometric function (the secant: see Appendix A3.1 and Figure A3.4) of z. Above 30° the curvature of the atmosphere (see Figure 13.25b) becomes important, and M is increasingly less than secant z, rising to a maximum of 38.

about 30°, however, the curvature of the atmosphere must be accounted for, otherwise the airmass would increase without bound toward the horizon. The maximum (Figure 13.25b) is 38 times zenithal. At this point, the spectral character of incoming light is severely altered, nearly all the blue light is eliminated, and the Sun (or the Moon) becomes noticeably red, especially if there is a bit of haze in the air. It is for this reason that the worst sunburns occur when the Sun is closest to the zenith. People are often painfully amazed at how quickly they will burn in the tropics, and will quite unsuccessfully wait for a tan in winter, when the Sun is always low. Figure 13.26 shows how the airmass actually changes with zenith angle.

The extinction is so great at the horizon that only the brightest stars can be seen.

Some 6000–8000 stars are visible to the naked eye over the whole sky, but on any night only about a third can be seen, even though we have half the celestial sphere above us, since the faintest ones can be viewed only well above the horizon. Atmospheric extinction is one of the twin banes of the astronomer (see Section 13.23 for the other). Its variability and capriciousness can make that task of determining stellar magnitudes quite difficult. Our blanket of air limits the typical accuracy of magnitudes to about 0.5%, or roughly 0.005 of a magnitude division.

13.16 Sunsets

As the Sun drops to – and then below – the horizon, the contrast of the blue sky and the red Sun becomes more intense. There seems to be little in nature quite as lovely as a sunset, unless perhaps its reverse, a sunrise. In a clear sky, just after the solar limb has disappeared, we will first see a rounded dome of light arching above the setting point, surrounded by the bright blue sky. The arch, produced by sunlight illuminating the atmosphere just above the darkening ground, can be colored yellow to red, depending on the degree of haze or dust in the air. This reddish glow – "Hercules' funeral pyre" – deepens in color and spreads along the horizon as the Sun drops and lights only higher and higher layers of atmosphere. At the same time, the sky at greater elevations darkens to a velvety blue, in part in response to absorption of light in the ozone layer. At maximum coloration, a fiery red strip a few degrees high shades upwards to yellow and even green before blending into the purplish, oncoming night.

Clouds can greatly heighten the effect. Even in what appear to be the clearest skies, we will often see thin, high, wispy cirrus strips near the horizon, the result of our looking at them nearly edge-on, enhancing their thickness to the line of sight. Indeed, they are overhead, but too thin to be seen. (There are few astronomers who have not left the dome after a night of observing to find these horizon clouds at sunrise, to their dismay knowing that they have also been present during the night, distorting the observations.) Lower cumulus or stratus clouds can vividly reflect the highly reddened light of the sub-horizon Sun, contrasting wildly with the blue celestial sphere. The variations in color seem almost infinite, depending as much as they do on local conditions such as cloud cover, humidity, the amount of wind-blown dust, or even pollution. Of considerable importance is the degree of any recent worldwide volcanic activity, which can launch immense globe-girdling clouds of high-altitude aerosols, and which in turn produce extraordinary sunsets. The evening sky colors following the great 1883 explosion of Krakatoa in Indonesia are legendary and we have had numerous displays in our own century as well. Description of a sunset is nearly impossible. Instead, go and watch.

13.17 Crepuscular radiation

And if you do watch, there are a great many other things to see. Clouds throw shadows into the air. Holes that lie in or between the clouds allow shafts of sunlight to penetrate

13.27. Crepuscular rays. The rays are caused by sunlight penetrating through holes in clouds and illuminating shafts of dusty air. (Author's photograph.)

through to illuminate ubiquitous airborne dust particles that reflect the light in all directions. We then see a sunbeam pierce the air or lance to ground, a *crepuscular ray* (Figure 13.27). An irregular cumulus cloud may throw many parallel shafts, which will appear to be radiating from a single point, an effect of perspective like converging railroad tracks. If the Sun is near the horizon, the crepuscular rays may radiate upward like a glorious celestial fan pinned by a brilliant light. The effect can be stunningly beautiful and is the basis for an unfortunate amount of bad art.

After the Sun sets, it may hide behind a cloud that is below the horizon. The parallel shafts of light then rise upward into the sky, diverging from the invisible Sun like a great surmounting corona that announces the grandeur of the event. If the Sun is not too dimmed by haze, these crepuscular rays may pass overhead, and then converge again to the opposite point on the horizon, an unforgettable sight.

Moonlight can create crepuscular radiation as well. The contrast against a dark sky is quite remarkable.

13.18 Noctilucent clouds

As the Sun drops below the horizon and darkness grows on the ground we may still see the lower-lying cumulus clouds directly illuminated. At two kilometers above the Earth, the horizon dip is 1°.3, so these billowy vapor-balls still catch sunlight several minutes past sunset. Cirrus clouds, at elevations of some 10 kilometers, will be lit when the Sun is 3° down, halfway to civil twilight, producing lovely contrast effects. But as the Sun descends, even these lose their light and disappear, blending into the deepening evening.

Imagine one's surprise then to see bright clouds in the sky well past the end of civil twilight when the sun is some 10° below the horizon and it is growing quite dark. This rare phenomenon of *noctilucent clouds* is visible only from far northern or southern climes and is caused by full sunlight shining on a layer of atmosphere 80 kilometers above the ground where thin clouds can form as a result of a temperature depression. This is the region where meteors burn and is two-thirds of the way to the aurora (see Section 13.30)!

13.19 Sun pillars

Directly after sunset (or before sunrise; or moonset, or rise) we may see a stark pillar of light – white or pink – standing straight up from the horizon above the illuminating body (Figure 13.28a). It is commonly caused by grazing reflection from ice particles suspended in cirrus clouds, or even by snowflakes in the atmosphere (Figure 13.28b). The pillar is quite common, and can occasionally be seen extending below the Sun as well. Sometimes it is also caused by refraction through elongated pencil-shaped crystals hanging in the air parallel to the horizon. The Moon can join this act too, producing lovely shafts of soft light against a deep background.

13.20 The Earth's shadow

Just after sunset, if the sky is very clear, you will see a distinct dark band began to rise in the east, the shadow cast by the Earth into the still-illuminated atmosphere (Figure 13.29). As the Sun sinks, the shadow rises, and because our layer of air is so thin, it creeps upward a bit faster than the Sun drops. We see it only as long as it is projected into the thick lower air layer, so that it contrasts well with the large amount of light still being scattered back to us from the Sun. As less light is sent earthward from above, the contrast is lost, and by the time the shadow climbs to a dozen or so degrees high, within roughly half an hour, it disappears. In the morning, we see the reverse with the shadow setting in the west, just before sunrise. The full Moon rising in Earth shadow is quite extraordinary.

13.21 The Moon illusion

The rising or setting Moon is not bright enough to cause much in the way of display. Yet it has its own mystique. Most astronomers have been asked the question: "Why is the Moon so large when it is on the horizon?" Nearly everyone has seen a huge coppery full Moon looming over the distant trees or houses, which then appears to shrink when it rises. The phenomenon is an optical illusion. As we saw from Section 9.8, the angular diameter of the Moon is actually slightly *smaller* when on the horizon than at meridian culmination, since it is some 6000 or so kilometers farther away owing to the Earth's diameter and rotation. It just seems larger at moonrise, probably because of a juxtapo-

(a)

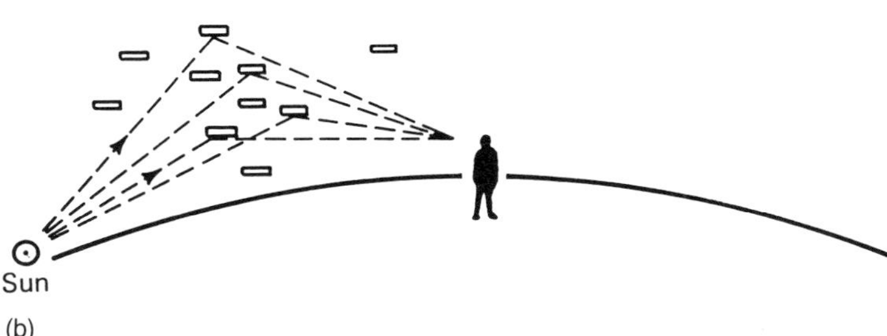

(b)

13.28. The Sun pillar. (a) Here, it is seen as a short spike sticking upward from the horizon above the already-set Sun. (b) It is caused by grazing reflection of sunlight from suspended ice crystals in clouds. The same effect can be produced by snowflakes and by refraction through hexagonal crystals. The Moon also can produce such pillars. ((a) Author's photograph.)

sition with everyday objects on the horizon. When it rises a substantial degree, we no longer have a reference with which to compare it, and it seems smaller. Constellations show the same effect: look at Orion the next time he rises, when the giant hunter appears enormous as he hovers low over the landscape.

13.22 Atmospheric refraction

We now turn, over the next several sections, to a wonderful variety of refractive (and diffractive!) atmospheric effects. We saw in Section 7.2 that refraction in the Earth's

13.29. The Earth's shadow, seen rising to the east after sunset from Kitt Peak. (Author's photograph.)

atmosphere lofts everything upwards in altitude (see Figure 7.3). The closer we look to the horizon, the greater the effect. Stars at 45° altitude are raised by only one minute of arc, but the setting Sun is actually lifted by its own angular diameter. Although not noticeable with the naked eye, refraction can cause difficulty for the astronomer in setting a telescope. Refraction also causes very noticeable distortion in the perceived image of the horizon Sun. The upper limb has a higher altitude, and is thereby less subject to the lofting effect; the lower limb is thus raised more than the upper, and the circular disk is severely flattened (Figure 13.30).

Atmospheric layers of varying temperature and density, particularly *inversions*, wherein a higher layer may have higher temperature and density rather than lower, cause the solar limb to appear jagged and the upper limb even to separate into pieces that seem to float off the Sun: the "*Chinese lantern effect* (Figure 13.30 inset). These phenomena are best seen from a seacoast where stronger inversions tend to develop, and obviously *only when there is enough haze in the air to make viewing comfortable*: caution, as always, is necessary when attempting to view the Sun. The Moon, of course, is also flattened when on the horizon and is completely safe to observe with binoculars (the Sun is not!), which renders the sight much clearer.

13.23 Seeing and scintillation

Adding to the beauty and luster of the nighttime sky is the twinkling of stars. Almost everyone knows of the effect, and has heard of it from childhood in the form of the familiar nursery rhyme. Twinkling is yet another example of refraction. Our atmosphere is not uniform and homogeneous, but contains cells of moving zones of differing density and temperature. As starlight passes through on the way to the ground, it is

13.30. Atmospherically distorted solar images. In the main photograph, the Sun is flattened by atmospheric refraction. (Note the large naked-eye sunspot down and to the left of center.) The inset shows the Chinese lantern effect caused by severe and anomalous refraction. (Main photograph by the author; inset by C. Treusch, Vatican Observatory, in *The Green Flash*, D. J. K. O'Connell, North-Holland, 1958.)

irregularly deflected by refraction within this lumpy medium, and since the cells are in motion, the path of the light is constantly changing. From the Earth's surface, the stars then appear to jump around, an aspect of twinkling called the *seeing*, and change in both brightness (called *scintillation*) and color.

The twinkling of stars is the other bane (the first being atmospheric absorption: see Section 13.15) of the observational astronomer. Refraction, as it smears the point of light that is a star into an apparent *seeing disk* (Figure 13.31a), markedly reduces our ability to see fine detail. For even a modestly sized telescope, the seeing disk is much larger than the diffraction disk (see Section 13.11). Scintillation can cause severe problems in the measurement of stellar brightnesses.

The degree of scintillation and the size of the seeing disk depend critically upon location, the amount of air that the starlight must pass through, and local weather conditions. The twinkling phenomenon is another of the reasons why astronomers place observatories on high mountain tops: there, the stellar disks can be as small as half a second of arc. This angle, though tiny, is still vastly larger than the actual angular sizes of stars, however, which range downward from about $0''.05$, and than the $0''.025$ diffraction disk of a 5-meter telescope. Stellar images are much worse at average locations: in

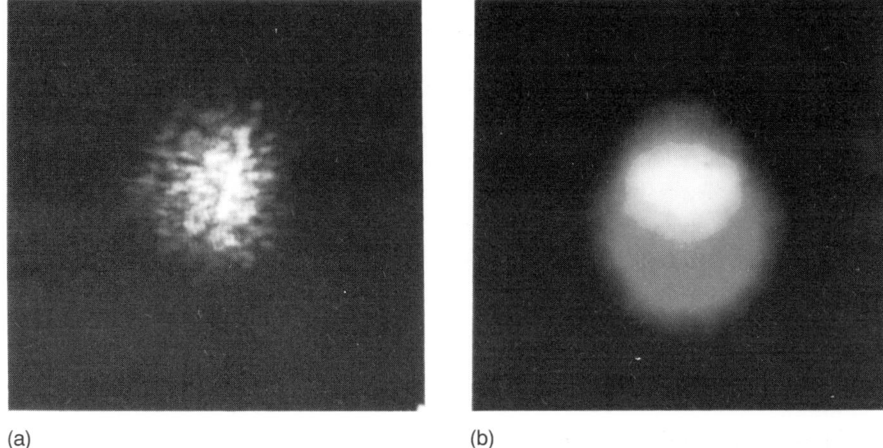

(a) (b)

13.31. Atmospheric effects on stars. (a) Capella is a close binary whose components are only 0.04 seconds of arc apart. Hundreds of individual images of the pair are spread out by turbulent atmospheric refraction to make the large blurred seeing disk seen at the telescope. (b) Atmospheric dispersion spreads the light of a star (here Venus) into a broad line perpendicular to the horizon with blue on top and red on the bottom. ((a) H. A. McAlister, National Optical Astronomy Observatories and Georgia State University; (b) C. Treusch, Vatican Observatory, in *The Green Flash*, D. J. K. O'Connell, North-Holland, 1958.)

the low flatlands they will be at best 4″ or 5″ across, and sometimes, especially after the passage of a weather front, can approach one minute of arc, rendering astronomical observations nearly impossible. The problem can be eliminated entirely by taking our telescopes into space. From the ground, however, we can overcome part of the problem by the use of "adaptive optics," in which zones of the reflecting mirrors are flexed by motors to compensate for the jumping image and steady it down.

The naked-eye observer can easily see the night-to-night weather-related variations in the degree of twinkling. Sometimes the stars seem to stare downward at you, at other times to jump madly about. The effect is clearly enhanced toward lower astronomical altitudes where the airmass is larger, and it is even worse in winter, when the atmosphere contains larger density variations. The champion twinkler in northern latitudes is Sirius, which satisfies both negative requirements, and is extremely bright as well. It can be a remarkable sight on a cold winter night, as refraction and the accompanying dispersion make it jitter and flash through the colors of the rainbow.

Even the casual observer will see, however, that the much brighter planets do not twinkle as much. They stay quite steady, while the stars shimmer around them. The reason is that planets present real surfaces that are generally larger than the seeing disk. Any one point within the planetary image will twinkle as much as a star, but the brightness and positional variations of all of the points average out, and the planet as a whole yields a fairly steady light. A telescopic view, however, shows the planetary image to be severely degraded. We can see little detail on its surface from the Earth with an ordinary telescope and must either employ sophisticated adaptive optics or go into space, preferably to the planet itself.

A good portion of the seeing problem is local to the observer, and for the astronomer is involved with the telescope housing and dome, which trap heat. The locally heated air then produces additional unwanted refraction along the light path. People who build and operate large observatories take considerable effort in controlling the telescope environment and in placing instruments where the proper wind flows minimize the effect. Small local observatories often have serious problems on visitors' nights. A crush of bodies in the telescope dome produce heat, and the warm air flowing out of the shutter destroys that which the visitor is trying to view. As so frequently happens in science, the act of observing is itself detrimental to the observation.

13.24 Atmospheric dispersion

Whenever we have refraction of light, we must also have dispersion (see Figure 13.4), and refraction through the atmosphere is no exception. One result, mentioned above, is erratic changes in star colors as they twinkle. In addition, shorter wavelengths of light are refracted upward by the air more than the longer, stretching the point image of a star into a short spectrum perpendicular to the horizon (Figure 13.31b). The effect is only visible with optical aid, and only when a star is below about 45° altitude, when the spectrum is then longer than a typical seeing disk of 2 seconds of arc. High telescopic power applied to stars near the horizon shows it very clearly. The phenomenon is quite lovely to look at, but as for twinkling, is very pesky to the professional observer, since it further distorts the images that the astronomer is attempting to observe and analyze.

13.25 The green flash

As the Sun sets, atmospheric dispersion becomes very pronounced: the Sun in a sense separates into many overlapping colored images, the violet one on top and the red one on the bottom. Below 1° altitude, however, scattering and absorption are so effective that the violet and blue images are removed, leaving green as the shortest wavelength. Just before the upper limb of the Sun disappears, the top edge develops a pronounced green rim of extraordinary color a few minutes of arc thick.

During sunset in a clear hazeless sky, the last thing to disappear will be this green rim whose light is then scattered and reflected by the atmosphere immediately above the Sun. The effect is a momentary flash of green light right at the point of sunset. If the air is sufficiently pure the light may even penetrate enough to allow a rare blue flash.

To see the flash, sunset must occur over a distant sharp horizon so as to allow for maximum dispersion. The best place would be an ocean sunset, viewed from a coastal mountain. It is commonly seen from aboard ship in the Pacific. Of course, we can also see it just before sunrise as well, if we know exactly where and when to look.

13.32. A double rainbow. The sky is darker between the bows and brighter inside the primary. (Author's photograph.)

The phenomenon is considerably more complex, however. It depends on the absorptive properties of the Earth's atmosphere and also seems to be greatly magnified by anomalous dispersion that is produced by temperature inversions (see Section 13.22) associated with atmospheric layering.

13.26 Rainbows

The sky presents few more glorious sights than the brilliantly colored arc of the rainbow following a sudden storm (Figure 13.32). How significant to humanity is this once mysterious sight: in biblical literature it announced God's promise to Noah that there be no more floods; the Greeks assigned it a goddess, Iris, a messenger to mortals; and there is the fabled pot of gold to be found at its end.

We know what causes them now, but that knowledge cannot remove their magic or their beauty. The cause is a combination of refraction, with the inevitable dispersion, and reflection. The Sun is low in the west, and an afternoon shower flees to the east, so that sunlight illuminates the still-falling raindrops. The drops act like tiny prisms that throw the dispersed and colored light – the solar spectrum – back to the watcher on the ground (Figure 13.33).

There is quite a bit more to the event than just a single bow. We will almost always see *two* bows, one above the other. The lower, the *primary bow*, is much the brighter, and always has red on the top and violet on the bottom. It is a segment of a full circle with an angular radius of 42° centered on the point in the sky opposite the Sun. The Sun must be up to create the bow, so this *anti-solar point* has to be below the horizon,

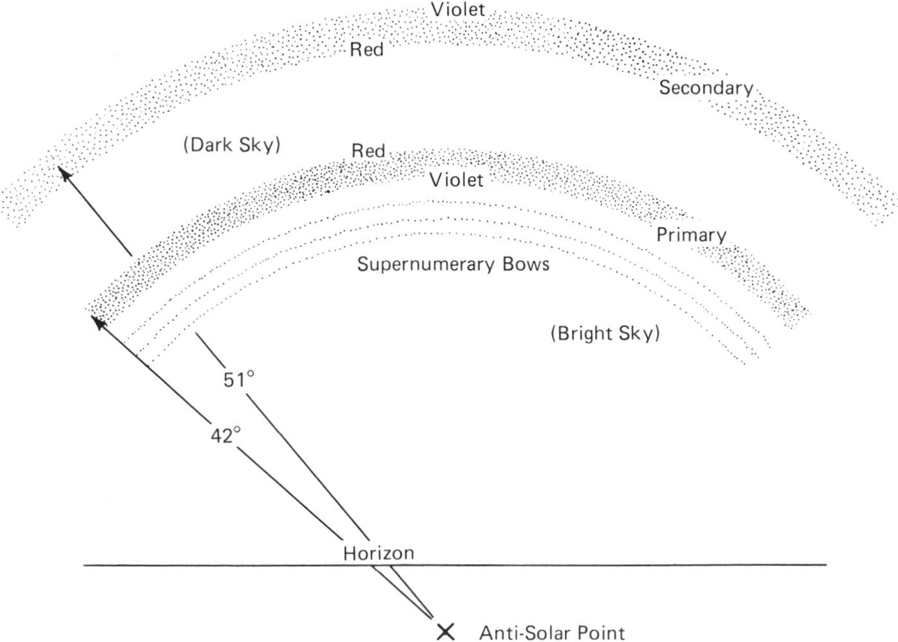

13.33. The phenomena of the rainbow. The bright primary bow, of 42° radius and the one commonly seen, has red on the top, violet on the bottom. The fainter secondary, 51° radius, has its colors reversed by an additional reflection in the raindrop, and is always present. Faint supernumerary bows are seen inside the primary.

always rendering the visible bow less than a semicircle as seen from the ground. Riding above the primary is the *secondary bow*, which is also centered on the anti-solar point, but with a radius of 51°, the colors reversed. If the rainbow is bright, and if we look carefully, we will also see a number of reddish *supernumerary bows* displayed below the primary. The brightness of the sky in the neighborhood of the bows varies as well: it is dark between the two, and bright inside the primary.

All this activity takes place on the level of the individual drops. To produce the primary bow (Figure 13.34a), a ray of sunlight is refracted upon entering a spherical falling drop and is dispersed into its component colors. The rays, represented in the figure only by red and violet, are then reflected from the back side of the drop, and further refracted and dispersed upon exit, whence they head toward the observer who then sees a spectrum. Light is actually refracted and reflected with a wide range of angles interior to 42°. But the rays are strongly concentrated to that angle, which defines a limit. The result is a bright colored bow surrounding a diffuse illuminated interior. The fainter secondary bow features two internal reflections (Figure 13.34b). The additional bounce both reverses the colors and changes the angle of exit relative to the Sun, resulting in a 51° radius.

The light rays that exit the drop at less than the 42° angle (those interior to the primary bow) interfere with one another like those in Figure 13.5. The result is a series

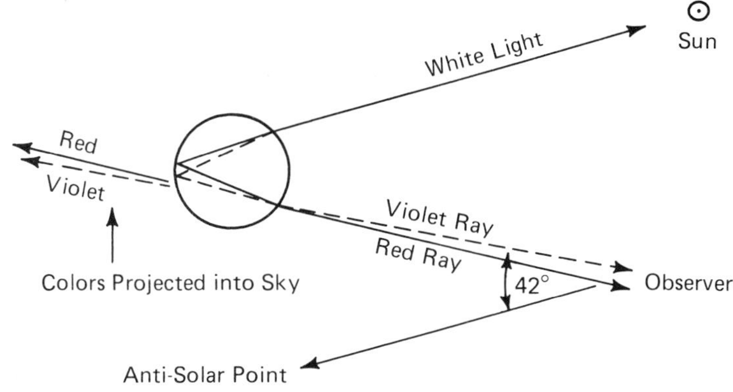

(a)

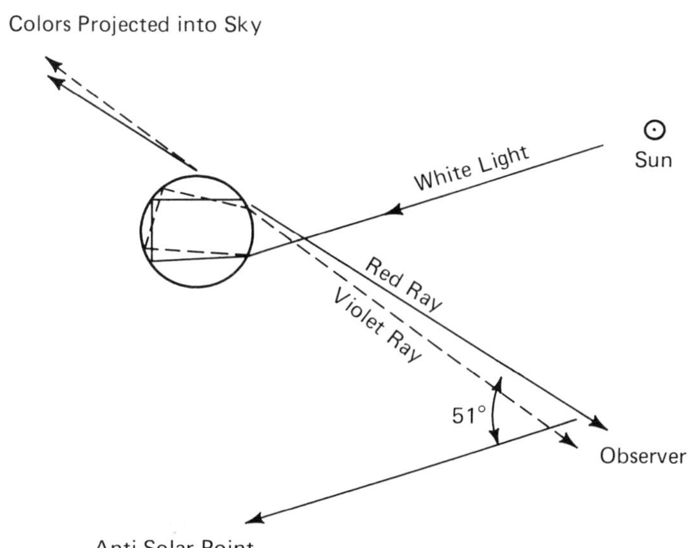

(b)

13.34. The formation of rainbows. The solid lines show the paths of the red rays, the dashed, violet. (a) Sunlight is refracted and dispersed upon entering the drop, reflects off its back side, then disperses further upon exiting. The observer sees the rays projected outward into the distance to form the primary bow. (b) The secondary rainbow is produced by two back-side reflections, so that violet appears on top rather than the bottom.

of light and dark fringes – the reddish supernumerary bows – fading away toward the anti-solar point. The internal reflections also cause the bows' light to be polarized, that from the inner one almost completely, from the outer less so.

Individual displays can differ markedly. The bows can appear only where sunlight illuminates rain, so that they are frequently seen as only partial arcs (as in Figure 13.32). If the drops are sparse, the primary will be faint, and the secondary undetectable, so that only one bow is apparent. The more uniform the drop sizes the more supernumeraries will be seen, and the smaller the drops the wider their spacing. The colors present also depend on the color of the Sun: at sunset or sunrise, the red Sun will produce a red bow.

Since the primary is 42° from the anti-solar point, it will be visible only if the Sun has an altitude less than 42°. As the bows are fragile, and must be reasonably well elevated to be seen, the effective limiting altitude will be somewhat less. Because the Sun transits the meridian in temperate latitudes with altitudes of the order of 50° or more during seasons when rain is most likely, the bows are only seen in early morning or late afternoon when the solar altitude is sufficiently low.

Since there could be a long interval between thunderstorms, make a bow by spraying a garden hose in the sunny air. Both bows will be visible, and the variations in the phenomena involved can be observed by changing the nozzle setting and thereby the droplet size.

We do not necessarily need the Sun. A rare and extraordinary sight is the *moonbow*, the rainbow created by the full or nearly full Moon. The illumination is so much less that color is nearly absent, and conditions must be just right to see it at all. But once viewed, this ghostly arc presented against a velvety sky is unforgettable.

13.27 Coronae

When thin clouds cover the sky, rainbows are replaced by other colored delights. The most common ones are the rings and halos around a bright Moon. The display of hues against a darkened sky attracts singular attention. There are a variety of forms, and all are caused by light transparent clouds. They are equally present around the Sun, but we tend to notice them less because the Sun is so dangerously bright that we cannot pause to look and admire it as we do the Moon, and we thus miss the associated phenomena.

The most common of all these sights is the *corona* or *aureole* around the Moon (Figure 13.35). We see this phenomenon as rings of color that are nearly contiguous to the lunar limb. It is caused by diffraction by the tiny water droplets or ice crystals, usually in high cirrus clouds. Because long wavelengths are displaced the most by diffraction (see Section 13.6), the outside of the corona will have a reddish cast and the inside, next to the Moon, will be bluish. The dimension of the corona – typically one to several degrees – depends on the size of the cloud particles. Sometimes we see nested coronae that correspond to the different spectral orders of Figure 13.6, where each sequence of colors begins with an inner blue band and ends with red: the author has seen as many as four around the Sun.

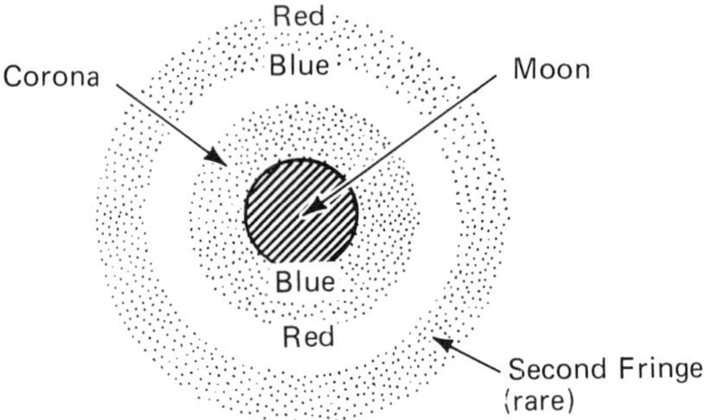

13.35. The corona. Coronae are seen around both the Moon and the Sun, and are caused by diffraction. Only the inner, first order, fringe is commonly seen. The outer one is a rare second order fringe.

The term "aureole" is sometimes used for the bright white light immediately surrounding the Sun. You can see it on a clear day if you first hide the Sun behind a barrier. This phenomenon is caused by scattering from aerosols in the air and is not produced by diffraction. The clearer the sky the smaller *this* aureole. Do not confuse the two.

13.28 Halos and sundogs

The common term "ring around the Moon" includes very different phenomena, which are discussed separately. The coronae described above are distinct from the lunar or solar *halos*. The most usual form is a circular ring well separated from the lunar or solar limb, with an inner radius of 22° (Figure 13.36), much larger than the corona. This beautiful sight is caused by refraction through ice crystals in high cold cirrus clouds. Water droplets will not produce it, which allows us immediately to identify the cloud's composition. The crystals must be shaped like hexagonal prisms, which are naturally produced as a by-product of the crystalline structure of ice: examine a snowflake, and you will see a six-sided figure. As the crystals tumble freely in the air, they will refract sunlight; the most common path of the rays will be through one of the six-sided faces, which will bend them through a 22° angle (Figure 13.37a), producing a halo of that radius. The blue light will be refracted the most, and thus the halo will exhibit red inner and blue outer edges. The halo light is again polarized, but not nearly so strongly as that from the inner rainbow. A less common refractive path for the light is through rarer cubic prisms or through the end faces of the hexagonal crystals (Figure 13.37b). This route produces the fainter and rarely seen 46° halo, which will be colored the same as the inner one.

We are not always aware of the solar halos because of our natural avoidance of the Sun. Be on the watch for one whenever you see a thin film of cirrus in the sky during the colder months when they are usually made of ice crystals. Just be cautious when seeking them so that you do not look directly at the Sun: hide it behind a building or

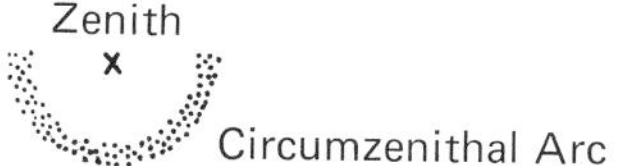

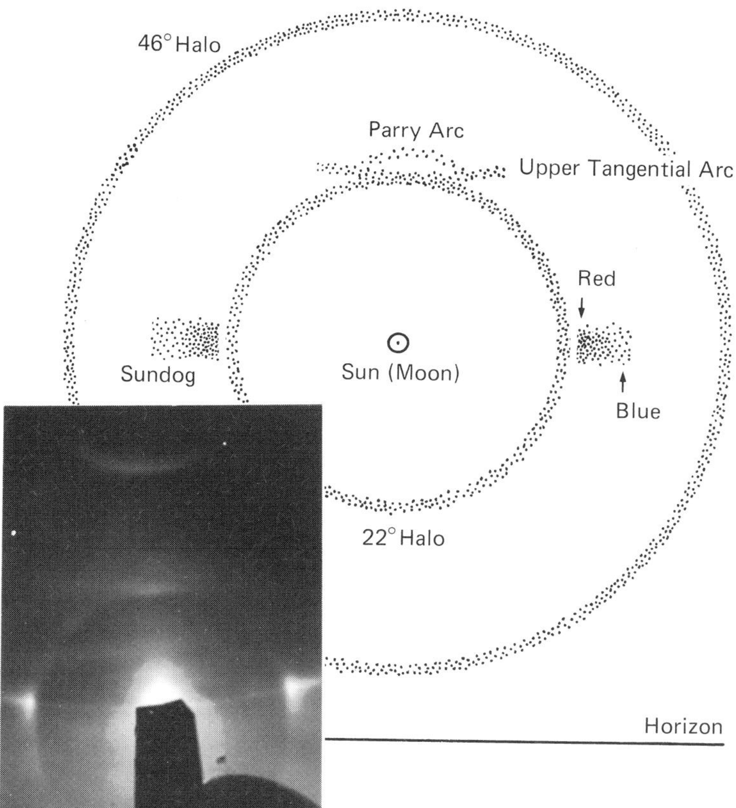

13.36. Halo phenomena caused by ice crystal clouds. The Sun is surrounded by the 22° and the much rarer 46° halos. The inner one has parhelia or sundogs attached on a line that runs through the Sun parallel to the horizon. The 22° halo also sports an upper tangential arc and a Parry arc. Around the zenith is a circumzenithal arc. The inset shows an example. (Inset: B. Morley, in *Rainbows, Halos, and Glories*, R. Greenler, Cambridge University Press, 1980.)

tree first. It is common to see only a portion of the halo because of the uneven distribution of clouds.

The lunar halo is a remarkable and lovely apparition. It is less frequently seen than the solar variety simply because the Moon must have a reasonably developed phase to be bright enough to produce it, but it seems more common because we are much more likely to look at the Moon. The colors also appear more vivid, perhaps because we see them against a darkened sky. When the lunar halo is present, we will also commonly

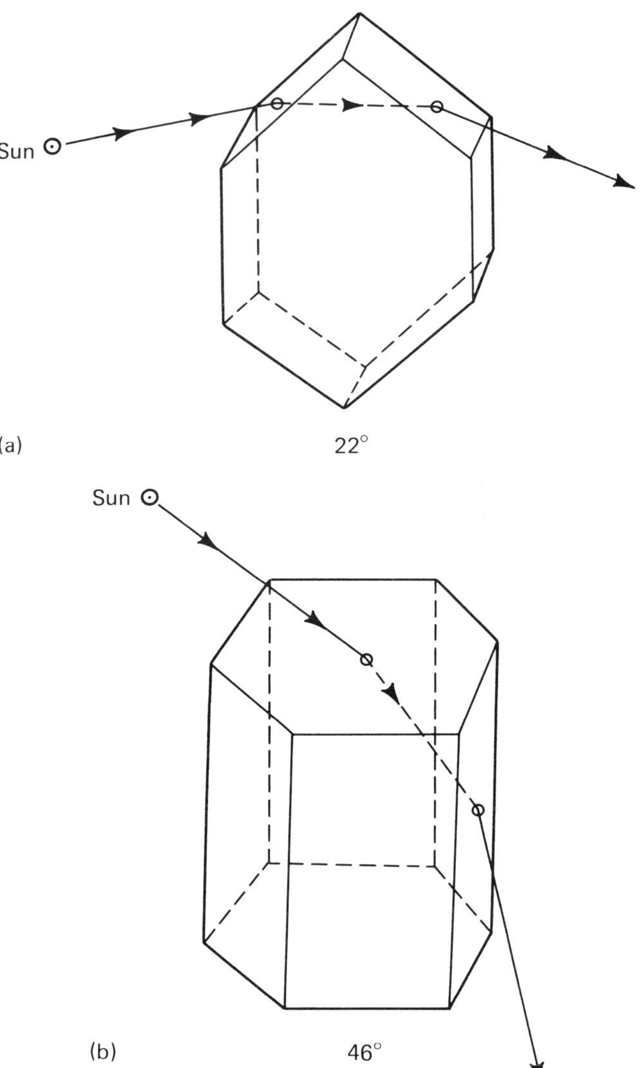

13.37. The origins of the halos. Ice crystals that produce solar and lunar halos are shaped like hexagonal prisms. (a) Sunlight refracted through the sides, which meet at a 60° angle, produces the 22° halo. (b) Light passing through the right-angled ends causes the 46° halo. The blue light bends more than the red and so to the observer seems to come from farther away from the Sun.

see the diffraction-produced corona nested inside: one of nature's more striking sights.

If the Sun is lower in the sky, the 22° halo is often significantly – and sometimes dramatically – brighter at points that are on a line running through the Sun parallel to the horizon (Figure 13.36). These luminous spots are called *mock suns*, or in the analogous case of moonlight, *mock moons*. Since they follow along with the Sun on its daily path, they are more familiarly known as *sundogs* (or, again, *moondogs*), and sometimes as *parhelia*.

The mock suns are not seen if the Sun is high enough in the sky that the light penetrates the clouds at a near-vertical angle. They develop as the Sun drops on its daily path below an altitude of about 40°, and get progressively brighter as sunset approaches. The colorful features (red on the inside, blue on the outside) are produced by sunlight as it enters the clouds at an oblique angle and is refracted through flat plate-shaped crystals that are preferentially oriented with their bases parallel to the horizon. Then (see Figure 13.37a) most of the halo's light will be concentrated to the sides, resulting in the horizontal positions of the sundogs. Although the crystals are positioned parallel to the ground, they are not placed that way to the line of sight unless the Sun is exactly setting or rising. The path of the light is then not at perfect right angles to the side of the prism, and as a result, the mock suns are displaced outside the halo. As altitude increases the separation gets larger, less light gets through, and the features become fainter. Above 40° they disappear.

Because of patchy cloud distribution, and a variety of crystal orientations and solar altitudes, we are constantly treated to entertainingly different patterns: one sundog (or moondog) or two, sundogs with no halo, all or partial halos with or without sundogs. Since snowflakes have the same hexagonal crystal structure as the little prisms in cirrus clouds, the halos and their sundogs are often brightly seen in fine blowing snow.

Numerous other features related to the halos depend upon variety in cloud crystal structure and orientation (see Figure 13.36). The most usual is the frequently seen *upper tangential arc* to the inner halo, caused by pencil-shaped crystals floating with their sides parallel to the ground, and the related *Parry arc*. The *circumzenithal arc* is also fairly common, though rarely noticed, since most people have little reason to look straight up. It is a brightly colored arc, a fragment of a halo that is centered on the zenith and is symmetrical about the vertical circle through the Sun. It is produced by light rays passing through the square ends of the plate crystals. Always look for it when you see a bright 22° halo. A remarkable array of other arcs are attached to both the large and small halos: be on the watch for them. Extensive descriptions are to be found in the books in the bibliography.

Much less common, but quite striking, is the *horizontal* or *parhelic* circle, a white band that runs around the sky through the Sun parallel to the ground. It is caused by reflection from the same ice crystals that produce the halo. Occasionally it will have white spots set into it at angles of 120° from the Sun.

13.29 Unidentified flying objects

Unidentified flying objects, or UFOs (Figure 13.38) have been the subject of hot popular debate for a good fraction of this century. People see strange lights and objects in the sky. Most have ready explanations. However, many observers choose to interpret these sightings – both real and false – as evidence for extraterrestrial life and visitors from space. The subject has produced a veritable industry for sensationalist newspapers.

The sightings fall into three categories: delusions, hoaxes, and real phenomena. Delusions have no explanations and can be dismissed out of hand. Hoaxes are remarkably easy to fabricate: an out-of-focus garbage-can cover does nicely. Photographs are

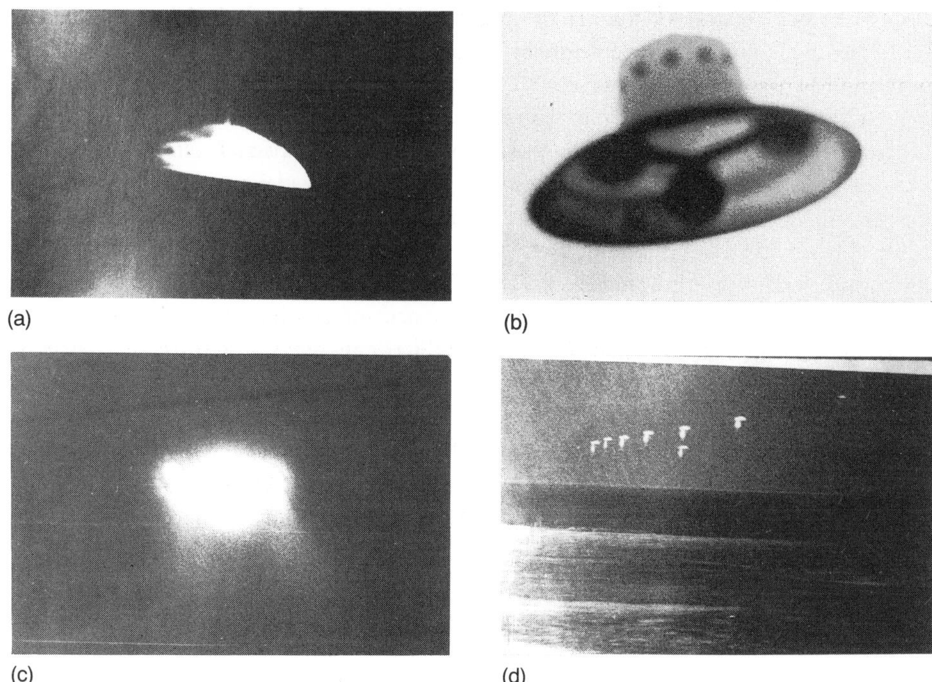

13.38. A variety of UFOs. On top are two are deliberate fabrications, on the bottom, real but misunderstood images: (a) an upside-down banana split dish hung on a wire; (b) a pie plate with ping-pong balls topped by a cottage-cheese carton; (c) an airplane with landing lights seen through clouds; (d) internal reflections in a camera. (Robert Sheaffer.)

so easily faked (Figures 13.38a and b) that they provide no real evidence at all. Other hoaxers have flown airplanes in tight formation to mimic an extraterrestrial spacecraft or have flown lighted candles in helium balloons (a dangerous trick at best).

The natural and real phenomena mistaken for UFOs are much more interesting. Of all the UFO candidates, Venus probably tops the list. As you drive along a country road it seems to follow you. Jet fighter pilots have tried to shoot it down and a sheriff and his deputy once chased it from Ohio into Pennsylvania, only to see it get away, probably as the Sun came up. Bright stars serve nicely too. Many people have never really watched a star twinkle and interpret the flashes of refracted light as running lights aboard a craft of some sort: brilliant Sirius has sent more than one person to the telephone to call the local paper or observatory. Altair and flanking companions β and γ Aquilae can appear like a winged aircraft flying slowly across the sky, much as they reminded our ancient ancestors of a celestial eagle. Real aircraft and Earth-orbiting satellites that move slowly and silently across the heavens also provide fodder for the UFO mill. The satellites in particular can appear odd since they will suddenly enter the Earth's shadow and disappear. Moreover, the eye can be fooled into "seeing" the satellite wander back and forth, away from its actual straight path.

Meteors are subject to misconception. A bright fireball (see Section 12.4) creates an illusion of closeness, so that a person might think that it has flashed just over his or her house or has fallen into the neighbor's yard, when it is really 100 kilometers or more away and has passed over the horizon. A fireball may also leave a tubular ionization train in its wake that might resemble a spaceship.

Mirages also produce UFOs. They are caused by anomalous and unusual refractions that can grossly distort an object and displace it from its normal position. Odd mechanical constructions are candidates as well. A university's radio telescope was illuminated at night by floodlights. A couple driving by on the road below were mystified by this contraption and when the floods were switched off, they reported to the local paper that it took off into the sky. Lightning, the northern and southern lights (to be explored in the next section), oddly shaped clouds, satellite launchings, explosive space experiments, high-flying balloons that catch sunlight after it is dark on the ground, all add fuel to the strange UFO fire.

There is no credible evidence that any of these UFOs are alien spacecraft. To be sure, there are unexplained sightings. These are commonly garbled accounts of what may have been seen: humans are notoriously bad witnesses, and descriptions are often confused and poor. Moreover, some few are hoaxes or delusions that can never be rationally explained by natural phenomena. To quote Philip Klass, an expert debunker of the subject, "Extraordinary events" (i.e. aliens from space) "require extraordinary evidence." And we have none at all.

13.30 The aurora

The stars present a backdrop for one of the most awesome of sights, the *northern* (or *southern*) *lights*, more properly the *aurora*, *borealis* in the north, *australis* in the south (Figure 13.39). They are not the results of optical effects or sunlight; indeed any light at all acts to diminish their glory. The Sun, however, is still responsible.

The aurora is principally a polar phenomenon and is associated with the northern sky in the northern hemisphere and vice versa. A display will commonly begin as a weak patch of red or green and can grow to a spectacular fireworks display that covers the entire heavens. The morphology, or structure, of the lights is highly complex, but there are a few basic forms that in a full display can sequentially change, one to another: we might see simple arcs that spread across the horizon; sometimes the arcs will be banded or consist of broad parallel sheets; the bands may break into narrow rays that align to the Earth's magnetic field; if the rays are long enough they may appear to converge to a point for a coronal effect much like that seen for crepuscular radiation; and in one of their more fascinating forms, rayed bands may be curved, giving us the appearance of draperies hanging in the sky.

The aurora is the a result of the interaction between the solar wind (see Section 12.3), the steady flow of particles (protons and electrons) from the Sun and the terrestrial and solar magnetic fields. Terrestrial magnetism traps the protons and electrons of the highly irregular solar wind into the broad *Van Allen radiation belts* that are far above the atmosphere and are symmetrical with the magnetic poles (Figure 13.40). At the same time, the wind grossly distorts the field, blowing it backwards, compressing the

13.39. An aurora. These lights in the upper atmosphere are most commonly seen during the maximum of a solar cycle. (D. Paprocki.)

part in front, and extending the portion in back into a long *magnetotail*. The Sun's magnetic field can connect with the terrestrial field, allowing the charged particles to ride into the planet. These interactions establish electrical current rings roughly 150 kilometers or so above the Earth that encircle the magnetic poles and have radii of about 20°. Associated with the current is the ionization of the upper atmosphere. When the free electrons rejoin their parent atoms, they emit a glow. The activity of different kinds of atoms, notably oxygen and nitrogen, produces the characteristic green and red colors.

We see the effects of the current as a broad, relatively permanent auroral ring centered on the magnetic poles. Disturbances in the solar wind and magnetic field called *magnetic substorms* will trigger irregularities and activity in the auroral zones that produce the visual displays seen on the ground. The display may last from a few minutes to several hours depending on the energy involved in the substorm. We will then see the aurora die away, possibly to be replaced with a new one a short time later. The aurorae are of course seen visually only when the sky is dark, and can be an endless source of amusement during long polar nights. They occur in the daytime as well, but cannot be observed from the ground because of the much brighter blue sky.

The Earth's overall magnetic axis is inclined to its rotational axis by roughly 11°, but near the surface is distorted by crustal ore deposits. The north magnetic pole (to which a compass needle points) is at a longitude of 101° W and a latitude of 76° N, which places it in far northern Canada, on Bathurst Island. The auroral zone, with its 20° radius, consequently passes nearly over the rotational pole, and through the rela-

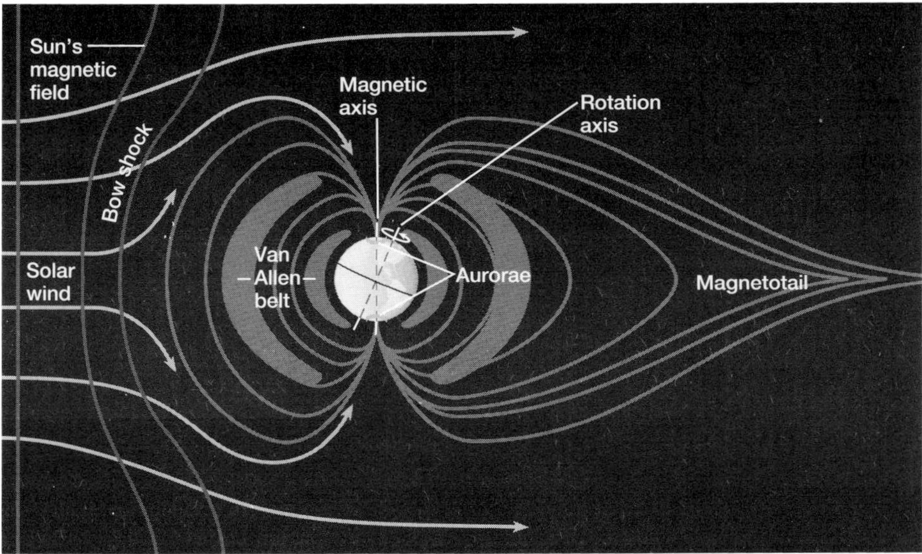

13.40. Earth's magnetism and the formation of aurorae. The dipole of the Earth's magnetic field is compressed on the sunward side by the solar wind. The shock wave where the solar wind impacts the Earth's field is actually about 15 Earth radii upstream. Opposite the Sun, the magnetic field is stretched into a long magnetotail extending 1000 Earth radii away. The solar magnetic field is coupled to the moving wind, is bent around the Earth, and occasionally connects with the Earth's field, allowing particles to enter. These electrons and protons are trapped in the two Van Allen belts. The magnetic and electrical interactions form huge current rings around the magnetic poles, causing the upper atmosphere to glow.

tively well populated northern parts of the western Canadian provinces, the Yukon, and Alaska, from which the fame of the lights has spread, then east across northern Quebec and Greenland. The south magnetic pole lies within Antarctica and the southern auroral zone passes across only frozen expanses and the far south Pacific Ocean.

The phenomenon of the aurora is not confined to pretty lights. The ionization and electrical activity associated with it can greatly enhance (or destroy, depending on wavelength) long-distance radio communication, and the electrical activity can cause currents in long-distance transmission lines that can blow circuit breakers and even shut down power grids.

The aurora is a creature of the *active Sun* and is modulated by the 11-year *sunspot* or *solar magnetic cycle*. A large number of dark spots can be seen on the solar surface in Figure 13.41. They are cooled areas that only look black in contrast to their surroundings and are ephemeral, lasting anywhere from a few days to a few weeks. The cycle is apparently caused by solar rotation. The Sun spins with a period of about 25 days at the equator, but it takes longer near the poles, so the gases shear past one another. One theory suggests that the shearing gases begin to wrap up and squeeze the solar magnetic field buried within them. Hot rising gases pop the field through the

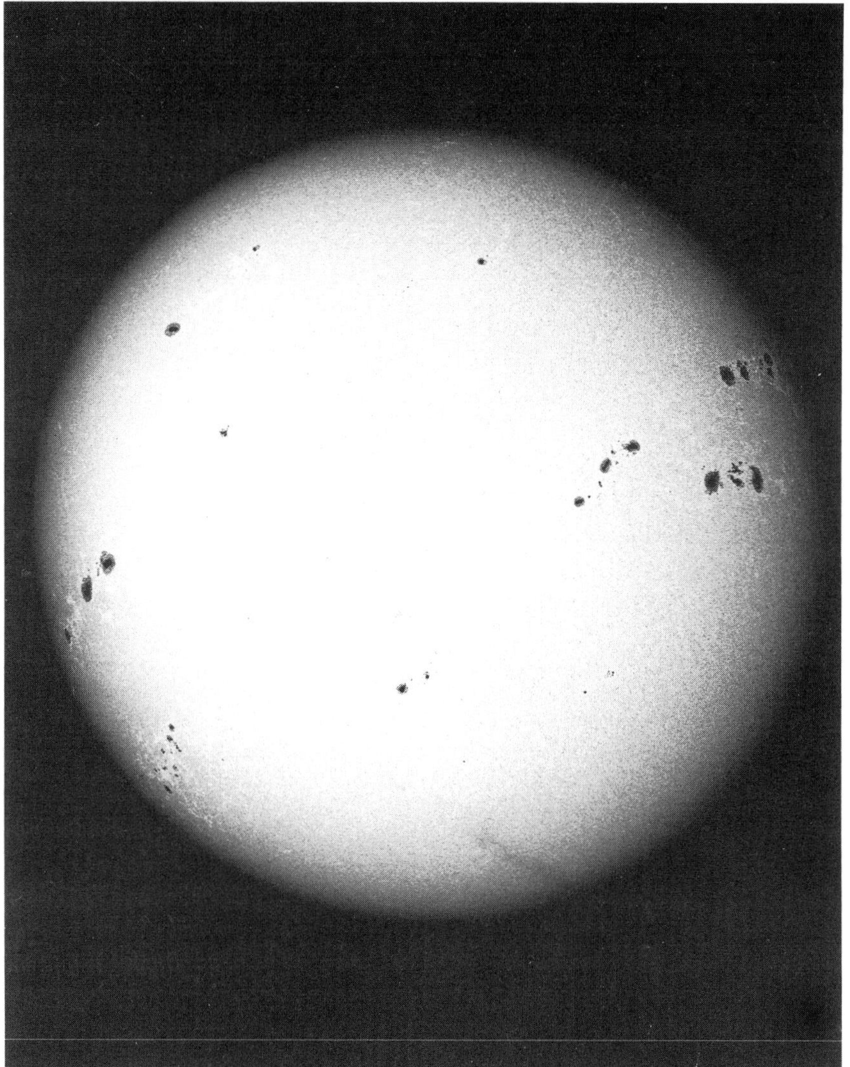

13.41. The spotted Sun. The bright photosphere (the apparent surface) of the Sun shines at a temperature of 5780 kelvin (5503 °C). Many dark sunspots, cooled (4500 K) areas symptomatic of solar magnetic activity, are visible. (Mt. Wilson and Las Campanas Observatories, Carnegie Institution of Washington.)

surface, slowing the flow of energy and producing a pair of sunspots, one where the field exits, the other where it re-enters. The spots have opposite magnetic polarities, or directions (one will be a north pole, the other a south pole), and the directions are reversed between hemispheres, the pairs commonly arranged in complex groups.

As the field starts to wind and the cycle develops, spot activity grows, and some 5 to 10 years after its onset the Sun may be quite covered with them. Occasionally one

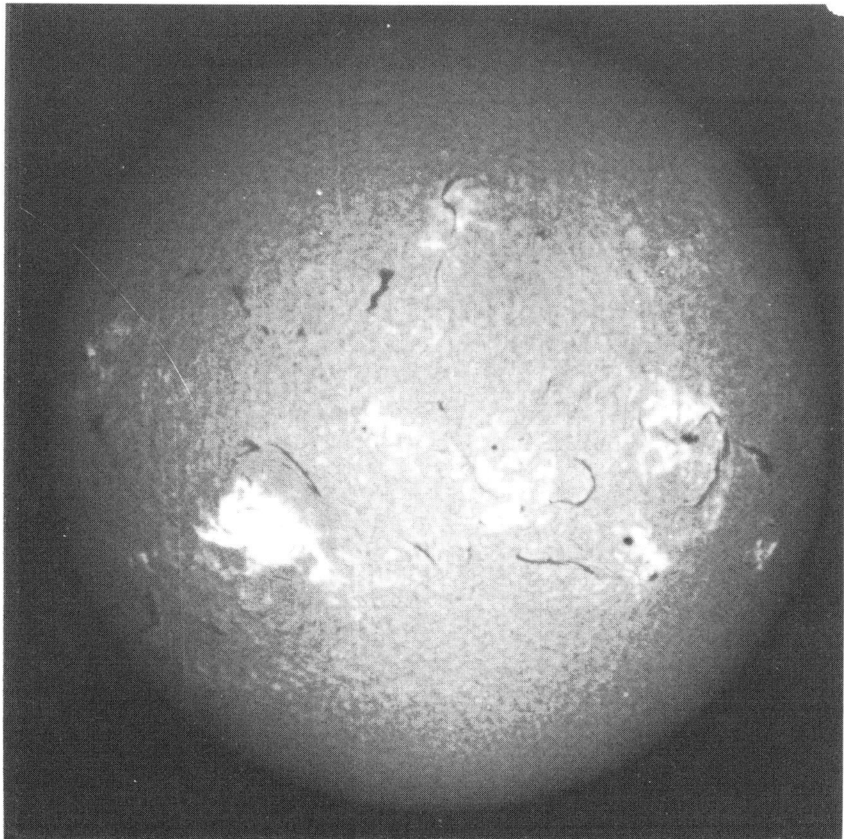

13.42. A solar flare. The flare is the bright patch seen to lower left in this photograph of the solar chromosphere. The flare was caused by an immense release of magnetic energy within a diameter much larger than Earth. Most are much smaller; they are very common near solar maximum. Flares produce X-rays and hurl particles into the solar wind. Prominences (see Figure 10.15) are also seen in relief against the Sun as thin, snake-like *filaments*. (McMath–Hulbert Observatory of the University of Michigan.)

may develop large enough, many times bigger than the Earth, to be seen even with the naked eye when the Sun is comfortably dimmed on the horizon by dust or fog, as in Figure 13.30. After 11 years (the range is about 8 to 15) the magnetic field breaks up and reorders itself, the spots disappear, and the whole cycle begins anew. Now for the next 11 years all the magnetic directions are reversed (including the overall solar field), so that the full cycle is actually 22 years long.

A variety of phenomena is associated with the spots. Most visible is a change in the brightness and shape of the solar corona as seen during a total solar eclipse (see Section 10.12 and compare Figures 10.7 and 10.15). At minimum, the corona pales considerably and develops long equatorial streamers; at maximum it brightens and becomes distinctly more round. Also associated with solar magnetism are red sheets of gas called *promi-*

13.43. Cascades of stars in Scorpius and Ophiuchus epitomize the glory of the sky. (From *An Atlas of the Milky Way*, F. E. Ross and M. R. Calvert, University of Chicago Press, Chicago, 1934.)

nences (see Figure 10.14) that extend from the solar chromosphere into the corona and whose numbers wax and wane with the solar cycle.

Most dramatic are the *solar flares* (Figure 13.42), and related *coronal mass ejections* (CMEs), magnetic releases that take place above the spots. They create shock waves in the solar wind, and produce huge numbers of high-speed protons and electrons that enrich the solar wind, disrupt the Earth's magnetic field, pour particles into the upper atmosphere, and trigger auroral displays. The larger a CME, the more intense will be the auroral light, and the wider the auroral zone, allowing it to be viewed farther from the central current ring.

The result is that aurorae are generally seen at middle latitudes, where most of us live, only near peak sunspot activity within the 11-year cycle. The cycle itself is highly variable: one maximum can be great, with a huge number of spots and flares, the next may be drab and relatively uninteresting. The reason for the variation is unknown. Intense activity, with wonderful displays, were seen in the late 1940s and especially in the 1959–60 solar peak. For a period of a few months the northern lights could be seen in the populated regions of southern Canada and the northern United States nearly every night. The early 1970 maximum was greatly reduced, however; 1980 was better, and some lovely displays were seen during the 1991–92 maximum.

The solar cycle produces only a modulation of the solar wind. It is always present and so is the aurora. Even though there may not be a display in progress, the sky always glows to some small degree as a result. This *permanent aurora* or *airglow* is one of the limiting factors in our ability to see faint stars with the telescope and one of the reasons why we wish to observe the heavens from the blackness of space. The airglow also limits the visibility of the subtle zodiacal light and gegenschein (see Section 12.6). If a solar minimum is especially quiet we can, from a dark mountain site, see the zodiacal light stretch across the sky along the entire ecliptic like a second, albeit much fainter, Milky Way.

13.31 The sky

In the telling of this story we watched the Earth turn and explored the intricate reflected motions of the heavens; looked to the past and wandered the sky, moving about the stars and constellations with our ancestral astronomers; admired the Milky Way in its grand circle around the heavens; found our way across the surface of the Earth; saw the Moon and planets wheel around us with mechanical exactness and examined something of their natures; and delighted at sunlight dancing through the air. The stage of the sky presents a free show. It will amaze, entertain, amuse, and intrigue for a lifetime.

Appendix 1

Graphs and tables

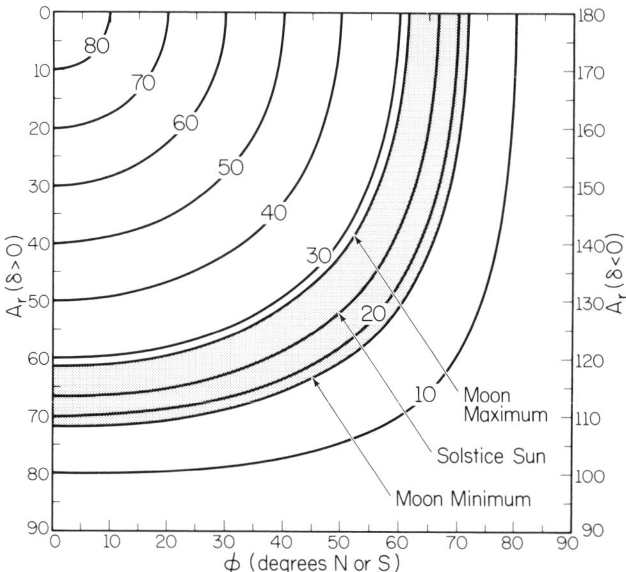

A1.1. Azimuth of the rising point, A_r, for bodies of different declination (δ) as a function of latitude (φ). Atmospheric refraction is not considered. If δ is positive use the left-hand scale; if δ is negative use the right-hand scale. The azimuth of the setting point is $A_s = 360° - A_r$. The stars are either circumpolar or do not rise for φ greater than the asymptotic limits of individual curves. Special curves indicate the solstitial Sun at δ = 23°27′ (Chapter 3) and solstitial Moon for the maximum and minimum celestial latitudes of ±5°09′ (Chapter 9).

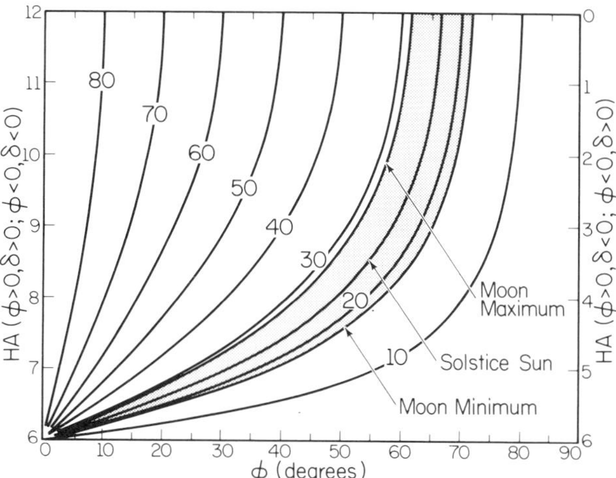

A1.2. Hour angle (HA) of the setting point for bodies of different declination (δ) as a function of latitude (ϕ). Atmospheric refraction is not considered. The duration above the horizon is twice the hour angle. Positive latitude: if δ is positive, use the left-hand scale, if δ is negative use the right-hand scale. Negative latitude: if δ is positive use the right-hand scale, if δ is negative use the left-hand scale. The stars are either circumpolar or do not rise for ϕ greater than the asymptotic limit of the individual curves. Special curves denote the solstitial Sun and Moon as in Figure A1.1. Hour angles are negative if the object is rising.

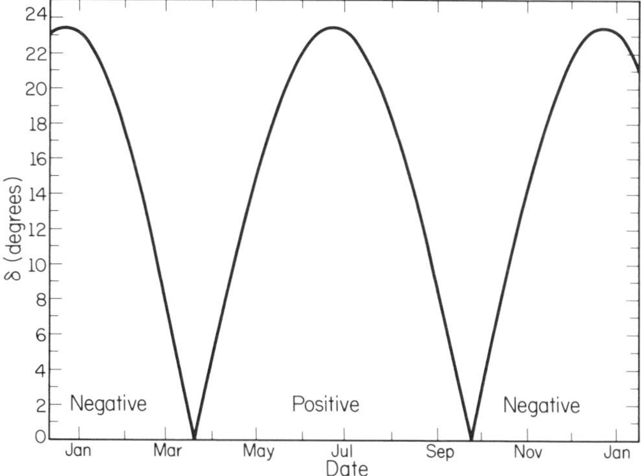

A1.3. The declination of the Sun throughout the year for the vernal equinox at 0^h on March 21. Declination is positive between March 21 and September 23, negative between September 23 and March 21, as indicated. The negative and positive curves are both drawn above the axis to enable maximum readability.

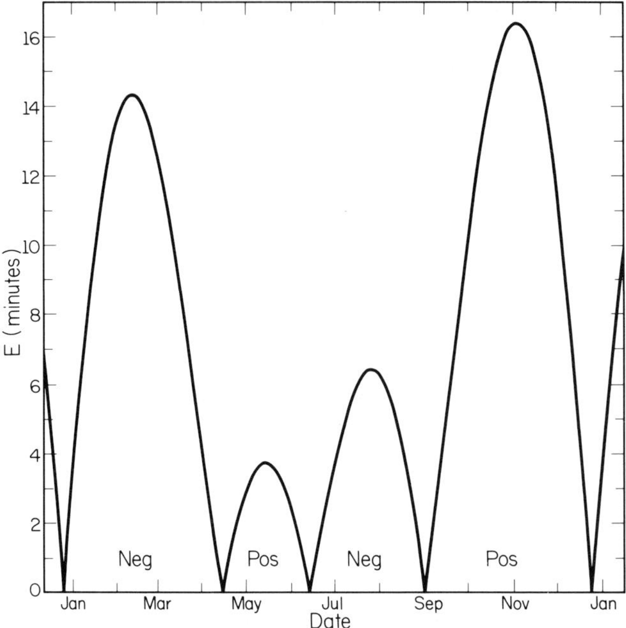

A1.4. The equation of time. E is the local apparent solar time minus the local mean solar time (E – LAST – LMST) in minutes for a year mid-way between leap years. The curve is accurate to within about 10 seconds for other years. For a leap year subtract one day from the horizontal axis after February 28. Whether the equation of time is positive or negative is indicated on the graph. The negative and positive curves are both drawn above the axis to enable maximum readability.

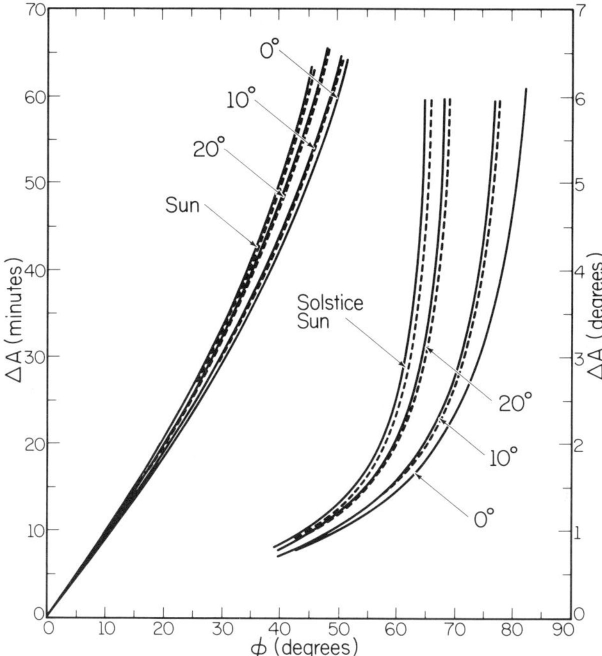

A1.5. Change in the azimuths of sunrise and sunset and moonrise and moonset in minutes of arc upon consideration of the upper limb (semidiameter) and atmospheric refraction compared with the azimuth for the center without refraction (Figure A1.1). ΔA is always toward the elevated pole: in the northern hemisphere, it is negative for sunrise and positive for sunset, vice versa in the southern hemisphere. The solid lines represent the Sun when its declination has the same sign as the latitude (i.e., positive for a northern hemisphere observer, negative for a southern), the dashed lines when the signs are opposite. Use the right-hand scale for the set of curves on the right, the left-hand scale for those on the left. Curves are given for the solstitial Sun.

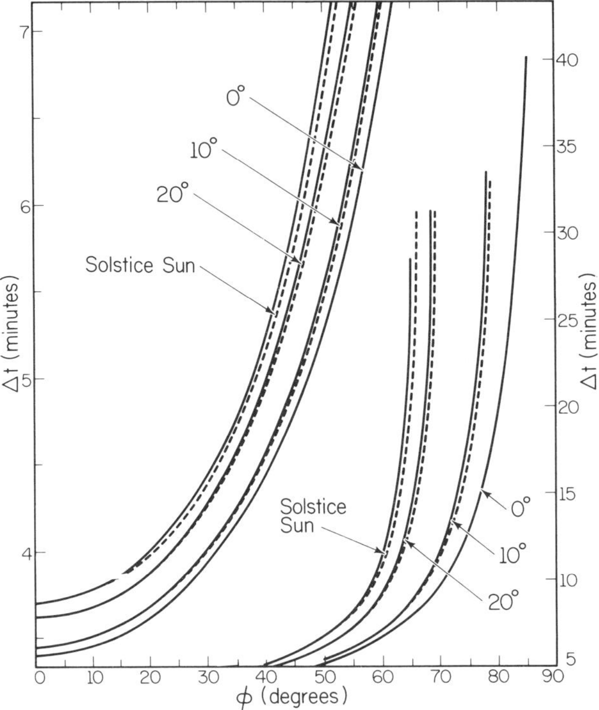

A1.6. Change in the times of sunrise and sunset and moonrise and moonset in minutes upon consideration of the upper limb (semidiameter) and atmospheric refraction, compared with the time for the center with no refraction (Figure A1.2). The value of Δt is positive at sunset, negative at sunrise. The solid lines represent the Sun when its declination has the same sign as the latitude, the dashed lines when the signs are opposite. Use the right-hand scale for the set of curves on the right, the left-hand scale for those on the left. Curves are given for the solstitial Sun.

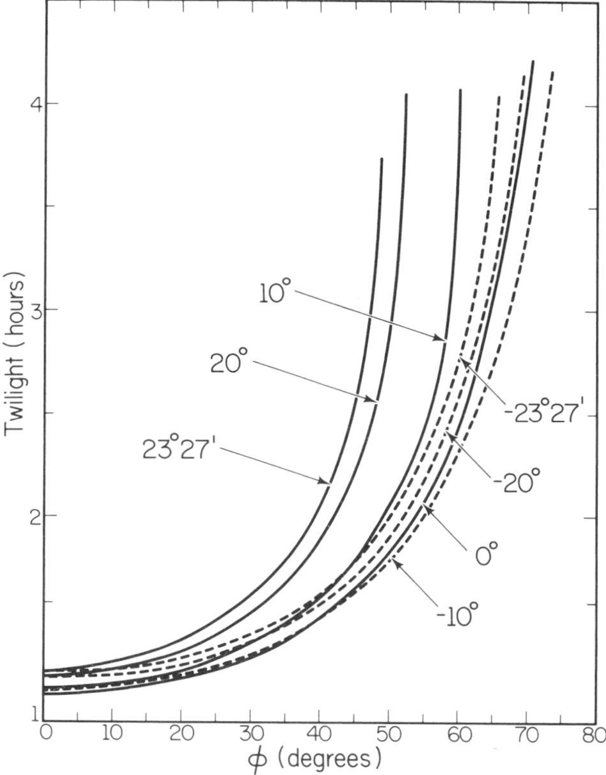

A1.7. The duration of astronomical twilight in hours as a function of latitude for different solar declinations. The solid lines represent the Sun when its declination has the same sign as the latitude, and the dashed lines when the signs are opposite. The extreme left-hand curves represent the solstitial Sun.

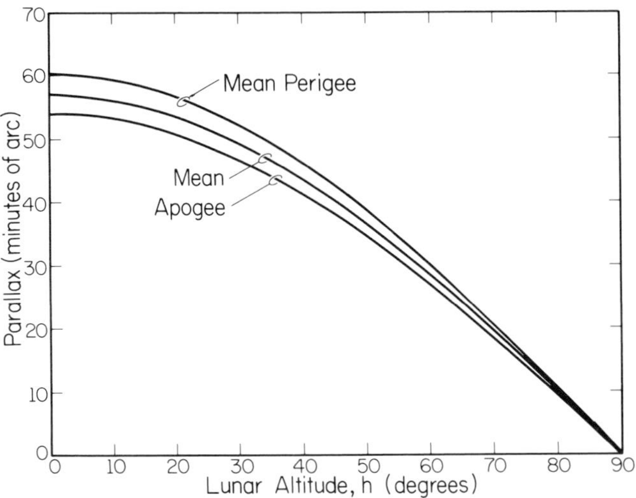

A1.8. Lunar parallax as a function of altitude for the mean distance and for mean perigee and apogee.

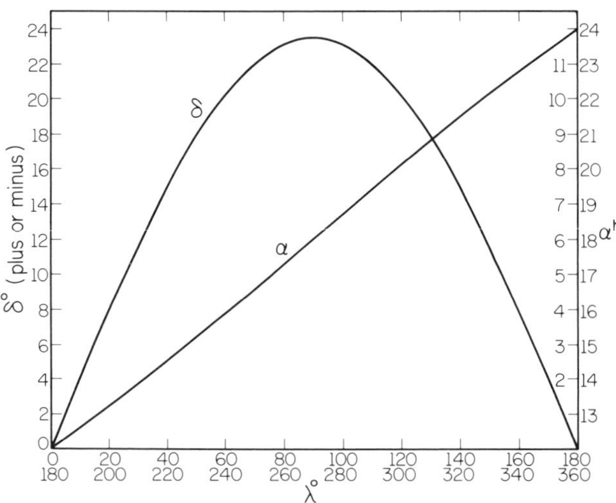

A1.9. Right ascension (α) and declination (δ) of the ecliptic as a function of celestial longitude (λ). For λ between 180° and 360°, δ is negative. Read the inside scale on the right-hand axis (right ascension) for λ between 0° and 180° and the outside scale for λ between 180° and 360°.

Table A1.1. *Day numbers. In the three columns Day 1 refers respectively to January 1, March 21, and September 23. Add 1 past February 28 for a leap year.*

Date	Begin January 1	Begin March 21	Begin Sept. 23	Date	Begin January 1	Begin March 21	Begin Sept. 23
Jan 1	1	287	101	Jul 1	182	103	282
10	10	296	110	10	191	112	291
20	20	306	120	20	201	122	301
31	31	316	131	31	212	133	312
Feb 1	32	317	132	Aug 1	213	134	313
10	41	327	141	10	222	143	322
20	51	337	151	20	232	153	332
28	59	345	159	31	243	164	343
Mar 1	60	346	160	Sep 1	244	165	344
10	69	355	169	10	253	174	353
20	79	365	179	20	263	184	363
31	90	11	190	30	273	194	8
Apr 1	91	12	191	Oct 1	274	195	9
10	100	21	200	10	283	204	18
20	110	31	210	20	293	214	28
30	120	41	220	31	304	225	39
May 1	121	42	221	Nov 1	305	226	40
10	130	51	230	10	314	235	49
20	140	61	240	20	324	245	59
31	151	72	251	30	334	255	69
Jun 1	152	73	252	Dec 1	335	256	70
10	161	82	261	10	344	265	79
20	171	92	271	20	354	275	89
30	181	102	281	31	365	286	100

Table A1.2. *Sunrise and sunset January–April*

UNIVERSAL TIME FOR MERIDIAN OF GREENWICH
SUNRISE

Lat.	−55°	−50°	−45°	−40°	−35°	−30°	−20°	−10°	0°	+10°	+20°	+30°	+35°	+40°	+42°	+44°	+46°	+48°	+50°	+52°	+54°	+56°	+58°	+60°	+62°	+64°	+66°
	h m	h m	h m	h m	h m	h m	h m	h m	h m	h m	h m	h m	h m	h m	h m	h m	h m	h m	h m	h m	h m	h m	h m	h m	h m	h m	h m
Jan −2	3 23	3 52	4 15	4 32	4 47	5 00	5 22	5 41	5 58	6 16	6 34	6 55	7 07	7 21	7 28	7 34	7 42	7 50	7 59	8 08	8 19	8 32	8 46	9 03	9 25	9 52	10 33
2	3 27	3 56	4 18	4 36	4 50	5 03	5 25	5 43	6 00	6 17	6 35	6 56	7 08	7 22	7 28	7 35	7 42	7 50	7 59	8 08	8 19	8 31	8 45	9 02	9 22	9 49	10 27
6	3 32	4 01	4 22	4 39	4 54	5 06	5 27	5 45	6 02	6 19	6 37	6 57	7 09	7 22	7 28	7 35	7 42	7 49	7 58	8 07	8 18	8 29	8 43	8 59	9 19	9 44	10 18
10	3 39	4 06	4 26	4 43	4 57	5 09	5 30	5 47	6 04	6 20	6 37	6 57	7 09	7 22	7 28	7 34	7 41	7 48	7 56	8 05	8 15	8 27	8 40	8 55	9 14	9 37	10 08
14	3 45	4 11	4 31	4 47	5 01	5 12	5 32	5 49	6 05	6 21	6 38	6 57	7 08	7 21	7 26	7 32	7 39	7 46	7 54	8 02	8 12	8 23	8 35	8 50	9 07	9 29	9 57
18	3 53	4 18	4 36	4 52	5 05	5 16	5 35	5 51	6 07	6 22	6 38	6 56	7 07	7 19	7 24	7 30	7 36	7 43	7 51	7 59	8 08	8 18	8 30	8 43	9 00	9 19	9 45
22	4 01	4 24	4 42	4 56	5 09	5 19	5 38	5 53	6 08	6 22	6 38	6 55	7 05	7 17	7 22	7 27	7 33	7 40	7 47	7 54	8 03	8 13	8 23	8 36	8 51	9 09	9 32
26	4 09	4 31	4 48	5 01	5 13	5 23	5 40	5 55	6 09	6 23	6 37	6 54	7 03	7 14	7 19	7 24	7 30	7 36	7 42	7 49	7 57	8 06	8 16	8 28	8 42	8 58	9 18
30	4 17	4 38	4 53	5 06	5 17	5 27	5 43	5 57	6 10	6 23	6 36	6 52	7 01	7 11	7 15	7 20	7 25	7 31	7 37	7 44	7 51	7 59	8 09	8 19	8 32	8 46	9 04
Feb. 3	4 26	4 45	4 59	5 11	5 21	5 30	5 45	5 58	6 10	6 22	6 35	6 49	6 58	7 07	7 11	7 16	7 20	7 26	7 31	7 37	7 44	7 52	8 00	8 10	8 21	8 34	8 50
7	4 35	4 52	5 05	5 16	5 25	5 34	5 47	6 00	6 11	6 22	6 33	6 47	6 54	7 03	7 07	7 11	7 15	7 20	7 25	7 31	7 37	7 44	7 51	8 00	8 10	8 22	8 36
11	4 44	4 59	5 11	5 21	5 29	5 37	5 50	6 01	6 11	6 21	6 32	6 44	6 51	6 58	7 02	7 06	7 09	7 14	7 18	7 23	7 29	7 35	7 42	7 50	7 59	8 09	8 21
15	4 52	5 06	5 17	5 26	5 33	5 40	5 52	6 02	6 11	6 20	6 29	6 40	6 46	6 53	6 57	7 00	7 03	7 07	7 11	7 16	7 21	7 26	7 32	7 39	7 47	7 56	8 07
19	5 01	5 13	5 23	5 30	5 37	5 43	5 53	6 02	6 11	6 19	6 27	6 37	6 42	6 48	6 51	6 54	6 57	7 00	7 04	7 08	7 12	7 17	7 22	7 28	7 35	7 43	7 52
23	5 09	5 20	5 28	5 35	5 41	5 46	5 55	6 03	6 10	6 17	6 24	6 33	6 37	6 43	6 45	6 48	6 50	6 53	6 56	7 00	7 03	7 07	7 12	7 17	7 23	7 29	7 37
27	5 18	5 27	5 34	5 40	5 45	5 49	5 57	6 03	6 09	6 15	6 22	6 29	6 33	6 37	6 39	6 41	6 43	6 46	6 48	6 51	6 54	6 57	7 01	7 05	7 10	7 16	7 22
Mar. 3	5 26	5 33	5 39	5 44	5 48	5 52	5 58	6 04	6 09	6 14	6 19	6 24	6 27	6 31	6 33	6 34	6 36	6 38	6 40	6 42	6 45	6 47	6 50	6 54	6 58	7 02	7 07
7	5 34	5 40	5 45	5 49	5 52	5 55	6 00	6 04	6 08	6 12	6 15	6 20	6 22	6 25	6 26	6 27	6 29	6 30	6 32	6 33	6 35	6 37	6 39	6 42	6 45	6 48	6 52
11	5 42	5 47	5 50	5 53	5 55	5 57	6 01	6 04	6 07	6 09	6 12	6 15	6 17	6 19	6 19	6 20	6 21	6 22	6 23	6 24	6 26	6 27	6 28	6 30	6 32	6 34	6 36
15	5 50	5 53	5 55	5 57	5 59	6 00	6 02	6 04	6 06	6 07	6 09	6 10	6 11	6 12	6 13	6 13	6 14	6 14	6 15	6 15	6 16	6 16	6 17	6 18	6 19	6 20	6 21
19	5 58	6 00	6 01	6 01	6 02	6 03	6 03	6 04	6 05	6 05	6 05	6 06	6 06	6 06	6 06	6 06	6 06	6 06	6 06	6 06	6 06	6 06	6 06	6 06	6 06	6 06	6 06
23	6 06	6 06	6 06	6 05	6 05	6 05	6 05	6 04	6 03	6 03	6 02	6 01	6 00	5 59	5 59	5 59	5 58	5 58	5 57	5 57	5 56	5 55	5 55	5 54	5 53	5 52	5 50
27	6 14	6 12	6 11	6 10	6 08	6 07	6 06	6 04	6 02	6 00	5 58	5 56	5 55	5 53	5 52	5 51	5 50	5 50	5 49	5 47	5 46	5 45	5 43	5 42	5 40	5 38	5 35
31	6 22	6 18	6 16	6 14	6 12	6 10	6 07	6 04	6 01	5 58	5 55	5 51	5 49	5 46	5 45	5 44	5 43	5 41	5 40	5 38	5 36	5 34	5 32	5 30	5 27	5 23	5 20
Apr. 4	6 29	6 25	6 21	6 18	6 15	6 12	6 08	6 04	6 00	5 56	5 51	5 46	5 43	5 40	5 38	5 37	5 35	5 33	5 31	5 29	5 27	5 24	5 21	5 17	5 14	5 09	5 04

SUNSET

		h m	h m	h m	h m	h m	h m	h m	h m	h m	h m	h m	h m	h m	h m	h m	h m	h m	h m	h m	h m	h m	h m	h m	h m	h m	h m	h m
Jan.	−2	20 41	20 12	19 49	19 32	19 17	19 04	18 42	18 23	18 06	17 49	17 30	17 09	16 57	16 43	16 37	16 30	16 23	16 15	16 06	15 56	15 45	15 33	15 18	15 01	14 40	14 12	13 32
	2	20 40	20 12	19 50	19 32	19 18	19 05	18 43	18 25	18 08	17 51	17 33	17 12	17 00	16 46	16 40	16 33	16 26	16 18	16 10	16 00	15 49	15 37	15 23	15 06	14 46	14 19	13 42
	6	20 39	20 11	19 49	19 32	19 18	19 05	18 44	18 26	18 10	17 53	17 35	17 15	17 03	16 50	16 44	16 37	16 30	16 22	16 14	16 05	15 54	15 43	15 29	15 13	14 53	14 28	13 54
	10	20 36	20 09	19 48	19 32	19 18	19 06	18 45	18 28	18 11	17 55	17 38	17 18	17 07	16 54	16 48	16 41	16 35	16 27	16 19	16 10	16 00	15 49	15 36	15 20	15 02	14 39	14 07
	14	20 32	20 06	19 46	19 30	19 17	19 05	18 46	18 29	18 13	17 57	17 40	17 21	17 10	16 58	16 52	16 46	16 40	16 32	16 25	16 16	16 07	15 56	15 43	15 29	15 12	14 50	14 22
	18	20 27	20 02	19 44	19 28	19 16	19 05	18 46	18 29	18 14	17 59	17 43	17 25	17 14	17 02	16 57	16 51	16 45	16 38	16 31	16 23	16 13	16 03	15 52	15 38	15 22	15 02	14 37
	22	20 21	19 58	19 41	19 26	19 14	19 03	18 45	18 30	18 15	18 01	17 46	17 28	17 18	17 07	17 02	16 56	16 50	16 44	16 37	16 29	16 21	16 11	16 00	15 48	15 33	15 15	14 52
	26	20 15	19 53	19 37	19 23	19 12	19 02	18 45	18 30	18 16	18 03	17 48	17 32	17 22	17 12	17 07	17 02	16 56	16 50	16 44	16 36	16 28	16 20	16 09	15 58	15 44	15 28	15 08
	30	20 08	19 48	19 32	19 20	19 09	19 00	18 44	18 30	18 17	18 04	17 51	17 35	17 26	17 16	17 12	17 07	17 02	16 56	16 50	16 44	16 36	16 28	16 19	16 08	15 56	15 41	15 23
Feb.	3	20 00	19 42	19 28	19 16	19 06	18 57	18 42	18 29	18 17	18 05	17 53	17 39	17 30	17 21	17 17	17 13	17 08	17 03	16 57	16 51	16 44	16 37	16 28	16 19	16 07	15 54	15 38
	7	19 52	19 36	19 22	19 12	19 02	18 54	18 41	18 29	18 18	18 07	17 55	17 42	17 34	17 26	17 22	17 18	17 14	17 09	17 04	16 58	16 52	16 45	16 38	16 29	16 19	16 07	15 53
	11	19 44	19 29	19 17	19 07	18 59	18 51	18 39	18 28	18 18	18 08	17 57	17 45	17 38	17 31	17 27	17 24	17 20	17 15	17 11	17 06	17 00	16 54	16 47	16 40	16 31	16 21	16 08
	15	19 35	19 21	19 11	19 02	18 54	18 48	18 36	18 27	18 18	18 09	17 59	17 48	17 42	17 35	17 32	17 29	17 26	17 22	17 18	17 13	17 08	17 03	16 57	16 50	16 42	16 34	16 23
	19	19 26	19 14	19 04	18 57	18 50	18 44	18 34	18 25	18 17	18 09	18 01	17 52	17 46	17 40	17 37	17 35	17 31	17 28	17 25	17 21	17 17	17 12	17 07	17 01	16 54	16 46	16 37
	23	19 16	19 06	18 58	18 51	18 45	18 40	18 31	18 24	18 17	18 10	18 03	17 55	17 50	17 45	17 42	17 40	17 37	17 34	17 31	17 28	17 25	17 21	17 16	17 11	17 05	16 59	16 51
	27	19 07	18 58	18 51	18 45	18 40	18 36	18 28	18 22	18 16	18 10	18 04	17 57	17 54	17 49	17 47	17 45	17 43	17 41	17 38	17 35	17 32	17 29	17 25	17 21	17 17	17 11	17 05
Mar.	3	18 57	18 50	18 44	18 39	18 35	18 31	18 25	18 20	18 15	18 11	18 06	18 00	17 57	17 54	17 52	17 50	17 49	17 47	17 45	17 43	17 40	17 38	17 35	17 31	17 28	17 24	17 19
	7	18 47	18 41	18 37	18 33	18 30	18 27	18 22	18 18	18 14	18 11	18 07	18 03	18 01	17 58	17 57	17 56	17 54	17 53	17 51	17 50	17 48	17 46	17 44	17 42	17 39	17 36	17 32
	11	18 37	18 33	18 29	18 27	18 24	18 22	18 19	18 16	18 13	18 11	18 08	18 06	18 04	18 02	18 01	18 01	18 00	17 59	17 58	17 57	17 56	17 54	17 53	17 51	17 50	17 48	17 45
	15	18 26	18 24	18 22	18 20	18 19	18 18	18 15	18 14	18 12	18 11	18 10	18 08	18 07	18 06	18 06	18 06	18 05	18 05	18 04	18 04	18 03	18 03	18 02	18 01	18 01	18 00	17 59
	19	18 16	18 15	18 14	18 14	18 13	18 13	18 12	18 12	18 11	18 11	18 11	18 11	18 11	18 11	18 11	18 11	18 11	18 11	18 11	18 11	18 11	18 11	18 11	18 11	18 11	18 12	18 12
	23	18 06	18 07	18 07	18 07	18 08	18 08	18 09	18 09	18 10	18 11	18 12	18 13	18 14	18 15	18 15	18 16	18 16	18 17	18 17	18 18	18 18	18 19	18 20	18 21	18 22	18 23	18 25
	27	17 56	17 58	17 59	18 01	18 02	18 03	18 05	18 07	18 09	18 11	18 13	18 15	18 17	18 19	18 20	18 20	18 21	18 22	18 23	18 25	18 26	18 27	18 29	18 31	18 33	18 35	18 38
	31	17 46	17 49	17 52	17 54	17 56	17 58	18 02	18 05	18 08	18 11	18 14	18 18	18 20	18 23	18 24	18 25	18 27	18 28	18 30	18 31	18 33	18 35	18 38	18 40	18 43	18 47	18 51
Apr.	4	17 36	17 41	17 45	17 48	17 51	17 54	17 58	18 02	18 06	18 10	18 15	18 20	18 23	18 27	18 28	18 30	18 32	18 34	18 36	18 38	18 41	18 44	18 47	18 50	18 54	18 59	19 04

Table A1.2. *Sunrise and sunset April–July*

UNIVERSAL TIME FOR MERIDIAN OF GREENWICH
SUNRISE

Lat.	−55°	−50°	−45°	−40°	−35°	−30°	−20°	−10°	0°	+10°	+20°	+30°	+35°	+40°	+42°	+44°	+46°	+48°	+50°	+52°	+54°	+56°	+58°	+60°	+62°	+64°	+66°
	h m	h m	h m	h m	h m	h m	h m	h m	h m	h m	h m	h m	h m	h m	h m	h m	h m	h m	h m	h m	h m	h m	h m	h m	h m	h m	h m
Mar. 31	6 22	6 18	6 16	6 14	6 12	6 10	6 07	6 04	6 01	5 58	5 55	5 51	5 49	5 46	5 45	5 44	5 43	5 41	5 40	5 38	5 36	5 34	5 32	5 30	5 27	5 23	5 20
Apr. 4	6 29	6 25	6 21	6 18	6 15	6 12	6 08	6 04	6 00	5 56	5 51	5 46	5 43	5 40	5 38	5 37	5 35	5 33	5 31	5 29	5 27	5 24	5 21	5 17	5 14	5 09	5 04
8	6 37	6 31	6 26	6 22	6 18	6 15	6 09	6 04	5 59	5 54	5 48	5 42	5 38	5 34	5 32	5 30	5 28	5 25	5 23	5 20	5 17	5 13	5 10	5 05	5 01	4 55	4 49
12	6 45	6 37	6 31	6 26	6 21	6 17	6 10	6 04	5 58	5 51	5 45	5 37	5 33	5 27	5 25	5 23	5 20	5 17	5 14	5 11	5 07	5 03	4 59	4 54	4 48	4 41	4 33
16	6 52	6 43	6 36	6 30	6 24	6 19	6 11	6 04	5 57	5 49	5 42	5 33	5 27	5 21	5 19	5 16	5 13	5 10	5 06	5 02	4 58	4 53	4 48	4 42	4 35	4 27	4 18
20	7 00	6 49	6 41	6 34	6 27	6 22	6 12	6 04	5 56	5 47	5 39	5 28	5 22	5 16	5 12	5 09	5 06	5 02	4 58	4 53	4 49	4 43	4 37	4 30	4 22	4 13	4 02
24	7 08	6 55	6 46	6 38	6 31	6 24	6 13	6 04	5 55	5 46	5 36	5 24	5 18	5 10	5 06	5 03	4 59	4 55	4 50	4 45	4 40	4 33	4 26	4 19	4 09	3 59	3 46
28	7 15	7 01	6 51	6 42	6 34	6 27	6 15	6 04	5 54	5 44	5 33	5 20	5 13	5 04	5 01	4 57	4 52	4 48	4 43	4 37	4 31	4 24	4 16	4 07	3 57	3 45	3 31
May 2	7 23	7 07	6 55	6 45	6 37	6 29	6 16	6 05	5 54	5 42	5 30	5 17	5 09	4 59	4 55	4 51	4 46	4 41	4 35	4 29	4 22	4 15	4 06	3 56	3 45	3 31	3 15
6	7 30	7 13	7 00	6 49	6 40	6 32	6 18	6 05	5 53	5 41	5 28	5 13	5 05	4 55	4 50	4 45	4 40	4 35	4 29	4 22	4 14	4 06	3 57	3 46	3 33	3 18	2 59
10	7 37	7 19	7 05	6 53	6 43	6 34	6 19	6 06	5 53	5 40	5 26	5 10	5 01	4 50	4 45	4 40	4 35	4 29	4 22	4 15	4 07	3 58	3 47	3 35	3 21	3 04	2 43
14	7 44	7 25	7 10	6 57	6 46	6 37	6 21	6 06	5 53	5 39	5 24	5 07	4 58	4 46	4 41	4 36	4 30	4 23	4 16	4 08	4 00	3 50	3 39	3 26	3 10	2 51	2 28
18	7 51	7 30	7 14	7 01	6 49	6 39	6 22	6 07	5 53	5 38	5 23	5 05	4 55	4 42	4 37	4 31	4 25	4 18	4 11	4 02	3 53	3 42	3 30	3 16	3 00	2 39	2 12
22	7 57	7 35	7 18	7 04	6 52	6 42	6 24	6 08	5 53	5 38	5 22	5 03	4 52	4 39	4 34	4 27	4 21	4 14	4 06	3 57	3 47	3 36	3 23	3 08	2 49	2 26	1 55
26	8 03	7 40	7 22	7 08	6 55	6 44	6 25	6 09	5 53	5 38	5 21	5 01	4 50	4 37	4 31	4 24	4 17	4 10	4 01	3 52	3 42	3 30	3 16	3 00	2 40	2 15	1 39
30	8 09	7 45	7 26	7 11	6 58	6 46	6 27	6 10	5 54	5 38	5 20	5 00	4 48	4 34	4 28	4 21	4 14	4 06	3 58	3 48	3 37	3 25	3 10	2 53	2 32	2 04	1 22
June 3	8 14	7 49	7 29	7 13	7 00	6 48	6 28	6 11	5 54	5 38	5 20	4 59	4 47	4 33	4 26	4 19	4 12	4 04	3 55	3 45	3 33	3 20	3 05	2 47	2 24	1 54	1 05
7	8 18	7 52	7 32	7 16	7 02	6 50	6 30	6 12	5 55	5 38	5 20	4 58	4 46	4 31	4 25	4 18	4 10	4 02	3 52	3 42	3 30	3 17	3 01	2 42	2 18	1 45	0 46
11	8 21	7 55	7 35	7 18	7 04	6 52	6 31	6 13	5 56	5 39	5 20	4 58	4 45	4 31	4 24	4 17	4 09	4 00	3 51	3 40	3 28	3 14	2 58	2 38	2 13	1 38	0 24
15	8 24	7 57	7 37	7 20	7 06	6 54	6 32	6 14	5 57	5 39	5 20	4 58	4 45	4 30	4 24	4 16	4 08	4 00	3 50	3 39	3 27	3 13	2 56	2 36	2 10	1 33	** **
19	8 26	7 59	7 38	7 21	7 07	6 55	6 34	6 15	5 58	5 40	5 21	4 59	4 46	4 31	4 24	4 17	4 09	4 00	3 50	3 39	3 27	3 13	2 56	2 35	2 09	1 31	** **
23	8 27	8 00	7 39	7 22	7 08	6 56	6 34	6 16	5 58	5 41	5 22	5 00	4 47	4 32	4 25	4 17	4 09	4 01	3 51	3 40	3 28	3 13	2 57	2 36	2 10	1 31	** **
27	8 27	8 00	7 39	7 23	7 09	6 56	6 35	6 17	5 59	5 42	5 23	5 01	4 48	4 33	4 26	4 19	4 11	4 02	3 52	3 42	3 29	3 15	2 58	2 38	2 12	1 35	** **
July 1	8 26	8 00	7 39	7 23	7 09	6 56	6 36	6 17	6 00	5 43	5 24	5 02	4 49	4 35	4 28	4 21	4 13	4 04	3 55	3 44	3 32	3 18	3 02	2 42	2 16	1 40	0 14
5	8 24	7 58	7 38	7 22	7 08	6 56	6 36	6 18	6 01	5 44	5 25	5 04	4 51	4 37	4 30	4 23	4 15	4 07	3 57	3 47	3 35	3 22	3 06	2 46	2 22	1 48	0 45

SUNSET

		h m	h m	h m	h m	h m	h m	h m	h m	h m	h m	h m	h m	h m	h m	h m	h m	h m	h m	h m	h m	h m	h m	h m	h m	h m	h m	h m
Mar.	31	17 46	17 49	17 52	17 54	17 56	17 58	18 02	18 05	18 08	18 11	18 14	18 18	18 20	18 23	18 24	18 25	18 27	18 28	18 30	18 31	18 33	18 35	18 38	18 40	18 43	18 47	18 51
Apr.	4	17 36	17 41	17 45	17 48	17 51	17 54	17 58	18 02	18 06	18 10	18 15	18 20	18 23	18 27	18 28	18 30	18 32	18 34	18 36	18 38	18 41	18 44	18 47	18 50	18 54	18 59	19 04
	8	17 26	17 32	17 37	17 42	17 45	17 49	17 55	18 00	18 05	18 10	18 16	18 23	18 27	18 31	18 33	18 35	18 37	18 40	18 42	18 45	18 48	18 52	18 56	19 00	19 05	19 11	19 17
	12	17 16	17 24	17 30	17 36	17 40	17 44	17 52	17 58	18 04	18 10	18 17	18 25	18 30	18 35	18 37	18 40	18 42	18 45	18 48	18 52	18 56	19 00	19 05	19 10	19 16	19 23	19 31
	16	17 06	17 16	17 23	17 30	17 35	17 40	17 48	17 56	18 03	18 10	18 18	18 28	18 33	18 39	18 42	18 45	18 48	18 51	18 55	18 59	19 03	19 08	19 13	19 20	19 27	19 35	19 44
	20	16 57	17 08	17 17	17 24	17 30	17 36	17 45	17 54	18 02	18 11	18 20	18 30	18 36	18 43	18 46	18 49	18 53	18 57	19 01	19 06	19 11	19 16	19 22	19 30	19 38	19 47	19 58
	24	16 48	17 00	17 10	17 18	17 25	17 32	17 43	17 52	18 02	18 11	18 21	18 33	18 39	18 47	18 51	18 54	18 58	19 03	19 07	19 12	19 18	19 24	19 31	19 40	19 49	20 00	20 13
	28	16 39	16 53	17 04	17 13	17 21	17 28	17 40	17 51	18 01	18 11	18 22	18 35	18 43	18 51	18 55	18 59	19 03	19 08	19 13	19 19	19 25	19 33	19 40	19 49	20 00	20 12	20 27
May	2	16 31	16 46	16 58	17 08	17 17	17 24	17 38	17 49	18 00	18 12	18 24	18 38	18 46	18 55	18 59	19 04	19 09	19 14	19 20	19 26	19 33	19 41	19 49	19 59	20 11	20 25	20 42
	6	16 23	16 39	16 52	17 03	17 13	17 21	17 35	17 48	18 00	18 12	18 25	18 40	18 49	18 59	19 04	19 09	19 14	19 19	19 26	19 33	19 40	19 49	19 58	20 09	20 22	20 38	20 57
	10	16 15	16 33	16 47	16 59	17 09	17 18	17 33	17 47	18 00	18 13	18 27	18 43	18 52	19 03	19 08	19 13	19 19	19 25	19 32	19 39	19 47	19 56	20 07	20 19	20 33	20 51	21 12
	14	16 08	16 27	16 43	16 55	17 06	17 15	17 32	17 46	18 00	18 14	18 28	18 45	18 56	19 07	19 12	19 18	19 24	19 30	19 37	19 45	19 54	20 04	20 15	20 29	20 44	21 04	21 28
	18	16 01	16 22	16 38	16 52	17 03	17 13	17 30	17 46	18 00	18 14	18 30	18 48	18 59	19 11	19 16	19 22	19 28	19 35	19 43	19 51	20 01	20 11	20 24	20 38	20 55	21 16	21 44
	22	15 55	16 17	16 35	16 49	17 01	17 11	17 29	17 45	18 00	18 15	18 32	18 50	19 02	19 14	19 20	19 26	19 33	19 40	19 48	19 57	20 07	20 19	20 32	20 47	21 06	21 29	22 01
	26	15 50	16 13	16 31	16 46	16 59	17 10	17 28	17 45	18 01	18 16	18 33	18 53	19 04	19 18	19 24	19 30	19 37	19 45	19 53	20 03	20 13	20 25	20 39	20 55	21 16	21 41	22 19
	30	15 46	16 10	16 29	16 44	16 57	17 08	17 28	17 45	18 01	18 17	18 35	18 55	19 07	19 21	19 27	19 34	19 41	19 49	19 58	20 08	20 19	20 31	20 46	21 03	21 25	21 53	22 36
June	3	15 42	16 07	16 27	16 42	16 56	17 07	17 28	17 45	18 02	18 18	18 36	18 57	19 10	19 24	19 30	19 37	19 45	19 53	20 02	20 12	20 23	20 37	20 52	21 10	21 33	22 04	22 55
	7	15 39	16 05	16 25	16 41	16 55	17 07	17 28	17 45	18 02	18 19	18 38	18 59	19 12	19 26	19 33	19 40	19 48	19 56	20 05	20 16	20 28	20 41	20 57	21 16	21 41	22 14	23 16
	11	15 37	16 04	16 24	16 41	16 55	17 07	17 28	17 46	18 03	18 21	18 39	19 01	19 14	19 29	19 35	19 42	19 50	19 59	20 08	20 19	20 31	20 45	21 01	21 21	21 47	22 22	23 44
	15	15 36	16 03	16 24	16 41	16 55	17 07	17 28	17 47	18 04	18 22	18 40	19 02	19 15	19 30	19 37	19 44	19 52	20 01	20 11	20 22	20 34	20 48	21 05	21 25	21 51	22 28	** **
	19	15 36	16 03	16 24	16 41	16 55	17 08	17 29	17 47	18 05	18 22	18 41	19 04	19 17	19 32	19 38	19 46	19 54	20 03	20 12	20 23	20 36	20 50	21 07	21 27	21 54	22 32	** **
	23	15 37	16 04	16 25	16 42	16 56	17 08	17 30	17 48	18 06	18 23	18 42	19 04	19 17	19 33	19 39	19 47	19 55	20 03	20 13	20 24	20 36	20 51	21 07	21 28	21 54	22 32	** **
	27	15 39	16 06	16 26	16 43	16 57	17 10	17 31	17 49	18 07	18 24	18 43	19 05	19 18	19 33	19 40	19 47	19 55	20 04	20 13	20 24	20 36	20 50	21 07	21 27	21 53	22 30	** **
July	1	15 41	16 08	16 28	16 45	16 59	17 11	17 32	17 50	18 07	18 25	18 43	19 05	19 18	19 33	19 39	19 47	19 54	20 03	20 13	20 23	20 35	20 49	21 05	21 25	21 50	22 26	23 43
	5	15 45	16 11	16 31	16 47	17 01	17 13	17 33	17 51	18 08	18 25	18 44	19 05	19 18	19 32	19 39	19 46	19 53	20 02	20 11	20 21	20 33	20 47	21 03	21 22	21 46	22 19	23 19

The symbols (-) and (*) indicate Sun continuously below and above horizon, respectively.

Table A1.2. *Sunrise and sunset July–October*

UNIVERSAL TIME FOR MERIDIAN OF GREENWICH
SUNRISE

Lat.		−55°	−50°	−45°	−40°	−35°	−30°	−20°	−10°	0°	+10°	+20°	+30°	+35°	+40°	+42°	+44°	+46°	+48°	+50°	+52°	+54°	+56°	+58°	+60°	+62°	+64°	+66°
		h m	h m	h m	h m	h m	h m	h m	h m	h m	h m	h m	h m	h m	h m	h m	h m	h m	h m	h m	h m	h m	h m	h m	h m	h m	h m	h m
July	1	8 26	8 00	7 39	7 23	7 09	6 56	6 36	6 17	6 00	5 43	5 24	5 02	4 49	4 35	4 28	4 21	4 13	4 04	3 55	3 44	3 32	3 18	3 02	2 42	2 16	1 40	0 14
	5	8 24	7 58	7 38	7 22	7 08	6 56	6 36	6 18	6 01	5 44	5 25	5 04	4 51	4 37	4 30	4 23	4 15	4 07	3 57	3 47	3 35	3 22	3 06	2 46	2 22	1 48	0 45
	9	8 22	7 56	7 37	7 21	7 08	6 56	6 36	6 18	6 01	5 45	5 27	5 06	4 53	4 39	4 33	4 26	4 18	4 10	4 01	3 51	3 39	3 26	3 11	2 52	2 29	1 58	1 06
	13	8 18	7 54	7 35	7 19	7 06	6 55	6 35	6 18	6 02	5 46	5 28	5 08	4 56	4 42	4 36	4 29	4 22	4 14	4 05	3 55	3 44	3 31	3 17	2 59	2 37	2 08	1 24
	17	8 13	7 50	7 32	7 17	7 05	6 54	6 35	6 18	6 02	5 47	5 30	5 10	4 58	4 45	4 39	4 32	4 25	4 18	4 09	4 00	3 49	3 37	3 23	3 07	2 46	2 20	1 42
	21	8 08	7 46	7 29	7 15	7 03	6 52	6 34	6 18	6 03	5 48	5 31	5 12	5 01	4 48	4 42	4 36	4 29	4 22	4 14	4 05	3 55	3 44	3 30	3 15	2 56	2 32	1 59
	25	8 02	7 41	7 25	7 12	7 00	6 50	6 33	6 17	6 03	5 48	5 33	5 14	5 04	4 52	4 46	4 40	4 34	4 27	4 19	4 11	4 01	3 50	3 38	3 24	3 06	2 45	2 16
	29	7 56	7 36	7 21	7 08	6 57	6 48	6 31	6 17	6 03	5 49	5 34	5 17	5 07	4 55	4 50	4 44	4 38	4 32	4 24	4 16	4 08	3 57	3 46	3 33	3 17	2 57	2 32
Aug.	2	7 49	7 31	7 16	7 04	6 54	6 45	6 29	6 16	6 03	5 50	5 36	5 19	5 10	4 59	4 54	4 49	4 43	4 37	4 30	4 23	4 14	4 05	3 54	3 42	3 27	3 10	2 48
	6	7 41	7 24	7 11	7 00	6 50	6 42	6 27	6 15	6 02	5 50	5 37	5 22	5 13	5 03	4 58	4 53	4 48	4 42	4 36	4 29	4 21	4 12	4 03	3 51	3 38	3 22	3 03
	10	7 33	7 18	7 05	6 55	6 46	6 39	6 25	6 13	6 02	5 51	5 38	5 24	5 16	5 06	5 02	4 58	4 53	4 47	4 42	4 35	4 28	4 20	4 11	4 01	3 49	3 35	3 18
	14	7 25	7 11	6 59	6 50	6 42	6 35	6 23	6 12	6 01	5 51	5 39	5 26	5 19	5 10	5 06	5 02	4 58	4 53	4 47	4 42	4 35	4 28	4 20	4 11	4 00	3 47	3 32
	18	7 16	7 03	6 53	6 45	6 37	6 31	6 20	6 10	6 00	5 51	5 41	5 29	5 22	5 14	5 10	5 07	5 03	4 58	4 53	4 48	4 42	4 36	4 29	4 20	4 11	4 00	3 46
	22	7 07	6 56	6 47	6 39	6 33	6 27	6 17	6 08	6 00	5 51	5 42	5 31	5 25	5 18	5 15	5 11	5 08	5 04	4 59	4 55	4 49	4 44	4 37	4 30	4 22	4 12	4 00
	26	6 57	6 48	6 40	6 33	6 28	6 23	6 14	6 06	5 59	5 51	5 43	5 33	5 28	5 22	5 19	5 16	5 13	5 09	5 05	5 01	4 57	4 52	4 46	4 40	4 32	4 24	4 14
	30	6 48	6 39	6 33	6 27	6 22	6 18	6 11	6 04	5 57	5 51	5 44	5 36	5 31	5 25	5 23	5 20	5 18	5 15	5 11	5 08	5 04	4 59	4 55	4 49	4 43	4 36	4 27
Sept	3	6 38	6 31	6 26	6 21	6 17	6 13	6 07	6 02	5 56	5 51	5 45	5 38	5 34	5 29	5 27	5 25	5 23	5 20	5 17	5 14	5 11	5 07	5 03	4 59	4 53	4 47	4 40
	7	6 28	6 23	6 18	6 15	6 11	6 09	6 04	5 59	5 55	5 50	5 46	5 40	5 37	5 33	5 31	5 29	5 28	5 25	5 23	5 21	5 18	5 15	5 12	5 08	5 04	4 59	4 53
	11	6 18	6 14	6 11	6 08	6 06	6 04	6 00	5 57	5 53	5 50	5 46	5 42	5 40	5 37	5 35	5 34	5 32	5 31	5 29	5 27	5 25	5 23	5 20	5 17	5 14	5 10	5 06
	15	6 08	6 05	6 03	6 02	6 00	5 59	5 56	5 54	5 52	5 50	5 47	5 44	5 42	5 40	5 40	5 39	5 37	5 36	5 35	5 34	5 32	5 31	5 29	5 27	5 24	5 22	5 19
	19	5 57	5 56	5 56	5 55	5 54	5 54	5 53	5 52	5 51	5 49	5 48	5 46	5 45	5 44	5 44	5 43	5 43	5 42	5 41	5 40	5 39	5 38	5 37	5 36	5 35	5 33	5 31
	23	5 47	5 48	5 48	5 48	5 49	5 49	5 49	5 49	5 49	5 49	5 49	5 49	5 48	5 48	5 48	5 48	5 48	5 47	5 47	5 47	5 47	5 46	5 46	5 46	5 45	5 45	5 44
	27	5 37	5 39	5 40	5 42	5 43	5 44	5 45	5 47	5 48	5 49	5 50	5 51	5 51	5 52	5 52	5 52	5 53	5 53	5 53	5 53	5 54	5 54	5 55	5 55	5 55	5 56	5 57
Oct.	1	5 26	5 30	5 33	5 35	5 37	5 39	5 42	5 44	5 46	5 49	5 51	5 53	5 54	5 56	5 56	5 57	5 58	5 58	5 59	6 00	6 01	6 02	6 03	6 04	6 06	6 07	6 09
	5	5 16	5 21	5 25	5 29	5 32	5 34	5 38	5 42	5 45	5 48	5 52	5 55	5 57	6 00	6 01	6 02	6 03	6 04	6 05	6 07	6 08	6 10	6 12	6 14	6 16	6 19	6 22

SUNSET

		h m	h m	h m	h m	h m	h m	h m	h m	h m	h m	h m	h m	h m	h m	h m	h m	h m	h m	h m	h m	h m	h m	h m	h m	h m	h m	h m
July	1	15 41	16 08	16 28	16 45	16 59	17 11	17 32	17 50	18 07	18 25	18 43	19 05	19 18	19 33	19 39	19 47	19 54	20 03	20 13	20 23	20 35	20 49	21 05	21 25	21 50	22 26	23 43
	5	15 45	16 11	16 31	16 47	17 01	17 13	17 33	17 51	18 08	18 25	18 44	19 05	19 18	19 32	19 39	19 46	19 53	20 02	20 11	20 21	20 33	20 47	21 03	21 22	21 46	22 19	23 19
	9	15 49	16 14	16 34	16 49	17 03	17 15	17 35	17 52	18 09	18 25	18 43	19 04	19 17	19 31	19 37	19 44	19 52	20 00	20 09	20 19	20 30	20 43	20 59	21 17	21 40	22 10	23 00
	13	15 54	16 18	16 37	16 52	17 05	17 17	17 36	17 53	18 09	18 26	18 43	19 03	19 15	19 29	19 35	19 42	19 49	19 57	20 06	20 16	20 27	20 39	20 54	21 11	21 32	22 01	22 43
	17	15 59	16 22	16 40	16 55	17 08	17 19	17 38	17 54	18 10	18 25	18 42	19 02	19 14	19 27	19 33	19 39	19 46	19 54	20 02	20 12	20 22	20 34	20 48	21 04	21 24	21 50	22 26
	21	16 05	16 27	16 44	16 58	17 10	17 21	17 39	17 55	18 10	18 25	18 41	19 00	19 11	19 24	19 30	19 36	19 43	19 50	19 58	20 07	20 17	20 28	20 41	20 56	21 15	21 38	22 10
	25	16 11	16 32	16 48	17 02	17 13	17 23	17 40	17 56	18 10	18 24	18 40	18 58	19 09	19 21	19 26	19 32	19 39	19 45	19 53	20 01	20 11	20 21	20 34	20 48	21 05	21 26	21 54
	29	16 18	16 37	16 52	17 05	17 16	17 25	17 42	17 56	18 10	18 24	18 39	18 56	19 06	19 17	19 22	19 28	19 34	19 40	19 48	19 55	20 04	20 14	20 25	20 39	20 54	21 13	21 38
Aug.	2	16 24	16 42	16 57	17 09	17 19	17 28	17 43	17 57	18 10	18 23	18 37	18 53	19 02	19 13	19 18	19 23	19 29	19 35	19 42	19 49	19 57	20 06	20 17	20 29	20 43	21 00	21 22
	6	16 31	16 48	17 01	17 12	17 22	17 30	17 45	17 57	18 09	18 21	18 35	18 50	18 58	19 09	19 13	19 18	19 23	19 29	19 35	19 42	19 49	19 58	20 08	20 19	20 31	20 47	21 06
	10	16 38	16 54	17 06	17 16	17 25	17 32	17 46	17 58	18 09	18 20	18 32	18 46	18 54	19 04	19 08	19 12	19 17	19 22	19 28	19 34	19 41	19 49	19 58	20 08	20 20	20 33	20 50
	14	16 46	16 59	17 11	17 20	17 28	17 35	17 47	17 58	18 08	18 18	18 30	18 42	18 50	18 59	19 02	19 07	19 11	19 16	19 21	19 27	19 33	19 40	19 48	19 57	20 07	20 20	20 34
	18	16 53	17 05	17 15	17 24	17 31	17 37	17 48	17 58	18 07	18 17	18 27	18 39	18 45	18 53	18 57	19 00	19 04	19 09	19 13	19 18	19 24	19 31	19 38	19 46	19 55	20 06	20 19
	22	17 00	17 11	17 20	17 27	17 34	17 39	17 49	17 58	18 06	18 15	18 24	18 34	18 40	18 47	18 51	18 54	18 57	19 01	19 05	19 10	19 15	19 21	19 27	19 34	19 42	19 52	20 03
	26	17 07	17 17	17 25	17 31	17 37	17 42	17 50	17 58	18 05	18 13	18 21	18 30	18 35	18 41	18 44	18 47	18 50	18 54	18 57	19 01	19 06	19 11	19 16	19 23	19 30	19 38	19 48
	30	17 15	17 23	17 29	17 35	17 39	17 44	17 51	17 58	18 04	18 10	18 17	18 25	18 30	18 35	18 38	18 40	18 43	18 46	18 49	18 53	18 56	19 01	19 05	19 11	19 17	19 24	19 32
Sept	3	17 22	17 29	17 34	17 38	17 42	17 46	17 52	17 57	18 03	18 08	18 14	18 21	18 25	18 29	18 31	18 33	18 35	18 38	18 41	18 44	18 47	18 50	18 54	18 59	19 04	19 10	19 16
	7	17 29	17 34	17 39	17 42	17 45	17 48	17 53	17 57	18 01	18 06	18 10	18 16	18 19	18 23	18 24	18 26	18 28	18 30	18 32	18 34	18 37	18 40	18 43	18 47	18 51	18 55	19 01
	11	17 37	17 40	17 43	17 46	17 48	17 50	17 54	17 57	18 00	18 03	18 07	18 11	18 13	18 16	18 17	18 19	18 20	18 22	18 23	18 25	18 27	18 29	18 32	18 35	18 38	18 41	18 45
	15	17 44	17 46	17 48	17 50	17 51	17 52	17 54	17 56	17 59	18 01	18 03	18 06	18 08	18 09	18 10	18 11	18 12	18 13	18 14	18 16	18 17	18 19	18 20	18 22	18 25	18 27	18 30
	19	17 52	17 52	17 53	17 53	17 54	17 54	17 55	17 56	17 57	17 58	17 59	18 01	18 02	18 03	18 03	18 04	18 04	18 05	18 06	18 06	18 07	18 08	18 09	18 10	18 11	18 13	18 15
	23	17 59	17 58	17 58	17 57	17 57	17 57	17 56	17 56	17 56	17 56	17 56	17 56	17 56	17 56	17 56	17 56	17 57	17 57	17 57	17 57	17 57	17 57	17 58	17 58	17 58	17 59	17 59
	27	18 07	18 04	18 03	18 01	18 00	17 59	17 57	17 56	17 54	17 53	17 52	17 51	17 50	17 50	17 49	17 49	17 49	17 48	17 48	17 48	17 47	17 47	17 46	17 46	17 45	17 45	17 44
Oct.	1	18 14	18 10	18 07	18 05	18 03	18 01	17 58	17 55	17 53	17 51	17 48	17 46	17 45	17 43	17 42	17 42	17 41	17 40	17 39	17 38	17 37	17 36	17 35	17 34	17 32	17 31	17 29
	5	18 22	18 17	18 12	18 09	18 06	18 03	17 59	17 55	17 52	17 48	17 45	17 41	17 39	17 37	17 36	17 34	17 33	17 32	17 31	17 29	17 28	17 26	17 24	17 22	17 19	17 17	17 13

Table A1.2. *Sunrise and sunset October–December*

UNIVERSAL TIME FOR MERIDIAN OF GREENWICH
SUNRISE

Lat.	−55°	−50°	−45°	−40°	−35°	−30°	−20°	−10°	0°	+10°	+20°	+30°	+35°	+40°	+42°	+44°	+46°	+48°	+50°	+52°	+54°	+56°	+58°	+60°	+62°	+64°	+66°
	h m	h m	h m	h m	h m	h m	h m	h m	h m	h m	h m	h m	h m	h m	h m	h m	h m	h m	h m	h m	h m	h m	h m	h m	h m	h m	h m
Oct. 1	5 26	5 30	5 33	5 35	5 37	5 39	5 42	5 44	5 46	5 49	5 51	5 53	5 54	5 56	5 56	5 57	5 58	5 58	5 59	6 00	6 01	6 02	6 03	6 04	6 06	6 07	6 09
5	5 16	5 21	5 25	5 29	5 32	5 34	5 38	5 42	5 45	5 48	5 52	5 55	5 57	6 00	6 01	6 02	6 03	6 04	6 05	6 07	6 08	6 10	6 12	6 14	6 16	6 19	6 22
9	5 06	5 13	5 18	5 22	5 26	5 29	5 35	5 40	5 44	5 48	5 53	5 58	6 01	6 04	6 05	6 07	6 08	6 10	6 12	6 14	6 16	6 18	6 21	6 24	6 27	6 31	6 35
13	4 56	5 04	5 11	5 16	5 21	5 25	5 32	5 38	5 43	5 48	5 54	6 00	6 04	6 08	6 10	6 12	6 14	6 16	6 18	6 20	6 23	6 26	6 30	6 33	6 38	6 43	6 48
17	4 46	4 56	5 04	5 10	5 15	5 20	5 28	5 36	5 42	5 49	5 55	6 03	6 07	6 12	6 14	6 16	6 19	6 22	6 24	6 27	6 31	6 35	6 39	6 43	6 49	6 55	7 02
21	4 37	4 48	4 57	5 04	5 11	5 16	5 26	5 34	5 41	5 49	5 57	6 06	6 11	6 16	6 19	6 22	6 24	6 28	6 31	6 34	6 38	6 43	6 48	6 53	7 00	7 07	7 15
25	4 27	4 40	4 50	4 59	5 06	5 12	5 23	5 32	5 41	5 49	5 58	6 08	6 14	6 21	6 24	6 27	6 30	6 34	6 37	6 42	6 46	6 51	6 57	7 03	7 11	7 19	7 29
29	4 18	4 33	4 44	4 53	5 01	5 08	5 20	5 31	5 40	5 50	6 00	6 11	6 18	6 25	6 28	6 32	6 36	6 40	6 44	6 49	6 54	7 00	7 06	7 14	7 22	7 32	7 43
Nov. 2	4 10	4 25	4 38	4 48	4 57	5 05	5 18	5 30	5 40	5 51	6 02	6 14	6 22	6 30	6 33	6 37	6 41	6 46	6 51	6 56	7 02	7 08	7 16	7 24	7 33	7 44	7 57
6	4 01	4 19	4 32	4 44	4 53	5 02	5 16	5 29	5 40	5 52	6 04	6 17	6 25	6 34	6 38	6 42	6 47	6 52	6 57	7 03	7 10	7 17	7 25	7 34	7 45	7 57	8 12
10	3 53	4 12	4 27	4 40	4 50	4 59	5 15	5 28	5 40	5 53	6 06	6 21	6 29	6 39	6 43	6 48	6 53	6 58	7 04	7 10	7 17	7 25	7 34	7 44	7 56	8 10	8 27
14	3 46	4 07	4 23	4 36	4 47	4 57	5 13	5 28	5 41	5 54	6 08	6 24	6 33	6 44	6 48	6 53	6 58	7 04	7 11	7 17	7 25	7 34	7 43	7 55	8 08	8 23	8 42
18	3 39	4 01	4 19	4 33	4 44	4 55	5 12	5 27	5 42	5 56	6 10	6 27	6 37	6 48	6 53	6 58	7 04	7 10	7 17	7 24	7 33	7 42	7 52	8 05	8 19	8 36	8 57
22	3 33	3 57	4 15	4 30	4 42	4 53	5 12	5 28	5 42	5 57	6 13	6 31	6 41	6 53	6 58	7 03	7 10	7 16	7 23	7 31	7 40	7 50	8 01	8 14	8 30	8 49	9 12
26	3 28	3 53	4 12	4 28	4 41	4 52	5 11	5 28	5 44	5 59	6 15	6 34	6 45	6 57	7 02	7 08	7 15	7 22	7 29	7 38	7 47	7 57	8 09	8 24	8 40	9 01	9 28
30	3 23	3 50	4 10	4 26	4 40	4 51	5 12	5 29	5 45	6 01	6 18	6 37	6 48	7 01	7 07	7 13	7 20	7 27	7 35	7 44	7 53	8 05	8 17	8 32	8 50	9 12	9 42
Dec. 4	3 20	3 47	4 08	4 25	4 39	4 51	5 12	5 30	5 46	6 03	6 20	6 40	6 52	7 05	7 11	7 17	7 24	7 32	7 40	7 49	7 59	8 11	8 24	8 40	8 59	9 23	9 56
8	3 17	3 46	4 07	4 25	4 39	4 52	5 13	5 31	5 48	6 05	6 23	6 43	6 55	7 09	7 15	7 22	7 29	7 36	7 45	7 54	8 05	8 17	8 31	8 47	9 07	9 33	10 09
12	3 16	3 45	4 07	4 25	4 39	4 52	5 14	5 33	5 50	6 07	6 25	6 46	6 58	7 12	7 18	7 25	7 32	7 40	7 49	7 59	8 10	8 22	8 36	8 53	9 14	9 41	10 20
16	3 15	3 45	4 08	4 26	4 41	4 53	5 15	5 34	5 52	6 09	6 28	6 49	7 01	7 15	7 21	7 28	7 36	7 44	7 52	8 02	8 13	8 26	8 41	8 58	9 19	9 47	10 29
20	3 16	3 46	4 09	4 27	4 42	4 55	5 17	5 36	5 54	6 11	6 30	6 51	7 03	7 18	7 24	7 31	7 38	7 46	7 55	8 05	8 16	8 29	8 44	9 01	9 23	9 51	10 34
24	3 18	3 48	4 11	4 29	4 44	4 57	5 19	5 38	5 56	6 13	6 32	6 53	7 05	7 20	7 26	7 33	7 40	7 48	7 57	8 07	8 18	8 31	8 46	9 03	9 25	9 53	10 36
28	3 21	3 51	4 14	4 31	4 46	4 59	5 21	5 40	5 58	6 15	6 34	6 55	7 07	7 21	7 27	7 34	7 41	7 50	7 58	8 08	8 19	8 32	8 46	9 04	9 25	9 53	10 34
32	3 25	3 55	4 17	4 34	4 49	5 02	5 24	5 42	6 00	6 17	6 35	6 56	7 08	7 22	7 28	7 35	7 42	7 50	7 59	8 08	8 19	8 31	8 46	9 03	9 23	9 50	10 29
36	3 31	3 59	4 21	4 38	4 52	5 05	5 26	5 45	6 02	6 18	6 36	6 57	7 09	7 22	7 28	7 35	7 42	7 50	7 58	8 08	8 18	8 30	8 44	9 00	9 20	9 45	10 21

SUNSET

		h m	h m	h m	h m	h m	h m	h m	h m	h m	h m	h m	h m	h m	h m	h m	h m	h m	h m	h m	h m	h m	h m	h m	h m	h m	h m	h m
Oct.	1	18 14	18 10	18 07	18 05	18 03	18 01	17 58	17 55	17 53	17 51	17 48	17 46	17 45	17 43	17 42	17 42	17 41	17 40	17 39	17 38	17 37	17 36	17 35	17 34	17 32	17 31	17 29
	5	18 22	18 17	18 12	18 09	18 06	18 03	17 59	17 55	17 52	17 48	17 45	17 41	17 39	17 37	17 36	17 34	17 33	17 32	17 31	17 29	17 28	17 26	17 24	17 22	17 19	17 17	17 13
	9	18 30	18 23	18 18	18 13	18 09	18 06	18 00	17 55	17 51	17 46	17 42	17 36	17 34	17 30	17 29	17 27	17 26	17 24	17 22	17 20	17 18	17 16	17 13	17 10	17 06	17 03	16 58
	13	18 38	18 29	18 23	18 17	18 13	18 08	18 01	17 55	17 50	17 44	17 38	17 32	17 28	17 24	17 22	17 20	17 18	17 16	17 14	17 11	17 08	17 05	17 02	16 58	16 54	16 49	16 43
	17	18 46	18 36	18 28	18 22	18 16	18 11	18 03	17 55	17 49	17 42	17 35	17 28	17 23	17 18	17 16	17 14	17 11	17 09	17 06	17 03	16 59	16 55	16 51	16 46	16 41	16 35	16 28
	21	18 54	18 43	18 33	18 26	18 19	18 14	18 04	17 56	17 48	17 40	17 32	17 23	17 18	17 12	17 10	17 07	17 04	17 01	16 58	16 54	16 50	16 46	16 41	16 35	16 29	16 21	16 13
	25	19 02	18 49	18 39	18 30	18 23	18 17	18 06	17 56	17 48	17 39	17 30	17 19	17 14	17 07	17 04	17 01	16 58	16 54	16 50	16 46	16 41	16 36	16 30	16 24	16 17	16 08	15 58
	29	19 11	18 56	18 44	18 35	18 27	18 20	18 07	17 57	17 47	17 38	17 27	17 16	17 09	17 02	16 59	16 55	16 51	16 47	16 43	16 38	16 33	16 27	16 20	16 13	16 05	15 55	15 43
Nov.	2	19 19	19 03	18 50	18 40	18 31	18 23	18 09	17 58	17 47	17 36	17 25	17 13	17 05	16 57	16 53	16 49	16 45	16 41	16 36	16 30	16 25	16 18	16 11	16 02	15 53	15 42	15 29
	6	19 27	19 10	18 56	18 44	18 34	18 26	18 11	17 59	17 47	17 36	17 23	17 10	17 02	16 53	16 49	16 44	16 40	16 35	16 29	16 23	16 17	16 10	16 02	15 52	15 42	15 29	15 14
	10	19 36	19 16	19 01	18 49	18 38	18 29	18 14	18 00	17 47	17 35	17 22	17 07	16 58	16 48	16 44	16 40	16 35	16 29	16 23	16 17	16 10	16 02	15 53	15 43	15 31	15 17	15 00
	14	19 44	19 23	19 07	18 54	18 42	18 33	18 16	18 01	17 48	17 35	17 21	17 05	16 55	16 45	16 40	16 35	16 30	16 24	16 18	16 11	16 03	15 54	15 45	15 34	15 21	15 05	14 46
	18	19 52	19 30	19 12	18 58	18 46	18 36	18 18	18 03	17 49	17 35	17 20	17 03	16 53	16 42	16 37	16 32	16 26	16 20	16 13	16 05	15 57	15 48	15 37	15 25	15 11	14 54	14 32
	22	20 00	19 36	19 18	19 03	18 50	18 39	18 21	18 05	17 50	17 35	17 19	17 01	16 51	16 39	16 34	16 28	16 22	16 16	16 08	16 01	15 52	15 42	15 30	15 17	15 02	14 43	14 19
	26	20 08	19 42	19 23	19 07	18 54	18 43	18 23	18 06	17 51	17 35	17 19	17 00	16 50	16 37	16 32	16 26	16 19	16 12	16 05	15 56	15 47	15 37	15 24	15 10	14 54	14 33	14 06
	30	20 15	19 48	19 28	19 12	18 58	18 46	18 26	18 08	17 52	17 36	17 19	17 00	16 49	16 36	16 30	16 24	16 17	16 10	16 02	15 53	15 43	15 32	15 19	15 05	14 47	14 24	13 54
Dec.	4	20 21	19 54	19 33	19 16	19 02	18 49	18 28	18 10	17 54	17 37	17 20	17 00	16 48	16 35	16 29	16 23	16 16	16 08	16 00	15 51	15 40	15 29	15 15	15 00	14 41	14 17	13 44
	8	20 27	19 58	19 37	19 19	19 05	18 52	18 31	18 13	17 56	17 39	17 21	17 00	16 48	16 35	16 28	16 22	16 15	16 07	15 59	15 49	15 38	15 26	15 13	14 56	14 36	14 10	13 34
	12	20 32	20 03	19 40	19 23	19 08	18 55	18 33	18 15	17 57	17 40	17 22	17 01	16 49	16 35	16 29	16 22	16 15	16 07	15 58	15 48	15 38	15 25	15 11	14 54	14 33	14 06	13 27
	16	20 36	20 06	19 44	19 26	19 11	18 58	18 36	18 17	17 59	17 42	17 23	17 02	16 50	16 36	16 30	16 23	16 15	16 07	15 58	15 49	15 38	15 25	15 10	14 53	14 32	14 04	13 22
	20	20 39	20 09	19 46	19 28	19 13	19 00	18 38	18 19	18 01	17 44	17 25	17 04	16 52	16 37	16 31	16 24	16 17	16 09	16 00	15 50	15 39	15 26	15 11	14 54	14 32	14 04	13 21
	24	20 41	20 11	19 48	19 30	19 15	19 02	18 40	18 21	18 03	17 46	17 27	17 06	16 54	16 39	16 33	16 26	16 19	16 11	16 02	15 52	15 41	15 28	15 13	14 56	14 34	14 06	13 23
	28	20 41	20 12	19 49	19 31	19 16	19 03	18 42	18 23	18 05	17 48	17 29	17 08	16 56	16 42	16 36	16 29	16 22	16 14	16 05	15 55	15 44	15 31	15 17	14 59	14 38	14 10	13 29
	32	20 41	20 12	19 50	19 32	19 17	19 05	18 43	18 24	18 07	17 50	17 32	17 11	16 59	16 45	16 39	16 32	16 25	16 17	16 08	15 59	15 48	15 36	15 21	15 04	14 44	14 17	13 38
	36	20 39	20 11	19 49	19 32	19 18	19 05	18 44	18 26	18 09	17 52	17 34	17 14	17 02	16 48	16 42	16 36	16 29	16 21	16 13	16 03	15 53	15 41	15 27	15 11	14 51	14 25	13 50

Table A1.3. *Twilight January–April*

UNIVERSAL TIME FOR MERIDIAN OF GREENWICH
BEGINNING ASTRONOMICAL TWILIGHT

Lat.	−55°	−50°	−45°	−40°	−35°	−30°	−20°	−10°	0°	+10°	+20°	+30°	+35°	+40°	+42°	+44°	+46°	+48°	+50°	+52°	+54°	+56°	+58°	+60°	+62°	+64°	+66°
	h m	h m	h m	h m	h m	h m	h m	h m	h m	h m	h m	h m	h m	h m	h m	h m	h m	h m	h m	h m	h m	h m	h m	h m	h m	h m	h m
Jan. −2	// //	// //	1 43	2 30	3 01	3 24	3 58	4 23	4 43	5 00	5 15	5 30	5 37	5 44	5 47	5 50	5 53	5 56	5 59	6 03	6 06	6 10	6 14	6 18	6 23	6 28	6 34
2	// //	// //	1 48	2 34	3 04	3 27	4 01	4 26	4 45	5 02	5 17	5 31	5 38	5 45	5 48	5 51	5 54	5 57	6 00	6 03	6 07	6 10	6 14	6 18	6 23	6 27	6 33
6	// //	// //	1 55	2 39	3 08	3 31	4 04	4 28	4 47	5 04	5 18	5 32	5 39	5 45	5 48	5 51	5 54	5 57	6 00	6 03	6 06	6 10	6 13	6 17	6 21	6 26	6 31
10	// //	// //	2 03	2 44	3 13	3 35	4 07	4 31	4 49	5 05	5 19	5 33	5 39	5 45	5 48	5 51	5 53	5 56	5 59	6 02	6 05	6 08	6 11	6 15	6 19	6 23	6 28
14	// //	0 51	2 11	2 50	3 18	3 39	4 10	4 33	4 51	5 07	5 20	5 33	5 39	5 45	5 47	5 50	5 52	5 55	5 57	6 00	6 03	6 06	6 09	6 12	6 15	6 19	6 23
18	// //	1 13	2 20	2 57	3 23	3 43	4 13	4 36	4 53	5 08	5 21	5 33	5 38	5 44	5 46	5 48	5 51	5 53	5 55	5 58	6 00	6 03	6 05	6 08	6 11	6 14	6 18
22	// //	1 32	2 30	3 04	3 29	3 48	4 17	4 38	4 55	5 09	5 21	5 32	5 37	5 42	5 44	5 46	5 48	5 50	5 52	5 55	5 57	5 59	6 01	6 03	6 06	6 08	6 11
26	// //	1 48	2 39	3 11	3 34	3 52	4 20	4 40	4 56	5 09	5 21	5 31	5 36	5 40	5 42	5 44	5 45	5 47	5 49	5 51	5 53	5 54	5 56	5 58	6 00	6 02	6 04
30	// //	2 03	2 48	3 18	3 40	3 57	4 23	4 42	4 58	5 10	5 20	5 29	5 34	5 38	5 39	5 41	5 42	5 43	5 45	5 46	5 48	5 49	5 50	5 52	5 53	5 54	5 55
Feb. 3	0 46	2 17	2 58	3 25	3 46	4 02	4 26	4 44	4 59	5 10	5 20	5 28	5 31	5 35	5 36	5 37	5 38	5 39	5 40	5 41	5 42	5 43	5 44	5 44	5 45	5 45	5 46
7	1 24	2 31	3 07	3 32	3 51	4 06	4 29	4 46	4 59	5 10	5 18	5 25	5 28	5 31	5 32	5 33	5 34	5 34	5 35	5 35	5 36	5 36	5 36	5 36	5 36	5 36	5 35
11	1 49	2 43	3 16	3 39	3 57	4 11	4 32	4 48	5 00	5 09	5 17	5 23	5 25	5 27	5 28	5 28	5 29	5 29	5 29	5 29	5 29	5 29	5 28	5 28	5 27	5 26	5 24
15	2 09	2 55	3 24	3 45	4 02	4 15	4 35	4 49	5 00	5 09	5 15	5 20	5 21	5 23	5 23	5 23	5 23	5 23	5 23	5 22	5 22	5 21	5 20	5 18	5 17	5 15	5 12
19	2 26	3 06	3 32	3 52	4 07	4 19	4 37	4 51	5 00	5 08	5 13	5 17	5 17	5 18	5 18	5 18	5 17	5 17	5 16	5 15	5 14	5 12	5 11	5 09	5 06	5 03	4 59
23	2 42	3 16	3 40	3 58	4 12	4 23	4 40	4 52	5 00	5 07	5 11	5 13	5 13	5 13	5 12	5 12	5 11	5 10	5 09	5 07	5 05	5 03	5 01	4 58	4 54	4 50	4 45
27	2 56	3 26	3 48	4 04	4 17	4 27	4 42	4 52	5 00	5 05	5 08	5 09	5 08	5 07	5 06	5 05	5 04	5 03	5 01	4 59	4 57	4 54	4 51	4 47	4 42	4 37	4 30
Mar. 3	3 09	3 36	3 55	4 10	4 21	4 30	4 44	4 53	5 00	5 04	5 05	5 05	5 04	5 01	5 00	4 59	4 57	4 55	4 53	4 50	4 47	4 44	4 40	4 35	4 29	4 23	4 14
7	3 21	3 45	4 02	4 15	4 25	4 34	4 46	4 54	4 59	5 02	5 02	5 00	4 58	4 55	4 54	4 52	4 50	4 47	4 44	4 41	4 37	4 33	4 28	4 23	4 16	4 08	3 58
11	3 32	3 53	4 09	4 20	4 30	4 37	4 47	4 54	4 58	5 00	4 59	4 56	4 53	4 49	4 47	4 44	4 42	4 39	4 35	4 32	4 27	4 22	4 16	4 10	4 01	3 52	3 40
15	3 43	4 01	4 15	4 25	4 33	4 40	4 49	4 54	4 57	4 58	4 56	4 51	4 47	4 42	4 40	4 37	4 34	4 30	4 26	4 22	4 17	4 11	4 04	3 56	3 46	3 35	3 21
19	3 53	4 09	4 21	4 30	4 37	4 43	4 50	4 54	4 56	4 55	4 52	4 46	4 41	4 35	4 32	4 29	4 26	4 21	4 17	4 12	4 06	3 59	3 51	3 41	3 30	3 17	3 00
23	4 02	4 17	4 27	4 35	4 41	4 45	4 51	4 54	4 55	4 53	4 49	4 41	4 35	4 28	4 25	4 21	4 17	4 12	4 07	4 01	3 54	3 46	3 37	3 26	3 13	2 57	2 37
27	4 11	4 24	4 33	4 39	4 44	4 48	4 52	4 54	4 53	4 51	4 45	4 36	4 29	4 21	4 17	4 13	4 08	4 03	3 57	3 50	3 42	3 33	3 23	3 10	2 55	2 36	2 10
31	4 20	4 30	4 38	4 44	4 48	4 50	4 54	4 54	4 52	4 48	4 41	4 30	4 23	4 14	4 10	4 05	3 59	3 53	3 47	3 39	3 30	3 20	3 08	2 53	2 35	2 12	1 38
Apr. 4	4 28	4 37	4 43	4 48	4 51	4 53	4 55	4 54	4 51	4 46	4 37	4 25	4 17	4 07	4 02	3 56	3 50	3 44	3 36	3 27	3 18	3 06	2 52	2 35	2 13	1 43	0 50

ENDING ASTRONOMICAL TWILIGHT

	h m	h m	h m	h m	h m	h m	h m	h m	h m	h m	h m	h m	h m	h m	h m	h m	h m	h m	h m	h m	h m	h m	h m	h m	h m	h m	h m
Jan. −2	// //	// //	22 21	21 34	21 03	20 40	20 06	19 41	19 21	19 04	18 49	18 35	18 28	18 20	18 17	18 14	18 11	18 08	18 05	18 02	17 58	17 54	17 50	17 46	17 42	17 36	17 31
2	// //	// //	22 19	21 34	21 04	20 41	20 07	19 42	19 23	19 06	18 51	18 37	18 30	18 23	18 20	18 18	18 15	18 12	18 08	18 05	18 02	17 58	17 54	17 50	17 46	17 41	17 36
6	// //	// //	22 15	21 32	21 03	20 41	20 08	19 43	19 24	19 08	18 53	18 40	18 33	18 26	18 24	18 21	18 18	18 15	18 12	18 09	18 06	18 02	17 59	17 55	17 51	17 46	17 41
10	// //	23 50	22 11	21 30	21 02	20 40	20 08	19 44	19 25	19 10	18 56	18 43	18 36	18 30	18 27	18 25	18 22	18 19	18 16	18 14	18 11	18 07	18 04	18 01	17 57	17 53	17 48
14	// //	23 22	22 05	21 27	21 00	20 39	20 08	19 45	19 27	19 11	18 58	18 46	18 40	18 34	18 31	18 29	18 26	18 24	18 21	18 18	18 16	18 13	18 10	18 07	18 03	18 00	17 56
18	// //	23 04	21 59	21 23	20 57	20 37	20 07	19 45	19 28	19 13	19 00	18 49	18 43	18 37	18 35	18 33	18 31	18 28	18 26	18 24	18 21	18 19	18 16	18 13	18 10	18 07	18 04
22	// //	22 48	21 52	21 18	20 54	20 35	20 06	19 45	19 28	19 14	19 02	18 52	18 46	18 41	18 39	18 37	18 35	18 33	18 31	18 29	18 27	18 25	18 23	18 20	18 18	18 16	18 13
26	// //	22 34	21 44	21 13	20 50	20 32	20 05	19 45	19 29	19 16	19 04	18 55	18 50	18 45	18 44	18 42	18 40	18 38	18 37	18 35	18 33	18 32	18 30	18 28	18 26	18 25	18 23
30	// //	22 20	21 37	21 07	20 46	20 29	20 03	19 44	19 29	19 17	19 06	18 57	18 53	18 50	18 48	18 47	18 45	18 44	18 42	18 41	18 40	18 39	18 37	18 36	18 35	18 34	18 33
Feb. 3	23 30	22 08	21 28	21 01	20 41	20 25	20 01	19 43	19 29	19 18	19 08	19 00	18 57	18 54	18 53	18 51	18 50	18 49	18 48	18 47	18 46	18 46	18 45	18 44	18 44	18 44	18 44
7	22 58	21 55	21 20	20 55	20 36	20 21	19 59	19 42	19 29	19 19	19 10	19 03	19 00	18 58	18 57	18 56	18 55	18 55	18 54	18 54	18 53	18 53	18 53	18 53	18 53	18 54	18 55
11	22 35	21 43	21 11	20 49	20 31	20 17	19 56	19 40	19 29	19 19	19 12	19 06	19 04	19 02	19 02	19 01	19 01	19 00	19 00	19 00	19 00	19 01	19 01	19 02	19 03	19 05	19 06
15	22 16	21 31	21 03	20 42	20 26	20 13	19 53	19 39	19 28	19 20	19 13	19 09	19 07	19 06	19 06	19 06	19 06	19 06	19 07	19 07	19 08	19 09	19 10	19 11	19 13	19 16	19 18
19	21 58	21 20	20 54	20 35	20 20	20 08	19 50	19 37	19 27	19 20	19 15	19 12	19 11	19 11	19 11	19 11	19 11	19 12	19 13	19 14	19 15	19 17	19 18	19 21	19 23	19 27	19 31
23	21 42	21 08	20 45	20 28	20 14	20 03	19 47	19 35	19 26	19 20	19 16	19 14	19 14	19 15	19 15	19 16	19 17	19 18	19 19	19 21	19 23	19 25	19 27	19 30	19 34	19 39	19 44
27	21 27	20 57	20 36	20 20	20 08	19 58	19 43	19 33	19 26	19 21	19 18	19 17	19 18	19 19	19 20	19 21	19 22	19 24	19 26	19 28	19 30	19 33	19 37	19 41	19 45	19 51	19 58
Mar. 3	21 13	20 46	20 28	20 13	20 02	19 53	19 40	19 31	19 24	19 21	19 19	19 20	19 21	19 24	19 25	19 26	19 28	19 30	19 32	19 35	19 38	19 42	19 46	19 51	19 57	20 04	20 12
7	20 59	20 36	20 19	20 06	19 56	19 48	19 36	19 28	19 23	19 21	19 20	19 22	19 25	19 28	19 30	19 32	19 34	19 36	19 39	19 42	19 46	19 51	19 56	20 02	20 09	20 17	20 27
11	20 46	20 25	20 10	19 59	19 50	19 43	19 33	19 26	19 22	19 21	19 21	19 25	19 28	19 32	19 34	19 37	19 39	19 43	19 46	19 50	19 54	20 00	20 06	20 13	20 21	20 31	20 43
15	20 33	20 15	20 02	19 52	19 44	19 38	19 29	19 24	19 21	19 21	19 23	19 28	19 32	19 37	19 39	19 42	19 45	19 49	19 53	19 58	20 03	20 09	20 16	20 24	20 34	20 46	21 01
19	20 21	20 05	19 53	19 45	19 38	19 33	19 25	19 21	19 20	19 21	19 24	19 30	19 35	19 41	19 44	19 48	19 51	19 56	20 00	20 06	20 12	20 19	20 27	20 37	20 48	21 02	21 20
23	20 09	19 55	19 45	19 38	19 32	19 27	19 22	19 19	19 19	19 21	19 25	19 33	19 39	19 46	19 49	19 53	19 58	20 02	20 08	20 14	20 21	20 29	20 38	20 50	21 03	21 20	21 41
27	19 58	19 46	19 37	19 31	19 26	19 22	19 18	19 17	19 17	19 21	19 26	19 36	19 42	19 51	19 55	19 59	20 04	20 09	20 15	20 22	20 30	20 40	20 51	21 03	21 19	21 39	22 06
31	19 47	19 37	19 30	19 24	19 20	19 18	19 15	19 14	19 16	19 21	19 28	19 39	19 46	19 56	20 00	20 05	20 10	20 17	20 23	20 31	20 40	20 51	21 03	21 18	21 37	22 02	22 39
Apr. 4	19 36	19 28	19 22	19 18	19 15	19 13	19 11	19 12	19 15	19 21	19 29	19 42	19 50	20 01	20 05	20 11	20 17	20 24	20 32	20 41	20 51	21 03	21 17	21 35	21 57	22 30	23 38

The symbols (*) and (/) indicate Sun continuously above horizon and continuous twilight, respectively.

Table A1.3. *Twilight April–July*

UNIVERSAL TIME FOR MERIDIAN OF GREENWICH
BEGINNING ASTRONOMICAL TWILIGHT

Lat.	-55°	-50°	-45°	-40°	-35°	-30°	-20°	-10°	0°	+10°	+20°	+30°	+35°	+40°	+42°	+44°	+46°	+48°	+50°	+52°	+54°	+56°	+58°	+60°	+62°	+64°	+66°
	h m	h m	h m	h m	h m	h m	h m	h m	h m	h m	h m	h m	h m	h m	h m	h m	h m	h m	h m	h m	h m	h m	h m	h m	h m	h m	h m
Mar. 31	4 20	4 30	4 38	4 44	4 48	4 50	4 54	4 54	4 52	4 48	4 41	4 30	4 23	4 14	4 10	4 05	3 59	3 53	3 47	3 39	3 30	3 20	3 08	2 53	2 35	2 12	1 38
Apr. 4	4 28	4 37	4 43	4 48	4 51	4 53	4 55	4 54	4 51	4 46	4 37	4 25	4 17	4 07	4 02	3 56	3 50	3 44	3 36	3 27	3 18	3 06	2 52	2 35	2 13	1 43	0 50
8	4 36	4 43	4 48	4 52	4 54	4 55	4 55	4 54	4 49	4 43	4 34	4 20	4 11	3 59	3 54	3 48	3 41	3 34	3 25	3 16	3 04	2 51	2 35	2 15	1 48	1 05	// //
12	4 44	4 49	4 53	4 55	4 57	4 57	4 56	4 53	4 48	4 40	4 30	4 15	4 04	3 52	3 46	3 39	3 32	3 24	3 14	3 03	2 51	2 36	2 17	1 52	1 16	// //	// //
16	4 51	4 55	4 58	4 59	5 00	5 00	4 57	4 53	4 47	4 38	4 26	4 09	3 58	3 44	3 38	3 31	3 23	3 13	3 03	2 51	2 37	2 19	1 57	1 25	0 16	// //	// //
20	4 58	5 01	5 02	5 03	5 03	5 02	4 58	4 53	4 46	4 36	4 23	4 04	3 52	3 37	3 30	3 22	3 13	3 03	2 51	2 38	2 22	2 01	1 34	0 48	// //	// //	// //
24	5 05	5 07	5 07	5 07	5 05	5 04	4 59	4 53	4 44	4 33	4 19	3 59	3 46	3 30	3 22	3 14	3 04	2 53	2 40	2 24	2 06	1 41	1 05	// //	// //	// //	// //
28	5 12	5 12	5 11	5 10	5 08	5 06	5 00	4 53	4 43	4 31	4 16	3 54	3 40	3 23	3 14	3 05	2 54	2 42	2 28	2 10	1 49	1 18	0 06	// //	// //	// //	// //
May 2	5 18	5 17	5 16	5 13	5 11	5 08	5 01	4 53	4 42	4 29	4 13	3 50	3 35	3 16	3 07	2 57	2 45	2 32	2 15	1 56	1 30	0 47	// //	// //	// //	// //	// //
6	5 24	5 22	5 20	5 17	5 14	5 10	5 02	4 53	4 41	4 27	4 10	3 45	3 29	3 09	2 59	2 48	2 36	2 21	2 03	1 40	1 07	// //	// //	// //	// //	// //	// //
10	5 30	5 27	5 24	5 20	5 16	5 12	5 03	4 53	4 41	4 26	4 07	3 41	3 24	3 03	2 52	2 40	2 27	2 10	1 50	1 23	0 37	// //	// //	// //	// //	// //	// //
14	5 36	5 32	5 27	5 23	5 19	5 14	5 04	4 53	4 40	4 24	4 04	3 37	3 19	2 56	2 45	2 33	2 18	1 59	1 36	1 03	// //	// //	// //	// //	// //	// //	// //
18	5 41	5 36	5 31	5 26	5 21	5 16	5 05	4 54	4 40	4 23	4 02	3 34	3 15	2 51	2 39	2 25	2 09	1 49	1 22	0 38	// //	// //	// //	// //	// //	// //	// //
22	5 46	5 40	5 35	5 29	5 24	5 18	5 07	4 54	4 40	4 22	4 00	3 31	3 11	2 45	2 33	2 18	2 01	1 38	1 07	// //	// //	// //	// //	// //	// //	// //	// //
26	5 51	5 44	5 38	5 32	5 26	5 20	5 08	4 55	4 40	4 22	3 59	3 28	3 07	2 41	2 27	2 12	1 53	1 28	0 50	// //	// //	// //	// //	// //	// //	// //	// //
30	5 55	5 48	5 41	5 34	5 28	5 22	5 09	4 55	4 40	4 21	3 58	3 26	3 05	2 37	2 23	2 06	1 45	1 18	0 27	// //	// //	// //	// //	// //	// //	// //	// //
June 3	5 59	5 51	5 44	5 37	5 30	5 24	5 10	4 56	4 40	4 21	3 57	3 24	3 02	2 33	2 19	2 01	1 39	1 08	// //	// //	// //	// //	// //	// //	// //	// //	// //
7	6 02	5 54	5 46	5 39	5 32	5 25	5 12	4 57	4 40	4 21	3 56	3 23	3 00	2 30	2 15	1 57	1 34	0 59	// //	// //	// //	// //	// //	// //	// //	// //	// //
11	6 05	5 56	5 48	5 41	5 34	5 27	5 13	4 58	4 41	4 21	3 56	3 22	2 59	2 29	2 13	1 54	1 29	0 51	// //	// //	// //	// //	// //	// //	// //	// //	// //
15	6 07	5 58	5 50	5 42	5 35	5 28	5 14	4 59	4 42	4 22	3 56	3 22	2 59	2 28	2 12	1 52	1 26	0 45	// //	// //	// //	// //	// //	// //	// //	// //	** **
19	6 08	5 59	5 51	5 44	5 36	5 29	5 15	5 00	4 42	4 22	3 57	3 22	2 59	2 27	2 11	1 51	1 25	0 42	// //	// //	// //	// //	// //	// //	// //	// //	** **
23	6 09	6 00	5 52	5 45	5 37	5 30	5 16	5 00	4 43	4 23	3 58	3 23	3 00	2 28	2 12	1 52	1 26	0 42	// //	// //	// //	// //	// //	// //	// //	// //	** **
27	6 10	6 01	5 53	5 45	5 38	5 31	5 16	5 01	4 44	4 24	3 59	3 24	3 01	2 30	2 14	1 54	1 28	0 46	// //	// //	// //	// //	// //	// //	// //	// //	** **
July 1	6 09	6 01	5 53	5 45	5 38	5 31	5 17	5 02	4 45	4 25	4 00	3 26	3 03	2 32	2 16	1 57	1 32	0 53	// //	// //	// //	// //	// //	// //	// //	// //	// //
5	6 08	6 00	5 52	5 45	5 38	5 31	5 17	5 03	4 46	4 26	4 02	3 28	3 05	2 35	2 20	2 01	1 37	1 02	// //	// //	// //	// //	// //	// //	// //	// //	// //

ENDING ASTRONOMICAL TWILIGHT

		h m	h m	h m	h m	h m	h m	h m	h m	h m	h m	h m	h m	h m	h m	h m	h m	h m	h m	h m	h m	h m	h m	h m	h m	h m	h m	h m
Mar.	31	19 47	19 37	19 30	19 24	19 20	19 18	19 15	19 14	19 16	19 21	19 28	19 39	19 46	19 56	20 00	20 05	20 10	20 17	20 23	20 31	20 40	20 51	21 03	21 18	21 37	22 02	22 39
Apr.	4	19 36	19 28	19 22	19 18	19 15	19 13	19 11	19 12	19 15	19 21	19 29	19 42	19 50	20 01	20 05	20 11	20 17	20 24	20 32	20 41	20 51	21 03	21 17	21 35	21 57	22 30	23 38
	8	19 26	19 19	19 15	19 11	19 09	19 08	19 08	19 10	19 14	19 21	19 31	19 45	19 54	20 06	20 11	20 17	20 24	20 32	20 40	20 50	21 02	21 15	21 32	21 53	22 22	23 11	// //
	12	19 17	19 11	19 08	19 06	19 04	19 04	19 05	19 08	19 14	19 21	19 32	19 48	19 58	20 11	20 17	20 24	20 31	20 40	20 49	21 00	21 13	21 29	21 49	22 14	22 54	// //	// //
	16	19 07	19 03	19 01	19 00	18 59	19 00	19 02	19 07	19 13	19 22	19 34	19 51	20 02	20 16	20 23	20 30	20 39	20 48	20 59	21 11	21 26	21 44	22 07	22 41	// //	// //	// //
	20	18 58	18 56	18 55	18 54	18 55	18 56	18 59	19 05	19 12	19 22	19 36	19 54	20 07	20 22	20 29	20 37	20 46	20 57	21 08	21 22	21 39	22 01	22 30	23 24	// //	// //	// //
	24	18 50	18 49	18 49	18 49	18 50	18 52	18 57	19 03	19 12	19 23	19 38	19 58	20 11	20 28	20 35	20 44	20 54	21 06	21 19	21 34	21 54	22 20	23 00	// //	// //	// //	// //
	28	18 42	18 42	18 43	18 44	18 46	18 49	18 55	19 02	19 12	19 24	19 40	20 01	20 16	20 33	20 42	20 51	21 02	21 15	21 30	21 47	22 10	22 43	// //	// //	// //	// //	// //
May	2	18 35	18 36	18 38	18 40	18 43	18 46	18 53	19 01	19 12	19 25	19 42	20 05	20 20	20 39	20 49	20 59	21 11	21 24	21 41	22 01	22 29	23 17	// //	// //	// //	// //	// //
	6	18 28	18 30	18 33	18 36	18 39	18 43	18 51	19 00	19 12	19 26	19 44	20 09	20 25	20 45	20 55	21 06	21 19	21 34	21 53	22 17	22 52	// //	// //	// //	// //	// //	// //
	10	18 22	18 25	18 29	18 32	18 36	18 40	18 49	19 00	19 12	19 27	19 46	20 12	20 29	20 51	21 02	21 14	21 28	21 45	22 06	22 34	23 27	// //	// //	// //	// //	// //	// //
	14	18 16	18 20	18 25	18 29	18 33	18 38	18 48	18 59	19 12	19 28	19 48	20 16	20 34	20 57	21 09	21 22	21 37	21 55	22 19	22 54	// //	// //	// //	// //	// //	// //	// //
	18	18 11	18 16	18 21	18 26	18 31	18 36	18 47	18 59	19 13	19 30	19 51	20 19	20 39	21 03	21 15	21 29	21 46	22 06	22 34	23 24	// //	// //	// //	// //	// //	// //	// //
	22	18 06	18 12	18 18	18 24	18 29	18 35	18 46	18 59	19 14	19 31	19 53	20 23	20 43	21 09	21 21	21 36	21 54	22 17	22 50	// //	// //	// //	// //	// //	// //	// //	// //
	26	18 03	18 09	18 16	18 22	18 28	18 34	18 46	18 59	19 14	19 32	19 55	20 26	20 47	21 14	21 28	21 43	22 03	22 28	23 09	// //	// //	// //	// //	// //	// //	// //	// //
	30	18 00	18 07	18 14	18 20	18 26	18 33	18 46	19 00	19 15	19 34	19 57	20 29	20 51	21 19	21 33	21 50	22 11	22 39	23 35	// //	// //	// //	// //	// //	// //	// //	// //
June	3	17 57	18 05	18 12	18 19	18 26	18 32	18 46	19 00	19 16	19 35	19 59	20 32	20 54	21 24	21 38	21 56	22 18	22 50	// //	// //	// //	// //	// //	// //	// //	// //	// //
	7	17 55	18 04	18 11	18 18	18 25	18 32	18 46	19 01	19 17	19 37	20 01	20 35	20 58	21 28	21 43	22 01	22 25	23 00	// //	// //	// //	// //	// //	// //	// //	// //	// //
	11	17 54	18 03	18 11	18 18	18 25	18 32	18 46	19 01	19 18	19 38	20 03	20 37	21 00	21 31	21 47	22 06	22 31	23 09	// //	// //	// //	// //	// //	// //	// //	// //	// //
	15	17 54	18 03	18 11	18 18	18 25	18 33	18 47	19 02	19 19	19 39	20 04	20 39	21 02	21 33	21 49	22 09	22 35	23 17	// //	// //	// //	// //	// //	// //	// //	// //	** **
	19	17 54	18 03	18 11	18 19	18 26	18 33	18 48	19 03	19 20	19 40	20 06	20 40	21 04	21 35	21 51	22 11	22 37	23 21	// //	// //	// //	// //	// //	// //	// //	// //	** **
	23	17 55	18 04	18 12	18 20	18 27	18 34	18 48	19 04	19 21	19 41	20 06	20 41	21 05	21 36	21 52	22 12	22 38	23 22	// //	// //	// //	// //	// //	// //	// //	// //	** **
	27	17 56	18 05	18 13	18 21	18 28	18 35	18 49	19 05	19 22	19 42	20 07	20 41	21 05	21 36	21 52	22 11	22 37	23 19	// //	// //	// //	// //	// //	// //	// //	// //	** **
July	1	17 58	18 07	18 15	18 22	18 29	18 36	18 50	19 05	19 22	19 42	20 07	20 41	21 04	21 35	21 51	22 10	22 34	23 13	// //	// //	// //	// //	// //	// //	// //	// //	// //
	5	18 01	18 09	18 17	18 24	18 31	18 38	18 52	19 06	19 23	19 42	20 07	20 40	21 03	21 33	21 48	22 07	22 30	23 05	// //	// //	// //	// //	// //	// //	// //	// //	// //

The symbols (*) and (/) indicate Sun continuously above horizon and continuous twilight, respectively.

Table A1.3. *Twilight July–October*

UNIVERSAL TIME FOR MERIDIAN OF GREENWICH
BEGINNING ASTRONOMICAL TWILIGHT

Lat.		−55°	−50°	−45°	−40°	−35°	−30°	−20°	−10°	0°	+10°	+20°	+30°	+35°	+40°	+42°	+44°	+46°	+48°	+50°	+52°	+54°	+56°	+58°	+60°	+62°	+64°	+66°
		h m	h m	h m	h m	h m	h m	h m	h m	h m	h m	h m	h m	h m	h m	h m	h m	h m	h m	h m	h m	h m	h m	h m	h m	h m	h m	h m
July	1	6 09	6 01	5 53	5 45	5 38	5 31	5 17	5 02	4 45	4 25	4 00	3 26	3 03	2 32	2 16	1 57	1 32	0 53	// //	// //	// //	// //	// //	// //	// //	// //	// //
	5	6 08	6 00	5 52	5 45	5 38	5 31	5 17	5 03	4 46	4 26	4 02	3 28	3 05	2 35	2 20	2 01	1 37	1 02	// //	// //	// //	// //	// //	// //	// //	// //	// //
	9	6 06	5 58	5 51	5 44	5 38	5 31	5 18	5 03	4 47	4 28	4 04	3 31	3 08	2 39	2 24	2 06	1 44	1 12	// //	// //	// //	// //	// //	// //	// //	// //	// //
	13	6 04	5 57	5 50	5 43	5 37	5 30	5 18	5 04	4 48	4 29	4 05	3 33	3 12	2 43	2 29	2 12	1 51	1 23	0 22	// //	// //	// //	// //	// //	// //	// //	// //
	17	6 01	5 54	5 48	5 42	5 36	5 30	5 17	5 04	4 49	4 30	4 07	3 36	3 15	2 48	2 35	2 19	1 59	1 33	0 52	// //	// //	// //	// //	// //	// //	// //	// //
	21	5 57	5 51	5 45	5 40	5 34	5 28	5 17	5 04	4 49	4 32	4 10	3 40	3 19	2 54	2 41	2 26	2 07	1 44	1 11	// //	// //	// //	// //	// //	// //	// //	// //
	25	5 53	5 47	5 42	5 37	5 32	5 27	5 16	5 04	4 50	4 33	4 12	3 43	3 24	2 59	2 47	2 33	2 16	1 55	1 27	0 33	// //	// //	// //	// //	// //	// //	// //
	29	5 48	5 43	5 39	5 34	5 30	5 25	5 15	5 03	4 50	4 34	4 14	3 46	3 28	3 05	2 53	2 40	2 25	2 06	1 41	1 05	// //	// //	// //	// //	// //	// //	// //
Aug.	2	5 42	5 38	5 35	5 31	5 27	5 23	5 13	5 03	4 50	4 35	4 16	3 50	3 32	3 10	3 00	2 47	2 33	2 16	1 55	1 26	0 24	// //	// //	// //	// //	// //	// //
	6	5 36	5 33	5 30	5 27	5 24	5 20	5 12	5 02	4 51	4 36	4 18	3 53	3 37	3 16	3 06	2 55	2 42	2 26	2 07	1 43	1 06	// //	// //	// //	// //	// //	// //
	10	5 29	5 27	5 25	5 23	5 20	5 17	5 10	5 01	4 51	4 37	4 20	3 57	3 41	3 22	3 13	3 02	2 50	2 36	2 19	1 58	1 30	0 36	// //	// //	// //	// //	// //
	14	5 21	5 21	5 20	5 19	5 16	5 14	5 08	5 00	4 50	4 38	4 22	4 00	3 46	3 28	3 19	3 09	2 58	2 45	2 30	2 12	1 48	1 14	// //	// //	// //	// //	// //
	18	5 14	5 14	5 14	5 14	5 12	5 11	5 05	4 59	4 50	4 39	4 24	4 03	3 50	3 33	3 25	3 16	3 06	2 54	2 41	2 25	2 05	1 38	0 54	// //	// //	// //	// //
	22	5 05	5 07	5 08	5 09	5 08	5 07	5 03	4 57	4 49	4 39	4 25	4 07	3 54	3 39	3 31	3 23	3 14	3 03	2 51	2 37	2 19	1 57	1 26	0 14	// //	// //	// //
	26	4 56	5 00	5 02	5 03	5 03	5 03	5 00	4 55	4 49	4 40	4 27	4 10	3 58	3 44	3 37	3 30	3 21	3 12	3 01	2 48	2 33	2 14	1 49	1 13	// //	// //	// //
	30	4 47	4 52	4 55	4 57	4 58	4 58	4 57	4 53	4 48	4 40	4 29	4 13	4 02	3 49	3 43	3 36	3 28	3 20	3 10	2 58	2 45	2 29	2 08	1 41	0 56	// //	// //
Sept	3	4 37	4 44	4 48	4 51	4 53	4 54	4 54	4 51	4 47	4 40	4 30	4 16	4 06	3 54	3 48	3 42	3 35	3 27	3 18	3 08	2 56	2 42	2 25	2 03	1 32	0 31	// //
	7	4 27	4 35	4 41	4 45	4 48	4 49	4 50	4 49	4 46	4 40	4 31	4 18	4 10	3 59	3 54	3 48	3 42	3 35	3 27	3 18	3 07	2 55	2 40	2 21	1 57	1 22	// //
	11	4 17	4 26	4 33	4 38	4 42	4 44	4 47	4 47	4 45	4 40	4 32	4 21	4 13	4 04	3 59	3 54	3 48	3 42	3 35	3 27	3 17	3 06	2 53	2 38	2 18	1 52	1 10
	15	4 06	4 17	4 25	4 32	4 36	4 39	4 43	4 44	4 43	4 40	4 34	4 24	4 17	4 08	4 04	4 00	3 55	3 49	3 43	3 35	3 27	3 17	3 06	2 53	2 36	2 15	1 45
	19	3 54	4 08	4 17	4 25	4 30	4 34	4 40	4 42	4 42	4 40	4 35	4 26	4 20	4 13	4 09	4 05	4 01	3 56	3 50	3 44	3 36	3 28	3 18	3 06	2 52	2 35	2 12
	23	3 43	3 58	4 09	4 18	4 24	4 29	4 36	4 39	4 41	4 39	4 36	4 29	4 24	4 17	4 14	4 10	4 07	4 02	3 57	3 52	3 45	3 38	3 29	3 19	3 07	2 52	2 33
	27	3 31	3 48	4 01	4 11	4 18	4 24	4 32	4 37	4 39	4 39	4 37	4 31	4 27	4 21	4 19	4 16	4 12	4 09	4 04	3 59	3 54	3 47	3 40	3 31	3 21	3 08	2 53
Oct.	1	3 18	3 38	3 52	4 03	4 12	4 19	4 28	4 34	4 38	4 39	4 38	4 34	4 30	4 26	4 23	4 21	4 18	4 15	4 11	4 07	4 02	3 56	3 50	3 43	3 34	3 23	3 10
	5	3 05	3 27	3 44	3 56	4 06	4 13	4 24	4 32	4 36	4 39	4 39	4 36	4 33	4 30	4 28	4 26	4 23	4 21	4 18	4 14	4 10	4 05	4 00	3 54	3 46	3 37	3 26

ENDING ASTRONOMICAL TWILIGHT

		h m	h m	h m	h m	h m	h m	h m	h m	h m	h m	h m	h m	h m	h m	h m	h m	h m	h m	h m	h m	h m	h m	h m	h m	h m	h m	h m
July	1	17 58	18 07	18 15	18 22	18 29	18 36	18 50	19 05	19 22	19 42	20 07	20 41	21 04	21 35	21 51	22 10	22 34	23 13	// //	// //	// //	// //	// //	// //	// //	// //	// //
	5	18 01	18 09	18 17	18 24	18 31	18 38	18 52	19 06	19 23	19 42	20 07	20 40	21 03	21 33	21 48	22 07	22 30	23 05	// //	// //	// //	// //	// //	// //	// //	// //	// //
	9	18 04	18 12	18 19	18 26	18 33	18 39	18 53	19 07	19 23	19 42	20 06	20 39	21 01	21 30	21 45	22 03	22 25	22 56	// //	// //	// //	// //	// //	// //	// //	// //	// //
	13	18 08	18 15	18 22	18 28	18 35	18 41	18 54	19 08	19 23	19 42	20 06	20 37	20 59	21 27	21 41	21 58	22 19	22 46	23 40	// //	// //	// //	// //	// //	// //	// //	// //
	17	18 12	18 18	18 25	18 31	18 37	18 43	18 55	19 08	19 23	19 42	20 04	20 35	20 56	21 23	21 36	21 52	22 11	22 36	23 16	// //	// //	// //	// //	// //	// //	// //	// //
	21	18 16	18 22	18 28	18 33	18 39	18 45	18 56	19 09	19 23	19 41	20 03	20 33	20 52	21 18	21 31	21 46	22 04	22 26	22 58	// //	// //	// //	// //	// //	// //	// //	// //
	25	18 21	18 26	18 31	18 36	18 41	18 46	18 57	19 09	19 23	19 40	20 01	20 29	20 49	21 13	21 25	21 39	21 55	22 16	22 43	23 30	// //	// //	// //	// //	// //	// //	// //
	29	18 26	18 30	18 35	18 39	18 44	18 48	18 58	19 09	19 23	19 38	19 59	20 26	20 44	21 07	21 18	21 31	21 47	22 05	22 28	23 03	// //	// //	// //	// //	// //	// //	// //
Aug.	2	18 31	18 35	18 38	18 42	18 46	18 50	18 59	19 10	19 22	19 37	19 56	20 22	20 39	21 01	21 12	21 24	21 38	21 54	22 15	22 43	23 33	// //	// //	// //	// //	// //	// //
	6	18 37	18 39	18 42	18 45	18 48	18 52	19 00	19 10	19 21	19 35	19 53	20 18	20 34	20 55	21 04	21 16	21 28	21 43	22 02	22 25	23 00	// //	// //	// //	// //	// //	// //
	10	18 43	18 44	18 46	18 48	18 51	18 54	19 01	19 10	19 20	19 33	19 50	20 13	20 29	20 48	20 57	21 07	21 19	21 33	21 49	22 09	22 36	23 23	// //	// //	// //	// //	// //
	14	18 49	18 49	18 50	18 51	18 53	18 56	19 02	19 09	19 19	19 31	19 47	20 09	20 23	20 41	20 49	20 59	21 10	21 22	21 37	21 54	22 17	22 49	// //	// //	// //	// //	// //
	18	18 55	18 54	18 54	18 55	18 56	18 58	19 03	19 09	19 18	19 29	19 44	20 04	20 17	20 33	20 41	20 50	21 00	21 11	21 25	21 40	22 00	22 25	23 05	// //	// //	// //	// //
	22	19 02	18 59	18 58	18 58	18 58	19 00	19 03	19 09	19 16	19 26	19 40	19 59	20 11	20 26	20 33	20 41	20 51	21 01	21 13	21 27	21 44	22 05	22 34	23 27	// //	// //	// //
	26	19 09	19 05	19 03	19 01	19 01	19 01	19 04	19 08	19 15	19 24	19 36	19 53	20 05	20 19	20 25	20 33	20 41	20 50	21 01	21 14	21 28	21 46	22 10	22 43	// //	// //	// //
	30	19 16	19 10	19 07	19 05	19 04	19 03	19 05	19 08	19 14	19 21	19 32	19 48	19 58	20 11	20 17	20 24	20 32	20 40	20 50	21 01	21 14	21 30	21 49	22 15	22 55	// //	// //
Sept	3	19 23	19 16	19 12	19 08	19 06	19 05	19 05	19 08	19 12	19 19	19 28	19 43	19 52	20 04	20 09	20 15	20 22	20 30	20 39	20 49	21 00	21 14	21 31	21 52	22 20	23 10	// //
	7	19 30	19 22	19 16	19 12	19 09	19 07	19 06	19 07	19 10	19 16	19 25	19 37	19 46	19 56	20 01	20 07	20 13	20 20	20 28	20 37	20 47	20 59	21 13	21 31	21 54	22 27	23 38
	11	19 38	19 28	19 21	19 16	19 12	19 10	19 07	19 07	19 09	19 13	19 21	19 32	19 39	19 49	19 53	19 58	20 04	20 10	20 17	20 25	20 34	20 45	20 57	21 12	21 31	21 56	22 34
	15	19 46	19 35	19 26	19 20	19 15	19 12	19 08	19 06	19 07	19 11	19 17	19 26	19 33	19 41	19 45	19 50	19 55	20 00	20 06	20 14	20 22	20 31	20 42	20 55	21 11	21 31	21 59
	19	19 55	19 41	19 31	19 24	19 18	19 14	19 08	19 06	19 06	19 08	19 13	19 21	19 27	19 34	19 38	19 41	19 46	19 51	19 56	20 03	20 10	20 18	20 27	20 39	20 52	21 09	21 31
	23	20 04	19 48	19 37	19 28	19 22	19 16	19 09	19 06	19 04	19 05	19 09	19 16	19 20	19 27	19 30	19 33	19 37	19 41	19 46	19 52	19 58	20 05	20 14	20 23	20 35	20 49	21 07
	27	20 13	19 56	19 42	19 33	19 25	19 19	19 10	19 05	19 03	19 03	19 05	19 10	19 14	19 20	19 22	19 25	19 29	19 32	19 37	19 41	19 47	19 53	20 00	20 09	20 19	20 31	20 46
Oct.	1	20 23	20 03	19 48	19 37	19 28	19 22	19 12	19 05	19 02	19 00	19 02	19 05	19 09	19 13	19 15	19 18	19 21	19 24	19 27	19 31	19 36	19 41	19 48	19 55	20 03	20 14	20 26
	5	20 34	20 11	19 55	19 42	19 32	19 24	19 13	19 05	19 01	18 58	18 58	19 00	19 03	19 06	19 08	19 10	19 13	19 15	19 18	19 22	19 26	19 30	19 35	19 42	19 49	19 57	20 08

The symbols (*) and (/) indicate Sun continuously above horizon and continuous twilight, respectively.

Table A1.3. *Twilight October–December*

UNIVERSAL TIME FOR MERIDIAN OF GREENWICH
BEGINNING ASTRONOMICAL TWILIGHT

Lat.	−55°	−50°	−45°	−40°	−35°	−30°	−20°	−10°	0°	+10°	+20°	+30°	+35°	+40°	+42°	+44°	+46°	+48°	+50°	+52°	+54°	+56°	+58°	+60°	+62°	+64°	+66°
	h m	h m	h m	h m	h m	h m	h m	h m	h m	h m	h m	h m	h m	h m	h m	h m	h m	h m	h m	h m	h m	h m	h m	h m	h m	h m	h m
Oct. 1	3 18	3 38	3 52	4 03	4 12	4 19	4 28	4 34	4 38	4 39	4 38	4 34	4 30	4 26	4 23	4 21	4 18	4 15	4 11	4 07	4 02	3 56	3 50	3 43	3 34	3 23	3 10
5	3 05	3 27	3 44	3 56	4 06	4 13	4 24	4 32	4 36	4 39	4 39	4 36	4 33	4 30	4 28	4 26	4 23	4 21	4 18	4 14	4 10	4 05	4 00	3 54	3 46	3 37	3 26
9	2 52	3 17	3 35	3 49	3 59	4 08	4 21	4 29	4 35	4 38	4 40	4 38	4 37	4 34	4 32	4 31	4 29	4 27	4 24	4 21	4 18	4 14	4 09	4 04	3 58	3 50	3 41
13	2 37	3 06	3 26	3 41	3 53	4 03	4 17	4 27	4 34	4 38	4 41	4 41	4 40	4 38	4 37	4 36	4 34	4 32	4 30	4 28	4 25	4 22	4 19	4 14	4 09	4 03	3 55
17	2 23	2 55	3 17	3 34	3 47	3 57	4 13	4 24	4 33	4 38	4 42	4 43	4 43	4 42	4 41	4 40	4 39	4 38	4 37	4 35	4 33	4 30	4 27	4 24	4 20	4 15	4 09
21	2 07	2 44	3 08	3 27	3 41	3 52	4 10	4 22	4 31	4 38	4 43	4 46	4 46	4 46	4 46	4 45	4 45	4 44	4 43	4 42	4 40	4 38	4 36	4 34	4 30	4 27	4 22
25	1 50	2 32	2 59	3 20	3 35	3 48	4 07	4 20	4 31	4 38	4 44	4 48	4 49	4 50	4 50	4 50	4 50	4 49	4 49	4 48	4 47	4 46	4 45	4 43	4 41	4 38	4 34
29	1 32	2 20	2 51	3 13	3 29	3 43	4 03	4 18	4 30	4 39	4 45	4 51	4 53	4 54	4 54	4 55	4 55	4 55	4 55	4 55	4 54	4 54	4 53	4 52	4 50	4 49	4 46
Nov. 2	1 11	2 08	2 42	3 06	3 24	3 38	4 00	4 17	4 29	4 39	4 47	4 53	4 56	4 58	4 59	4 59	5 00	5 00	5 01	5 01	5 01	5 01	5 01	5 01	5 00	4 59	4 58
6	0 45	1 56	2 33	2 59	3 19	3 34	3 58	4 15	4 29	4 40	4 48	4 56	4 59	5 02	5 03	5 04	5 05	5 06	5 07	5 07	5 08	5 08	5 09	5 09	5 09	5 09	5 09
10	// //	1 43	2 25	2 53	3 14	3 30	3 56	4 14	4 28	4 40	4 50	4 59	5 02	5 06	5 07	5 09	5 10	5 11	5 12	5 13	5 15	5 16	5 17	5 17	5 18	5 19	5 20
14	// //	1 30	2 17	2 47	3 09	3 27	3 53	4 13	4 28	4 41	4 52	5 01	5 06	5 10	5 12	5 13	5 15	5 16	5 18	5 19	5 21	5 22	5 24	5 25	5 27	5 28	5 30
18	// //	1 16	2 09	2 42	3 05	3 24	3 52	4 12	4 29	4 42	4 54	5 04	5 09	5 14	5 16	5 18	5 20	5 21	5 23	5 25	5 27	5 29	5 31	5 33	5 35	5 37	5 40
22	// //	1 02	2 02	2 37	3 02	3 21	3 51	4 12	4 29	4 43	4 56	5 07	5 12	5 18	5 20	5 22	5 24	5 26	5 28	5 31	5 33	5 35	5 38	5 40	5 43	5 46	5 49
26	// //	0 45	1 55	2 32	2 59	3 19	3 50	4 12	4 30	4 45	4 58	5 10	5 16	5 22	5 24	5 26	5 29	5 31	5 33	5 36	5 39	5 41	5 44	5 47	5 50	5 54	5 57
30	// //	0 24	1 49	2 29	2 56	3 18	3 49	4 12	4 31	4 46	5 00	5 13	5 19	5 25	5 28	5 30	5 33	5 35	5 38	5 41	5 44	5 47	5 50	5 53	5 57	6 01	6 05
Dec. 4	// //	// //	1 44	2 26	2 55	3 16	3 49	4 13	4 32	4 48	5 02	5 16	5 22	5 29	5 31	5 34	5 37	5 40	5 42	5 45	5 49	5 52	5 55	5 59	6 03	6 07	6 12
8	// //	// //	1 40	2 24	2 54	3 16	3 49	4 14	4 33	4 50	5 05	5 18	5 25	5 32	5 35	5 37	5 40	5 43	5 46	5 50	5 53	5 56	6 00	6 04	6 08	6 13	6 18
12	// //	// //	1 37	2 23	2 53	3 16	3 50	4 15	4 35	4 52	5 07	5 21	5 28	5 35	5 38	5 41	5 44	5 47	5 50	5 53	5 57	6 00	6 04	6 09	6 13	6 18	6 23
16	// //	// //	1 35	2 23	2 54	3 17	3 51	4 17	4 37	4 54	5 09	5 23	5 30	5 38	5 41	5 44	5 47	5 50	5 53	5 57	6 00	6 04	6 08	6 12	6 17	6 22	6 28
20	// //	// //	1 36	2 24	2 55	3 18	3 53	4 18	4 39	4 56	5 11	5 25	5 33	5 40	5 43	5 46	5 49	5 52	5 56	5 59	6 03	6 07	6 11	6 15	6 20	6 25	6 31
24	// //	// //	1 37	2 26	2 57	3 20	3 55	4 20	4 41	4 58	5 13	5 27	5 35	5 42	5 45	5 48	5 51	5 54	5 58	6 01	6 05	6 09	6 13	6 17	6 22	6 27	6 33
28	// //	// //	1 41	2 28	3 00	3 23	3 57	4 23	4 43	5 00	5 15	5 29	5 36	5 43	5 46	5 49	5 53	5 56	5 59	6 02	6 06	6 10	6 14	6 18	6 23	6 28	6 34
32	// //	// //	1 46	2 32	3 03	3 26	4 00	4 25	4 45	5 02	5 17	5 31	5 38	5 45	5 47	5 50	5 53	5 57	6 00	6 03	6 07	6 10	6 14	6 18	6 23	6 28	6 33
36	// //	// //	1 53	2 37	3 07	3 29	4 03	4 27	4 47	5 03	5 18	5 32	5 38	5 45	5 48	5 51	5 54	5 57	6 00	6 03	6 06	6 10	6 13	6 17	6 22	6 26	6 31

ENDING ASTRONOMICAL TWILIGHT

		h m	h m	h m	h m	h m	h m	h m	h m	h m	h m	h m	h m	h m	h m	h m	h m	h m	h m	h m	h m	h m	h m	h m	h m	h m	h m	h m
Oct.	1	20 23	20 03	19 48	19 37	19 28	19 22	19 12	19 05	19 02	19 00	19 02	19 05	19 09	19 13	19 15	19 18	19 21	19 24	19 27	19 31	19 36	19 41	19 48	19 55	20 03	20 14	20 26
	5	20 34	20 11	19 55	19 42	19 32	19 24	19 13	19 05	19 01	18 58	18 58	19 00	19 03	19 06	19 08	19 10	19 13	19 15	19 18	19 22	19 26	19 30	19 35	19 42	19 49	19 57	20 08
	9	20 46	20 20	20 01	19 47	19 36	19 27	19 14	19 06	19 00	18 56	18 55	18 56	18 57	19 00	19 01	19 03	19 05	19 07	19 09	19 12	19 16	19 19	19 24	19 29	19 35	19 42	19 51
	13	20 58	20 28	20 08	19 52	19 40	19 31	19 16	19 06	18 59	18 54	18 52	18 51	18 52	18 54	18 55	18 56	18 58	18 59	19 01	19 03	19 06	19 09	19 13	19 17	19 22	19 28	19 35
	17	21 11	20 38	20 15	19 58	19 45	19 34	19 18	19 07	18 58	18 53	18 49	18 47	18 47	18 48	18 49	18 50	18 51	18 52	18 53	18 55	18 57	18 59	19 02	19 05	19 09	19 14	19 20
	21	21 26	20 48	20 23	20 04	19 49	19 38	19 20	19 07	18 58	18 51	18 46	18 43	18 43	18 43	18 43	18 43	18 44	18 45	18 46	18 47	18 48	18 50	18 52	18 54	18 57	19 01	19 06
	25	21 42	20 58	20 30	20 10	19 54	19 41	19 22	19 08	18 58	18 50	18 44	18 40	18 38	18 38	18 38	18 38	18 38	18 38	18 38	18 39	18 40	18 41	18 42	18 44	18 46	18 49	18 52
	29	22 00	21 10	20 38	20 16	19 59	19 45	19 25	19 09	18 58	18 49	18 42	18 36	18 34	18 33	18 32	18 32	18 32	18 32	18 32	18 32	18 32	18 33	18 33	18 34	18 36	18 37	18 40
Nov.	2	22 22	21 21	20 47	20 23	20 04	19 49	19 27	19 11	18 58	18 48	18 40	18 34	18 31	18 29	18 28	18 27	18 27	18 26	18 26	18 25	18 25	18 25	18 25	18 25	18 26	18 27	18 28
	6	22 50	21 34	20 56	20 29	20 09	19 54	19 30	19 12	18 59	18 48	18 39	18 31	18 28	18 25	18 24	18 23	18 22	18 21	18 20	18 19	18 18	18 18	18 17	18 17	18 17	18 17	18 17
	10	// //	21 47	21 05	20 36	20 15	19 58	19 33	19 14	18 59	18 48	18 37	18 29	18 25	18 21	18 20	18 19	18 17	18 16	18 15	18 14	18 12	18 11	18 10	18 09	18 08	18 08	18 07
	14	// //	22 02	21 14	20 43	20 20	20 03	19 36	19 16	19 00	18 48	18 37	18 27	18 23	18 18	18 17	18 15	18 14	18 12	18 10	18 09	18 07	18 06	18 04	18 03	18 01	17 59	17 58
	18	// //	22 17	21 23	20 50	20 26	20 07	19 39	19 18	19 02	18 48	18 36	18 26	18 21	18 16	18 14	18 12	18 10	18 08	18 06	18 05	18 03	18 01	17 59	17 56	17 54	17 52	17 50
	22	// //	22 34	21 32	20 57	20 31	20 12	19 42	19 20	19 03	18 49	18 36	18 25	18 19	18 14	18 12	18 10	18 08	18 05	18 03	18 01	17 59	17 56	17 54	17 51	17 49	17 46	17 43
	26	// //	22 54	21 41	21 03	20 37	20 16	19 45	19 23	19 05	18 50	18 36	18 24	18 18	18 13	18 10	18 08	18 06	18 03	18 01	17 58	17 55	17 53	17 50	17 47	17 44	17 40	17 36
	30	// //	23 20	21 50	21 09	20 42	20 20	19 48	19 25	19 06	18 51	18 37	18 24	18 18	18 12	18 09	18 07	18 04	18 01	17 59	17 56	17 53	17 50	17 47	17 43	17 40	17 36	17 31
Dec.	4	// //	// //	21 58	21 15	20 46	20 24	19 51	19 27	19 08	18 52	18 38	18 25	18 18	18 11	18 09	18 06	18 03	18 00	17 57	17 54	17 51	17 48	17 44	17 41	17 37	17 32	17 28
	8	// //	// //	22 05	21 20	20 51	20 28	19 54	19 30	19 10	18 54	18 39	18 25	18 18	18 12	18 09	18 06	18 03	18 00	17 57	17 54	17 50	17 47	17 43	17 39	17 35	17 30	17 25
	12	// //	// //	22 11	21 25	20 54	20 31	19 57	19 32	19 12	18 55	18 40	18 26	18 19	18 12	18 09	18 06	18 03	18 00	17 57	17 54	17 50	17 47	17 43	17 38	17 34	17 29	17 23
	16	// //	// //	22 16	21 29	20 58	20 34	20 00	19 34	19 14	18 57	18 42	18 28	18 21	18 13	18 10	18 07	18 04	18 01	17 58	17 54	17 51	17 47	17 43	17 39	17 34	17 29	17 23
	20	// //	// //	22 20	21 32	21 00	20 37	20 02	19 37	19 16	18 59	18 44	18 30	18 22	18 15	18 12	18 09	18 06	18 03	17 59	17 56	17 52	17 48	17 44	17 40	17 35	17 30	17 24
	24	// //	// //	22 21	21 33	21 02	20 38	20 04	19 39	19 18	19 01	18 46	18 32	18 24	18 17	18 14	18 11	18 08	18 05	18 01	17 58	17 54	17 50	17 46	17 42	17 37	17 32	17 26
	28	// //	// //	22 21	21 34	21 03	20 40	20 05	19 40	19 20	19 03	18 48	18 34	18 27	18 20	18 17	18 14	18 11	18 07	18 04	18 01	17 57	17 53	17 49	17 45	17 40	17 35	17 30
	32	// //	// //	22 20	21 34	21 04	20 41	20 07	19 42	19 22	19 05	18 50	18 36	18 29	18 22	18 19	18 17	18 14	18 10	18 07	18 04	18 00	17 57	17 53	17 49	17 44	17 39	17 34
	36	// //	// //	22 17	21 33	21 03	20 41	20 07	19 43	19 24	19 07	18 53	18 39	18 32	18 25	18 23	18 20	18 17	18 14	18 11	18 08	18 04	18 01	17 57	17 53	17 49	17 45	17 39

The symbols (*) and (/) indicate Sun continuously above horizon and continuous twilight, respectively.

Appendix 2

Star maps[1]

The following six maps locate the constellations and brighter stars. The first shows the north polar region down to about 50° N declination. The next four are seasonal equatorial maps that display stars between 60° N and 60° S declination, and the last shows the south polar region. Declinations are given along a central hour circle. Right ascensions are noted around the peripheries of the polar maps and along the celestial equator on the equatorial maps.

Stars are generally selected to indicate constellation positions and outlines. They are shown through fourth magnitude, although their census is not complete. A few fainter stars are included to indicate the locations of obscure constellations. All 88 constellations are represented, although several are indicated by only their brightest stars. Brighter stars that display color (non-white) are rendered gray. A small number of non-stellar objects, clusters, nebulae, and galaxies are also included.

The maps show the broad outline of the Milky Way. However, much of its intricate detail is omitted. The galactic poles are indicated, and the galactic equator, the mid-line of the Galaxy, is marked with galactic longitude starting at the galactic center in Sagittarius.

The complete paths of precession of the north and south celestial poles are shown on Maps 1 and 6. Partial paths are outlined for the NCP on Maps 2 and 5 and for the SCP on Maps 3 and 4. These paths are distorted from circles as a result of distortions inherent in mapping a sphere onto a plane.

The dates along the edges of the polar maps and along the tops and bottoms of the equatorial maps indicate the appearance of the sky at approximately 8:30 PM ($20^h\ 30^m$) local time. To use the north polar map face north and rotate the map so that the current month appears at the top. In the southern hemisphere use the south polar map similarly. The celestial pole will have an elevation in degrees equal to your latitude. To use the equatorial maps in the northern hemisphere face south and line up the current

[1]Adapted from the color maps in *Astronomy!*, James B. Kaler, HarperCollins, New York, 1994.

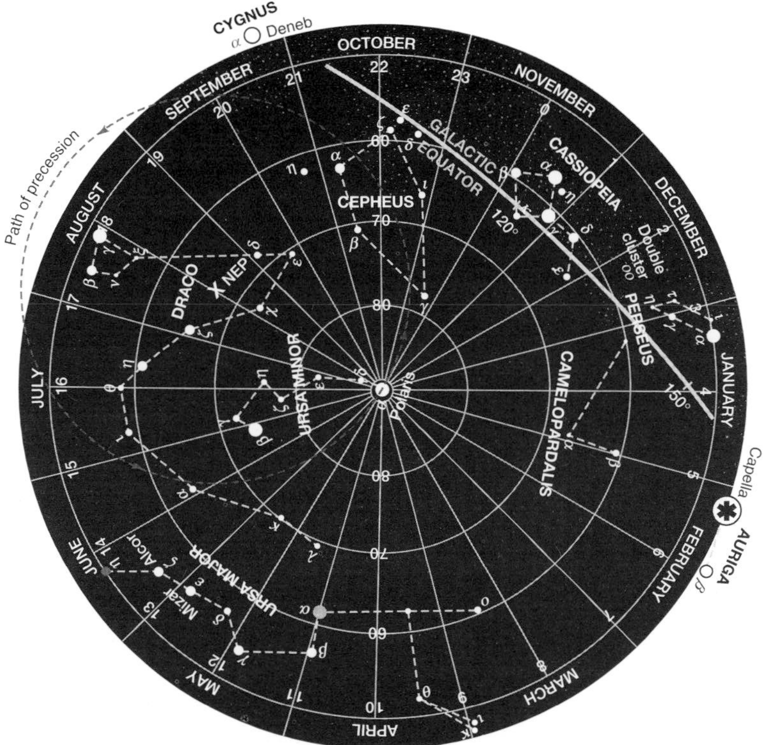

Map 1. *The North Polar Constellations*

month with the celestial meridian. To use them in the southern hemisphere, face north and turn them upside down. The equator point (the intersection between the celestial equator and the meridian) will have an elevation in degrees equal to 90° minus the latitude.

For each hour past 20^{h} 30^{m} shift or rotate the map one hour to the west, that is, align an hour circle that is one additional hour to the east. For every 2 hours past 20^{h} 30^{m} add one month to your current month. For example, if it is March 15 at 20^{h} 30^{m}, you would align "March" on Map 4 with the celestial meridian. If it is 2^{h} 30^{m}, set "June" (Map 5) on the meridian.

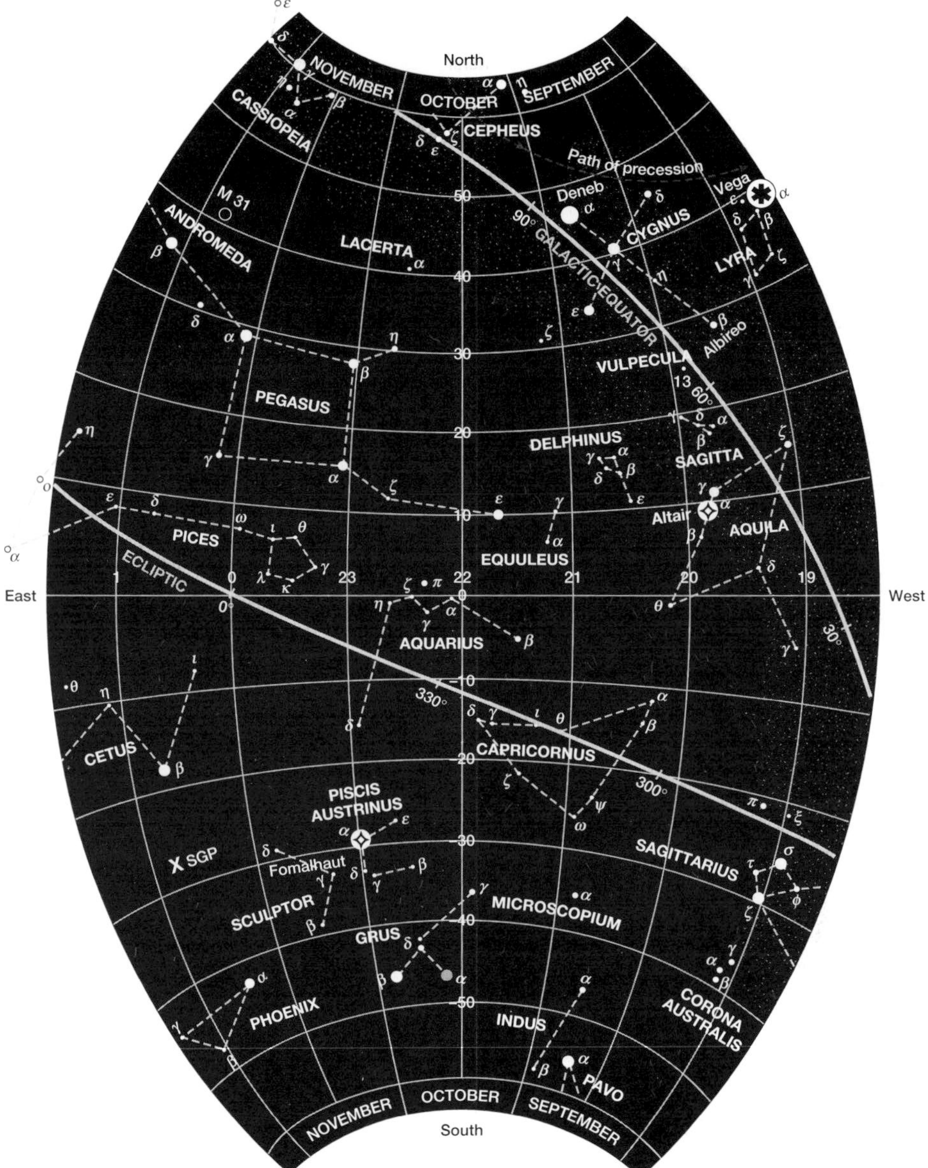

Map 2. *The Constellations of Northern Autumn, Southern Spring*

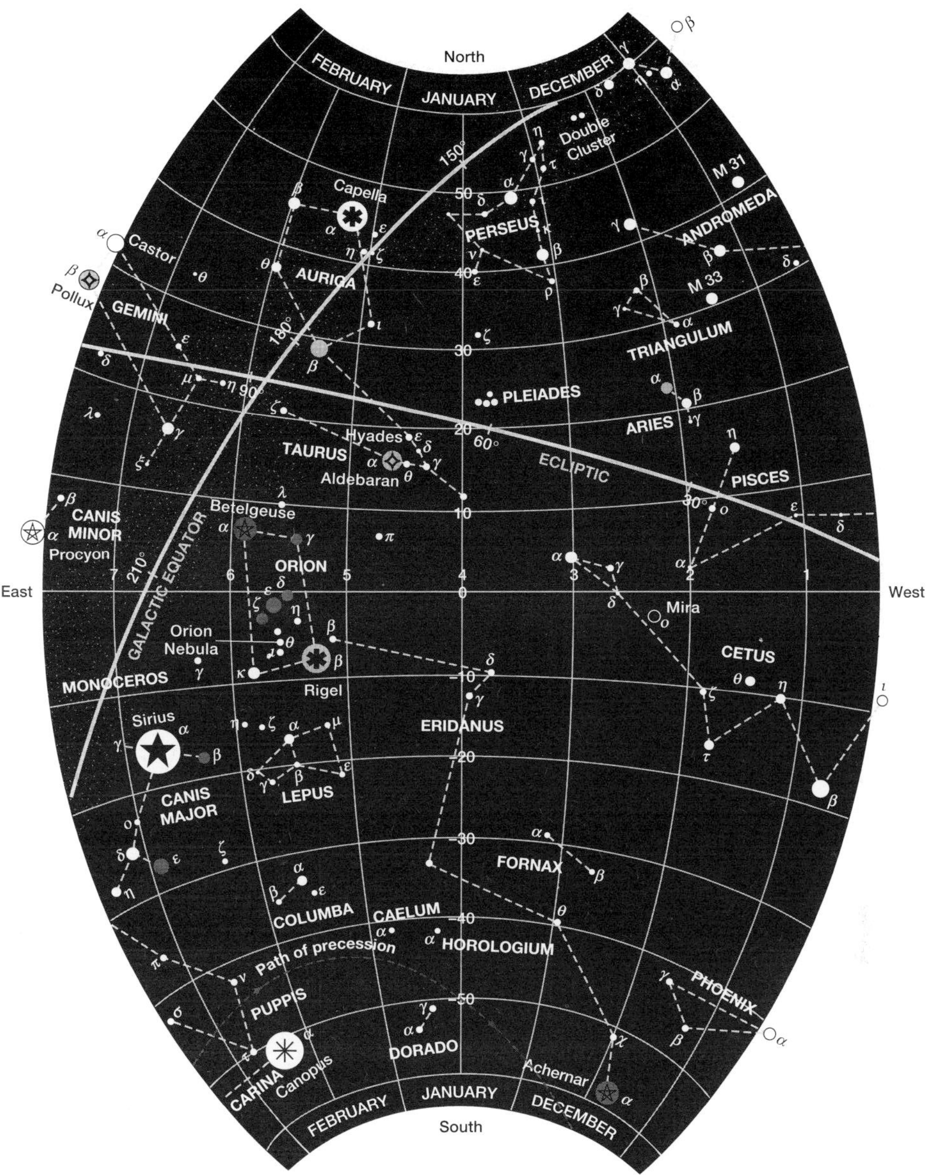

Map 3. *The Constellations of Northern Winter, Southern Summer*

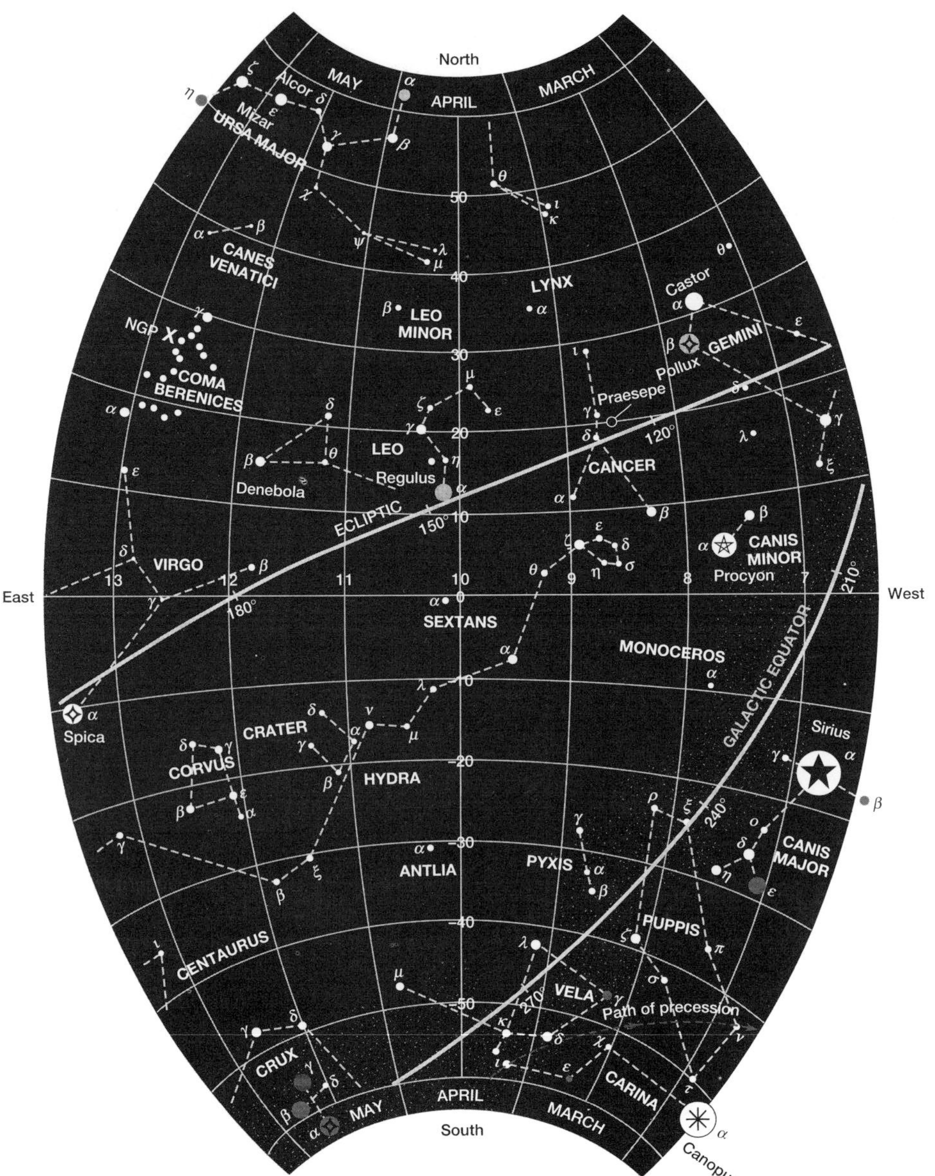

Map 4. *The Constellations of Northern Spring, Southern Autumn*

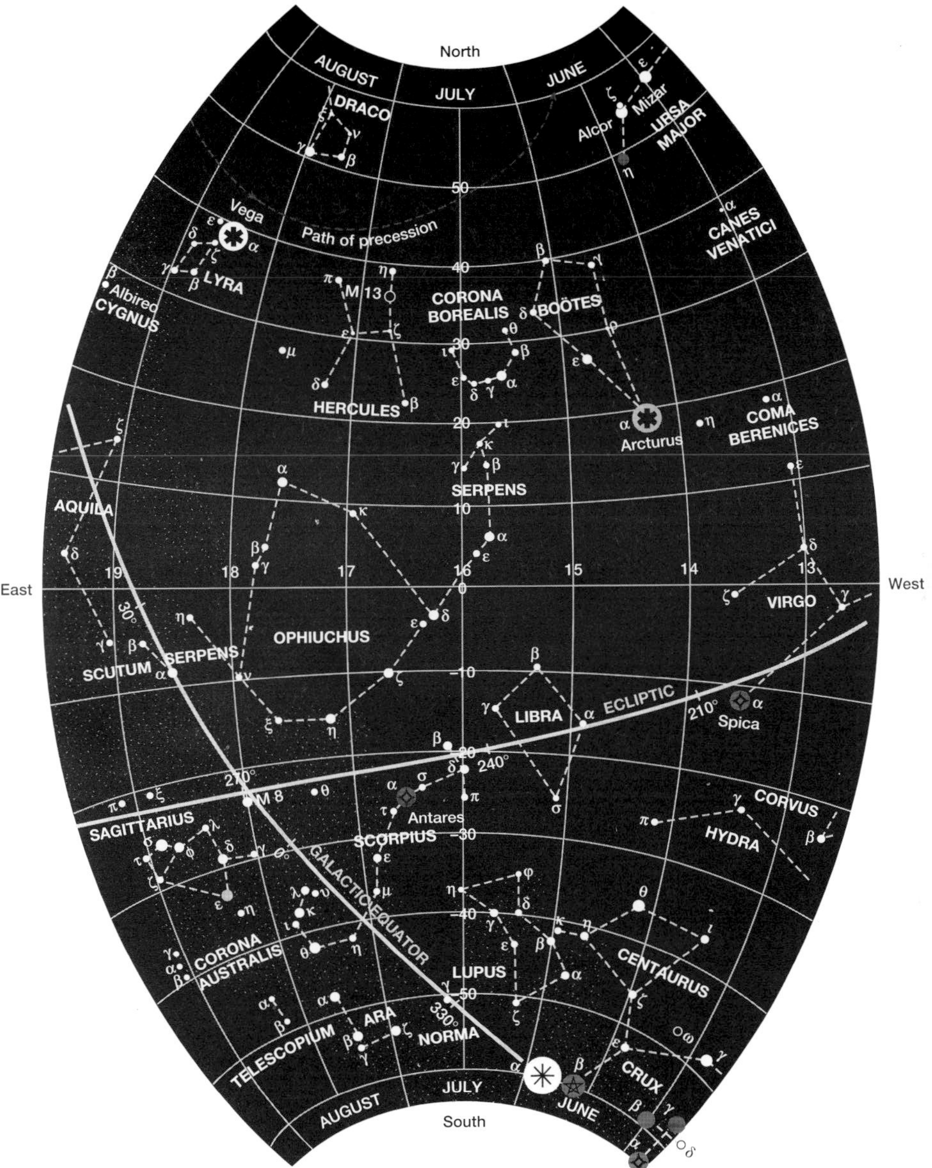

Map 5. *The Constellations of Northern Summer, Southern Winter*

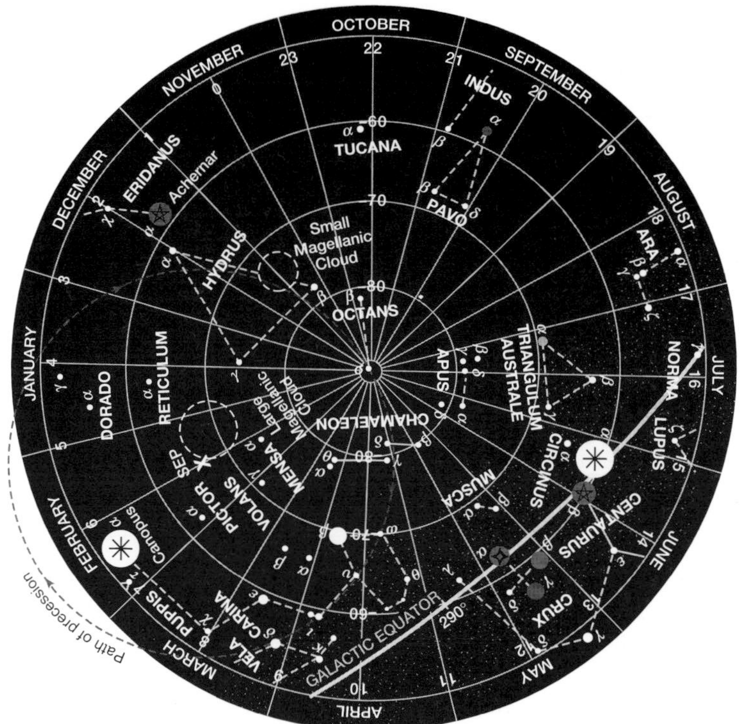

Map 6. *The South Polar Constellations*

Appendix 3

Trigonometric relationships

The direction of this book has been toward qualitative geometric relationships, presenting the results of quantitative calculations graphically, but eschewing the equations required to produce them in favor of non-mathematical demonstrations. The philosophy has been to develop a fundamental understanding of the various aspects of the sky. However, it clear that the celestial sphere is subject to mathematically rigorous treatment. Other books listed in the bibliography will take you through that adventure. This appendix provides an introduction to the subject to demonstrate some of the simpler calculations that can be performed and to provide a bridge to more technical texts.

A3.1 Basic plane trigonometry

Trigonometry is the study of triangles and the mathematical relationships found among their sides and angles. This appendix can hardly present a complete examination of the subject, but a brief review of the most basic and useful concepts, specifically definitions of the trigonometric functions and how to use them, is in order.

Figure A3.1 shows a right triangle with sides a, b, and c; opposite each are angles, A, B, and C. Angle B is a right angle of 90°, the necessary ingredient of a right triangle. We can define six quantities called the *trigonometric functions* that interconnect these sides and angles (note, *only* for right triangles). The first example is the *sine* of an angle. The term is abbreviated sin and the "sine of angle A" is written as sin A. The sine of an angle is defined as the ratio of the length of the opposite side to that of the hypotenuse (the side opposite the right angle), or using Figure A3.1,

$$\sin A = a/b\ .$$

The *cosine* (cos) of an angle is the ratio of the lengths of the side *adjacent* to the angle (that not the hypotenuse) and the hypotenuse, and the *tangent* (tan) the ratio of the opposite and adjacent sides. The other three, the *secant* (sec), *cosecant* (csc), and the *cotangent* (cot) are the inverses of the cosine, sine, and tangent, that is, the cosecant of

Table A3.1. *Definitions of the trigonometric functions (based on Figure A1.1)*

sine of A = sin $A = a/b$
cosine of A = cos $A = c/b$
tangent of A = tan $A = a/c$
secant of A = sec $A = b/c = 1/\cos A$
cosecant of A = csc $A = b/a = 1/\sin A$
cotangent of A = cot $A = c/a = 1/\tan A$

A is $1/\sin A$ or b/a, etc. These functions and their abbreviations are summarized in Table A3.1. Similarly, sin $C = c/b$, cos $C = a/b$, tan $C = c/a$, etc.

The values of these functions vary continuously with the values of the angle. They can be found in mathematical texts, which give tables for each degree, and sometimes for each minute of arc, between 0° and 90°. The values from 90° to 360° are symmetrically related to those between 0° and 90° and are readily found from relations discussed below. Values of the functions for more refined angles, say to the second of arc, can be found by interpolation. Since the secant, cosecant, and tangent are inverses of cosine, sine, and tangent, only the values of the latter three are usually given. The more sophisticated hand calculators will generate them for any angle, so such tables will not be presented here.

They are, however, graphed in Figure A3.2 for all angles from 0° to 360°. Note how the sine passes through unity (1) at 90° and the cosine through zero; the tangent goes off the graph to infinity (∞) at 90°, then returns from $-\infty$ as we proceed from 90° to 180°, where it passes zero and repeats the behavior. The functions are all positive in the first quadrant of the circle (0° to 90°), and take on negative values in various other quadrants, as seen in Figure A3.2 and as summarized in Figure A3.3. For angle A of Figure A3.1 the quadrants are connected according to the relations given in Table A3.2 below. Knowledge of these sign relationships is critical in the solutions of triangles.

Trigonometry pervades all of science and engineering. You know the values of the angles and have measured one side, say c in Figure A3.1; you can then calculate the lengths of a and b from the above rules, and find $a = c \tan A$ and $b = c/\cos A = c \sec A$. One of the trigonometric functions has already been used, in Section 13.15, where

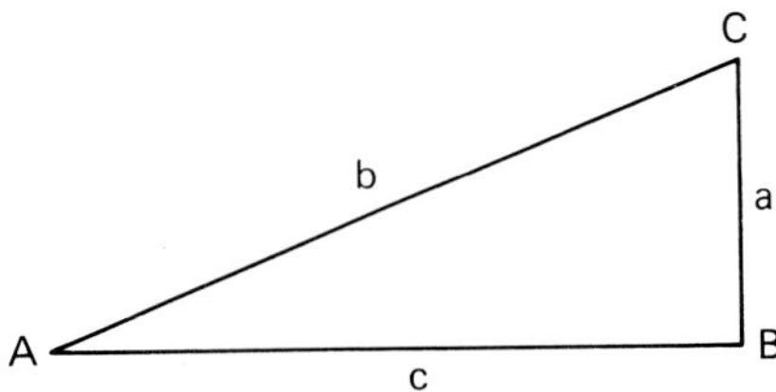

A3.1. A right triangle with side lengths a, b, c and opposing angles A, B, and C. Angle B is 90°.

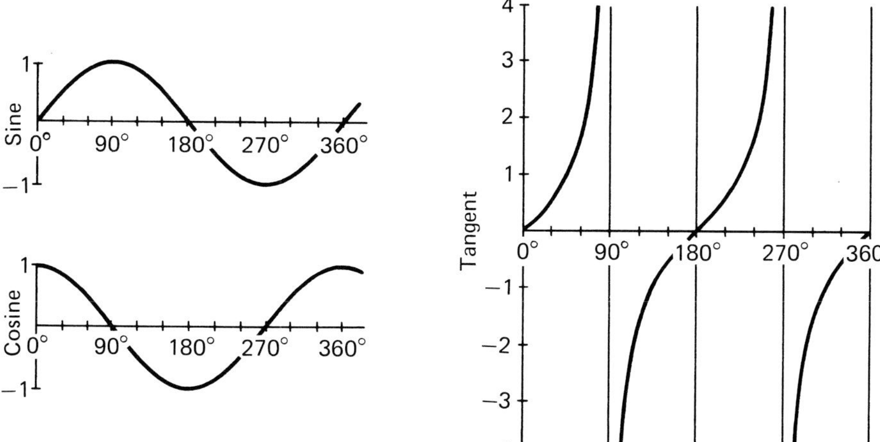

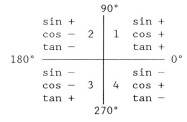

A3.2. The graphical representations of the sine, cosine, and tangent.

		90°		
	sin + cos − tan −	2	1	sin + cos + tan +
180°				0°
	sin − cos − tan +	3	4	sin − cos + tan −
		270°		

A3.3. The signs of the trigonometric functions within the four numbered quadrants.

Table A3.2. *The relations among the quadrants*

Quadrant 2: $\sin(180° - A) = \sin A$;
$\cos(180° - A) = -\cos A$; $\tan(180° - A) = -\tan A$
Quadrant 3: $\sin(180° + A) = -\sin A$;
$\cos(180° + A) = -\cos A$; $\tan(180° + A) = \tan A$
Quadrant 4: $\sin(360° - A) = -\sin A$;
$\cos(360° - A) = \cos A$; $\tan(360° - A) = -\tan A$

we explored the subject of airmass and atmospheric extinction. Look again at Figure 13.25a as redrawn in Figure A3.4. We use a simple atmosphere with its top drawn as a plane parallel to the ground. Toward the zenith we look out into space through the minimum unit atmosphere. The airmass through which we will look at any zenith angle is a function of z, i.e., $\cos z = 1/M$ or $M = \sec z$ (remember, however, that for high z the secant of z is larger than the true airmass: see Figure 13.25b).

The functions are intimately related to one another. For example, from their defi-

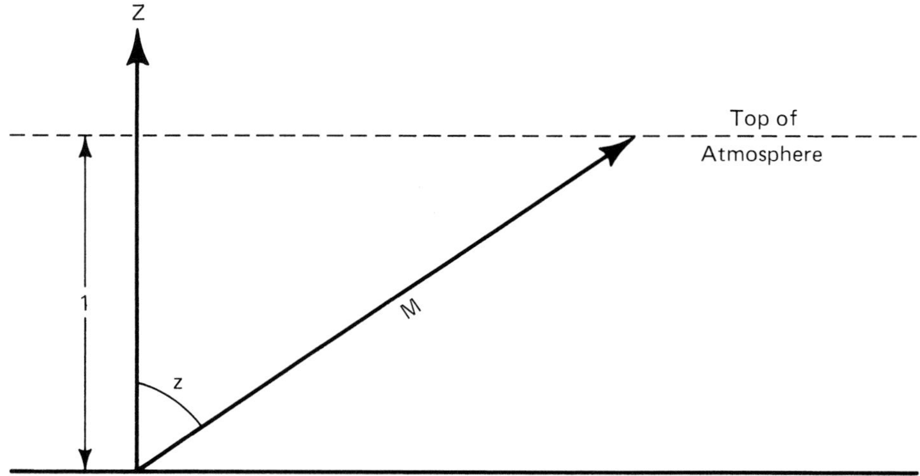

A3.4. The calculation of airmass. We use a simple plane-parallel approximation to the atmosphere, with unit atmosphere toward the zenith. The airmass, M, is thus the secant of z.

nitions in Table A3.1, we see that the tangent of an angle is just its sine divided by its cosine. The most useful relations are summarized in Table A3.3.

The above discussion concerns the right triangle, which is a special case of triangles in general. Two laws (not proven here) are useful in the solution of the general triangle of Figure A3.5. The first, the *law of sines*, relates the sines of the angles to their opposite sides:

$$(1)\quad \frac{\sin A}{a} = \frac{\sin B}{b} = \frac{\sin C}{c} \ .$$

Thus $a = b\,(\sin A/\sin B)$, and so forth. The *law of cosines* relates a side with the cosine of the opposite angle:

$$(2)\quad \begin{aligned} a^2 &= b^2 + c^2 - 2bc \cos A \\ b^2 &= a^2 + c^2 - 2ac \cos B \\ c^2 &= a^2 + b^2 - 2ab \cos C \ . \end{aligned}$$

The centerpieces of this appendix involve the analogues of these laws for triangles drawn on a sphere.

Table A3.3. *Some relationships among the trigonometric functions*

$\sin(90° - A) = \cos A$	$\cos(90° - A) = \sin A$
$\sin(90° + A) = \cos A$	$\cos(90° + A) = -\sin A$
$\tan A = \sin A/\cos A$	

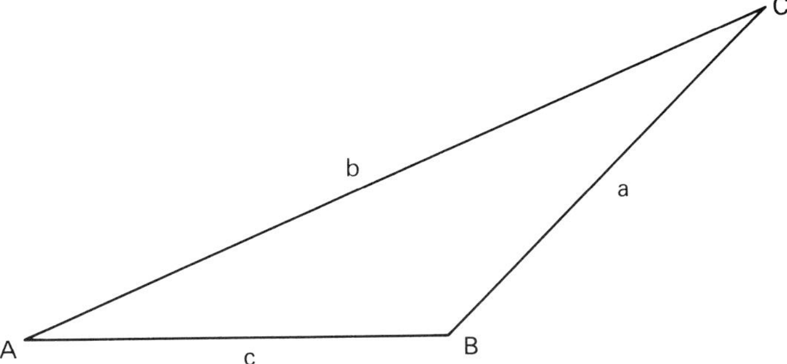

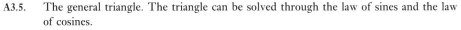

A3.5. The general triangle. The triangle can be solved through the law of sines and the law of cosines.

A3.2 Spherical trigonometry

Similar rules connect the sides and angles of the *spherical triangle* (Figure A3.6). Draw three great circles on the sphere. They must intersect at three points, creating angles A, B, and C (to be identified later with real angles in the sky) and defining a triangle. Opposite each angle are now, not sides measured in physical units (like centimeters), but arcs measured in degrees, referred to respectively as a, b, and c.

The sides and angles of the spherical triangle are related by three rules. From the law of cosines of plane trigonometry, equation (2), we can derive (though not here) the *law of cosines* of spherical trigonometry:

$$\begin{aligned} \cos a &= \cos b \cos c + \sin b \sin c \cos A, \\ \cos b &= \cos a \cos c + \sin a \sin c \cos B, \qquad (3) \\ \cos c &= \cos a \cos b + \sin a \sin b \cos C. \end{aligned}$$

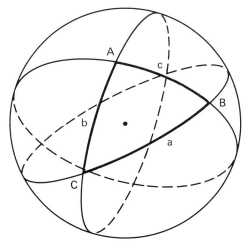

A3.6. The spherical triangle. Spherical triangle ABC is constructed from the arcs of three intersecting great circles, the angles labelled A, B, and C and the sides opposite them a, b, and c. All six quantities are measured in degrees.

Knowledge of two sides and the enclosed angle allows the computation of the other side, or given all the sides, we can derive any of the angles from

$$\cos A = (\cos a - \cos b \cos c)/\sin b \sin c,$$

etc.

The law of cosines, equation (3), is the fundamental rule. From it we derive another useful formula, the *law of sines*:

$$\frac{\sin A}{\sin a} = \frac{\sin B}{\sin b} = \frac{\sin C}{\sin c} \tag{4}$$

so that

$$\sin A = \frac{\sin a}{\sin b} \sin B,$$

etc. Given two sides and an opposite angle, we can always find the other opposite angle, or given two angles and an opposite side, we can find the other opposite side.

The law of cosines also generates a third, occasionally useful, relation called the *five parts rule*:

$$\begin{aligned} &\sin c \cos A = \cos a \sin b - \cos b \sin a \cos C, \\ (5)\quad &\sin c \cos B = \cos b \sin a - \cos a \sin b \cos C \\ &\sin a \cos B = \cos b \sin c - \cos c \sin b \cos A, \end{aligned}$$

etc.

The next step is application.

A3.3. The astronomical triangle

The commonest and simplest use of these rules is for the solution of the various parts of the astronomical triangle (see Section 2.5 and Figure A3.7). This configuration is used to translate between the equatorial coordinates hour angle and declination and horizon coordinates altitude and azimuth, and vice versa. It was employed earlier to calculate such things as the rising and setting positions of the Sun and Moon (Chapter 3), and is the fundamental tool used in the determination of terrestrial position from the stars (Chapter 8).

The astronomical triangle (*PZS* in Figure A3.7) is composed of segments of the celestial meridian (pole to zenith), the star's vertical circle or (zenith to star), and the hour circle (pole to star). The sides are therefore (in degrees) $(90° - \phi)$, $(90° - h) = z$, and $(90° - \delta)$, where ϕ, h, z, and δ are respectively the latitude, altitude, zenith distance, and declination (refer to Chapters 1 and 2.) The angles depend upon whether the star is in the eastern or western hemisphere. In the drawing, with the star in the west, they are t, the hour angle (equal to HA, the usual hour angle measured along the celestial equator), Z, the *zenith angle*, here equal to $360° - A$ where A is the azimuth, and p, the *paralactic angle*. If the star is in the eastern hemisphere, the angles would then be hour

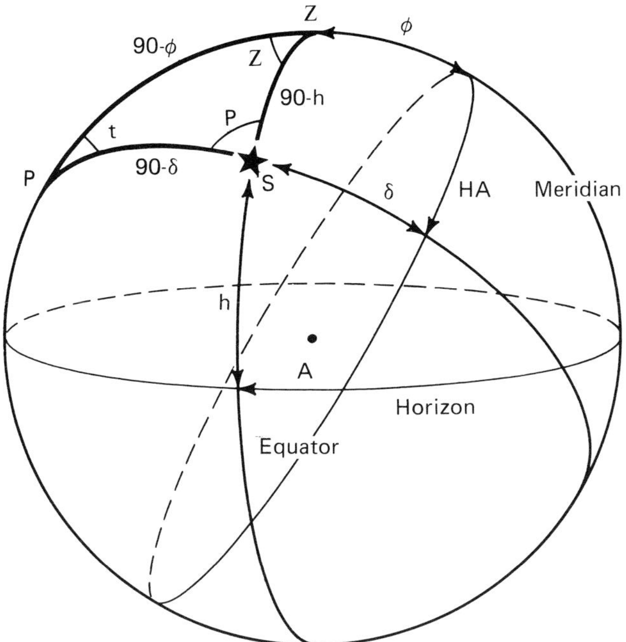

A3.7. The astronomical triangle, formed by the arcs of the celestial meridian from the zenith (*Z*) to the pole (*P*), the vertical circle from the zenith to the star, *S*, and the hour circle from the pole to the star.

angle east, or $t = 360° - HA$, and azimuth, A. Hour angle, of course, must here be expressed in degrees.

If we apply the law of cosines, we see that

$$(6)\quad \cos(90° - h) = \cos(90° - \delta)\cos(90° - \phi) + \sin(90° - \delta)\sin(90° - \phi)\cos t.$$

From the relations in Table A3.3 we can rewrite the equation as

$$(7)\quad \sin h = \sin \delta \sin \phi + \cos \delta \cos \phi \cos t.$$

Equation (7) allows us to solve for altitude if we know the declination and hour angle of the star plus the observer's latitude. To get the azimuth, we apply the law of sines to the astronomical triangle to derive

$$(8)\quad \frac{\sin Z}{\sin(90° - \delta)} = \frac{\sin t}{\sin(90° - h)},$$

or

$$\frac{\sin Z}{\cos \delta} = \frac{\sin t}{\cos h},$$

whence

$$(9)\quad \sin Z = \cos \delta \frac{\sin t}{\cos h}.$$

We may also apply the five parts rule and derive

$$\sin(90° - h)\cos Z = \sin(90° - \phi)\cos(90° - \delta)$$
$$-\cos(90° - \phi)\sin(90° - \delta)\cos t,$$

or

$$(10)\quad \cos Z \cos h = \cos \phi \sin \delta - \sin \phi \cos \delta \cos t\ .$$

We have now completed a *transformation of coordinates* from the equatorial system to the horizon system, which takes us from (HA, δ), or (α, δ) if the sidereal time is known, to (h, A), where again $Z = 360° - A$ and $t = HA$ if the star is west of the meridian and $Z = A$ and $t = (360° - HA)$ if the star is east of it.

We can just as easily transform in the other direction, from the horizon system back into equatorial coordinates by applying the analogous relations to Figure A3.7 to derive

$$(11)\quad \sin \delta = \sin \phi \sin h + \cos \phi \cos h \cos Z,$$

$$(12)\quad \sin t = \cos h\ \frac{\sin Z}{\cos \delta}\ .$$

and

$$(13)\quad \cos t \cos \delta = \cos \phi \sin h - \sin \phi \cos h \cos Z.$$

The signs of the functions provide the necessary information on the angle's quadrant (see Figure A3.3). For example, if $\sin A$ is positive we know that A is in the first or second quadrant, between 0° and 180°; if it is negative, then A is in the third or fourth quadrants, between 180° and 360°. There is no difficulty with the solution of h and δ from equations (7) and (11). If the sine of the angle is negative we know that the star is at a negative altitude, that is, below the horizon, or that it is at a negative declination, below the equator. Neither do we have any problem determining whether azimuth (from equation (9)) is above or below 180°, since all we need know is whether t is east or west, which will be given; or for that matter whether t (from equation (12)) is east or west, as it will be specified by the value of azimuth.

A problem does arise if A or t are near 90° or 270° (or Z is near 90°), close enough that one cannot tell by a simple diagram like Figure A3.7; for example, the sine of 91° is the same as that of 89°, i.e. there is a "sign ambiguity." We can discriminate by calculating $\cos Z$ or $\cos t$ from equations (10) and (13), since the cosine switches signs in passing from the first to second quadrants, whereas the sine does not (i.e. if the angle is 89° the cosine is positive, but if it is 91° the cosine is negative): see Figure A3.3. The same thing happens at 270°, as the sine stays negative in going from the third to the fourth quadrant, whereas the cosine reverses back from negative to positive.

A pair of examples shows how these equations can manipulated. First, what is the altitude and azimuth of the star Arcturus at 18 hours local time on June 1 at Paris, latitude 48° N? We will work here only to the nearest degree. From Table A1.1 (Appendix 1), June 1 is 253 days past September 21, when local mean solar and sidereal time are assumed to be equal, so that sidereal time is $253 \times 3^{m}\ 57^{s}$ or $16^{h}\ 39^{m}$ ahead of solar, whence local sidereal time is $10^{h}\ 39^{m}$. The right ascension of Arcturus is $14^{h}\ 15^{m}$, so that its hour angle at the moment is $-03^{h}\ 36^{m}$ or 54° east, and the declination is +19°.

From equation (7)

$$\sin h = \sin 19^\circ \sin 48^\circ + \cos 19^\circ \cos 48^\circ \cos 54^\circ.$$

From a calculator, or from trigonometric tables readily available in a variety of reference books,

$$\sin h = (0.326)(0.743) + (0.946)(0.669)(0.588) = 0.614.$$

Taking the inverse of sin h, we find $h = 38^\circ$. From equation (9) (note that the hour angle is east),

$$\sin Z = \sin A = \cos 19^\circ \sin 54^\circ/\cos 38^\circ = (0.946)(0.809)/0.788 = \quad 0.971,$$

so that A (since the sine is positive in both the first and second quadrants) is either 76° or 104°. These values are really too close to discriminate by a diagram, so use equation (10) instead, which gives the cosine of Z for which the quadrant is unambiguous. Substituting, we have

$$\begin{aligned}\cos Z = \cos A &= (\cos 48^\circ \sin 19^\circ - \sin 48^\circ \cos 19^\circ \cos 54^\circ)/\cos 38^\circ \\ &= [(0.669)(0.326) - (0.743)(0.946)(0.588)]/0.788 = -0.248,\end{aligned}$$

from which $A = 104^\circ$. It would obviously have been simpler to have used this equation in the first place.

Next, mix the transformations some and calculate the time and azimuth of sunset on May 1 at Chicago, latitude 42° N. On this date the solar declination (Figure A1.3) is +15° (which for simplicity we assume to be exact). At sunset, the altitude of the Sun is actually −51′ (see Section 7.2). From equation (11) we can write

$$\cos Z = (\sin \delta - \sin \phi \sin h)/\cos \phi \cos h,$$

so that

$$\cos Z = [\sin 15^\circ - \sin 42^\circ \sin(-51')]/\cos 42^\circ \cos(-51').$$

From a calculator, or trigonometric tables,

$$\cos Z = [0.2588 - (0.6691)(-0.01483)]/(0.7431)(0.9999) = 0.3617,$$

from which $Z = 68^\circ\ 48'$ and $A = 291^\circ\ 12'$.

We can calculate t from a variety of equations. First use equation (12) again to demonstrate the sign ambiguity discussed above. We have

$$\begin{aligned}\sin t &= \cos(-51') \sin (68^\circ\ 48')/\cos 15^\circ \\ &= (0.9999)(0.9323)/0.9659 = 0.9651.\end{aligned}$$

The hour angle t could be either 74° 49′ or 105° 11′. Since the date is past the vernal equinox (so that the Sun sets north of west) we know that the latter must be correct.

For the sake of completeness we can also use equations (7), (10), or (13) to find the cosine of t. From (10),

$$\cos t = (\cos \phi \sin \delta - \cos Z \cos h)/\sin \phi \cos \delta =$$

$$[\cos 42° \sin 15° - \cos (68°48') \cos(-51')]/\sin 42° \cos 15° =$$
$$[(0.7431)(0.2588) - (0.3616)(0.9999)]/(0.6691)(0.9659) =$$
$$-0.2619.$$

The cosine is negative, so that we know unambiguously that t is in the second quadrant. The solution above yields 105° 11′, the same answer as before. This value of t translates into $07^h\ 01^m$, rounded to the nearest minute. Adding 12^h gives a local apparent solar time of sunset of $19^h\ 01^m$.

From Figure A1.4, the equation of time on May 1 is $+3^m$, so that local mean solar time is $18^h\ 58^m$, which agrees perfectly with the prediction in Table A1.2. The longitude of Chicago is $05^h\ 51^m$ W, hence Central Standard Time, that of the 90th meridian west, is $18^h\ 49^m$ = 6:49 PM, the clock time of sunset.

A3.4 Equatorial and celestial coordinates

Next use the rules of spherical trigonometry to develop the equations that allow the transformation of the equatorial coordinates of right ascension (α) and declination (δ) into the ecliptic coordinates, celestial longitude (α) and latitude (β): see Section 3.18. The coordinate systems are connected through the spherical triangle defined by the north ecliptic pole, the north celestial pole, and the star, as drawn in Figure A3.8. The sides of this triangle are $90° - \delta$, $90° - \beta$, and ϵ, the obliquity of the ecliptic, which has a value of 23° 26′ (it changes only very slowly with perturbations of the Earth's orbit). The angles, as drawn here, are $90° + \alpha$ at the NCP and $90° - \lambda$ at the NEP.

We can first write, applying the law of cosines,

$$\cos(90° - \beta) = \cos\epsilon \cos(90° - \delta) + \sin\epsilon \sin(90° - \delta) \cos(90° + \alpha),$$

or from Table A3.3,

$$\sin\beta = \cos\epsilon \sin\delta - \sin\epsilon \cos\delta \sin\alpha.$$

Substituting for $\epsilon = 23°\ 26'$, we can write

$$(14)\quad \sin\beta = 0.9175 \sin\delta - 0.3977 \cos\delta \sin\alpha.$$

From the law of sines,

$$\sin(90° - \lambda) = \sin(90° - \delta) \sin(90° + \alpha)/\sin(90° - \beta),$$

or

$$(15)\quad \cos\lambda = \cos\delta \cos\alpha/\cos\beta.$$

Transforming in the other direction, again using the cosine law,

$$\cos(90° - \delta) = \cos\epsilon \cos(90° - \beta)$$
$$+ \sin\epsilon \sin(90° - \beta) \cos(90° - \lambda),$$

whence

$$(16)\quad \sin\delta = 0.9175 \sin\beta + 0.3977 \cos\beta \sin\lambda\ ;$$

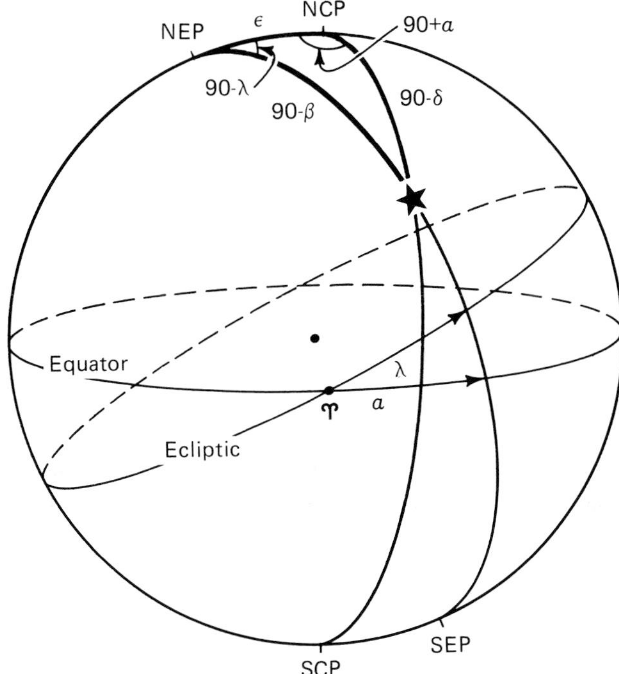

A3.8. Equatorial and celestial coordinates of a star. The sides of the triangle are the obliquity of the ecliptic, $\epsilon = 23° 27'$, $90° - \delta$, and $90° - \beta$. The angles are $90° - \lambda$ at the north ecliptic pole (NEP) and $90° + \alpha$ at the north celestial pole.

and from equation (15)

(17) $\cos \alpha = \cos \beta \cos \lambda / \cos \delta$

The calculation of β and δ presents no sign ambiguities, as they run only between plus and minus 90°, and the sine changes from positive to negative across zero. However, α and λ run the full circle, so that we will not necessarily know from equations (15) and (17) whether they are in the first or fourth quadrants, for which the cosine is positive (see Figure A3.3), or in the second or third, for which it is negative. We can resolve the dilemma by solving for $\sin \alpha$ or $\sin \lambda$ from equations (14) or (16), which yield respectively

(18) $\sin \lambda = (\sin \delta - 0.9175 \sin \beta)/0.3977 \cos \beta$,

and

(19) $\sin \alpha = (0.9175 \sin \delta - \sin \beta)/0.3977 \cos \delta$.

Alternatively, we can apply the five parts rule to Figure A3.8 to obtain

$$\sin(90° - \beta)\cos(90° - \lambda) = \cos(90° - \delta)\sin \epsilon$$
$$- \cos \epsilon \sin(90° - \delta) \cos(90° + \alpha).$$

This equation reduces to

(20) $\cos\beta \sin\lambda = 0.3977 \sin\delta + 0.9175 \cos\delta \sin\alpha$,

which we can solve for either $\sin\lambda$ or $\sin\alpha$.

Look at a pair of examples. First, what are the celestial coordinates of our star Arcturus, $\alpha = 14^h\ 15^m = 214°$, $\delta = +19°$? Again, for simplicity, round to the nearest degree. From equation (14),

$$\sin\beta = 0.9175 \sin 19° - 0.3977 \cos 19° \sin 214° =$$
$$(0.9175)(0.3256) + (0.3977)(0.9455)(0.5592) = 0.5090,$$

(from Table A3.3 the sine of 214° is negative). The celestial latitude, β, is then $+30°.60$ (we temporarily retain the decimals to avoid round-off errors below: always carry at least one more decimal than you ultimately want). From equation (15)

$$\cos\lambda = \cos 19° \cos 214°/\cos 30°.60$$
$$- (0.9455)(0.8290)/0.8607 = -0.9107.$$

The cosine is negative in quadrants two and three, so that λ could be either 156° or 204°. We could discriminate with a carefully drawn diagram, but it is better to solve for $\sin\lambda$. From equation (18) we find

$$\sin\lambda = (\sin 19° - 0.9175 \sin 30°.60)/0.3977 \cos 30°.6 =$$
$$(0.3256 - 0.4670)/(0.3977)(0.8607) = -0.4131,$$

from which the longitude must be in the third quadrant, or 204°. Equation (20) gives the same answer.

As a second example, calculate the right ascension and declination of the Sun at noon on February 25. The Sun is a special case: by definition, $\beta_\odot = 0°$, since the Sun is always on the ecliptic. Consequently, equations (16) and (17) reduce to

(21) $\sin\delta_\odot = 0.3977 \sin\lambda_\odot$

and

(22) $\cos\alpha_\odot = \cos\lambda_\odot/\cos\delta_\odot$.

These relationships are graphed in Appendix Figure A1.9. Since $\sin\delta/\cos\delta = \tan\delta$ (Table A3.3) we can also manipulate equation (14) to read

(23) $\tan\delta_\odot = 0.4335 \sin\alpha_\odot$

Assume that the equinox passage occurs at March 21.0. The equation of time from Figure A1.4 is $-7^m\ 20^s$, so that the mean Sun is 1°.83 west of the apparent Sun. At a rate of 0.9856 degrees per day (360°/365.24 days) it will take the mean Sun 1.86 days, to March 22.86, to reach the equinox, $\alpha = 0^h$, $\lambda = 0°$. From Table A1.1, February 25.5 is 339.64 days past this moment, so that the right ascension of the mean Sun must be $339.64 \times 0.9856 = 334°.75$. The equation of time on February 25.5 is $-13^m\ 00^s$, so the right ascension of the apparent Sun is 3°.25 east of that of the mean Sun, or $338°.0 = 22^h\ 32^m$. Therefore from equation (23)

$$\tan \delta_\odot = (0.4335)(\sin 338.0) = (0.4335)(-0.3746) = -0.1625,$$

and $\delta_\odot = -9°.23 = -9°14'$, which is in agreement with the result fround from Figure A1.3.

Another common transformation involves the conversion of equatorial coordinates into galactic coordinates (galactic longitude and latitude, l and b) and back again. For that we need to know that the 1950 right ascension and declination of the north galactic pole are $12^h\ 49^m.0$ and $+27°\ 24'.0$. Galactic longitude starts at the galactic center ($\alpha = 17^h\ 42^m.4$, $\delta = -28°55'$, 1950), and proceeds northward. Try developing the equations as an exercise.

The basic concepts outlined here also allow us to develop more complex formulations, such as those that govern celestial navigation, precession, aberration, nutation and the like. Armed with these fundamentals, you may now be able to proceed to more quantitative, mathematically oriented texts.

Bibliography

ARCHAEOASTRONOMY

Aveni, A. F. 1989. *Empires of Time: Calendars, Clocks, and Cultures.* New York: Basic Books.

Hawkins, G. S. 1965. *Stonehenge Decoded.* Garden City, N.Y.: Doubleday.

Urton, G. 1981. *At the Crossroads of the Earth and the Sky: An Andean Cosmology.* Austin: University of Texas Press.

ASTROLOGY

Bok, B. and Jerome, L. 1975. *Objections to Astrology.* Buffalo, New York: Prometheus Books.

Culver, R. B. and Ianna, P. A. 1979. *The Gemini Syndrome.* Tucson: Pachart.

Jerome, L. E. 1977. *Astrology Disproved.* Buffalo, New York: Prometheus.

CONSTELLATIONS

Allen, R. H. 1963. *Star Names, Their Lore and Meaning.* Reprint. New York: Dover.

Krupp, E. C. 1991. *Beyond the Blue Horizon.* New York: HarperCollins.

Kunitzsch, P. and Smart, T. 1986. *Short Guide to Modern Star Names and their Derivations.* Wiesbaden, Germany: Otto Harrassowitz.

Menzel, D. H. and Pasachoff, J. 1982. *A Field Guide to the Stars and Planets.* 3rd. ed. New York: Houghton-Mifflin.

Ridpath, I. 1988. *Star Tales.* New York: Universe Books.

Warner, D. 1979. *The Sky Explored.* New York: Alan R. Liss

Williamson, R. A. 1984. *Living the Sky: The Cosmos of the American Indian.* Boston: Houghton-Mifflin.

ECLIPSES

Meeus, J., Grosjean, C. C., and Vanderleen, W. 1966. *Canon of Solar Eclipses.* New York: Pergamon.

Mitchell, S. A. 1951. *Eclipses of the Sun*. New York: Columbia University Press.

Newton, R. R. 1970. *Ancient Astronomical Observations of the Accelerations of the Earth and Moon*. Baltimore: Johns Hopkins Press.

Oppolzer, T. 1962. *Canon of Eclipses*. Reprint. New York: Dover.

Schaeffer, B. E. 1992. Lunar Eclipses that Changed the World. *Sky and Telescope*, **84**:639 (December).

Zirker, J. B. 1984. *Total Eclipses of the Sun*. New York: Prentice-Hall.

HISTORY

Abetti, G. 1952. *The History of Astronomy*. New York: Henry Schuman.

Berry, A. 1961. *A Short History of Astronomy*. Reprint. New York: Dover.

Gingerich, O. 1993. *The Eye of Heaven: Ptolemy, Copernicus, Kepler*. New York: American Institute of Physics.

Hodson, D. G. ed. 1974. *The Place of Astronomy in the Ancient World*. Oxford: Oxford University Press.

Pannekoek, A. 1989. *A History of Astronomy*. Reprint. New York: Dover.

Pedersen, O. 1974. *A Survey of the Almagest*. Odense, Denmark: Odense University Press.

LIGHT, COLOR, AND THE ATMOSPHERE

Davis, N. 1992. *The Aurora Watcher's Handbook*. College, Alaska: The University of Alaska Press.

Greenler, R. 1980. *Rainbows, Halos, and Glories* . Cambridge: Cambridge University Press.

Meinel, A. and Meinel, M. 1983. *Sunsets, Twilights, and Evening Skies*. Cambridge, England: Cambridge University Press.

Minnaert, M. 1954. *The Nature of Light and Color in the Open Air*. Reprint. New York: Dover.

Minnaert, M. 1993. *Light and Color in the Outdoors*. New York: Springer

Overheim, R. D. and Wagner, E. L. 1982. *Light and Color*. New York: Wiley.

Sobel, M. I. 1987. *Light*. Chicago, University of Chicago Press.

Williamson, S. J. and Cummins, H. Z. *Light and Color in Nature and Art*. New York: Wiley.

SOLAR SYSTEM

Beatty, J. K., and Chaikin, A. eds. 1990. *The New Solar System*. 3rd ed. Cambridge, England: Cambridge University Press.

Binzel, R. P., Gehrels, T., and Matthews, M. S. 1989. *Asteroids II*. Tucson: University of Arizona Press.

Brandt, J. C. 1992. *Rendezvous in Space: The Science of Comets*. New York: Freeman.

Dodd, R. T. 1981. *Meteorites*. Cambridge, England: Cambridge University Press.

Gropman, D. 1985. *Comet Fever: A Popular History of Halley's Comet*. New York: Simon & Schuster.

Hodge, P. 1994. *Meteorite Craters and Impact Structures of the Earth*. Cambridge: Cambridge University Press.

Hunt, G. and Moore, P. 1981. *Jupiter*. New York: Rand McNally.

Hunt, G. and Moore, P. 1982. *Saturn*. New York: Rand McNally.

Kaufmann, W. J. 1979. *Planets and Moons*. New York: Freeman.

Kronk, G. W. 1988. *Meteor Showers: A Descriptive Catalogue*. Hillside, N.J.: Enslow.

Moore, P. 1981. *The Moon*. New York: Rand McNally.

Morrison, D. 1993. *Exploring Planetary Worlds*. Scientific American Library. New York: Freeman.

Morrison, D. and Owen, T. 1988. *The Planetary System*. New York: Addison-Wesley.

Narlikar, I. V. 1982. *The Lighter Side of Gravity*. San Francisco: Freeman.

Rothery, D. A. 1992. *The Satellites of the Outer Planets*. Oxford: Oxford University Press.

Whipple, F. L. 1985. *The Mystery of Comets*. Washington, D.C.: Smithsonian Institution Press.

Wilhelms, D. E. 1987. *The Geologic History of the Moon*. US Geological Survey Professional Paper 1348. Washington, D.C.: US Government Printing Office.

SPHERICAL ASTRONOMY

Green, R. M. 1985. *Spherical Astronomy*. Cambridge, England: Cambridge University Press.

McNally, D. 1973. *Positional Astronomy*. New York: Wiley.

Nassau, J. J. 1948. *Practical Astronomy*. New York: McGraw-Hill

Smart, W. M. 1977. *Spherical Astronomy*. 6th ed. Cambridge, England: Cambridge University Press.

Woolard, E. W. and Clemence, G. M. 1966. *Spherical Astronomy*. New York: Academic Press.

Astronomical Almanac. Issued annually and jointly by the Nautical Almanac Office of the United States Naval Observatory and Her Majesty's Nautical Almanac Office, Royal Greenwich Observatory.

STAR MAPS AND ATLASES

Becvar, A. 1949. *Skalnate Pleso Atlas of the Heavens*. Cambridge, Massachusetts: Sky Publishing Corp.

Ridpath, I. ed. 1989. *Norton's Star Atlas*. 19th ed. Cambridge, Mass.: Sky Publishing Corp. (All naked-eye stars and many telescopic objects.)

Tirion, W. 1981. *Atlas 2000*. Cambridge, Mass.: Sky Publishing Corp. (Deep atlas to 8th magnitude includes a large number of non-stellar objects.)

Tirion, W., Rappaport, B., and Lovi, G. 1988 *Uranometria 2000*. Richmond, Va.:Willmann-Bell. (Includes stars to 10th magnitude and non-stellar objects.)

Vehrenberg, H. 1970. *Atlas Stellarum*. Düsseldorf: Treugesell-Verlag.

SC1, SC2, SC3 Star Charts, Cambridge, Mass.: Sky Publishing Corp. (Simple charts for beginning constellation study.)

SUN, STARS, AND GALAXIES

Bok, B. and Bok, P. 1981. *The Milky Way*, Cambridge, Mass.: Harvard University Press.

Burnham, R. R. 1978. *Burnham's Celestial Handbook*. 3 vols. New York: Dover.

Ferris, T. 1980. *Galaxies*. New York: Stewart, Tabori, and Chang.

Hodge, P. 1986. *Galaxies*. Cambridge, Mass.: Harvard University Press.

Kaler, J. B. 1992. *Stars*. Scientific American Library. New York: Freeman.

Marshall, L. A. 1988. *The Supernova Story*. New York: Plenum.

Nicolson, I. 1982. *The Sun*. New York: Rand McNally.

Phillips, K. J. H. 1992. *Guide to the Sun*. Cambridge, England: Cambridge University Press.

Sandage, A. 1961. *The Hubble Atlas of Galaxies*. Washington, D.C.: Carnegie Institution of Washington.

Petit, M. 1987. *Variable Stars*. New York: Wiley.

Verschuur, G. 1989. *Interstellar Matters*. New York: Springer.

Wentzel, D. G. 1989. *The Restless Sun*. Washington, D. C.: Smithsonian Institution Press.

TELESCOPES

King, H. C. 1979. *The History of the Telescope*. Reprint. New York: Dover.

Tucker, W. and Tucker, K. 1986. *The Cosmic Enquirers*. Cambridge, Mass.: Harvard University Press.

Verschuur, G. 1987. *The Invisible Universe Revealed: The Story of Radio Astronomy*. New York: Springer.

UNIDENTIFIED FLYING OBJECTS

Condon, E. U. 1968. *Scientific Study of Unidentified Flying Objects*. Boulder, Colorado: University of Colorado Press.

Klass, P. J. 1974. *UFOs Explained*. New York: Random House.

Klass, P. J. 1983. *UFOs: The Public Deceived*. Buffalo, New York: Prometheus.

Sheaffer, R. 1981. *The UFO Verdict: Examining the Evidence*. Buffalo, New York: Prometheus.

Index